Epigenetics and Human Health

For further volumes:
http://www.springer.com/series/8561

Constanze Bonifer • Peter N. Cockerill
Editors

Transcriptional and Epigenetic Mechanisms Regulating Normal and Aberrant Blood Cell Development

Springer

Editors
Constanze Bonifer
Peter N. Cockerill
University of Birmingham
Medical and Dental Sciences
Institute of Biomedical Research
Birmingham
United Kingdom

ISSN 2191-2262 ISSN 2191-2270 (electronic)
ISBN 978-3-642-45197-3 ISBN 978-3-642-45198-0 (eBook)
DOI 10.1007/978-3-642-45198-0
Springer Heidelberg New York Dordrecht London

Library of Congress Control Number: 2014933585

Printed on acid-free paper

Springer is part of Springer Science+Business Media (www.springer.com)

Preface

During vertebrate hematopoiesis many specialized cells types are formed with vastly different functions, such as B cells, T cells, granulocytes, macrophages, erythrocytes, and megakaryocytes. The complex blood cell system found in humans has evolved from a few simple cell types mainly involved in oxygen transport and phagocytosis to a highly efficient cell production facility, which constantly replenishes cells involved in oxygen transport, wound healing, the removal of cellular debris by phagocytosis, as well as providing a highly efficient innate and adaptive immune system designed to protect the body from infectious diseases, parasites, and tumor cells. To tightly control the enormous proliferative potential of developing blood cells, an intricately balanced signaling and transcription network has evolved that ensures that the different blood cell types are formed at the right time and in the right numbers. Finely tuned regulatory mechanisms ensure that blood cells function properly and have a determined life span. Moreover, in the adaptive immune system, long-lived memory cells have evolved that ensure that when pathogens have been seen once they will never cause a problem again.

All of these features of the hematopoietic system are under transcriptional and epigenetic control. Failures in this control cause incomplete differentiation, a dysfunctional immune system, problems with wound healing, as well as uncontrolled proliferation of blood cells and cancer. As the principles of differentiation control are similar in all multicellular organisms, the hematopoietic system has served as an excellent model system to study the principles of the epigenetic and transcriptional control of cell fate decisions in general. In this book we will therefore make a journey from first asking how very primitive organisms use the epigenetic regulatory machinery to balance growth with differentiation control, towards digging deep into what controls the function of specialized cells of the human immune system.

To introduce a general readership into the molecular basis of gene expression control in a chromatin context, Peter Cockerill and Constanze Bonifer will first introduce the general principles of chromatin structure and gene expression control. The next two chapters will then describe non-mammalian hematopoiesis and we will discover that flies make blood but exist without blood vessels, that fish make blood cells in the kidney, and which precise genetic circuitries are required for these developmental pathways. The first by Paul Badenhorst introduces *Drosophila melanogaster* hematopoiesis and demonstrates why this model system is highly informative for mammalian blood cell development and human leukemia. The second by Xiaoxing Bai informs us about blood cell development in zebra fish as an important genetic model for vertebrate hematopoiesis and tells us why this model is important for drug screening.

The remainder of the book focuses primarily on mammalian hematopoiesis. We start with five chapters outlining general principles of gene regulation and development, beginning with Valerie Kouskoff and coworkers who describe the regulatory processes that drive the development of hematopoietic stem cells in the mammalian embryo. We then will make a detour into the realm of Polycomb complexes in the chapter written by Miguel Vidal which describes one of the most fundamental mechanisms used by all tissues in all animals to establish patterns of development and differentiation. Although this chapter will not focus on hematopoiesis, it will explain in comprehensive detail the general principles of the biochemical nature of Polycomb complexes and how they regulate gene expression and outline the breathtaking complexity of this system that we are only now beginning to understand. The following chapter by Vincent Van den Boom et al. then makes it abundantly why knowledge of the Polycomb system is essential for our understanding of normal blood cell development and it illustrates how aberrations in this pathway lead to abnormal blood cell development and contribute to diseases such as leukemia. To gain a more complete understanding of the regulatory network controlling development we next learn that the repressive activity of Polycomb proteins is counterbalanced by the Trithorax family of activating factors. This is the theme of the chapter written by Robert Slany. He introduces the MLL family of transcriptional activators, and he outlines why chromosomal translocations involving MLL disturb the balance between Polycomb and Trithorax activities at *HOX* genes and why such a disturbance causes leukemia. The final chapter on basic mechanisms is written by Grant Challen and Jenny Trowbridge and it explains the role of DNA methylation in reinforcing the decisions of stem cells to differentiate into all mature blood cell lineages and how this process is disturbed in malignant hematopoiesis.

The next major theme of the book focuses on mechanisms of hematopoietic differentiation and includes five articles describing how the development of the myeloid, erythroid/megakaryocytic, and lymphoid lineages is controlled. Peter Laslo and Thomas Stopka explain the control of myelopoiesis by transcription factors and epigenetic regulators and how mutation of their respective genes causes myeloid malignancies. Doug Vernimmen describes the network of transcription factors and the epigenetic regulators that control development and gene expression in the erythroid and megakaryocytic lineages. We then focus on the adaptive immune system and learn the fundamentals of how we acquire the ability to recognize millions of foreign antigens. We start with basic concepts of how the T-cell lineage develops from lymphoid progenitors in a chapter by Will Bailis and Warren Pear which describes the transcription factor networks and selection processes controlling T-cell differentiation in the thymus. In this chapter we also learn that the mechanisms driving T-cell development can also be diverted to induce T-cell leukemia. To complete the story of T-cell development, Cristina Hernandez-Munain and coworkers describe the intricate mechanisms that lead to the huge diversity of T-cell antigen receptors (TCR) as a result of TCR gene rearrangements. This chapter is perfectly balanced by a parallel description by Kirkham et al. of the processes controlling immunoglobulin gene rearrangements in the B-cell lineage. This chapter also explores how B cells develop from lymphoid progenitors and informs of the ways how leukemia or immune deficiency arises as a result of defects in the gene rearrangement process.

The final two chapters will shed light on the molecular mechanisms that regulate immune cell function and describe processes that establish normal cells or lead to the development of cells with impaired function. Here we concentrate on two lineages: T cells and macrophages. First we learn from the group of Toshinori Nakayama of how we maintain adaptive immunity in T cells once an existing infection has been resolved. This introduces the concept of molecular memory in memory T cells which allows these cells to respond rapidly to subsequent exposure to the same pathogens. We also learn that T cells have the ability to differentiate along different pathways in response to intrinsic and extrinsic signals, allowing them to tailor their responses to different types of pathogens. The chapter by Poletti et al. describes the role of macrophages in regulating an inflammatory response, outlining in fine molecular detail recent genome-wide studies that shed light on how this response is controlled at the level of gene regulation.

At the end of this journey, we hope that the scientist/science student/health professional reader will understand general principles of cellular differentiation control at the molecular level and what is actually meant by epigenetic and transcriptional regulation. We also hope that this book will help readers to develop a clear picture of how gene regulatory processes function in a chromatin context and how their deregulation causes blood cell development to go astray.

Birmingham, UK

Constanze Bonifer
Peter N. Cockerill

Contents

List of Contributors

Úrsula Angulo Department of Cellular Biology and Immunology, Instituto de Parasitología y Biomedicina "López-Neyra" (IPBLN-CSIC), Consejo Superior de Investigaciones Científicas, Armilla, Granada, Spain

Paul Badenhorst School of Infection and Immunity, College of Medical and Dental Sciences, Institute of Biomedical Research, University of Birmingham, Edgbaston, Birmingham, UK

Xiaoying Bai Department of Obstetrics and Gynecology, Cecil H. and Ida Green Center for Reproductive Biology Sciences, The University of Texas Southwestern Medical Center, Dallas, TX, USA

Will Bailis Department of Pathology and Laboratory Medicine, Abramson Family Cancer Research Institute, and Institute for Immunology, The Perelman School of Medicine at the University of Pennsylvania, Philadelphia, PA, USA

Sarah Bevington School of Immunity and Infection, Institute of Biomedical Research, University of Birmingham, Edgbaston, Birmingham, UK

Constanze Bonifer School of Cancer Sciences, College of Medical and Dental Sciences, Institute of Biomedical Research, University of Birmingham, Birmingham, UK

Joan Boyes School of Molecular and Cellular Biology, Faculty of Biological Sciences, University of Leeds, Leeds, UK

Annet Z. Brouwers-Vos Department of Experimental Hematology, University Medical Center Groningen, University of Groningen, Groningen, The Netherlands

Grant A. Challen Division of Oncology, Department of Internal Medicine, Washington University in St. Louis, St. Louis, MO, USA

Peter N. Cockerill School of Immunity and Infection, College of Medical and Dental Sciences, Institute of Biomedical Research, University of Birmingham, Birmingham, UK

Guilherme Costa Stem Cell Biology Group, Cancer Research UK Manchester Institute, The University of Manchester, Manchester, UK

Alessia Curina Department of Experimental Oncology, European Institute of Oncology (IEO), Milan, Italy

Beatriz del Blanco Department of Cellular Biology and Immunology, Instituto de Parasitología y Biomedicina "López-Neyra" (IPBLN-CSIC), Consejo Superior de Investigaciones Científicas, Armilla, Granada, Spain

Maud Fleury Stem Cell Haematopoiesis Group, Cancer Research UK Manchester Institute, The University of Manchester, Manchester, UK

Serena Ghisletti Department of Experimental Oncology, European Institute of Oncology (IEO), Milan, Italy

Cristina Hernández-Munain Department of Cellular Biology and Immunology, Instituto de Parasitología y Biomedicina "López-Neyra" (IPBLN-CSIC), Consejo Superior de Investigaciones Científicas, Armilla, Granada, Spain

Christopher M. Kirkham School of Molecular and Cellular Biology, Faculty of Biological Sciences, University of Leeds, Leeds, UK

Valerie Kouskoff Stem Cell Haematopoiesis Group, Cancer Research UK Manchester Institute, The University of Manchester, Manchester, UK

Georges Lacaud Stem Cell Biology Group, Cancer Research UK Manchester Institute, The University of Manchester, Manchester, UK

Peter Laslo Section of Experimental Haematology, Leeds Institute of Cancer and Pathology, St. James' University Hospital, University of Leeds, Leeds, UK

Toshinori Nakayama Department of Immunology, Graduate School of Medicine, Chiba University, Chiba, Japan

Gioacchino Natoli Department of Experimental Oncology, European Institute of Oncology (IEO), Milan, Italy

Atsushi Onodera Department of Immunology, Graduate School of Medicine, Chiba University, Chiba, Japan

Warren S. Pear Department of Pathology and Laboratory Medicine, Abramson Family Cancer Research Institute, and Institute for Immunology, The Perelman School of Medicine at the University of Pennsylvania, Philadelphia, PA, USA

Flor M. Perez-Campo Stem Cell Biology Group, Cancer Research UK Manchester Institute, The University of Manchester, Manchester, UK

Department of Internal Medicine, Hospital U.M. Valdecilla, IFIMAV, University of Cantabria, Santander, Spain

Sara Polletti Department of Experimental Oncology, European Institute of Oncology (IEO), Milan, Italy

Hein Schepers Department of Experimental Hematology, University Medical Center Groningen, University of Groningen, Groningen, The Netherlands

Jan Jacob Schuringa Department of Experimental Hematology, University Medical Center Groningen, University of Groningen, Groningen, The Netherlands

James N. Scott School of Molecular and Cellular Biology, Faculty of Biological Sciences, University of Leeds, Leeds, UK

Robert K. Slany Department of Genetics, Friedrich Alexander Universitaet Erlangen, Erlangen, Germany

Tomas Stopka Institute of Pathophysiology and First Department of Medicine-Haematology, First Faculty of Medicine, Charles University in Prague, Prague, Czech Republic

Jennifer J. Trowbridge The Jackson Laboratory, Bar Harbor, ME, USA

Damon J. Tumes Department of Immunology, Graduate School of Medicine, Chiba University, Chiba, Japan

Vincent van den Boom Department of Experimental Hematology, University Medical Center Groningen, University of Groningen, Groningen, The Netherlands

Douglas Vernimmen The Roslin Institute, Developmental Biology Division, University of Edinburgh, Easter Bush, Midlothian, UK

Miguel Vidal Department of Cellular and Molecular Biology, Centro de Investigaciones Biológicas (CSIC), Madrid, Spain

Chapter 1
The Epigenetic Regulatory Machinery

Constanze Bonifer and Peter N. Cockerill

Abstract The processes of eukaryotic development and cellular differentiation are under transcriptional and epigenetic control by essentially the same mechanisms in all multi-cellular organisms. Here we briefly summarise the general principles of chromatin structure and gene regulation. Because the bulk of the chromatin in the nucleus exists in a highly condensed state, the main level at which gene expression is controlled is at the level of the accessibility of genes and their regulatory elements to the transcription apparatus. In this article we will describe the complex machinery that covalently modifies the DNA and histones, remodels chromatin structure and allows transcription factors to find their targets within regulatory elements. We will establish the concept that epigenetic and transcriptional regulation involves a finely tuned balance between activators and repressors, which function via a huge variety of mechanisms to either introduce or erase modifications to the basic chromatin template.

Keywords Chromatin • Epigenetics • Nucleosome • Histone • Transcription factor • Gene regulation • DNA methylation

The entire body plan of an individual is encoded in its genome. However, because each cell type expresses a different gene expression programme, only a subset of this information is accessed within each specific lineage. Alterations to these gene expression programmes are under epigenetic and transcriptional control and are the underlying basis for the differentiation of multiple cell types from a pluripotent fertilised oocyte. But how do we get from one cell to many diverse cell types? Epigenetic research

---

C. Bonifer (✉)
School of Cancer Sciences, College of Medical and Dental Sciences, Institute of Biomedical Research, University of Birmingham, Birmingham, UK
e-mail: c.bonifer@bham.ac.uk

P.N. Cockerill (✉)
School of Immunity and Infection, College of Medical and Dental Sciences, Institute of Biomedical Research, University of Birmingham, Birmingham, UK
e-mail: p.n.cockerill@bham.ac.uk

C. Bonifer and P.N. Cockerill (eds.), *Transcriptional and Epigenetic Mechanisms Regulating Normal and Aberrant Blood Cell Development*, Epigenetics and Human Health, DOI 10.1007/978-3-642-45198-0_1, © Springer-Verlag Berlin Heidelberg 2014

describes the molecular mechanisms by which tissue-specific gene expression patterns are established and maintained during multiple rounds of cell division. In recent years, great progress has been made in the identification of the molecular players involved in epigenetic control of gene expression during development. The work of many laboratories has established that regulating the interplay of transcription factors with chromatin components is the major driver of the cellular differentiation process.

In this introduction we will give a general overview of basic chromatin structure and the molecular principles that govern transitions between active and inactive transcriptional states. This brief introduction is not intended to be comprehensive, but it will allow the non-expert reader to gain a first insight into gene expression control at the level of the epigenome and will hopefully make it easier to understand the next chapters. We have, however, provided a more comprehensive description of the basic features of chromatin structure and function in a recent review article (Cockerill 2011).

1.1 The Chromatin Template

The basic building block of chromatin is the nucleosome which consists of 146 bp of DNA wrapped around a histone protein octamer containing two molecules each of histones H2A, H2B, H3 and H4. Each histone molecule consists of a globular domain closely associated with the nucleosomal DNA and highly basic unfolded N- and C-terminal tail domains which protrude outwards from the nucleosome. The structure of the nucleosome was defined at high resolution by X-ray crystallography in 1997 (Luger et al. 1997). A simplified model of this structure, showing just the N-terminal tails, is depicted in Fig. 1.1a, which also illustrates the order of assembly of the nucleosome. At its heart are two dimers of H3 and H4 which contact the central ~70 bp of DNA, with dimers of H2A and H2B binding above and below this inner core and interacting with an additional 30–40 bp of DNA on either side.

Individual nucleosomes are separated by linker regions to give an overall average repeat length of around 180–200 bp. These linkers are in most cases bound by the linker histone H1 which enhances the higher order folding of nucleosomes together in the form of a more condensed 30 nm diameter chromatin fibre and maintains a slightly longer nucleosomal repeat length (Woodcock et al. 2006; Robinson et al. 2008). Histone H1 also stabilises the nucleosome and locks it in place by simultaneously binding to DNA sequences located at the dyad axis on the surface of the nucleosome and contacting the flanking 10 bp of each linker region (Fig. 1.1b) (Syed et al. 2010). This creates a particle called the chromatosome in which histones closely contact a total of 166 bp of DNA. Histone H1 is typically thought of as a repressive factor associated with condensed inactive gene loci (Wolffe 1989). However, this distinction is far from absolute as significant levels of H1 are also found within active genes (Kamakaka and Thomas 1990). Furthermore, even inactive condensed chromatin is a highly dynamic structure whereby there is rapid exchange of histone H1 within the chromatin fibre (Lever et al. 2000; Misteli et al. 2000).

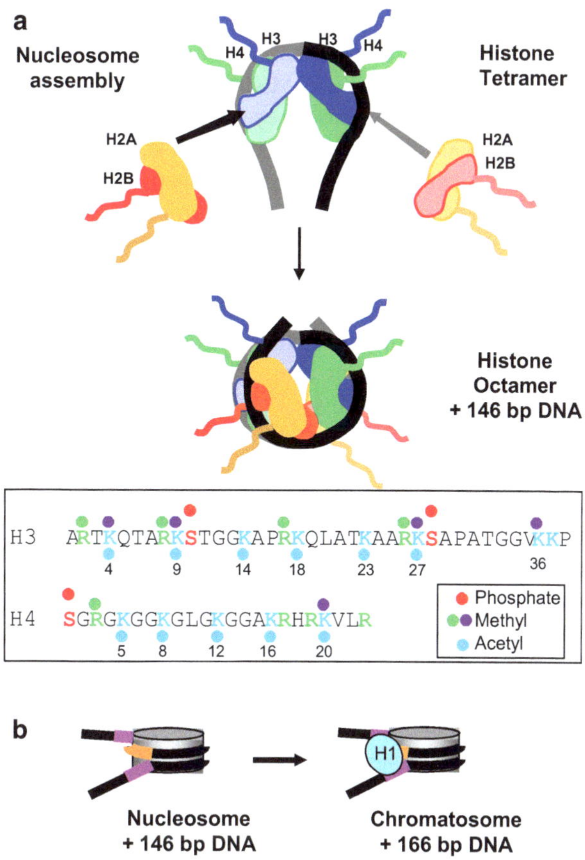

Fig. 1.1 Composition of nucleosomes. (**a**) The nucleosome is the basic building block of chromatin and consists of an inner core of two molecules each of H3 and H4, plus dimers of H2A and H2B above and below this core, and with 1.7 turns of DNA equalling 146 bp wrapped around the outside. For simplicity, the C-terminal tails of the histones are not shown. The N-terminal tails of histone H3 and H4 are shown in the *box* below, plus the major post-translational modifications that control nucleosome functions. (**b**) Within native chromatin, most nucleosomes are also associated with histone H1, forming a particle called the chromatosome which incorporates 166 bp of DNA. H1 engages DNA at the dyad axis of the nucleosome, shown in *orange*, plus 10 bp of each linker at the point where they exit the nucleosome, shown in *purple*

Within the interphase nucleus the chromatin exists mostly at an even higher level of condensation than the 30 nm fibre (Kireev et al. 2008; Hu et al. 2009), meaning that chromatin accessibility represents the most stringent level at which gene expression is controlled. In the absence of transcription, the chromatin of genes adopts a heritable silent state by default which is characterised by a number of distinct structural features. This includes the modification of the DNA itself via the methylation of CG dinucleotides and the modification of the histones by repressive marks.

1.2 Chromatin Modifications

The histone tail domains in particular are subject to a wide variety of covalent modifications that directly alter the structure and function of chromatin. The N terminal tails of H3 and H4 are the histone domains that stand out as being subject to the greatest number of modifications which significantly influence gene regulation. Figure 1.1a summarises the most widely studied modifications affecting these two specific tail domains. Modifications affecting the H3 N-terminal tail are of particular interest because they generate docking sites for a great number of both activators and repressors of transcription. H4 N-terminal tail modifications are of interest because in addition to their specific regulatory roles, they directly impact upon nucleosome:nucleosome interactions and the higher order folding of chromatin (Shogren-Knaak et al. 2006; Robinson et al. 2008). The most commonly encountered modifications to these tails include (1) acetyl lysine, which recruits bromodomain proteins usually associated with gene activation, (2) methyl lysine, which recruits chromo domain and PHD family proteins and (3) phospho-serine and methyl arginine, which can influence the activities of chromatin modifiers on adjacent lysines. However, lysine methylation at different sites in H3 has very different outcomes on transcription: K4 methylation is associated with activators, while H9 methylation is associated with repression by heterochromatin protein HP1, and K27 methylation is associated with repression by Polycomb complexes (Shilatifard 2008; Cockerill 2011). The same lysine is often subject to competition between different modifiers that are trying to establish opposing functional states. In the case of H3, for example acetyl K9 and K27 recruit activators, whereas trimethyl K9 and K27 recruit repressors. This whole process is controlled by the balance between histone acetyl transferases (HATs) and histone deacetylases (HDACs) controlling acetylation versus histone methyl transferases (HMTs) and histone demethylases (HDMs) controlling methylation of the very same amino acids (Fischle et al. 2005; Shahbazian and Grunstein 2007; Suganuma and Workman 2008). Some of these modifications on adjacent or nearby amino acids are mutually exclusive or antagonistic. For example, H3 R2 methylation suppresses trimethylation of H3 K4(Guccione et al. 2007; Kirmizis et al. 2007), while H3 S28 phosphorylation antagonises binding of polycomb proteins to methylated K27 (Gehani et al. 2010) and H3 S10 phosphorylation blocks binding of HP1 to methylated K9 (Fischle et al. 2005). In the case of the H3 tail the amino acid sequence ARKS is repeated at both K9 and K27, allowing in each case a complex interplay between all of the above four different types of modifications.

These pathways are frequent targets for dysregulation in diseases of blood cells. For example, mutations in the kinase Jak2 lead to constitutive phosphorylation of H3 Y41, which blocks the binding of HP1a and leads to activation of key target genes such as Lmo2 (Dawson et al. 2009). Oncogenic activation of other kinases can lead to gene activation via H3 S10 or S28 phosphorylation.

It is also generally accepted that histone acetylation has a non-specific role in gene activation because the neutralisation of highly charged lysine side chains by acetylation will have the added effect of loosening up the ionic interactions between the histone tails and the DNA. Hence, on active genes, chromatin is more dynamic and much less compact. Some specific modifications, such as acetylation of the H4 tail at K16, lead to decompaction of the 30-nm chromatin fibre by loosening the interactions between adjacent nucleosomes (Shogren-Knaak et al. 2006) or between the tails and the linker DNA. Other modifications, such as acetylation of H3 K56 and K122, occur within the globular domain and can boost transcription simply by assisting in the process of unravelling nucleosomes during transcription (Williams et al. 2008; Tropberger et al. 2013).

1.3 Maintaining Inactive Chromatin

DNA methylation at CG dinucleotides plays a major role during development in maintaining specific patterns of gene expression in mammals and is introduced by DNA-methyltransferases (DNMTs). DNMT3a and DNMT3b are capable of methylating cytosines de novo on unmethylated DNA. DNMT1 requires a methylated cytosine at one strand of a newly replicated CG residue to maintain previously installed methylation by modifying the cytosine residue on the other strand. Methylated DNA is recognised by specific methyl-binding proteins, such as MeCP2, which cooperate with other enzymatic activities to maintain the inactive transcriptional state. These include, for example, HDACs which ensure that chromatin at sites of high DNA methylation is not acetylated and exists in its most compact state. Repressive histone modifications play a role in maintaining heterochromatin via interacting with other proteins. As mentioned above, di- or trimethyl histone H3 lysine 9 serves as a docking site for heterochromatin protein 1 (HP1) which also interacts with HDACs. In this way, heterochromatin is maintained by a self-sustaining process that involves the cooperation between many different protein complexes. HP1-dependent heterochromatin is also capable of self-propagation, and in one study its rate of spread along chromatin was measured at about one nucleosome every 6 h (Hathaway et al. 2012).

The above examples of the transcriptionally active, "open" chromatin state and the inactive, highly compacted state are only two extremes in the spectrum of mechanisms that control gene expression. Genes can adopt a number of intermediate inactive or "primed" conformations which are characterised by the binding of transcription factors to enhancers and promoters, but are nevertheless associated with an absence of mRNA production. One such mechanism is characterised by the binding of Polycomb group complexes (PcG) which keep genes in a repressed state where they can still be easily reactivated. The molecular mechanism of how these highly heterogeneous complexes function and their precise nature will be explained in later chapters of this book this book by Vidal and by van den Boom et al. In this introduction it is sufficient to say that depending on the type of complex, the PcG complexes PRC2 and PRC1 work together to deposit a trimethyl mark on lysine 27 of H3 and to ubiquinate the C terminal tail of histone H2A. The presence of these modifications is compatible with the binding of RNA Polymerase II, but is incompatible with productive transcriptional elongation. To be activated, such genes need to receive a stimulus, either in the form of developmental cues, signal transduction, or the binding of additional transcription factors. In turn, this intermediate state can also characterise genes that are on their way to be silenced. It is this property of holding genes "in limbo" which is responsible for the fact that PcG complexes are important regulators of self-renewal in stem cells, but are also important for making sure that genes are switched on and off at the right developmental stage.

1.4 Changing Chromatin Structure: Transcription Factors and Co-activators

Compact chromatin presents a formidable barrier to the transcription apparatus. The activation of eukaryotic genes therefore requires an alteration of their chromatin structure. However, it should be noted that the interaction of chromatin proteins with their targets is a highly dynamic process with the half-lives of even heterochromatin proteins such as HP1 in the range of seconds (Schmiedeberg et al. 2004), indicating that other factors such as transcriptional activators can slip in if they are present at high enough concentrations. Transcription factors recognise a specific DNA sequence and, most importantly, recruit non-DNA binding co-factors to initiate the establishment of an active chromatin state. Transcriptional co-activators function by a wide variety of mechanisms and include, for example HATs such as CBP and p300 and chromatin remodelers such as SWI/SNF family Brg1 complexes that are capable of remodelling nucleosomes. It is likely that such a mechanism is also responsible for recruiting enzymatic activities such as TET proteins that are involved in the removal of methylated cytosines. TET proteins and their role in changing transcriptional states will be described in more detail in the chapter by Challen and Trowbridge. This concerted action leads to a less compact chromatin with enhanced accessibility for other transcription factors which can interact with each other and form large multi-protein assemblies bringing together proteins from close-by or from several kilobases away. The final result of this assembly process is the onset of productive mRNA transcription by the elongating form of RNA Polymerase II.

Clearly, the first big obstacle presented to the regulatory apparatus is simply the ability of transcription factors to gain entry to their binding sites, which are typically occluded by chromatin proteins when genes are inactive. As summarised in the examples shown in Fig. 1.2, there are different means by which regulatory proteins cooperate to gain entry to sites occupied by nucleosomes. As depicted in panel a, DNA-binding proteins have an intrinsic ability to cooperate in the process of sequentially peeling nucleosomal DNA away from the surface of the nucleosome (Adams and Workman 1995). This merely requires the presence of sufficient levels and numbers of factors and binding sites to progressively dissociate DNA from the nucleosome core, beginning with sites close to the point at which the DNA exits the nucleosome. A more invasive mechanism is depicted in panel b, whereby many transcription factors recruit ATP-dependent remodelers such as Brg1, which have the ability to reel out the DNA from one side and thereby translocate nucleosomes along the DNA to reveal the underlying binding sites (Narlikar et al. 2002). This may be the main mechanism that creates DNase I hypersensitive sites. In all of these

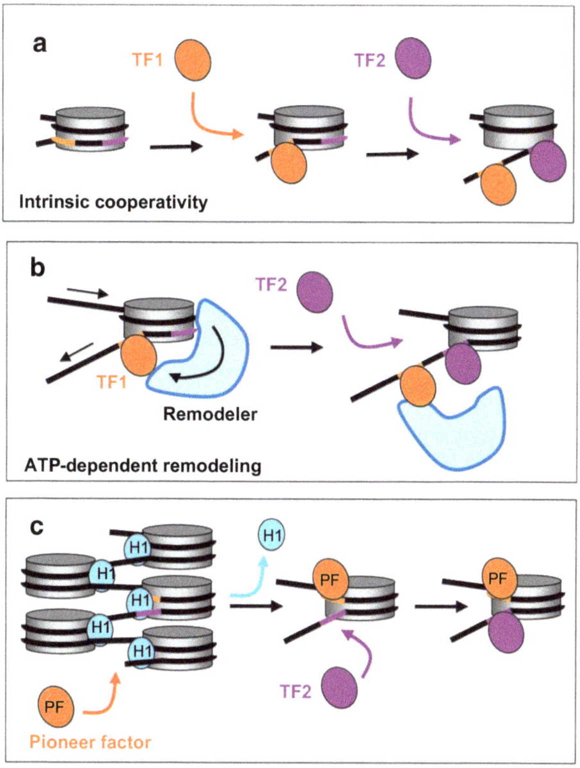

Fig. 1.2 Mechanisms of action employed by transcription factors (TF) to disrupt chromatin and bind to DNA. (**a**) Most TFs have an intrinsic ability to compete with histones for DNA sequences located at the point where DNA exits the nucleosome where the contacts are the most dynamic. Furthermore, TFs can progressively peel the DNA away from the nucleosome surface by binding sequentially to adjacent binding sites, starting at the boundaries. This leads to an intrinsic cooperativity in the binding of TFs. (**b**) Chromatin remodelers such as the SWI/SNF and ISWI families use energy from ATP to translocate DNA around the nucleosome, and thereby either freeing up TF-binding sites that were previously occluded, or covering up sites that were previously free. (**c**) Pioneer factors (PF) such as FoxA1 and FoxO1 utilise a specialised mechanism to open up regions of condensed chromatin. These proteins have a histone-like fold that is able to mimic histone H1 and bind to the same sites at the nucleosome dyad (depicted in *orange*) normally occupied by H1. By creating localised openings in the chromatin fibre, pioneer factors render binding sites for other TFs more accessible

cases, transcription factor interactions will be assisted by histone acetylation which acts indirectly by increasing nucleosome mobility and directly by targeting the recruitment of remodelers containing bromodomains.

An alternative means of activating chromatin (panel c) is utilised by factors that are capable of interacting with compact chromatin and opening it without co-factors, the so-called pioneer factors, which pave the way for other factors to bind (Cirillo et al. 2002; Hatta and Cirillo 2007; Zaret and Carroll 2011). They overcome the problems of steric hindrance faced by most other transcription factors which find it difficult to bind efficiently to DNA within nucleosomes because some of the points of interaction are strongly bound by the histones. In contrast, pioneer factors such as FoxA1 and FoxO1 have a histone-like domain that is able to bind to just one face of DNA sequences located at the very midpoint of nucleosomes, known as the dyad axis (Cirillo and Zaret 2007; Sekiya et al. 2009; Zaret and Carroll 2011). This concept will be described in more detail in the chapter by Poletti et al. Last, but not least, recent results indicate that interactions between transcription factors and the nucleosomal templates are even more complex than previously thought (Ballare et al. 2013). Hence, we now know that in some instances, nucleosomes actually aid efficient recruitment of transcription factors.

Even after a transcription factor has found its binding site, it still faces stiff competition with both nucleosomes and other transcription factors (Voss et al. 2011). In some cases a newly bound factor can recruit a remodeler which then leads to its own eviction by repositioning a nucleosome over its own binding site (Voss et al. 2011). Alternatively, transcription factors can lose their access if a passing polymerase directs the relocation of a nucleosome over its binding site (Lefevre et al. 2008).

Transcription factors are not all activators, and many transcription factors act as repressors of gene expression. Repressors function by recruiting co-repressors which include HDACs and chromatin remodelers such as NuRD which evict activating factors by sliding nucleosomes over their DNA-binding sites. However, some factors can repress transcription by simply interfering with the binding of RNA-Polymerase II (Pol II) to the promoter by either direct competition or by driving anti-sense transcription across the promoter (Ingram et al. 2011). Moreover, it was recently shown that the distinction between "inactive" and "active" chromatin is more blurred than previously thought. The maintenance of heterochromatin requires the transcription of non-coding RNAs, at least in yeast, and in mammals the binding of sequence-specific transcription factors (Zofall et al. 2012; Bulut-Karslioglu et al. 2012).

1.5 Maintaining Active Chromatin

We learned above that the maintenance of silent chromatin is a highly dynamic, but self-sustaining process. Its heritability during cell division is mediated by the fact that many of the enzymatic activities involved in inactive chromatin maintenance associate with the DNA-replication machinery and faithfully copy the epigenetic template as shown for polycomb complexes (Follmer et al. 2012). In contrast, the mechanisms that propagate the active transcriptional state through cell division have been much less clear. Moreover, most transcriptional activators are stripped off their binding sites during DNA replication and when chromosomes become highly compacted during mitosis (Martinez-Balbas et al. 1995). However, in recent years, several mechanisms have been described that function as "mitotic bookmarks" facilitating the re-assembly of productive transcription complexes

(Zaidi et al. 2011). This included the mitotically stable binding of transcription factors such as GATA1 (Kadauke et al. 2012) and FOXA1 (Caravaca et al. 2013). The number of such factors that can be retained during mitosis is constantly increasing (Kadauke and Blobel 2012). For example, the histone H3 lysine 4 methylase MLL, which is associated with active promoters, is also retained at active genes during mitosis (Blobel et al. 2009). In all described cases, "bookmarked" genes were activated with accelerated kinetics after cell division compared to non-marked genes.

1.6 Deregulation of Gene Expression in Disease

The different mechanisms described above ensure that specific genes are transcribed only when needed: at the right developmental stage, in the right environmental context, in the right cell type and at the right level. For a few model genes their developmental control has been studied in great depth (Higgs et al. 2012). Such studies show a mind-boggling complexity of different mechanisms ensuring that such genes are correctly regulated in development. Billions of years of evolution have fine-tuned DNA-coded transcriptional networks to an extent that they have an inherent ability to use a few initial materials, (i.e. deposition of maternal proteins in the egg) and external signals (i.e. blastocyst implantation) to kick off a process that culminates in the development of an adult organism which can maintain its structure for many years and is even capable of reproducing. However, even the most perfect system is not immune to being derailed. In the next chapters, a number of authors will describe the mechanisms of epigenetic control of normal blood cell development and how the intricate balance between gene activation and silencing is disturbed in disease. Let the journey begin.

References

Adams CC, Workman JL (1995) Binding of disparate transcriptional activators to nucleosomal DNA is inherently cooperative. Mol Cell Biol 15:1405–1421

Ballare C, Castellano G, Gaveglia L et al (2013) Nucleosome-driven transcription factor binding and gene regulation. Mol Cell 49:67–79

Blobel GA, Kadauke S, Wang E, Lau AW, Zuber J, Chou MM, Vakoc CR (2009) A reconfigured pattern of MLL occupancy within mitotic chromatin promotes rapid transcriptional reactivation following mitotic exit. Mol Cell 36:970–983

Bulut-Karslioglu A, Perrera V, Scaranaro M, de la Rosa-Velazquez IA et al (2012) A transcription factor-based mechanism for mouse heterochromatin formation. Nat Struct Mol Biol 19:1023–30

Caravaca JM, Donahue G, Becker JS, He X, Vinson C, Zaret KS (2013) Bookmarking by specific and nonspecific binding of FoxA1 pioneer factor to mitotic chromosomes. Genes Dev 27:251–260

Cirillo LA, Zaret KS (2007) Specific interactions of the wing domains of FOXA1 transcription factor with DNA. J Mol Biol 366:720–724

Cirillo LA, Lin FR, Cuesta I, Friedman D, Jarnik M, Zaret KS (2002) Opening of compacted chromatin by early developmental transcription factors HNF3 (FoxA) and GATA-4. Mol Cell 9:279–289

Cockerill PN (2011) Structure and function of active chromatin and DNase I Hypersensitive Sites. FEBS J 278(13):2182–2210

Dawson MA, Bannister AJ, Gottgens B, Foster SD, Bartke T, Green AR, Kouzarides T (2009) JAK2 phosphorylates histone H3Y41 and excludes HP1alpha from chromatin. Nature 461:819–822

Fischle W, Tseng BS, Dormann HL et al (2005) Regulation of HP1-chromatin binding by histone H3 methylation and phosphorylation. Nature 438:1116–1122

Follmer NE, Wani AH, Francis NJ (2012) A polycomb group protein is retained at specific sites on chromatin in mitosis. PLoS Genet 8:e1003135

Gehani SS, Agrawal-Singh S, Dietrich N, Christophersen NS, Helin K, Hansen K (2010) Polycomb group protein displacement and gene activation through MSK-dependent H3K27me3S28 phosphorylation. Mol Cell 39:886–900

Guccione E, Bassi C, Casadio F et al (2007) Methylation of histone H3R2 by PRMT6 and H3K4 by an MLL complex are mutually exclusive. Nature 449:933–937

Hathaway NA, Bell O, Hodges C, Miller EL, Neel DS, Crabtree GR (2012) Dynamics and memory of heterochromatin in living cells. Cell 149:1447–1460

Hatta M, Cirillo LA (2007) Chromatin opening and stable perturbation of core histone: DNA contacts by FoxO1. J Biol Chem 282:35583–35593

Higgs DR, Engel JD, Stamatoyannopoulos G (2012) Thalassaemia. Lancet 379:373–383

Hu Y, Kireev I, Plutz M, Ashourian N, Belmont AS (2009) Large-scale chromatin structure of inducible genes: transcription on a condensed, linear template. J Cell Biol 185:87–100

Ingram RM, Valeaux S, Wilson N et al (2011) Differential regulation of sense and antisense promoter activity at the Csf1R locus in B cells by the transcription factor PAX5. Exp Hematol 39(730–740):e731–e732

Kadauke S, Blobel GA (2012) "Remembering" tissue-specific transcription patterns through mitosis. Cell Cycle 11:3911–3912

Kadauke S, Udugama MI, Pawlicki JM et al (2012) Tissue-specific mitotic bookmarking by hematopoietic transcription factor GATA1. Cell 150:725–737

Kamakaka RT, Thomas JO (1990) Chromatin structure of transcriptionally competent and repressed genes. EMBO J 9:3997–4006

Kireev I, Lakonishok M, Liu W, Joshi VN, Powell R, Belmont AS (2008) In vivo immunogold labeling confirms large-scale chromatin folding motifs. Nat Methods 5:311–313

Kirmizis A, Santos-Rosa H, Penkett CJ et al (2007) Arginine methylation at histone H3R2 controls deposition of H3K4 trimethylation. Nature 449:928–932

Lefevre P, Witham J, Lacroix CE, Cockerill PN, Bonifer C (2008) The LPS-induced transcriptional upregulation of the chicken lysozyme locus involves CTCF eviction and noncoding RNA transcription. Mol Cell 32:129–139

Lever MA, Th'ng JP, Sun X, Hendzel MJ (2000) Rapid exchange of histone H1.1 on chromatin in living human cells. Nature 408:873–876

Luger K, Mader AW, Richmond RK, Sargent DF, Richmond TJ (1997) Crystal structure of the nucleosome core particle at 2.8 A resolution. Nature 389:251–260

Martinez-Balbas MA, Dey A, Rabindran SK, Ozato K, Wu C (1995) Displacement of sequence-specific transcription factors from mitotic chromatin. Cell 83:29–38

Misteli T, Gunjan A, Hock R, Bustin M, Brown DT (2000) Dynamic binding of histone H1 to chromatin in living cells. Nature 408:877–881

Narlikar GJ, Fan HY, Kingston RE (2002) Cooperation between complexes that regulate chromatin structure and transcription. Cell 108:475–487

Robinson PJ, An W, Routh A, Martino F, Chapman L, Roeder RG, Rhodes D (2008) 30 nm chromatin fibre decompaction requires both H4-K16 acetylation and linker histone eviction. J Mol Biol 381:816–825

Schmiedeberg L, Weisshart K, Diekmann S, Meyer Zu Hoerste G, Hemmerich P (2004) High- and low-mobility populations of HP1 in heterochromatin of mammalian cells. Mol Biol Cell 15: 2819–2833
Sekiya T, Muthurajan UM, Luger K, Tulin AV, Zaret KS (2009) Nucleosome-binding affinity as a primary determinant of the nuclear mobility of the pioneer transcription factor FoxA. Genes Dev 23:804–809
Shahbazian MD, Grunstein M (2007) Functions of site-specific histone acetylation and deacetylation. Annu Rev Biochem 76:75–100
Shilatifard A (2008) Molecular implementation and physiological roles for histone H3 lysine 4 (H3K4) methylation. Curr Opin Cell Biol 20:341–348
Shogren-Knaak M, Ishii H, Sun JM, Pazin MJ, Davie JR, Peterson CL (2006) Histone H4-K16 acetylation controls chromatin structure and protein interactions. Science 311:844–847
Suganuma T, Workman JL (2008) Crosstalk among histone modifications. Cell 135:604–607
Syed SH, Goutte-Gattat D, Becker N et al (2010) Single-base resolution mapping of H1-nucleosome interactions and 3D organization of the nucleosome. Proc Natl Acad Sci USA 107:9620–9625
Tropberger P, Pott S, Keller C et al (2013) Regulation of transcription through acetylation of H3K122 on the lateral surface of the histone octamer. Cell 152:859–872
Voss TC, Schiltz RL, Sung MH et al (2011) Dynamic exchange at regulatory elements during chromatin remodeling underlies assisted loading mechanism. Cell 146:544–554
Williams SK, Truong D, Tyler JK (2008) Acetylation in the globular core of histone H3 on lysine-56 promotes chromatin disassembly during transcriptional activation. Proc Natl Acad Sci USA 105:9000–9005
Wolffe AP (1989) Dominant and specific repression of Xenopus oocyte 5S RNA genes and satellite I DNA by histone H1. EMBO J 8:527–537
Woodcock CL, Skoultchi AI, Fan Y (2006) Role of linker histone in chromatin structure and function: H1 stoichiometry and nucleosome repeat length. Chromosome Res 14:17–25
Zaidi SK, Young DW, Montecino M, van Wijnen AJ, Stein JL, Lian JB, Stein GS (2011) Bookmarking the genome: maintenance of epigenetic information. J Biol Chem 286:18355–18361
Zaret KS, Carroll JS (2011) Pioneer transcription factors: establishing competence for gene expression. Genes Dev 25:2227–2241
Zofall M, Yamanaka S, Reyes-Turcu FE, Zhang K, Rubin C, Grewal SI (2012) RNA elimination machinery targeting meiotic mRNAs promotes facultative heterochromatin formation. Science 335:96–100

Part I
Nonmammalian Hematopoiesis

Chapter 2
What Can We Learn from Flies: Epigenetic Mechanisms Regulating Blood Cell Development in *Drosophila*

Paul Badenhorst

Abstract *Drosophila* (fruit flies) possess a highly effective innate immune system that provides defence against pathogens that include bacteria, fungi and parasites. Pathogens are neutralised by mechanisms that include phagocytosis, encapsulation and melanisation. Circulating cells called haemocytes are a key component of the innate immune system and include cells that resemble the granulocyte–macrophage lineages of mammals. The mechanisms that regulate *Drosophila* haematopoietic progenitor specification and differentiation are highly conserved, allowing *Drosophila* to be used as a useful model to understand transcriptional regulation of haematopoiesis. In this review I will summarise the mesodermal origin of *Drosophila* haemocyte precursors and describe parallels with mammalian haemangioblast precursors. I will discuss key signalling pathways and transcription factors that regulate differentiation of the three principal haemocyte cell types. There are significant parallels with the transcriptional circuitry that controls mammalian haematopoiesis, with transcription factors such as GATA factors, RUNX family members and STAT proteins influencing the specification and differentiation of *Drosophila* haemocytes. These transcription factors recruit co-repressor or co-activator complexes that alter chromatin structure to regulate gene expression. I will discuss how the *Drosophila* haematopoietic compartment has been used to explore function of ATP-dependent chromatin remodelling complexes and histone modifying complexes. As key regulators of haematopoiesis are conserved, the great genetic amenability of *Drosophila* offers a powerful system to dissect function of leukaemogenic fusion proteins such as RUNX1-ETO. In the final section of the review the use of genetic screens to identify novel RUNX1-ETO interacting factors will be discussed.

Keywords Drosophila innate immunity • Haemocyte • Plasmatocyte • Lamellocyte • Chromatin remodelling • NURF

P. Badenhorst (✉)
School of Infection and Immunity, College of Medical and Dental Sciences, Institute of Biomedical Research, University of Birmingham, Edgbaston, Birmingham, UK
e-mail: p.w.badenhorst@bham.ac.uk

C. Bonifer and P.N. Cockerill (eds.), *Transcriptional and Epigenetic Mechanisms Regulating Normal and Aberrant Blood Cell Development*, Epigenetics and Human Health, DOI 10.1007/978-3-642-45198-0_2, © Springer-Verlag Berlin Heidelberg 2014

2.1 *Drosophila* Cellular Innate Immune Function

Leukocytes are key mediators of the innate immune responses of both humans and invertebrates. *Drosophila* possess leukocyte-like cells (called haemocytes) that are able to neutralise fungal and bacterial pathogens and parasites. Extensive work by the Rizki and colleagues in the 1950s identified three circulating haemocyte cell types in *Drosophila* larvae (Rizki 1957a). The most abundant are *plasmatocytes*, which account for approximately 95 % of circulating haemocytes. Plasmatocytes can function as macrophages to remove bacteria, foreign material and apoptotic cells by phagocytosis (Salt 1970; Rizki and Rizki 1980; Tepass et al. 1994; Franc et al. 1996). Plasmatocytes have additional functions in tissue remodelling through their ability to secrete components of the extracellular matrix (Fessler and Fessler 1989). The plasmatocyte appears to be a plastic cell type and, like monocytes, has the ability to differentiate into a number of activated cell types that include macrophages, podocytes and *lamellocytes* [See Fig. 2.1 and also (Rizki 1957a; Gateff 1978b)]. Lamellocytes are large flattened cells that are responsible for encapsulating foreign material or aberrant/damaged host tissue that is recognised as "non-self" (Salt 1970; Rizki and Rizki 1974). Lamellocytes occur rarely in larval haemolymph in the absence of immune challenge. However, large numbers differentiate either upon infestation by parasitic wasps (Nappi and Streams 1969; Rizki and Rizki 1992) or in a number of so-called melanotic "tumour" mutant strains (Rizki 1957b; Sparrow 1978). The third cell type that is detected is the *crystal cell*, which constitutes approximately 5 % of larval haemocytes (Gateff 1978a). Crystal cells contain a variable number of large paracrystalline inclusions (Rizki 1957a) that contain precursors of melanin that can be oxidised by phenoloxidase (PO) located in the cytoplasm of crystal cells (Rizki and Rizki 1959).

Drosophila larvae and adults have an open circulatory system. Haemocytes are circulated in the haemolymph via contractions of a primitive single chambered heart (the dorsal vessel) and by peristaltic contractions of the body in larvae (Lanot et al. 2001). It is important to note that *Drosophila* are devoid of oxygen transporting blood cells; oxygen transport is mediated by direct contact with a branching network of trachea (Poulson 1950). The three *Drosophila* haemocytes cell types are solely responsible for innate immune function of *Drosophila* and mediate three key responses that are respectively *phagocytosis*, *encapsulation* and *melanisation*.

2.1.1 *Phagocytosis*

Targeted ablation of plasmatocytes by induced apoptosis confirms that plasmatocytes are responsible for the removal of microorganisms and apoptotic material by phagocytosis. Depletion of plasmatocytes in adults reduces bacterial clearing and decreases survival after infection (Charroux and Royet 2009; Defaye et al. 2009) and in embryos causes lethality due to defects in CNS morphology as a

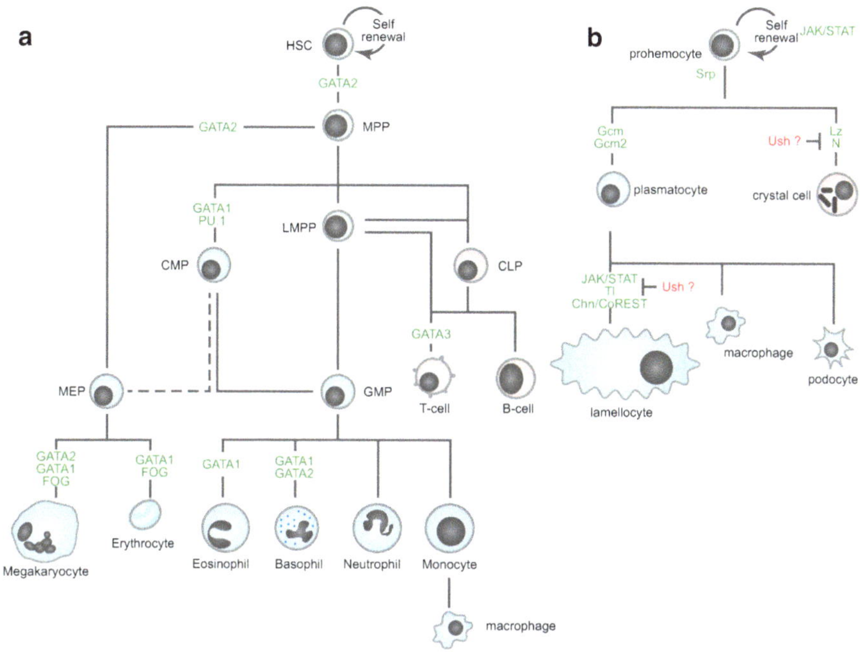

Fig. 2.1 Comparison of human and *Drosophila* haematopoietic lineages. (**a**) Human haematopoietic lineages showing origin of granulocyte/macrophage, erythroid and lymphoid lineages. GATA factors play key roles in maintenance of haematopoietic precursors and differentiation of major haematopoietic cell types. (**b**) *Drosophila* haematopoiesis. Three major differentiated cell types are detected: plasmatocytes, crystal cells and lamellocytes. The GATA factor Srp plays a key role in specifying haematopoietic progenitors (prohaemocytes). Transcription factors implicated in lineage differentiation are indicated (*red* antagonises, *green* confers fates). No lymphoid adaptive immune cells or erythroid cells are detected in *Drosophila*. Only granulocyte/macrophage-type innate immune effectors are present. *HSC* haematopoietic stem cell, *CMP* common myeloid progenitor, *CLP* common lymphoid progenitor, *MEP* megakaryocytic/erythroid progenitor, *MPP* multipotent progenitor, *LMPP* lymphoid-restricted multipotent progenitor, *GMP* granulocyte–monocyte progenitor

result of failure to clear apoptotic cells (Defaye et al. 2009). A particular advantage of the *Drosophila* system is the ease of both forward and reverse genetic approaches to identify factors required for recognition of bacterial and fungal pathogens and apoptotic cells by plasmatocytes (Franc et al. 1996, 1999; Ramet et al. 2002; Philips et al. 2005; Stuart et al. 2005; Stroschein-Stevenson et al. 2006). These screens have identified conserved proteins that are required both for the recognition of particles to be engulfed and for subsequent internalisation in a specialised vesicle compartment the phagosome.

Recognition factors include cell surface receptors that bind directly to particles to be engulfed and opsonins that coat the particle and serve as a signal for recognition by cell surface receptors. In the case of apoptotic cells the key mediator of recognition is the CD36 homologue Croquemort (Franc et al. 1996, 1999),

However, CD36 is a multi-ligand receptor that is also able to recognise *Staphylococcus aureus* (Stuart et al. 2005). CD36 is a class B scavenger receptor (SR), and other scavenger receptors including the SR-BI homologue Peste and the class C scavenger receptor (SR-CI) have been shown to bind microbes (Ramet et al. 2001; Philips et al. 2005). A second group of receptors include the EGF repeat containing proteins Eater (Kocks et al. 2005)and Nimrod C1 (Kurucz et al. 2007) that are able to bind to bacterial surfaces via the EGF repeats, and Draper that is required for removal of apoptotic glial cells (Freeman et al. 2003). Opsonins include the thioester containing proteins (TEPs) that are related to mammalian α_2 macroglobulin and C3 (Lagueux et al. 2000). TEPs are secreted into the haemolymph and up-regulated after microbial challenge (Lagueux et al. 2000; Johansson et al. 2005) and have been shown to bind microbes and enhance phagocytosis (Stroschein-Stevenson et al. 2006).

2.1.2 Encapsulation

Particles that are too large to be engulfed during phagocytosis are neutralised by encapsulation that effectively walls off particles in inert masses coated with a dense layer of melanin. Lamellocytes are primarily responsible for the encapsulation response and recognise both foreign material, such as parasites, and aberrant/damaged tissue (Salt 1970; Rizki and Rizki 1974). A normal pathogen target of lamellocytes is the egg and larval forms of parasitoid wasps such as *Leptopilina*. Female parasitoid wasps use an ovipositor to inject eggs into the body cavity of larvae of another host insect species. These eggs hatch into larvae that complete the initial stages of their life cycles inside the host, consuming the host to sustain their development. Lamellocytes are seldom detected in larval haemolymph in the absence of immune challenge, but large numbers differentiate upon infestation by parasitoid wasps (Nappi and Streams 1969; Rizki and Rizki 1992) in an attempt to encapsulate and neutralise the injected wasp eggs (Russo et al. 1996; Williams 2009). Lamellocyte differentiation is accompanied by up-regulation of cell adhesion molecules such as integrins (Irving et al. 2005; Kwon et al. 2008), up-regulation of markers of actin polymerisation (Stofanko et al. 2008) and factors that link integrins to cytoskeleton such as Vinculin (Wertheim et al. 2005; Kwon et al. 2008) and changes in the distribution of the *Drosophila* L1CAM homologue Neuroglian (Williams 2009). These changes are potentially required for adhesion to the wasp egg, but also homotypic adhesion of lamellocytes to form a capsule surrounding particles. The capsule is subsequently melanised to generate an inert nodule that neutralises the pathogen. It had been speculated that crystal cells participate in the melanisation of these capsules (Rizki and Rizki 1980); however, it has subsequently been shown that lamellocytes may also express phenoloxidase enzymes required for melanisation (Kwon et al. 2008; Nam et al. 2008). During the process of melanisation, cytotoxic reactive oxygen and nitrogen species can potentially be generated and function in pathogen killing (Christensen et al. 2005), as evidenced by rises in the levels of NO radicals during the response to parasitisation (Carton et al. 2009).

Lamellocytes also differentiate in response to aberrant or damaged tissue or dysregulation of haematopoiesis to produce so-called "melanotic tumours" (Rizki and Rizki 1974; Sparrow 1978). These are not true neoplasms as they are incapable of autonomous growth or invasion but are more appropriately termed melanotic pseudotumours (Barigozzi 1969). Melanotic tumours arise either as free-floating aggregates of lamellocytes in the haemocoel or as fixed accumulations of lamellocytes, typically near the caudal fat body, in which lamellocytes appear to encapsulate host tissue. It is speculated that these occur as a result of recognition of tissue as "non-self" through disruption of the basement membrane of tissue or appearance of fat body contents in the haemocoel (Rizki and Rizki 1974). Plasmatocytes are known to secrete components of the extracellular matrix (Fessler et al. 1994) and it has been proposed that this normally renders them neutral to surfaces covered by the proteins they secrete. Removal of these surfaces would allow lamellocyte reaction. As during the normal response to parasitoid wasp eggs, these lamellocyte aggregates subsequently melanise to generate blackened masses that can be readily observed both in larva and in adults (See Fig. 2.8a). The ease of visualising melanotic tumours has allowed both traditional genetic screens and inducible RNAi screens to identify melanotic tumour suppressor genes (Barigozzi 1969; Sparrow 1978; Watson et al. 1991; Garzino et al. 1992; Hanratty and Dearolf 1993; Harrison et al. 1995; Rodriguez et al. 1996; Avet-Rochex et al. 2010). As shall be discussed later this has provided a convenient assay and tool to explore functions of epigenetic regulators in the control of *Drosophila* haematopoietic function.

2.1.3 Melanisation

The final innate immune response mediated by haemocytes is the process of melanisation that is required during wound healing and coagulation (Galko and Krasnow 2004; Bidla et al. 2007). Crystal cells are key mediators of melanisation responses. They have long been recognised to be exquisitely sensitive to changes in the haemolymph, releasing paracrystalline inclusions of melanin precursors and phenoloxidase (PO) into the surrounding medium when activated (Rizki and Rizki 1980). It is understood that PO is produced as an inactive precursor (prophenoloxidase, proPO) that is converted to active PO by haemolymph (humoral) serine proteinase cascades allowing integration of the cellular and humoral innate responses [reviewed in Cerenius et al. (2008), Cerenius et al. (2010)]. Although melanin is not toxic, cytotoxic reactive oxygen and nitrogen species are generated as by-products of the melanisation cascade and can function in bacterial and pathogen killing (Christensen et al. 2005). Thus, while morphologically quite distinct from mammalian granulocytes, crystal cells may be functionally related to granulocytes that release cytotoxic agents during degranulation that accompanies granulocyte activation.

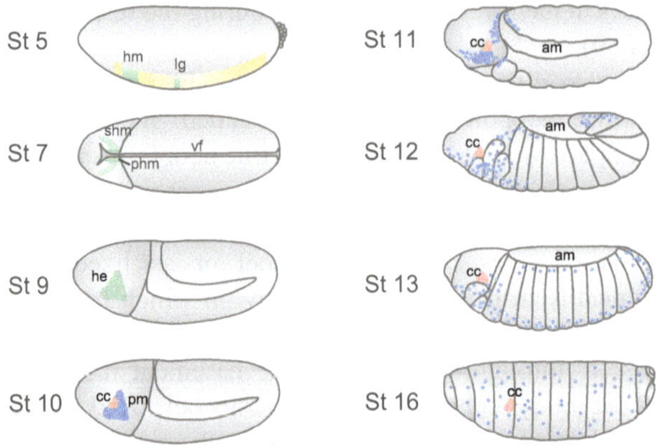

Fig. 2.2 *Drosophila* embryonic haematopoiesis. Schematic showing origin of *Drosophila* haematopoietic precursors and development of the embryonic haematopoietic system. Embryo-derived haemocytes (he) originate from the procephalic mesoderm which delaminates from the blastoderm surface in two waves, either invaginating through the ventral furrow (vf) during gastrulation to form the primary head mesoderm (phm) or delaminating from the ectoderm as a result of vertically orientated divisions to generate the secondary head mesoderm (shm). Haemocyte precursors from both populations fuse to form a cluster of Srp-expressing haemocytes in the procephalic region on either side of the embryo by embryonic stage 9. Prohaemocytes then differentiate into either crystal cells (cc) or mainly plasmatocytes (pm). During subsequent embryonic stages plasmatocytes disperse through the embryo along well-characterised migration pathways until shortly before hatching they are uniformly spread throughout the embryo. Crystal cell clusters from either side of the embryo will eventually form a single cluster centred on the proventriculus. At larval hatching both plasmatocyte and crystal cell populations disperse into the circulating haemolymph. Embryonic stages are according to (Campos-Ortega and Hartenstein 1985)

2.2 *Drosophila* Haematopoiesis

As in mammals two distinct waves of haematopoiesis can be detected in *Drosophila*. The first occurs in embryonic stages and corresponds loosely with primitive haematopoiesis. The second phase of haematopoiesis commences during larval stages in the lymph glands and is speculated to correspond to definitive haematopoiesis. As summarised in Fig. 2.2, cell fate mapping studies have revealed that haemocytes originate from two distinct anlagens in the mesoderm of blastoderm stage embryos (Holz et al. 2003). The first that generates embryonic haemocytes corresponds to a portion of the head mesoderm (Fig. 2.2, hm). The second anlagen is present in the trunk mesoderm and exclusively generates the lymph gland lobes that are responsible for definitive haematopoiesis (Fig. 2.2, lg). In the following section I describe how these cells give rise to the different types of haematopoietic cells.

2.2.1 Embryonic Haematopoiesis

The head mesoderm that will generate the embryonic haemocytes from originates two phases. As shown in Fig. 2.2, during gastrulation a part of the head mesoderm, the primary head mesoderm (phm), invaginates as the anterior portion of the ventral furrow (de Velasco et al. 2006). Additional head mesoderm is also generated during a secondary process of delamination events to generate the secondary head mesoderm (shm). The secondary head mesoderm is generated in part by division of cells of the surface epithelium in a plane vertical to the epithelium. This results in the generation of inner daughter cells which become the secondary head mesoderm and outer cells that remain ectoderm (de Velasco et al. 2006).

The secondary and primary head mesoderm cells intermingle to form two monolayered sheets of cells on either side of the midline of the embryo. These migrate dorsally and by stage 9 of embryogenesis form two plates of cells that can be recognised as haemocyte precursors (prohaemocytes) that express the GATA factor Serpent (Srp) (Rehorn et al. 1996). By stage 10 of embryogenesis, these prohaemocytes differentiate into either plasmatocytes (pm) or between 20 and 30 crystal cells (cc) (Lebestky et al. 2000; Fossett et al. 2003; Waltzer et al. 2003). In the embryo only these two haemocyte cell types are generated; lamellocytes are never observed prior to larval stages. The crystal cells remain localised as bilateral clusters on either side of the embryo. However by embryonic stage 11 the plasmatocytes disperse and follow a number of highly stereotyped migration pathways through the embryo (Tepass et al. 1994; Cho et al. 2002; Bruckner et al. 2004). Plasmatocytes migrate across the amnioserosa (Fig. 2.2, am) towards the caudal end of the germband-extended embryo, forming a distinct cluster of plasmatocytes once germband retraction commences (Fig. 2.2, stage 12). Subsequently, plasmatocytes migrate through the developing nerve cord, the gut and dorsal epidermis eventually becoming uniformly dispersed prior to larval hatching. By this stage the two bilateral clusters of crystal cells merge to form a loose aggregate surrounding part of the gut, the proventriculus (Lebestky et al. 2000).

Both plasmatocytes and crystal cells persist into larval stages and constitute the circulating haemocytes found in larval stages (Lanot et al. 2001; Holz et al. 2003). It is important to stress that haemocytes generated in the lymph glands during the second wave of haematopoiesis are not liberated into circulation under normal circumstances (Holz et al. 2003; Grigorian et al. 2011) so that all cells in circulation in larvae derive from embryonic haematopoiesis. At the end of embryogenesis there are approximately 700 plasmatocytes (Tepass et al. 1994), but these increase by division to generate in excess of 5,000 plasmatocytes by the end of larval stages (Lanot et al. 2001). This is largely due to increases in plasmatocyte numbers as these are the only haemocyte types that have been observed to undergo cell division (Rizki 1978; Lanot et al. 2001). In third instar larva approximately two-thirds of haemocytes freely circulate in the haemolymph; the remainder attach to the inner surface of the cuticle to form a number of segmentally repeated sessile compartments that contain both plasmatocytes and crystal cells (Lanot et al. 2001; Stofanko et al. 2008; Makhijani et al. 2011). The function of these sessile compartments is unclear, although it has been proposed that they provide a progenitor pool for lamellocytes (Markus et al. 2009), immune sentinels or a depot function that is liberated upon infection (Stofanko et al. 2010).

2.2.2 *Post-Embryonic Haematopoiesis*

The second wave of haematopoiesis is initiated in the lymph glands during larval stages. Haemocytes generated in the lymph gland are not liberated into circulation until after metamorphosis and together with haemocytes of embryonic origin will contribute to the circulating pupal and adult haemocyte pool (Lanot et al. 2001; Holz et al. 2003; Grigorian et al. 2011). Development of the lymph gland initiates during embryonic stages although haemocytes only start to differentiate in the lymph gland during larval stages. The development of the lymph gland is intimately associated with that of the cardioblasts of the primitive heart (the dorsal vessel) and the associated pericardial cells. Indeed lineage tracing experiments demonstrate the existence of a common precursor for both the lymph gland and cardioblasts, a linkage that parallels the common vascular and blood haemangioblast precursors found in the aorta-gonad-mesonephros region of vertebrate embryos (Medvinsky et al. 1993; Medvinsky and Dzierzak 1996; Mandal et al. 2004).

2.2.2.1 Development of the Lymph Gland

During gastrulation in *Drosophila* embryos the ventral part of the blastoderm invaginates through the ventral furrow (Fig. 2.2, vf) to form mesoderm that then spreads dorsally as a monolayer of cells along the inner surface of the ectoderm. The dorsal mesoderm (Fig. 2.3, dm), the dorsal-most strip of this mesoderm, generates cardioblast and lymph gland precursors (Bodmer 1993). Potential to form the lymph gland and cardioblasts becomes restricted to clusters of cells in each segment (Fig. 2.3, cm). This restriction is mediated through the co-ordinate action of the BMP-4 (Dpp), FGF (Htl), Wnt (Wg) and Notch signalling pathways on the cardiogenic mesoderm. BMP-4, FGF and Wnt favour while Notch antagonises cardiogenic mesoderm development (Frasch 1995; Wu et al. 1995; Beiman et al. 1996; Mandal et al. 2004; Stathopoulos et al. 2004). These pathways cooperate to turn on expression of the GATA-4, -5, -6 homologue Pannier (Pnr) (Klinedinst and Bodmer 2003) and the Nkx2.5 homologue Tinman (Tin) (Bodmer 1993) in the cardiogenic mesoderm. At the start of germband retraction, the cardiogenic mesoderm can be observed as a row segmentally repeated clusters of cells in close juxtaposition to the amnioserosa on either side of the embryo (Fig. 2.3, stage 12).

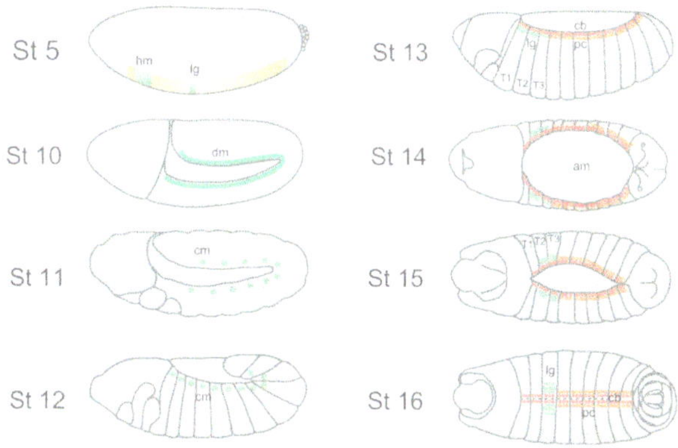

Fig. 2.3 Developmental origin of the larval lymph gland and dorsal vessel. The haematopoietic precursors of the larval lymph gland and cardioblasts that generate the dorsal vessel derive from cardiogenic mesoderm progenitors (cm) located in the dorsal mesoderm (dm) of the embryo. These divide to generate medially cardioblasts (cb) or laterally either lymph gland (lg) or pericardial nephrocyte precursors (pc). In thoracic segments (T1–T3) lymph precursors are generated while in abdominal segments pericardial nephrocyte precursors (pc) are formed. Initially lymph gland precursor populations on either side of the embryo form three spatially distinct populations along the anterior–posterior axis, but these fuse by embryonic stage 16 to form a single cluster located in segment T3. At the same time cardioblast, lymph gland and pericardial nephrocyte precursors from either side of the embryo move towards the dorsal midline of the embryo during the process of dorsal closure. This involves the dorsally directed migration of the lateral mesoderm and epidermis from either side of the embryo, during which the two flanks move over the amnioserosa and fuse along the dorsal midline. The dorsal vessel is formed from two rows of cardioblasts that run the length of the embryo. Lymph gland clusters from either side of the embryo remain separated and form the two primary lobes of the larval lymph gland. These express the GATA factor Srp and are composed of prohaemocyte precursors. Embryos in stages 5–13 are shown in *lateral view*. Embryos in stages 14–16 are shown in *dorsal view*. Embryonic stages are according to Campos-Ortega and Hartenstein (1985)

During germband retraction (Fig. 2.3, stage 13) the cardiogenic mesoderm divides to produce two cell lineages—medial cardioblasts (cb) that maintain expression of Pnr and Tin and precursors of the lymph gland (lg) and the pericardial nephrocytes (pc) that express the zinc finger transcription factor Odd skipped (Odd) and down-regulate expression of Pnr and Tin (Ward and Skeath 2000; Mandal et al. 2004). Restriction of cardioblast versus lymph gland and pericardial nephrocyte fate requires a second function of Notch to inhibit cardioblast development. Selective activation of Notch in the lymph gland and pericardial nephrocyte precursors appears to be achieved by asymmetric division of the cardiogenic mesoderm precursors and unequal partitioning of determinants such as Numb (Ward and Skeath 2000). Subsequently, lymph gland fate is restricted to the anterior of the embryo as a result of regulatory input from the HOX genes that are differentially expressed along the anterior–posterior axis of the embryo. In particular Ultrabithorax (Ubx) that is expressed in abdominal segments inhibits lymph gland development and allows development of pericardial nephrocyte fate (Mandal et al. 2004). As a result three clusters of lymph gland precursors are generated in thoracic segments T1–T3, while in abdominal segments pericardial nephrocytes develop (Fig. 2.3, stage 12). Lymph gland precursors then express the GATA transcription factor Serpent (Srp) that confers haemocyte fate, as during embryonic haematopoiesis.

Initially the lymph gland clusters are well separated, but during the process of dorsal closure they move posteriorly and coalesce into a single cluster in segment T3 that will form the primary lobe of the lymph gland (Fig. 2.3, stage 16). Moreover, during the process of dorsal closure the lateral edges of the epidermis together with the cardiogenic mesoderm also migrate towards the dorsal midline of the embryo and fuses to bring together cardioblast and lymph gland precursors that were initially on opposite sides of the embryo (Fig. 2.3, compare stage 14 and stage 16). This generates the final structure of the lymph gland with two lobes of 20–30 prohaemocytes on either side of the future dorsal vessel that runs the length of the embryo.

Within the Srp-expressing lymph gland cells a distinct compartment is generated towards the posterior of the primary lobe (Mandal et al. 2007). This region expresses Serrate, a ligand of the Notch pathway (Lebestky et al. 2003), the *Drosophila* early B-cell factor Collier (Col) (Crozatier et al. 2004), Hedgehog (Mandal et al. 2007) and ligands of the JAK/STAT pathway (Jung et al. 2005; Krzemien et al. 2007). This region, termed the posterior signalling centre (PSC), is speculated to function as a haematopoietic niche that regulates self-renewal and differentiation of flanking prohaemocytes in the lymph gland (Krzemien et al. 2007; Mandal et al. 2007).

2.2.2.2 Lymph Gland Haemematopoiesis

During larval stages the primary lobes of the embryonic lymph gland expand and additional pairs of smaller secondary lobes develop posterior to the primary lobes (Jung et al. 2005). By second instar larval stages there are approximately 200 prohaemocytes in each primary lobe and this number increases tenfold by late third larval instar stages such that prior to pupariation the primary lobes are considerably expanded. Under normal circumstances the secondary lobes remain small and do not contribute significant numbers of haemocytes, but these can be triggered to expand in response to immune challenge (Lanot et al. 2001). The lymph gland is not surrounded by a cellular capsule (Lanot et al. 2001), but exhibits a clear branching network of extracellular matrix (Jung et al. 2005) that maintains structure of the lymph gland and is left behind when differentiated haemocytes are liberated at pupariation (Grigorian et al. 2011).

During early larval stages there is no evidence of differentiation of prohaemocytes. During second larval instar stages markers of mature plasmatocytes begin to be detected (Jung et al. 2005), but these are detected at the periphery of the lobes that are still predominantly composed of replicating prohaemocytes. However, as shown in Fig. 2.4a, during third larval instar stages significant numbers of differentiated haemocyte types, including plasmatocytes,

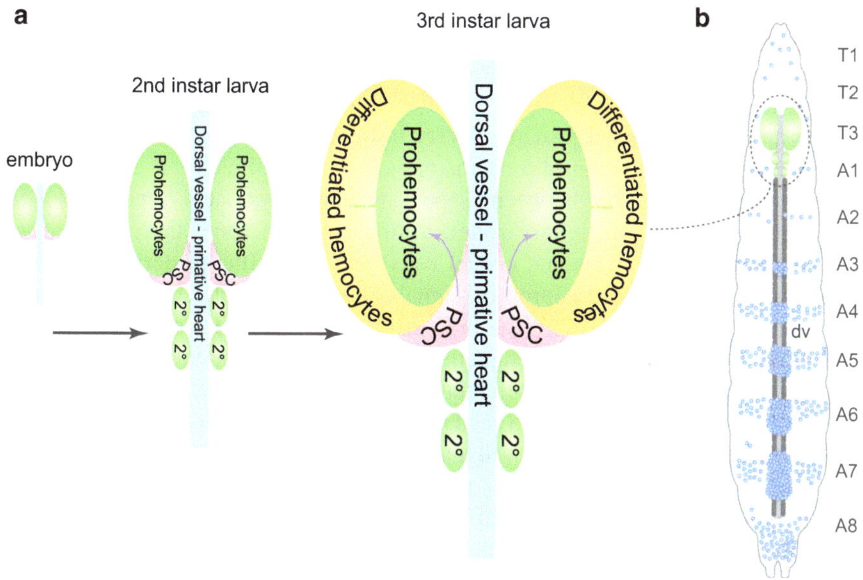

Fig. 2.4 Larval haematopoiesis. (**a**) The second wave of haematopoiesis or definitive haematopoiesis takes place in the paired lymph glands that flank the dorsal vessel. At the end of embryogenesis two regions can be distinguished within the lymph gland, the prohaemocytes (*green*) that give rise to blood cells and the posterior signalling centre (PSC, in *pink*) that acts as a hub to control prohaemocyte self-renewal and differentiation. During early larval stages the primary lobes of the lymph gland increase in size and secondary lobes develop posterior to the primary lobes flanking the dorsal vessel. By third larval instar prohaemocytes within the primary lobes start to differentiate into either plasmatocytes or crystal cells. At this stage regional organisation of the lymph gland into a medullary zone that contains prohaemocytes (*green*) and a cortical zone that contains differentiating haemocytes (*yellow*) can be detected. Under normal circumstances, haemocytes are not liberated from the lymph gland into circulation during larval stages, but are released at pupariation. Under normal conditions secondary lobes remain reduced and show no evidence of haemocyte differentiation until after pupariation when cells are released. (**b**) Haemocytes in circulation during larval stages are embryo-derived haemocytes that persist and continue to replicate after larval hatching. Haemocytes can be detected freely circulating in the haemolymph as well as attached to the inner surface of the integument in stereotyped locations in sessile haematopoietic compartments. Thoracic (T1–T3) and abdominal (A1–A8) segments are indicated

crystal cells and a few lamellocytes can be detected. At this stage the primary lymph gland lobe shows a clear distinction between a medullary zone (MZ) that contains prohaemocytes and a peripheral cortical zone that contains differentiated haemocytes (Jung et al. 2005; Mandal et al. 2007). The two zones can be distinguished by a number of reporters and markers; in particular the medullary zone expresses Domeless and Upd3, receptors and ligands that activate that JAK/STAT pathway (Jung et al. 2005; Krzemien et al. 2007), Wingless the ligand of the Wnt pathway (Sinenko et al. 2009) and the differentiation-regulating translational repressor Bam (Tokusumi et al. 2011). Under normal circumstances the smaller secondary lobes do not show a distinction between medullary and cortical zones and appear to consist of prohaemocytes (Jung et al. 2005) until after pupariation, when the remaining cells appear to differentiate into plasmatocytes (Grigorian et al. 2011).

The larval lymph gland provides a very powerful and experimentally tractable model to explore regulation of a haematopoietic stem cell niche. It exhibits clear ultrastructural distinction between a pool of undifferentiated precursors (the prohaemocytes in the medullary zone), a differentiation zone (the cortical zone that contains plasmatocytes and crystal cells) and a hub [the posterior signalling centre (PSC)] that is the source of signals that regulate the self-renewal and differentiation of the prohaemocyte precursors (Fig. 2.4a). This has already been exploited to define intercellular signalling pathways that can control the balance between self-renewal and differentiation (Lebestky et al. 2003; Krzemien et al. 2007; Mandal et al. 2007; Sinenko et al. 2009). However, it has also begun to be exploited to understand how signals such as oxidative stress (Owusu-Ansah and Banerjee 2009), energy status (Dragojlovic-Munther and Martinez-Agosto 2012), hypoxia (Mukherjee et al. 2011) and insulin signalling (Shim et al. 2012) affect the haematopoietic niche. The challenge now is to exploit this system to understand differences in chromatin structure between progenitors and committed cells within the haematopoietic niche, and how the signals identified above act on the chromatin landscape.

2.3 Transcriptional Control of *Drosophila* Haematopoiesis

The regulatory circuitry that controls *Drosophila* blood cell development is well characterised and demonstrates significant similarity to that governing myeloid differentiation in vertebrates, with transcription factors such as GATA factors, RUNX family members and STAT proteins influencing the specification and differentiation of *Drosophila* haemocytes (Fig. 2.1). As described in preceding sections and shown in Fig. 2.1b, the specification of haemocytes and precursors, the prohaemocytes, requires the expression of the GATA factor Srp (Rehorn et al. 1996; Bernardoni et al. 1997; Lebestky et al. 2000; Mandal et al. 2004). This has obvious parallels to vertebrate haematopoiesis where GATA-1, -2, -3 are required for development of specific haematopoietic lineages (Orkin 1995). Indeed it was initially suggested that the Srp amino acid sequence is more closely related to vertebrate GATA-1, -2, -3 than to GATA-4, -5, -6 (Rehorn et al. 1996). Maintenance of prohaemocytes appears to require activation of the JAK/STAT pathway. In larval lymph glands the medullary zone that contains undifferentiated prohaemocytes expresses Domeless and Upd3, receptors and ligands that activate the JAK/STAT pathway (Krzemien et al. 2007). In mutants that lack the sole *Drosophila* STAT (Stat92E), prohaemocytes prematurely differentiate, suggesting that JAK/STAT is required for prohaemocyte self-renewal (Krzemien et al. 2007). In contrast, activating mutants in the sole *Drosophila* JAK Hopscotch (Hop), which is most closely related to human JAK3, trigger hypertrophy of the larval lymph glands (Harrison et al. 1995; Luo et al. 1995).

The subsequent differentiation of prohaemocytes into plasmatocytes requires the action of both Glial Cells Missing (Gcm) and Gcm2 (Bernardoni et al. 1997; Lebestky et al. 2000; Alfonso and Jones 2002; Bataille et al. 2005). Homologues of both Gcm and Gcm2 are present in mammals but to date have not demonstrated role in haematopoiesis, although the Gcm homologue GCMB has been implicated in parathyroid adenoma (Mannstadt et al. 2011).

In contrast, the development of crystal cells requires the function of the Runx1/AML1 homologue Lozenge (Lz) (Lebestky et al. 2000; Fossett et al. 2003; Waltzer et al. 2003). In loss-of-function Lz mutants crystal cells are lost (Lebestky et al. 2000) while over-expression of Lz in prohaemocytes is sufficient to drive supernumerary crystal cell formation although this only occurs in tissues that express Srp indicating collaboration between GATA factors and Runx1/AML1 (Waltzer et al. 2003). In addition to Lz, activation of the Notch pathway has been shown to be required for crystal cell differentiation both during embryonic and larval haematopoiesis (Duvic et al. 2002; Lebestky et al. 2003). Recent *ch*romatin *i*mmuno*p*recipitation-coupled *seq*uencing (ChIP-Seq) analysis of the Notch transducer Suppressor of Hairless [Su(H)] indicates that Notch enforces crystal cell fates, but that binding to enhancers of target genes requires flanking GATA and Lz sites. Lz binding appears to be required to allow enhancers to respond to Notch (Terriente-Felix et al. 2013).

Lozenge is one of two Runx family members in flies, the other being the class-defining Runt transcription factor (Kania et al. 1990). Runt has no discernable function in *Drosophila* haematopoiesis, but its activity in other tissues has been exploited to characterise mechanisms of function of Runx transcription factors. In the embryo, Runt acts both as a transcriptional repressor of the pair-rule genes *hairy* (*h*) and *even-skipped* (*eve*) (Manoukian and Krause 1993; Aronson et al. 1997) and activator of the sex-determining gene *Sex-lethal* (*Sxl*) (Kramer et al. 1999). Lz shows similar dichotomy and in the fly eye, where Lz is also expressed, can either activate *dPax2* or repress *Deadpan* (*Dpn*) expression (Canon and Banerjee 2003). Repression by both Runt and Lz can be mediated by recruitment of the Groucho [in humans Transducin-Like Enhancer of split (TLE)] repressor protein (Aronson et al. 1997; Canon and Banerjee 2003), a feature conserved in vertebrate Runx1/AML1 (Levanon et al. 1998). Groucho (Gro) is a dedicated co-repressor first shown to be recruited by WRPW motifs on target proteins (Paroush et al. 1994). The domain bound by Gro on Runx proteins is the related conserved peptide VWRPY (Aronson et al. 1997). Although both VWRPY and WRPW motifs are required for Gro-mediated repression in vivo (Aronson et al. 1997; Canon and Banerjee 2003), there are some distinctions between the mechanisms of action of these peptides. Gro binding to VWRPY is weaker than that observed with WRPW (Jennings et al. 2006) and the VPRWY motif appears to function as a regulatable repressor domain unlike WRPW, which is a constitutive repressor. Thus, in the fly eye, while VPRWY-containing Lz rescue constructs both activate *dPax2* and repress *Dpn*, mutated VWRPY constructs only activate *dPax2* but fail to repress *Dpn*. In

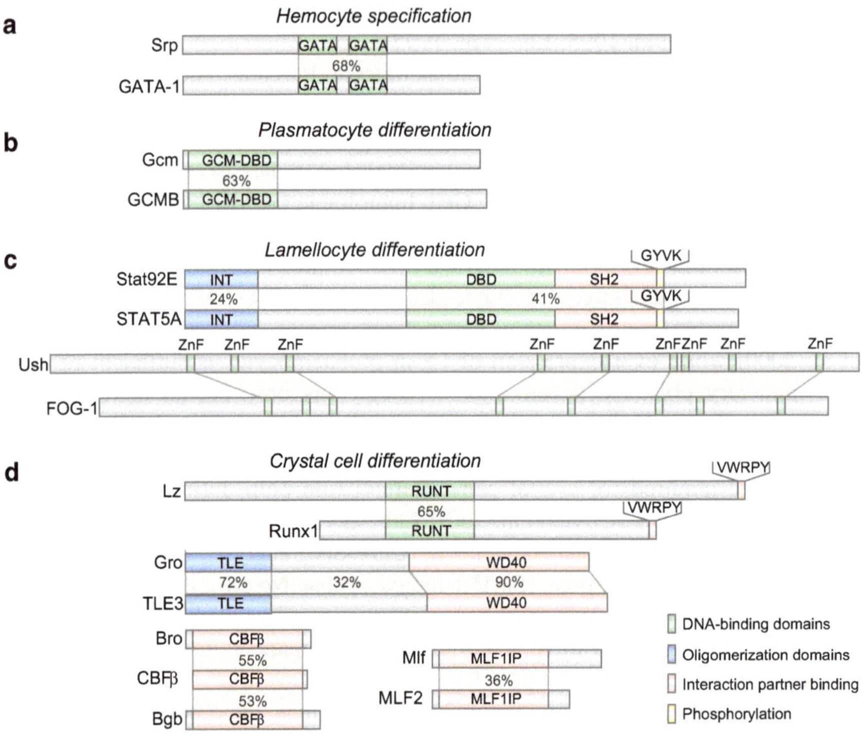

Fig. 2.5 Transcription factors that regulate *Drosophila* haematopoiesis are conserved. The key transcription factors that regulate (**a**) haemocyte specification and (**b**) plasmatocyte, (**c**) lamellocyte and (**d**) crystal cell differentiation are shown together with known human homologues. Conserved domains and regions of homology are indicated. Percentage amino acid identity in regions of homology is denoted. Conserved domains are *colour coded* according to function as shown in the *key*

contrast, WRPW substitution constructs fail to activate *dPax2* but repress *Dpn*. It appears that the VWRPY motif may be regulated through the binding of co-factor proteins like Cut (the homologue of CCAAT displacement protein (CDP) which has been shown to enhance binding of Lz to Gro (Canon and Banerjee 2003). However, it is equally feasible that the VPRWY motif provides a platform for integrating signal inputs from kinases.

Additional co-factors of Runt and Lz were identified by two-hybrid screen using the Runt homology domain (Fig. 2.5). These included two *Drosophila* homologues of core binding factor-Beta (CBFβ), Brother [*B*eta for *R*unt and *other*s (Bro)] and Big-brother (Bgb) (Golling et al. 1996). These are non DNA-binding cofactors of Runt and Lz that increase the affinity of Runx proteins for target sites and are redundantly required for repression and activation by Runt and Lz (Li and Gergen 1999; Kaminker et al. 2001). Exhaustive characterisation of Bro or Bgb function in haemocyte development has not been performed although it has been shown that over-expression of Bro or Bgb in haemocytes triggers increased haemocyte number and is also able to suppress effects of AML1-ETO fusion protein over-expression in haemocytes (Sinenko et al. 2010).

An additional factor that has been identified as required for crystal cell development is the *Drosophila* homologue of myeloid leukaemia factor 1 (MLF1). MLF1 is a translocation partner detected in a number of myelodysplasia (MDS) and acute myeloid leukaemia (AML) cases (Arber et al. 2003). *Drosophila* Mlf is expressed in crystal cells and appears to be required for crystal cell differentiation as markers of mature crystal cell fate such as prophenoloxidases are absent from *mlf* mutant embryos (Bras et al. 2012). Mlf is required for activation of Lz reporter cells in haemocyte-derived cell lines and appears to be required to stabilise levels of nuclear Lz in crystal cell precursors (Bras et al. 2012). Intriguingly Mlf also appears to be required for function of the RUNX1-ETO fusion protein in crystal cells (Bras et al. 2012).

While Notch, Lz, Srp and Mlf are positively acting factors that are required for crystal cell differentiation, the Friend of GATA (FOG) homologue U-shaped (Ush) has been suggested to prevent crystal cell differentiation. In embryos, Ush is expressed in haemocyte precursors and plasmatocytes but is down-regulated in crystal cells (Fossett et al. 2001). As over-expression of Ush was able to decrease crystal cell number while crystal cell numbers were increased in *Ush* mutants, it was proposed that Ush is a repressor of crystal cell development (Fossett et al. 2001). This is similar to observed functions of vertebrate FOG, in maintaining multipotent haematopoietic progenitors and antagonising eosinophil differentiation (Querfurth et al. 2000).

Lamellocyte differentiation can be induced by activation of signalling pathways that include the JAK/STAT (Luo et al. 1995; Kwon et al. 2008), Toll (Qiu et al. 1998) and JNK pathways (Zettervall et al. 2004). In addition to triggering lymph gland hypertrophy by controlling prohaemocyte self-renewal, gain-of-function activating mutants *Drosophila* JAK mutations trigger the differentiation of haemocytes into lamellocytes and the development of melanotic tumours (Harrison et al. 1995; Luo et al. 1995). This effect is transduced through STAT as deletion of the sole *Drosophila* STAT (Stat92E—homologue of STAT5A) suppresses these effects (Luo et al. 1997). It has been suggested that the JAK/STAT pathway acts in part by targeting the Friend of GATA protein U-Shaped (Ush). In *ush* mutants lamellocyte numbers are increased suggesting that a normal function of Ush is also to repress lamellocyte development from plasmatocytes (Sorrentino et al. 2007; Frandsen et al. 2008).

In the course of a gain-of-function genetic screen to identify regulators of haemocyte development, we identified the *Drosophila* NRSF/REST-like transcription factor Chn (Stofanko et al. 2008). Over-expression of Chn is able to induce plasmatocytes to differentiate into lamellocytes both in circulation and in lymph glands (Stofanko et al. 2010). Chn is able to bind to CoREST (Tsuda et al. 2006), suggesting that recruitment of the CoREST complex and associated histone deacetylase (HDAC) and histone demethylase components is required for lamellocyte differentiation. Finally, we have identified the ATP-dependent chromatin remodelling enzyme NURF as a repressor of lamellocyte development.

NURF is required to repress that JAK/STAT pathway and in NURF mutants the JAK/STAT pathway is activated leading to lamellocyte differentiation and melanotic tumours (Badenhorst et al. 2002; Kwon et al. 2008). These results emphasise the key role of chromatin modifying and remodelling enzymes in controlling lamellocyte development, but also illustrate a simple assay that can be used to identify function of epigenetic regulators in haematopoiesis—screening for the development of melanotic tumours. In the following section we discuss how this has been used to identify epigenetic factors required for haematopoiesis.

2.4 Epigenetic Regulation of Haemocyte Development

The great advantage of *Drosophila* as a model system to study haematopoiesis is the genetic amenability of *Drosophila*. Traditionally flies have been used in genetic screens in which males are randomly mutated using mutagens such as ethyl methanesulfonate (EMS) and progeny screened for mutants that disrupt biological processes of interest. Such so-called "forward" genetic screens have the advantage of identifying novel unanticipated components of developmental pathways like haematopoiesis. To this arsenal have been added the tools of systematic targeted protein over-expression (for example EP lines) and RNAi screens (Rorth 1996; Rorth et al. 1998; Dietzl et al. 2007) that allow tissue-specific gain-of-function and loss-of-function screens. These tools also allow the over-expression and targeted ablation of defined genes of interest and supplement extensive P-element-induced mutant collections for "reverse" genetic approaches to determine haematopoietic functions of known proteins or protein complexes such as ATP-dependent chromatin remodelling enzymes.

2.4.1 Genetic Screens for New Regulators of Haematopoiesis

The conspicuous appearance of melanotic tumours in *Drosophila* third instar larvae has provided a convenient phenotype to use to identify new regulators of haematopoiesis in *Drosophila*. Melanotic tumours were first reported by Bridges (Bridges 1916) and since then extensive collections have been generated (Barigozzi 1969; Gateff 1978a; Sparrow 1978). Many of these relied on the identification of spontaneous mutants; however, mutant screens using EMS have also been performed to identify melanotic tumour suppressors (Watson et al. 1991; Rodriguez et al. 1996; Braun et al. 1997). The usefulness of this approach is highlighted by the identification of the *Drosophila* JAK (Hanratty and Dearolf 1993), the *Drosophila* TIP60 complex subunit Domino (Ruhf et al. 2001), the *Drosophila* Toll (Tl) pathway including the Tl receptor and the *Drosophila* IκBα homologue Cactus (Braun et al. 1997; Qiu et al. 1998) and Escargot (Esg) the *Drosophila* homologue of the epithelial–mesenchyme transition regulator Slug/SNAI2 (Rodriguez et al. 1996), all of which play an important role in blood cell development and function.

More recently both gain-of-function genetic screens and targeted inducible RNAi screens have been performed to identify additional regulators of haematopoiesis. In an effort to identify novel factors that control larval haemocyte migration and differentiation, my laboratory has performed a modular misexpression screen to over-expresses ~20 % of *Drosophila* genes specifically in *Drosophila* circulating and lymph gland plasmatocytes using the GAL4-UAS system (Rorth 1996). To conduct this screen, a *Drosophila* strain that expresses the yeast transcriptional activator GAL4 in haemocytes using a blood-specific promoter (*Pxn-GAL4*) was crossed to a library of GAL4 responder (EP/EY) lines. These lines were generated by randomly mobilising a transposon that contains a GAL4-responsive promoter throughout the genome. Genes adjacent to the EP/EY transposon can be over-expressed using GAL4. The *Pxn-GAL4* driver also contained a *UAS-GFP* transgene that allowed haemocytes to be observed live in the transparent third instar larvae (Fig. 2.6). 3,412 insertions were screened to identify 101 candidate regulators of fly haematopoiesis (Stofanko et al. 2008). These included *Drosophila* homologues of CBP, JARID2 a component of the Polycomb repressive complex, the H3K9 and H3K36 demethylase KDM4/JMJD2, c-Fos, Slug/SNAI2 and the REST/NRSF homologue Chn.

Targeted RNAi knock-down screens have also been performed to identify new factors required for function of the posterior signalling centre (PSC), the hub that maintains the lymph gland haematopoietic niche (Tokusumi et al. 2012), and to identify additional melanotic tumour suppressors (Avet-Rochex et al. 2010). These screens identified the *Drosophila* SWI/SNF ATP-dependent chromatin remodelling complex BAP as a key regulator of PSC function and collaborating with the GATA factor Srp to control prohaemocyte self-renewal and differentiation (Tokusumi et al. 2012). Melanotic tumour suppressors identified include expected candidates that have previously been shown to cause melanotic tumours like Ush and Cactus, and novel chromatin associated components such as Tip60, WDR5, a component of the MLL and COMPASS histone H3 Lys4 (H3K4) methyltransferase complexes, and the histone chaperone Spt6 (Avet-Rochex et al. 2010).

2.4.2 Regulation of Haematopoiesis by ATP-Dependent Chromatin Remodelling Enzymes

ATP dependent chromatin remodelling complexes are large multisubunit protein complexes that use the energy of ATP hydrolysis to alter the dynamic properties of nucleosomes, the basic units of chromatin. As shown in Fig. 2.7 ATP-dependent chromatin remodelling enzymes can be divided into four broad categories depending on the energy utilising ATPase subunit at the core of the complex. These ATPases have broad homology to the SWI2/SNF2 subunit of the yeast

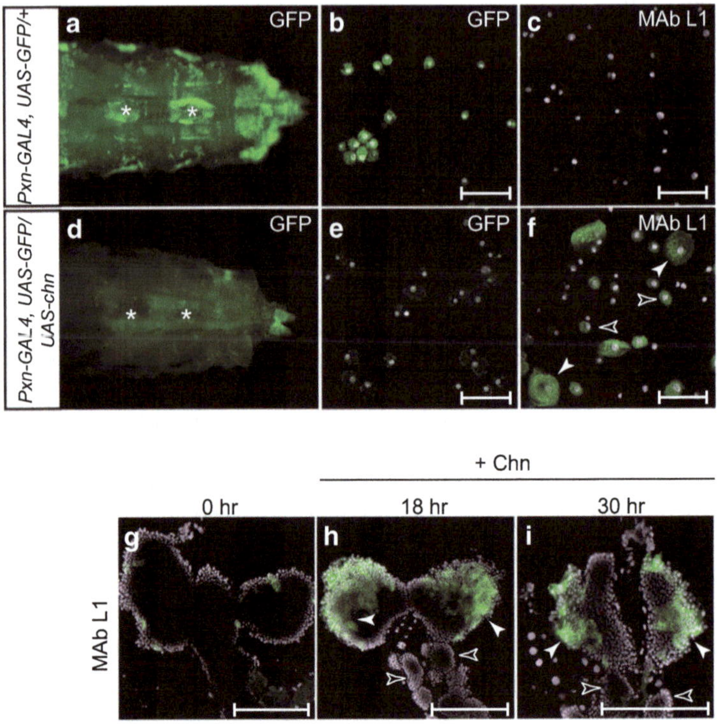

Fig. 2.6 Chn controls lamellocyte differentiation. Over-expression of Chn decreases numbers of (**a**, **b**) sessile (*asterisk*) and (**b**, **e**) circulating plasmatocytes. (**c**, **f**) MAb L1 staining indicates that Chn over-expression transforms plasmatocytes into lamellocytes. Circulating haemocytes were isolated from (**b**, **c**) *Pxn-GAL4*, *UAS-GFP x w1118* and (**e**, **f**) *Pxn-GAL4*, *UAS-GFP x UAS-chn* third instar larvae. (**g–i**) Chn over-expression increases lamellocyte number in primary lymph glands. Lamellocytes are not detected in the secondary lobes. In all panels GFP expression or antibody staining is shown in *green* and DAPI-stained nuclei in *purple*. *Scale bars* indicate 50 μm

SWI/SNF chromatin remodelling complex, but have some unique features that dictate individual activities and the ancillary subunits that are associated with the ATPase to form large multisubunit remodelling complexes [reviewed in Choudhary and Varga-Weisz (2007), Hota and Bartholomew (2011)]. The four nominal groupings of the ATP-dependent chromatin remodelling enzymes are the SWI/SNF, ISWI, CHD (Mi-2) and INO80/SWR1 complexes. The principal activities associated with the SWI/SNF2 complexes are nucleosome sliding and disruption, while ISWI and Mi-2 remodellers mediate nucleosome sliding *in cis* with no displacement from the nucleosome template. The INO80 and SWR1 subtypes catalyse histone variant exchange, either inserting or replacing histone variant dimers H2A.Z/H2B for/with canonical H2A/H2B dimers. In addition to their ability to slide nucleosomes the Mi-2 complexes like NURF are associated with histone deacetylases HDAC-1 and HDAC-2 (Rpd3 in *Drosophila*) that mediate removal of active histone acetylation marks and thus have a repressive function.

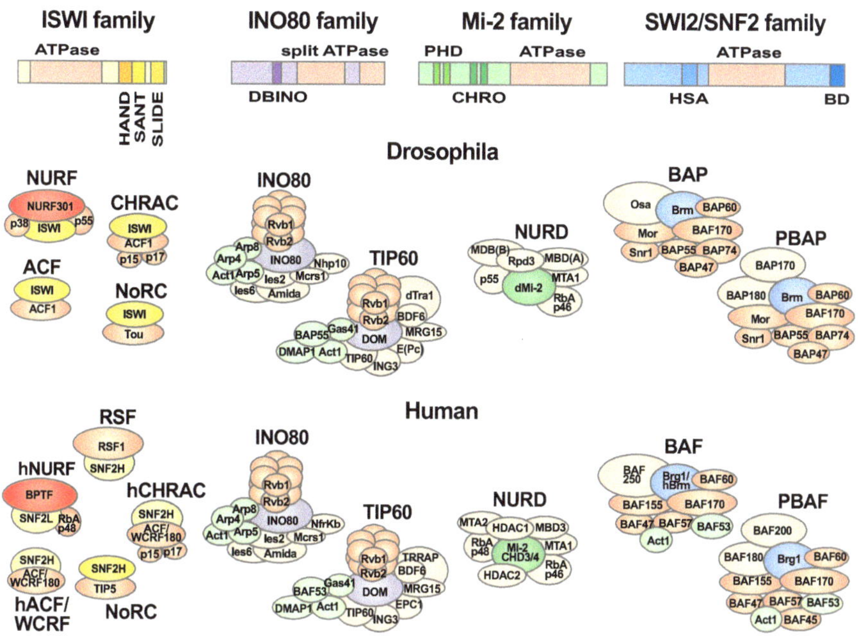

Fig. 2.7 ATP-dependent chromatin remodelling factors. ATP-dependent chromatin remodelling factors can be divided into four broad families depending on the catalytic ATPase subunit that is at the core of each complex. The four main groupings are the ISWI, INO80, Mi-2 and SWI/SNF families. The members of these classes of enzymes in both *Drosophila* and humans are shown along with the subunit composition of the complexes. Core catalytic subunits are *colour coded*, as are signature subunits for each complex

The functions of NURD-type complexes in mammalian haematopoiesis are well established, both via interactions with FOG-1 (Gao et al. 2010; Miccio et al. 2010) and the lymphoid system regulator Ikaros (Kim et al. 1999). There is also evidence from mammalian systems implicating the SWI/SNF subtype complexes BAF and PBAF in haematopoiesis (Bultman et al. 2005) and that these SWI/SNF complexes may be involved in facilitating binding of TAL1 to chromatin (Bultman et al. 2005; Hu et al. 2011). This is consistent with RNAi screens that identify the *Drosophila* BAP complex (BAF in humans; see Fig. 2.8) as a regulator of prohaemocyte self-renewal and differentiation (Tokusumi et al. 2012). The best evidence for roles of ISWI and INO80/SWR1 complexes in blood cell development is provided by studies of fly haematopoiesis.

Domino (Dom), which encodes the catalytic ATPase subunit of the fly and human TIP60 complex (Kusch et al. 2004), was one of the first ATP-dependent chromatin remodelling complexes to be implicated in early haematopoiesis. Enhancer traps in the *domino* gene are expressed in haemocytes and *dom* mutants develop melanotic tumours (Braun et al. 1997). Unlike tumours that derive from circulating lamellocyte aggregates, the tumours in *dom* mutants are in fact melanised lymph glands containing necrotic prohaemocytes, suggesting that

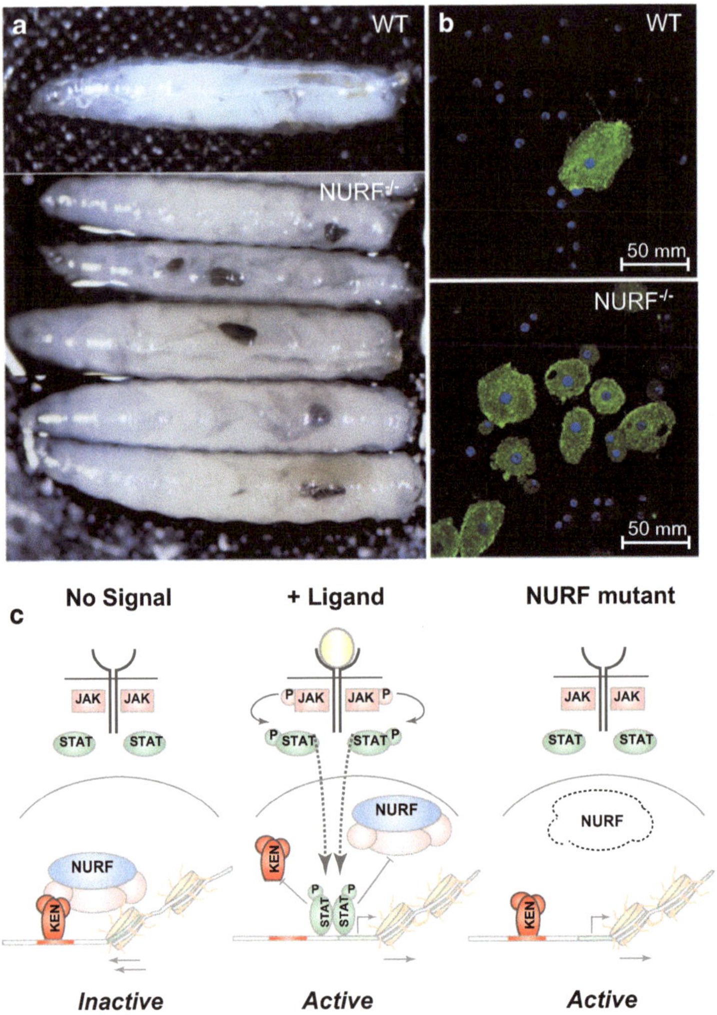

Fig. 2.8 NURF is a melanotic tumour suppressor. (**a**) Mutants lacking the NURF ATP-dependent chromatin remodelling complex NURF display melanotic tumours. (**b**) MAb L1 staining indicates that melanotic tumours are caused by ectopic differentiation of lamellocytes in NURF mutants (compare wild-type and NURF mutant haemolymph). (**c**) NURF is a repressor of JAK/STAT target genes. In unstimulated conditions (No signal) NURF binds to and is recruited by the Bcl6 homologue Ken to JAK/STAT target promoters. NURF slides/positions a nucleosome over the promoter to block transcription. After stimulation (+Ligand), Stat92E enters the nucleus and binds promoters, displacing both Ken and NURF. The repressive nucleosome position is not maintained. The repressive nucleosome position also cannot be maintained in NURF mutants and JAK/STAT targets are not silenced. As a result activation in the absence of STAT nuclear entry occurs

Domino-containing complexes like TIP60 are required for prohaemocyte survival (Braun et al. 1997). Destruction of the prohaemocyte compartment is accompanied by loss of circulating haemocytes which impairs response to pathogens (Braun et al. 1998). The Dom locus expresses two isoforms Dom-A and Dom-B (Ruhf et al. 2001). Dom-A is a subunit of the TIP60 complex that mediates both acetylation and exchange of histone H2A variants and is required for DNA-damage repair (Kusch et al. 2004), suggesting that loss of prohaemocytes may be due to impaired double-strand break repair. Prohaemocytes are known to contain elevated levels of reactive oxygen species (Owusu-Ansah and Banerjee 2009) and may be sensitised to loss of DNA-damage repair enzymes. Alternatively, *dom* phenotypes could be due to altered transcription programmes. Yeast complexes containing the Dom homologue Swr1 mediate incorporation of the histone variant H2A.Z at 5′ ends of genes that is required for transcription (Mizuguchi et al. 2004; Raisner et al. 2005; Zhang et al. 2005). It seems feasible that Dom-containing complexes may be targeted to specific promoters and enhancers to mediate H2A.Z histone variant incorporation which alters nucleosome structure to allow for subsequent binding of other DNA-binding factors (Jin et al. 2009; Hu et al. 2013). Certainly, there is evidence that the myeloid zinc finger protein 2A (MZF-2A) can bind to the mouse Dom-A homologue (Ogawa et al. 2003).

Work in our laboratory has demonstrated that the ISWI class chromatin remodelling complex NURF (the *nu*cleosome *r*emodelling *f*actor) is involved in haematopoiesis. NURF was one of the first ATP-dependent chromatin remodelling enzymes identified. NURF is composed of four subunits of which the largest subunit, NURF301, is NURF specific. NURF catalyses energy-dependent nucleosome sliding (Xiao et al. 2001; Barak et al. 2003). By sliding nucleosomes, NURF can alternatively expose or block transcription factor binding sites, and has been shown to be required for both transcription activation and repression (Badenhorst et al. 2002, 2005; Barak et al. 2003; Kwon et al. 2008). We have shown by microarray profiling that *Drosophila* NURF is a co-repressor of a large number of JAK/STAT target genes in haemocytes (Kwon et al. 2008). In NURF mutants, JAK/STAT target genes are precociously activated. As has been observed with gain-of-function JAK mutants, NURF mutants exhibit hypertrophy of the larval lymph glands, increases in haemocyte number and the transformation of plasmatocytes into lamellocytes leading to the production of melanotic tumours (Fig. 2.8) (Badenhorst et al. 2002; Kwon et al. 2008).

In silico analysis of promoters regulated by NURF identifies a consensus regulatory element consisting of a STAT-binding sequence overlapped by a recognition sequence for a transcriptional repressor, the *Drosophila* Bcl6 homologue Ken (Kwon et al. 2008). NURF and Ken interact physically and genetically, and NURF and Ken co-localise at target sites in haemocytes, suggesting that NURF is recruited by Ken to repress STAT responders. We have speculated that in unstimulated conditions NURF-mediated nucleosome sliding represses targets by positioning a nucleosome over the transcription start site. When the JAK/STAT pathway is activated, however, Ken and NURF are displaced by Stat92E switching promoters from a repressive to active state. In NURF mutants, these repressive nucleosome positions are not established and thus precocious activation of STAT target genes occurs resulting in the haematological transformations observed.

NURF recruitment and activity at JAK/STAT targets may be regulated by changes in its nucleosomal substrate induced by post-translational modification of the histone tails or histone variant exchange. The largest subunit of NURF (NURF301 in *Drosophila*, BPTF in humans) contains three PHD (Plant Homeo Domain) fingers and a C-terminal Bromodomain. These motifs have the ability to bind to modified histone tails and it has been shown that the C-terminal PHD finger of NURF301/BPTF binds histone H3 trimethylated at lysine position 4 (H3K4 $(Me)_3$) (Wysocka et al. 2006; Kwon et al. 2009). It is proposed that H3K4$(Me)_3$ recruits NURF to sites of action in the genome, with NURF acting as the ultimate effector of this modification. Significantly, the MLL/COMPASS enzyme complex that establishes the H3K4$(Me)_3$ mark in humans is a major factor in haematopoietic malignancy [reviewed in Muntean and Hess (2012)], making it a priority to investigate functions of NURF-type complexes in mammalian haematopoiesis. In flies knock-down of the fly homologue of WDR5—a component of the MLL/COMPASS complex—results in melanotic tumours like NURF mutants (Avet-Rochex et al. 2010), reinforcing the notion that ATP-dependent chromatin remodelling and histone post-translational modifications (HPTMs) do not act independently but rather that HPTMs provide molecular rheostats to control chromatin binding and function of "readers" like the chromatin remodelling enzyme NURF. By controlling the distribution and combinations of HPTMs, chromatin binding of remodelling complexes can be regulated.

2.4.3 Regulation by Histone Modifying Complexes

The distribution of histone post-translational modifications (HPTMs) is controlled by the balancing activities of families of "writers" such as histone acetyltransferases (HATs) and histone methyltransferases (HMTs), which establish acetylation and methylation marks, respectively, and "erasers" such as histone deacetylases (HDACs) and histone demethylases that remove these marks. These do not exist as isolated proteins but are often present in present in large multisubunit co-activator and co-repressor assemblies The activity of the MLL/COMPASS complex in generating the activating H3K4$(Me)_3$ mark and its role in haematopoietic malignancy in flies and humans are well defined as discussed above. Components of other co-activator complexes such as p300/CBP have also been identified in genetic screens for perturbed haematopoiesis in flies (Stofanko et al. 2008).

However, the most significant advances provided by *Drosophila* have been in the identification of histone modifying co-repressor complexes that regulate haematopoiesis. The Gro/TLE family of co-repressors that were first identified in flies as binding partners of the Runx proteins Runt and Lz (Aronson et al. 1997), and confirmed as binding to AML1(Levanon et al. 1998), have been shown to repress transcription either by oligomerising on chromatin (Song et al. 2004), but also to be associated with the histone deacetylase Rpd3 (HDAC1) (Chen et al. 1999). More recently the Gro homologue TLE4 has been shown to be part of a complex that contains the histone arginine methyltransferase PRMT5 (Patel et al. 2012). This TLE4 complex displaces activating MLL H3K4 methyltransferase complexes from the Pax2 transcription factor and methylates H3R3 residues allowing for subsequent recruitment of the Polycomb proteins Ezh2 and Suz12 that mediate repression.

Genetic screens have also identified histone H4K20 monomethylase Pr-set7 as a regulator of haematopoiesis. Pr-set7 was identified as a factor required to maintain PSC hub cells of the larval haematopoietic niche (Tokusumi et al. 2012) and Pr-set7 mutants develop melanotic tumours like gain-of-function JAK/STAT mutants (Minakhina and Steward 2006). Pr-set7 has also been identified as a regulator of JAK/STAT function in the haemocyte-derived Kc167 cell line (Fisher et al. 2012). The H4K20(Me)$_1$ mark functions by allowing the recruitment of binding partners such as the tumour suppressor L(3)mbt. L(3)mbt is in complex with HP1 and H1 and is speculated to act as a "chromatin lock" to negatively regulate gene transcription (Trojer et al. 2007).

Finally, data from our laboratory point to the role of the co-repressor complex CoREST in *Drosophila* haematopoiesis. We have shown that the *Drosophila* REST/NRSF homologue Chn is a key regulator of lamellocyte development. As shown in Fig. 2.6, over-expression of Chn in plasmatocytes is sufficient to trigger differentiation into lamellocytes (Stofanko et al. 2010). This is associated with repression of plasmatocyte-determinant Gcm and onset of expression of lamellocyte markers. Chn has been shown to associate with the *Drosophila* CoREST complex (Dallman et al. 2004; Tsuda et al. 2006). The Mammalian CoREST complex includes the scaffold protein CoREST and both the histone deacetylase Rdp3 (HDAC1) and lysine-specific demethylase-1 (Lsd1) (You et al. 2001; Shi et al. 2005), one of the first histone lysine demethylases identified (Shi et al. 2004). We have shown that RNAi knockdown of Rpd3 and Lsd1 prevents Chn-dependent lamellocyte differentiation, as does treatment with Lsd1 and HDAC chemical inhibitors, confirming that Chn acts via the CoREST complex. Haematopoietic functions of CoREST in mammals are confirmed by the observation that the CoREST complex is associated with TAL1 (Hu et al. 2009) and the transcription factors Gfi-1/1b and that inhibition of CoREST and Lsd1 affects erythroid, megakaryocyte and granulocyte differentiation (Saleque et al. 2007).

2.5 *Drosophila* as a Tool to Investigate Function of Leukaemogenic Fusion Proteins

An example of the effectiveness of the *Drosophila* model system has been the use of both the fly haematopoietic system and eye to dissect mechanism of action of the leukaemogenic fusion protein RUNX1-ETO. RUNX1-ETO is a fusion transcription factor generated by the t(8;21) translocation, and is present in adult (4–12 %) and paediatric (12–30 %) AML patients. It contains the RUNT homology domain of RUNX1 and most of the ETO gene [reviewed in Hatlen et al. (2012)]. The fly RUNX1 homologue Lz is expressed both in crystal cells, as described above, and also in the fly eye where it specifies lens-secreting cone cells in the ommatidial units that compose the compound eye (Daga et al. 1996; Canon and Banerjee 2003). Mann and colleagues have exploited cone cell differentiation to investigate function of the RUNX1-ETO fusion protein (Wildonger and Mann 2005). In particular the eye system was used to explore whether RUNX1-ETO interferes with normal Runx (Lz) function either by acting as a constitutive repressor of Lz target genes or by acting as a dominant-negative activity that competes with Lz for co-factors that are required for Lz functions—both gene activation and repression. Interestingly, the data suggest that RUNX1-ETO does not function as a dominant negative as phenotypes generated by over-expressing RUNX-ETO or by removing Lz are distinct. Moreover, over-expression of Bro or Bgb, the CBF homologues that enhance Lz binding and would be expected to counteract a dominant-negative action of RUNX1-ETO, did not suppress its phenotype (Wildonger and Mann 2005). However, reduction in Bgb levels suppresses the RUNX1-ETO over-expression phenotype (Wildonger and Mann 2005) as it enhances Lz loss-of-function phenotypes (Li and Gergen 1999; Kaminker et al. 2001), suggesting that RUNX1-ETO binding to targets is required for function. In support of the idea that RUNX1-ETO functions as a constitutive repressor, RUNX1-ETO over-expression was able to repress expression of *dPax2* a target that is normally activated by Lz in the eye, and an analogous Lz fusion protein with the Engrailed repressor domain generated similar over-expression phenotypes in the eye as RUNX1-ETO (Wildonger and Mann 2005).

Subsequently RUNX1-ETO has also been over-expressed in haemocytes and used as the basis of a modifier screen to isolate factors that are required for fusion protein function. RUNX1-ETO has been over-expressed both in crystal cells that normally express the Runx protein Lz (Osman et al. 2009) as well as plasmatocytes that do not express Lz (Sinenko et al. 2010). Haematopoietic phenotypes are induced in both cases that have been used to isolate modifiers of function. Over-expression of RUNX1-ETO in crystal cells under the control of the Lz promoter leads to increased numbers of committed crystal cells but appears to block terminal differentiation of crystal cells as prophenoloxidases fail to be expressed in these cells (Osman et al. 2009). Over-expression of RUNX1-ETO also leads to lethality at the pupal stage (Lz is also expressed in other tissues in addition to the eye) and this lethality has been used to isolate suppressors of RUNX1-ETO activity by simultaneous inducible RNAi. These experiments have identified CalpainB (CalpB), a member of a large family of Ca2-dependent proteases as a RUNX1-ETO suppressor (Osman et al. 2009). Knock-down of CalpB restores crystal cell differentiation in RUNX1-ETO over-expressing animals and also appears capable of selectively decreasing viability of Kasumi-1 cells that carry the RUNX1-ETO expressing t(8;21) translocation (Higuchi et al. 2002).

Experiments in mouse models have shown that over-expression of RUNX1-ETO alone is insufficient to trigger AML unless secondary mutations are present (Yuan et al. 2001; Higuchi et al. 2002). In humans, approximately 70 % of t(8;21) patient samples contain addition mutations in tyrosine kinases such as c-KIT and FLT3 (Beghini et al. 2000; Care et al. 2003; Kuchenbauer et al. 2006). The *Drosophila* RUNX1-ETO over-expressing model provides a potentially powerful system to identify collaborating mutations that can enhance leukaemogenesis. When over-expressed in plasmatocytes, RUNX1-ETO triggers the production of melanotic tumours (Sinenko et al. 2010). By screening for mutations that either increase or inhibit melanotic tumour production 22 modifiers of RUNX1-ETO were selected. Amongst these are components of the Wnt signalling pathway, the ligand Wnt4 and the receptors Frizzled (Fz) and Frizzled-2 (Fz2) (Sinenko et al. 2010). The interaction of these candidates with RUNX1-ETO remains to be characterised; however, it is known that Wnt signalling is required to for prohaemocyte self-renewal (Sinenko et al. 2009) as has been observed for self-renewal of vertebrate haematopoietic stem cells [reviewed in Staal and Clevers (2005)]. Significantly, the initial enhancer screen only utilised a panel of 231 chromosomal deficiencies that do not completely cover the *Drosophila* genome, and there is potential that many interactors may have been missed. Saturation EMS mutagenesis or inducible RNAi knockdown could be used to identify additional enhancers. EMS mutagenesis in particular is an attractive tool given its ability to generate both loss-of-function but also activating or neomorphic mutations that may more accurately reflect the mutation load of leukaemic cells.

2.6 Outlook

The great genetic amenability of *Drosophila* and the ability easily to conduct rapid forward and reverse genetic screens offer a powerful model system in which to identify new components of developmental pathways. This system has already been exploited to clarify mechanisms of action and partners of the RUNX1-ETO leukaemogenic fusion but has great potential to be used in similar genetic screens to identify collaborating factors for other leukaemogenic fusions. This is especially true of fusions involving chromatin modifying or associated proteins where well-established biochemical methods using *Drosophila* extracts allow identification and in vitro functional characterisation of complexes. A good example of the power of these techniques are studies showing AF4, AF9, ELL and EAF participation in the super elongation complex (Lin et al. 2010; Smith et al. 2011). The genetic amenability of *Drosophila* can also be used to generate transgenic fluorescent strains that allow in vivo characterisation of haematopoiesis. For example, we have developed a simplified screening assay, which uses a combination of GFP (green) and mCherry (red) fluorescent reporters for plasmatocytes and lamellocytes, respectively, to identify additional factors required for Chn/CoREST-induced lamellocytes differentiation. As *Drosophila* larvae are transparent, expression of these reporters can be visualised in live third instar larvae, and the effect of systematic inducible RNAi mediated knock-down of other genes examined. These types of approaches illustrate the great advantage of the *Drosophila* system as a tool to identifying new components of conserved pathways and processes.

References

Alfonso TB, Jones BW (2002) gcm2 promotes glial cell differentiation and is required with glial cells missing for macrophage development in Drosophila. Dev Biol 248:369–383

Arber DA, Chang KL, Lyda MH, Bedell V, Spielberger R, Slovak ML (2003) Detection of NPM/MLF1 fusion in t(3;5)-positive acute myeloid leukemia and myelodysplasia. Hum Pathol 34:809–813

Aronson BD, Fisher AL, Blechman K, Caudy M, Gergen JP (1997) Groucho-dependent and -independent repression activities of Runt domain proteins. Mol Cell Biol 17:5581–5587

Avet-Rochex A, Boyer K, Polesello C, Gobert V, Osman D, Roch F et al (2010) An in vivo RNA interference screen identifies gene networks controlling Drosophila melanogaster blood cell homeostasis. BMC Dev Biol 10:65

Badenhorst P, Voas M, Rebay I, Wu C (2002) Biological functions of the ISWI chromatin remodeling complex NURF. Genes Dev 16:3186–3198

Badenhorst P, Xiao H, Cherbas L, Kwon SY, Voas M, Rebay I et al (2005) The Drosophila nucleosome remodeling factor NURF is required for Ecdysteroid signaling and metamorphosis. Genes Dev 19:2540–2545

Barak O, Lazzaro MA, Lane WS, Speicher DW, Picketts DJ, Shiekhattar R (2003) Isolation of human NURF: a regulator of Engrailed gene expression. EMBO J 22:6089–6100

Barigozzi C (1969) Genetic control of melanotic tumors in Drosophila. Natl Cancer Inst Monogr 31:277–290

Bataille L, Auge B, Ferjoux G, Haenlin M, Waltzer L (2005) Resolving embryonic blood cell fate choice in Drosophila: interplay of GCM and RUNX factors. Development 132:4635–4644

Beghini A, Peterlongo P, Ripamonti CB, Larizza L, Cairoli R, Morra E et al (2000) C-kit mutations in core binding factor leukemias. Blood 95:726–728

Beiman M, Shilo BZ, Volk T (1996) Heartless, a Drosophila FGF receptor homolog, is essential for cell migration and establishment of several mesodermal lineages. Genes Dev 10:2993–3002

Bernardoni R, Vivancos V, Giangrande A (1997) glide/gcm is expressed and required in the scavenger cell lineage. Dev Biol 191:118–130

Bidla G, Dushay MS, Theopold U (2007) Crystal cell rupture after injury in Drosophila requires the JNK pathway, small GTPases and the TNF homolog Eiger. J Cell Sci 120:1209–1215

Bodmer R (1993) The gene tinman is required for specification of the heart and visceral muscles in Drosophila. Development 118:719–729

Bras S, Martin-Lanneree S, Gobert V, Auge B, Breig O, Sanial M et al (2012) Myeloid leukemia factor is a conserved regulator of RUNX transcription factor activity involved in hematopoiesis. Proc Natl Acad Sci U S A 109:4986–4991

Braun A, Lemaitre B, Lanot R, Zachary D, Meister M (1997) Drosophila Immunity: analysis of Larval Hemocytes by P-Element-Mediated Enhancer Trap. Genetics 147:623–634

Braun A, Hoffmann JA, Meister M (1998) Analysis of the Drosophila host defense in domino mutant larvae, which are devoid of hemocytes. Proc Natl Acad Sci U S A 95:14337–14342

Bridges CB (1916) Non-disjunction as proof of the chromosome theory of heredity. Genetics 1:1–52

Bruckner K, Kockel L, Duchek P, Luque CM, Rorth P, Perrimon N (2004) The PDGF/VEGF receptor controls blood cell survival in Drosophila. Dev Cell 7:73–84

Bultman SJ, Gebuhr TC, Magnuson T (2005) A Brg1 mutation that uncouples ATPase activity from chromatin remodeling reveals an essential role for SWI/SNF-related complexes in beta-globin expression and erythroid development. Genes Dev 19:2849–2861

Campos-Ortega JA, Hartenstein V (1985) The embryonic development of *Drosophila* melanogaster. Springer, Berlin

Canon J, Banerjee U (2003) In vivo analysis of a developmental circuit for direct transcriptional activation and repression in the same cell by a Runx protein. Genes Dev 17:838–843

Care RS, Valk PJ, Goodeve AC, Abu-Duhier FM, Geertsma-Kleinekoort WM, Wilson GA et al (2003) Incidence and prognosis of c-KIT and FLT3 mutations in core binding factor (CBF) acute myeloid leukaemias. Br J Haematol 121:775–777

Carton Y, Frey F, Nappi AJ (2009) Parasite-induced changes in nitric oxide levels in Drosophila paramelanica. J Parasitol 95:1134–1141

Cerenius L, Lee BL, Soderhall K (2008) The proPO-system: pros and cons for its role in invertebrate immunity. Trends Immunol 29:263–271

Cerenius L, S-i K, Lee BL, Nonaka M, Söderhäll K (2010) Proteolytic cascades and their involvement in invertebrate immunity. Trends Biochem Sci 35:575–583

Charroux B, Royet J (2009) Elimination of plasmatocytes by targeted apoptosis reveals their role in multiple aspects of the Drosophila immune response. Proc Natl Acad Sci U S A 106:9797–9802

Chen G, Fernandez J, Mische S, Courey AJ (1999) A functional interaction between the histone deacetylase Rpd3 and the corepressor groucho in Drosophila development. Genes Dev 13:2218–2230

Cho NK, Keyes L, Johnson E, Heller J, Ryner L, Karim F et al (2002) Developmental control of blood cell migration by the drosophila VEGF pathway. Cell 108:865–876

Choudhary P, Varga-Weisz P (2007) ATP-dependent chromatin remodelling: action and reaction. Subcell Biochem 41:29–43

Christensen BM, Li J, Chen CC, Nappi AJ (2005) Melanization immune responses in mosquito vectors. Trends Parasitol 21:192–199

Crozatier M, Ubeda JM, Vincent A, Meister M (2004) Cellular immune response to parasitization in Drosophila requires the EBF orthologue collier. PLoS Biol 2:E196

Daga A, Karlovich CA, Dumstrei K, Banerjee U (1996) Patterning of cells in the Drosophila eye by Lozenge, which shares homologous domains with AML1. Genes Dev 10:1194–1205

Dallman JE, Allopenna J, Bassett A, Travers A, Mandel G (2004) A conserved role but different partners for the transcriptional corepressor CoREST in fly and mammalian nervous system formation. J Neurosci 24:7186–7193

de Velasco B, Mandal L, Mkrtchyan M, Hartenstein V (2006) Subdivision and developmental fate of the head mesoderm in Drosophila melanogaster. Dev Genes Evol 216:39–51

Defaye A, Evans I, Crozatier M, Wood W, Lemaitre B, Leulier F (2009) Genetic ablation of Drosophila phagocytes reveals their contribution to both development and resistance to bacterial infection. J Innate Immun 1:322–334

Dietzl G, Chen D, Schnorrer F, Su KC, Barinova Y, Fellner M et al (2007) A genome-wide transgenic RNAi library for conditional gene inactivation in Drosophila. Nature 448:151–156

Dragojlovic-Munther M, Martinez-Agosto JA (2012) Multifaceted roles of PTEN and TSC orchestrate growth and differentiation of Drosophila blood progenitors. Development 139:3752–3763

Duvic B, Hoffmann JA, Meister M, Royet J (2002) Notch signaling controls lineage specification during Drosophila larval hematopoiesis. Curr Biol 12:1923–1927

Fessler JH, Fessler LI (1989) Drosophila extracellular matrix. Annu Rev Cell Biol 5:309–339

Fessler LI, Nelson RE, Fessler JH (1994) Drosophila extracellular matrix. Methods Enzymol 245:271–294

Fisher KH, Wright VM, Taylor A, Zeidler MP, Brown S (2012) Advances in genome-wide RNAi cellular screens: a case study using the Drosophila JAK/STAT pathway. BMC Genomics 13:506

Fossett N, Tevosian SG, Gajewski K, Zhang Q, Orkin SH, Schulz RA (2001) The Friend of GATA proteins U-shaped, FOG-1, and FOG-2 function as negative regulators of blood, heart, and eye development in Drosophila. Proc Natl Acad Sci U S A 98:7342–7347
Fossett N, Hyman K, Gajewski K, Orkin SH, Schulz RA (2003) Combinatorial interactions of Serpent, Lozenge, and U-shaped regulate crystal cell lineage commitment during Drosophila hematopoiesis. Proc Natl Acad Sci 100:11451–11456
Franc NC, Dimarcq JL, Lagueux M, Hoffmann J, Ezekowitz RA (1996) Croquemort, a novel Drosophila hemocyte/macrophage receptor that recognizes apoptotic cells. Immunity 4:431–443
Franc NC, Heitzler P, Ezekowitz RA, White K (1999) Requirement for croquemort in phagocytosis of apoptotic cells in Drosophila. Science 284:1991–1994
Frandsen JL, Gunn B, Muratoglu S, Fossett N, Newfeld SJ (2008) Salmonella pathogenesis reveals that BMP signaling regulates blood cell homeostasis and immune responses in Drosophila. Proc Natl Acad Sci U S A 105:14952–14957
Frasch M (1995) Induction of visceral and cardiac mesoderm by ectodermal Dpp in the early Drosophila embryo. Nature 374:464–467
Freeman MR, Delrow J, Kim J, Johnson E, Doe CQ (2003) Unwrapping glial biology: Gcm target genes regulating glial development, diversification, and function. Neuron 38:567–580
Galko MJ, Krasnow MA (2004) Cellular and genetic analysis of wound healing in Drosophila larvae. PLoS Biol 2:E239
Gao Z, Huang Z, Olivey HE, Gurbuxani S, Crispino JD, Svensson EC (2010) FOG-1-mediated recruitment of NuRD is required for cell lineage re-enforcement during haematopoiesis. EMBO J 29:457–468
Garzino V, Pereira A, Laurenti P, Graba Y, Levis RW, Le Parco Y et al (1992) Cell lineage-specific expression of modulo, a dose-dependent modifier of variegation in Drosophila. EMBO J 11:4471–4479
Gateff E (1978a) The genetics and epigenetics of neoplasms in Drosophila. Biol Rev Camb Philos Soc 53:123–168
Gateff E (1978b) Malignant neoplasms of genetic origin in Drosophila melanogaster. Science 200:1448–1459
Golling G, Li L, Pepling M, Stebbins M, Gergen JP (1996) Drosophila homologs of the proto-oncogene product PEBP2/CBF beta regulate the DNA-binding properties of Runt. Mol Cell Biol 16:932–942
Grigorian M, Mandal L, Hartenstein V (2011) Hematopoiesis at the onset of metamorphosis: terminal differentiation and dissociation of the Drosophila lymph gland. Dev Genes Evol 221:121–131
Hanratty WP, Dearolf CR (1993) The Drosophila tumorous-lethal hematopoietic oncogene is a dominant mutation in the hopscotch locus. Mol Gen Genet 238:33–37
Harrison DA, Binari R, Nahreini TS, Gilman M, Perrimon N (1995) Activation of a Drosophila Janus kinase (JAK) causes hematopoietic neoplasia and developmental defects. EMBO J 14:2857–2865
Hatlen MA, Wang L, Nimer SD (2012) AML1-ETO driven acute leukemia: insights into pathogenesis and potential therapeutic approaches. Front Med 6:248–262
Higuchi M, O'Brien D, Kumaravelu P, Lenny N, Yeoh EJ, Downing JR (2002) Expression of a conditional AML1-ETO oncogene bypasses embryonic lethality and establishes a murine model of human t(8;21) acute myeloid leukemia. Cancer Cell 1:63–74
Holz A, Bossinger B, Strasser T, Janning W, Klapper R (2003) The two origins of hemocytes in Drosophila. Development 130:4955–4962
Hota SK, Bartholomew B (2011) Diversity of operation in ATP-dependent chromatin remodelers. Biochim Biophys Acta 1809:476–487
Hu X, Li X, Valverde K, Fu X, Noguchi C, Qiu Y et al (2009) LSD1-mediated epigenetic modification is required for TAL1 function and hematopoiesis. Proc Natl Acad Sci U S A 106:10141–10146

Hu G, Schones DE, Cui K, Ybarra R, Northrup D, Tang Q et al (2011) Regulation of nucleosome landscape and transcription factor targeting at tissue-specific enhancers by BRG1. Genome Res 21:1650–1658
Hu G, Cui K, Northrup D, Liu C, Wang C, Tang Q et al (2013) H2A.Z facilitates access of active and repressive complexes to chromatin in embryonic stem cell self-renewal and differentiation. Cell Stem Cell 12:180–192
Irving P, Ubeda JM, Doucet D, Troxler L, Lagueux M, Zachary D et al (2005) New insights into Drosophila larval haemocyte functions through genome-wide analysis. Cell Microbiol 7:335–350
Jennings BH, Pickles LM, Wainwright SM, Roe SM, Pearl LH, Ish-Horowicz D (2006) Molecular recognition of transcriptional repressor motifs by the WD domain of the Groucho/TLE corepressor. Mol Cell 22:645–655
Jin C, Zang C, Wei G, Cui K, Peng W, Zhao K et al (2009) H3.3/H2A.Z double variant-containing nucleosomes mark 'nucleosome-free regions' of active promoters and other regulatory regions. Nat Genet 41:941–945
Johansson KC, Metzendorf C, Soderhall K (2005) Microarray analysis of immune challenged Drosophila hemocytes. Exp Cell Res 305:145–155
Jung SH, Evans CJ, Uemura C, Banerjee U (2005) The Drosophila lymph gland as a developmental model of hematopoiesis. Development 132:2521–2533
Kaminker JS, Singh R, Lebestky T, Yan H, Banerjee U (2001) Redundant function of runt domain binding partners, big brother and brother, during drosophila development. Development 128:2639–2648
Kania MA, Bonner AS, Duffy JB, Gergen JP (1990) The Drosophila segmentation gene runt encodes a novel nuclear regulatory protein that is also expressed in the developing nervous system. Genes Dev 4:1701–1713
Kim J, Sif S, Jones B, Jackson A, Koipally J, Heller E et al (1999) Ikaros DNA-binding proteins direct formation of chromatin remodeling complexes in lymphocytes. Immunity 10:345–355
Klinedinst SL, Bodmer R (2003) Gata factor Pannier is required to establish competence for heart progenitor formation. Development 130:3027–3038
Kocks C, Cho JH, Nehme N, Ulvila J, Pearson AM, Meister M et al (2005) Eater, a transmembrane protein mediating phagocytosis of bacterial pathogens in Drosophila. Cell 123:335–346
Kramer SG, Jinks TM, Schedl P, Gergen JP (1999) Direct activation of sex-lethal transcription by the Drosophila runt protein. Development 126:191–200
Krzemien J, Dubois L, Makki R, Meister M, Vincent A, Crozatier M (2007) Control of blood cell homeostasis in Drosophila larvae by the posterior signalling centre. Nature 446:325–328
Kuchenbauer F, Schnittger S, Look T, Gilliland G, Tenen D, Haferlach T et al (2006) Identification of additional cytogenetic and molecular genetic abnormalities in acute myeloid leukaemia with t(8;21)/AML1-ETO. Br J Haematol 134:616–619
Kurucz E, Markus R, Zsamboki J, Folkl-Medzihradszky K, Darula Z, Vilmos P et al (2007) Nimrod, a putative phagocytosis receptor with EGF repeats in Drosophila plasmatocytes. Curr Biol 17:649–654
Kusch T, Florens L, Macdonald WH, Swanson SK, Glaser RL, Yates JR 3rd et al (2004) Acetylation by Tip60 is required for selective histone variant exchange at DNA lesions. Science 306:2084–2087
Kwon SY, Xiao H, Glover BP, Tjian R, Wu C, Badenhorst P (2008) The nucleosome remodeling factor (NURF) regulates genes involved in Drosophila innate immunity. Dev Biol 316:538–547
Kwon SY, Xiao H, Wu C, Badenhorst P (2009) Alternative splicing of NURF301 generates distinct NURF chromatin remodeling complexes with altered modified histone binding specificities. PLoS Genet 5:e1000574
Lagueux M, Perrodou E, Levashina EA, Capovilla M, Hoffmann JA (2000) Constitutive expression of a complement-like protein in toll and JAK gain-of-function mutants of Drosophila. Proc Natl Acad Sci U S A 97:11427–11432

Lanot R, Zachary D, Holder F, Meister M (2001) Postembryonic hematopoiesis in Drosophila. Dev Biol 230:243–257
Lebestky T, Chang T, Hartenstein V, Banerjee U (2000) Specification of Drosophila hematopoietic lineage by conserved transcription factors. Science 288:146–149
Lebestky T, Jung SH, Banerjee U (2003) A Serrate-expressing signaling center controls Drosophila hematopoiesis. Genes Dev 17:348–353
Levanon D, Goldstein RE, Bernstein Y, Tang H, Goldenberg D, Stifani S et al (1998) Transcriptional repression by AML1 and LEF-1 is mediated by the TLE/Groucho corepressors. Proc Natl Acad Sci 95:11590–11595
Li LH, Gergen JP (1999) Differential interactions between brother proteins and Runt domain proteins in the Drosophila embryo and eye. Development 126:3313–3322
Lin C, Smith ER, Takahashi H, Lai KC, Martin-Brown S, Florens L et al (2010) AFF4, a component of the ELL/P-TEFb elongation complex and a shared subunit of MLL chimeras, can link transcription elongation to leukemia. Mol Cell 37:429–437
Luo H, Hanratty WP, Dearolf CR (1995) An amino acid substitution in the Drosophila hopTum-l Jak kinase causes leukemia-like hematopoietic defects. EMBO J 14:1412–1420
Luo H, Rose P, Barber D, Hanratty WP, Lee S, Roberts TM et al (1997) Mutation in the Jak kinase JH2 domain hyperactivates Drosophila and mammalian Jak-Stat pathways. Mol Cell Biol 17:1562–1571
Makhijani K, Alexander B, Tanaka T, Rulifson E, Bruckner K (2011) The peripheral nervous system supports blood cell homing and survival in the Drosophila larva. Development 138:5379–5391
Mandal L, Banerjee U, Hartenstein V (2004) Evidence for a fruit fly hemangioblast and similarities between lymph-gland hematopoiesis in fruit fly and mammal aorta-gonadal-mesonephros mesoderm. Nat Genet 36:1019–1023
Mandal L, Martinez-Agosto JA, Evans CJ, Hartenstein V, Banerjee U (2007) A Hedgehog- and Antennapedia-dependent niche maintains Drosophila haematopoietic precursors. Nature 446:320–324
Mannstadt M, Holick E, Zhao W, Juppner H (2011) Mutational analysis of GCMB, a parathyroid-specific transcription factor, in parathyroid adenoma of primary hyperparathyroidism. J Endocrinol 210:165–171
Manoukian AS, Krause HM (1993) Control of segmental asymmetry in Drosophila embryos. Development 118:785–796
Markus R, Laurinyecz B, Kurucz E, Honti V, Bajusz I, Sipos B et al (2009) Sessile hemocytes as a hematopoietic compartment in Drosophila melanogaster. Proc Natl Acad Sci U S A 106:4805–4809
Medvinsky A, Dzierzak E (1996) Definitive hematopoiesis is autonomously initiated by the AGM region. Cell 86:897–906
Medvinsky AL, Samoylina NL, Muller AM, Dzierzak EA (1993) An early pre-liver intraembryonic source of CFU-S in the developing mouse. Nature 364:64–67
Miccio A, Wang Y, Hong W, Gregory GD, Wang H, Yu X et al (2010) NuRD mediates activating and repressive functions of GATA-1 and FOG-1 during blood development. EMBO J 29:442–456
Minakhina S, Steward R (2006) Melanotic mutants in Drosophila: pathways and phenotypes. Genetics 174:253–263
Mizuguchi G, Shen X, Landry J, Wu WH, Sen S, Wu C (2004) ATP-driven exchange of histone H2AZ variant catalyzed by SWR1 chromatin remodeling complex. Science 303:343–348
Mukherjee T, Kim WS, Mandal L, Banerjee U (2011) Interaction between Notch and Hif-alpha in development and survival of Drosophila blood cells. Science 332:1210–1213
Muntean AG, Hess JL (2012) The pathogenesis of mixed-lineage leukemia. Annu Rev Pathol 7:283–301
Nam HJ, Jang IH, Asano T, Lee WJ (2008) Involvement of pro-phenoloxidase 3 in lamellocyte-mediated spontaneous melanization in Drosophila. Mol Cells 26:606–610

Nappi AJ, Streams FA (1969) Haemocytic reactions of Drosophila melanogaster to the parasites Pseudocoila mellipes and P. bochei. J Insect Physiol 15:1551–1566
Ogawa H, Ueda T, Aoyama T, Aronheim A, Nagata S, Fukunaga R (2003) A SWI2/SNF2-type ATPase/helicase protein, mDomino, interacts with myeloid zinc finger protein 2A (MZF-2A) to regulate its transcriptional activity. Genes Cells 8:325–339
Orkin SH (1995) Transcription factors and hematopoietic development. J Biol Chem 270:4955–4958
Osman D, Gobert V, Ponthan F, Heidenreich O, Haenlin M, Waltzer L (2009) A Drosophila model identifies calpains as modulators of the human leukemogenic fusion protein AML1-ETO. Proc Natl Acad Sci U S A 106:12043–12048
Owusu-Ansah E, Banerjee U (2009) Reactive oxygen species prime Drosophila haematopoietic progenitors for differentiation. Nature 461:537–541
Paroush Z, Finley RL Jr, Kidd T, Wainwright SM, Ingham PW, Brent R et al (1994) Groucho is required for Drosophila neurogenesis, segmentation, and sex determination and interacts directly with hairy-related bHLH proteins. Cell 79:805–815
Patel SR, Bhumbra SS, Paknikar RS, Dressler GR (2012) Epigenetic mechanisms of Groucho/Grg/TLE mediated transcriptional repression. Mol Cell 45:185–195
Philips JA, Rubin EJ, Perrimon N (2005) Drosophila RNAi screen reveals CD36 family member required for mycobacterial infection. Science 309:1251–1253
Poulson DF (1950) Histogenesis, organogenesis and differentiation in the embryo of Drosophila melanogaster Meigen. In: Demerec M (ed) Biology of *Drosophila*. Hafner, New York, NY, pp 168–274
Qiu P, Pan PC, Govind S (1998) A role for the Drosophila Toll/Cactus pathway in larval hematopoiesis. Development 125:1909–1920
Querfurth E, Schuster M, Kulessa H, Crispino JD, Doderlein G, Orkin SH et al (2000) Antagonism between C/EBPbeta and FOG in eosinophil lineage commitment of multipotent hematopoietic progenitors. Genes Dev 14:2515–2525
Raisner RM, Hartley PD, Meneghini MD, Bao MZ, Liu CL, Schreiber SL et al (2005) Histone variant H2A.Z marks the 5′ ends of both active and inactive genes in euchromatin. Cell 123:233–248
Ramet M, Pearson A, Manfruelli P, Li X, Koziel H, Gobel V et al (2001) Drosophila scavenger receptor CI is a pattern recognition receptor for bacteria. Immunity 15:1027–1038
Ramet M, Manfruelli P, Pearson A, Mathey-Prevot B, Ezekowitz RA (2002) Functional genomic analysis of phagocytosis and identification of a Drosophila receptor for E. coli. Nature 416:644–648
Rehorn KP, Thelen H, Michelson AM, Reuter R (1996) A molecular aspect of hematopoiesis and endoderm development common to vertebrates and Drosophila. Development 122:4023–4031
Rizki MTM (1957a) Alterations in the haemocyte population of Drosophila melanogaster. J Morphol 100:437–458
Rizki MTM (1957b) Tumor formationin relation to metamorphosis in Drosophila melanogaster. J Morphol 100:459–472
Rizki TM (1978) The circulatory system and associated cells and tissues. In: Wright TRF, Ashburner M (eds) Genetics and biology of *Drosophila*, vol 2b. Academic, London, pp 398–451
Rizki MT, Rizki RM (1959) Functional significance of the crystal cells in the larva of Drosophila melanogaster. J Biophys Biochem Cytol 5:235–240
Rizki TM, Rizki RM (1974) Basement membrane abnormalities in melanotic tumour formation in *Drosophila*. Experientia 30:543–546
Rizki TM, Rizki RM (1980) Properties of the larval hemocytes of Drosophila melanogaster. Experientia 36:1223–1226
Rizki TM, Rizki RM (1992) Lamellocyte differentiation in Drosophila larvae parasitized by Leptopilina. Dev Comp Immunol 16:103–110

Rodriguez A, Zhou Z, Tang ML, Meller S, Chen J, Bellen H et al (1996) Identification of immune system and response genes, and novel mutations causing melanotic tumor formation in Drosophila melanogaster. Genetics 143:929–940
Rorth P (1996) A modular misexpression screen in Drosophila detecting tissue-specific phenotypes. Proc Natl Acad Sci U S A 93:12418–12422
Rorth P, Szabo K, Bailey A, Laverty T, Rehm J, Rubin GM et al (1998) Systematic gain-of-function genetics in Drosophila. Development 125:1049–1057
Ruhf ML, Braun A, Papoulas O, Tamkun JW, Randsholt N, Meister M (2001) The domino gene of Drosophila encodes novel members of the SWI2/SNF2 family of DNA-dependent ATPases, which contribute to the silencing of homeotic genes. Development 128:1429–1441
Russo J, Dupas S, Frey F, Carton Y, Brehelin M (1996) Insect immunity: early events in the encapsulation process of parasitoid (Leptopilina boulardi) eggs in resistant and susceptible strains of Drosophila. Parasitology 112(Pt 1):135–142
Saleque S, Kim J, Rooke HM, Orkin SH (2007) Epigenetic regulation of hematopoietic differentiation by Gfi-1 and Gfi-1b is mediated by the cofactors CoREST and LSD1. Mol Cell 27:562–572
Salt GW (1970) The cellular defence reactions of insects, vol 16, Cambridge monographs in experimental biology. Cambridge University Press, Cambridge
Shi Y, Lan F, Matson C, Mulligan P, Whetstine JR, Cole PA et al (2004) Histone demethylation mediated by the nuclear amine oxidase homolog LSD1. Cell 119:941–953
Shi YJ, Matson C, Lan F, Iwase S, Baba T, Shi Y (2005) Regulation of LSD1 histone demethylase activity by its associated factors. Mol Cell 19:857–864
Shim J, Mukherjee T, Banerjee U (2012) Direct sensing of systemic and nutritional signals by haematopoietic progenitors in Drosophila. Nat Cell Biol 14:394–400
Sinenko SA, Mandal L, Martinez-Agosto JA, Banerjee U (2009) Dual role of wingless signaling in stem-like hematopoietic precursor maintenance in Drosophila. Dev Cell 16:756–763
Sinenko SA, Hung T, Moroz T, Tran QM, Sidhu S, Cheney MD et al (2010) Genetic manipulation of AML1-ETO-induced expansion of hematopoietic precursors in a Drosophila model. Blood 116:4612–4620
Smith ER, Lin C, Garrett AS, Thornton J, Mohaghegh N, Hu D et al (2011) The little elongation complex regulates small nuclear RNA transcription. Mol Cell 44:954–965
Song H, Hasson P, Paroush Z, Courey AJ (2004) Groucho oligomerization is required for repression in vivo. Mol Cell Biol 24:4341–4350
Sorrentino RP, Tokusumi T, Schulz RA (2007) The friend of GATA protein U-shaped functions as a hematopoietic tumor suppressor in Drosophila. Dev Biol 311:311–323
Sparrow JC (1978) Melanotic tumours. In: Ashburner M, Wright TRF (eds) The genetics and biology of *Drosophila*, vol 2b. Academic, London, pp 277–313
Staal FJ, Clevers HC (2005) WNT signalling and haematopoiesis: a WNT-WNT situation. Nat Rev Immunol 5:21–30
Stathopoulos A, Tam B, Ronshaugen M, Frasch M, Levine M (2004) pyramus and thisbe: FGF genes that pattern the mesoderm of Drosophila embryos. Genes Dev 18:687–699
Stofanko M, Kwon SY, Badenhorst P (2008) A misexpression screen to identify regulators of Drosophila larval hemocyte development. Genetics 180:253–267
Stofanko M, Kwon SY, Badenhorst P (2010) Lineage tracing of lamellocytes demonstrates Drosophila macrophage plasticity. PLoS One 5:e14051
Stroschein-Stevenson SL, Foley E, O'Farrell PH, Johnson AD (2006) Identification of Drosophila gene products required for phagocytosis of Candida albicans. PLoS Biol 4:e4
Stuart LM, Deng J, Silver JM, Takahashi K, Tseng AA, Hennessy EJ et al (2005) Response to Staphylococcus aureus requires CD36-mediated phagocytosis triggered by the COOH-terminal cytoplasmic domain. J Cell Biol 170:477–485
Tepass U, Fessler LI, Aziz A, Hartenstein V (1994) Embryonic origin of hemocytes and their relationship to cell death in Drosophila. Development 120:1829–1837

Terriente-Felix A, Li J, Collins S, Mulligan A, Reekie I, Bernard F et al (2013) Notch cooperates with Lozenge/Runx to lock haemocytes into a differentiation programme. Development 140:926–937

Tokusumi T, Tokusumi Y, Hopkins DW, Shoue DA, Corona L, Schulz RA (2011) Germ line differentiation factor bag of marbles is a regulator of hematopoietic progenitor maintenance during Drosophila hematopoiesis. Development 138:3879–3884

Tokusumi Y, Tokusumi T, Shoue DA, Schulz RA (2012) Gene regulatory networks controlling hematopoietic progenitor niche cell production and differentiation in the Drosophila lymph gland. PLoS One 7:e41604

Trojer P, Li G, Sims RJ 3rd, Vaquero A, Kalakonda N, Boccuni P et al (2007) L3MBTL1, a histone-methylation-dependent chromatin lock. Cell 129:915–928

Tsuda L, Kaido M, Lim YM, Kato K, Aigaki T, Hayashi S (2006) An NRSF/REST-like repressor downstream of Ebi/SMRTER/Su(H) regulates eye development in Drosophila. EMBO J 25:3191–3202

Waltzer L, Ferjoux G, Bataille L, Haenlin M (2003) Cooperation between the GATA and RUNX factors serpent and lozenge during Drosophila hematopoiesis. EMBO J 22:6516–6525

Ward EJ, Skeath JB (2000) Characterization of a novel subset of cardiac cells and their progenitors in the Drosophila embryo. Development 127:4959–4969

Watson KL, Johnson TK, Denell RE (1991) Lethal(1) aberrant immune response mutations leading to melanotic tumor formation in Drosophila melanogaster. Dev Genet 12:173–187

Wertheim B, Kraaijeveld AR, Schuster E, Blanc E, Hopkins M, Pletcher SD et al (2005) Genome-wide gene expression in response to parasitoid attack in Drosophila. Genome Biol 6:R94

Wildonger J, Mann RS (2005) The t(8;21) translocation converts AML1 into a constitutive transcriptional repressor. Development 132:2263–2272

Williams MJ (2009) The Drosophila cell adhesion molecule Neuroglian regulates Lissencephaly-1 localisation in circulating immunosurveillance cells. BMC Immunol 10:17

Wu X, Golden K, Bodmer R (1995) Heart development in Drosophila requires the segment polarity gene wingless. Dev Biol 169:619–628

Wysocka J, Swigut T, Xiao H, Milne TA, Kwon SY, Landry J et al (2006) A PHD finger of NURF couples histone H3 lysine 4 trimethylation with chromatin remodelling. Nature 442:86–90

Xiao H, Sandaltzopoulos R, Wang HM, Hamiche A, Ranallo R, Lee KM et al (2001) Dual functions of largest NURF subunit NURF301 in nucleosome sliding and transcription factor interactions. Mol Cell 8:531–543

You A, Tong JK, Grozinger CM, Schreiber SL (2001) CoREST is an integral component of the CoREST- human histone deacetylase complex. Proc Natl Acad Sci U S A 98:1454–1458

Yuan Y, Zhou L, Miyamoto T, Iwasaki H, Harakawa N, Hetherington CJ et al (2001) AML1-ETO expression is directly involved in the development of acute myeloid leukemia in the presence of additional mutations. Proc Natl Acad Sci U S A 98:10398–10403

Zettervall CJ, Anderl I, Williams MJ, Palmer R, Kurucz E, Ando I et al (2004) A directed screen for genes involved in Drosophila blood cell activation. Proc Natl Acad Sci U S A 101:14192–14197

Zhang H, Roberts DN, Cairns BR (2005) Genome-wide dynamics of Htz1, a histone H2A variant that poises repressed/basal promoters for activation through histone loss. Cell 123:219–231

Chapter 3
Epigenetic and Transcriptional Mechanisms Regulating Blood Cell Development in Zebrafish

Xiaoying Bai

Abstract The remarkably conserved hematopoietic programs between zebrafish and mammals make zebrafish an excellent model to study vertebrate blood development. Studies on zebrafish from the past 2 decades have greatly contributed to our understanding of human hematopoiesis and associated blood disorders. This chapter summarizes the important genetic and epigenetic factors involved in transcriptional regulation of blood development in zebrafish embryos. Recent advances using zebrafish for blood disease modeling and therapeutic discovery will also be discussed.

Keywords Hematopoiesis • Stem cell • Leukemia • Anemia • Zebrafish

3.1 Introduction

Over the past 20 years, zebrafish has proven to be a powerful vertebrate model for large-scale genetic investigations. The unique advantages of zebrafish, such as external fertilization and embryogenesis, large brood size, and short developmental time, greatly facilitate forward genetic screens. In addition, its transparent embryo helps to detect and analyze blood mutants. To date, more than 26 complementation groups with hematopoietic defects have been isolated (Ransom et al. 1996). Moreover, available reverse genetic approaches to study specific genes of interest make this organism an excellent model for the control of vertebrate development. Gene knockdown by morpholino antisense oligonucleotides can be used to inhibit specific gene function through microinjection into 1- to 4-cell stage embryos. Although morpholino activity only lasts a few days, it is a quick and reliable way to assess

X. Bai (✉)
Department of Obstetrics and Gynecology, Cecil H. and Ida Green Center for Reproductive Biology Sciences, The University of Texas Southwestern Medical Center, 5323 Harry Hines Boulevard, J7.130B, Dallas, TX 75390, USA
e-mail: Xiaoying.Bai@UTSouthwestern.edu

C. Bonifer and P.N. Cockerill (eds.), *Transcriptional and Epigenetic Mechanisms Regulating Normal and Aberrant Blood Cell Development*, Epigenetics and Human Health, DOI 10.1007/978-3-642-45198-0_3, © Springer-Verlag Berlin Heidelberg 2014

gene function during development, as most organs, including the blood system, are already formed and functional during the first 5 days after fertilization. Targeted genetic mutants can be obtained by target induced local lesions in genome (TILLING) in which genomic DNA from thousands of *N*-ethyl-N nitrosourea (ENU)-mutagenized zebrafish is analyzed by exon-sequencing to identify mutations in specific genes (Kettleborough et al. 2013). More recently, targeted gene knockout has been established in zebrafish using new gene-targeting technologies including zinc finger endonucleases (ZFNs), transcription activator-like effector nucleases (TALENs), and CRISPR-CAS systems (Ekker 2008; Huang et al. 2011; Sander et al. 2011; Chang et al. 2013). In addition to genetic tools, transgenic methods are well established in zebrafish; multiple transgenic lines labeling the blood system ranging from early precursors to specific lineages have been generated and greatly facilitate lineage tracing and fate mapping studies.

This chapter will review the general process of hematopoiesis during zebrafish development with the highlight of important findings from zebrafish studies that have advanced our understanding of normal and malignant hematopoiesis in general.

3.2 Overview of Zebrafish Hematopoiesis

Like in other vertebrates, hematopoiesis in zebrafish is characterized by two sequential waves taking place at anatomically distinct sites during embryonic development (Table 3.1). Primitive hematopoiesis generates mainly erythroid cells and some macrophages and megakaryocytes. Zebrafish primitive erythroid cells are generated intraembryonically within the posterior lateral mesoderm (PML), from two strips of precursor cells expressing the transcription factor genes *scl*, *lmo2*, and *gata1*. These cells migrate toward the trunk midline to form the intermediate cell mass (ICM) (Fig. 3.1a), an equivalent to the mammalian yolk sac island (Detrich et al. 1995). At 15 h post-fertilization (hpf), these cells begin to express embryonic globin (Willett et al. 1999). Between 24 and 26 hpf, the heart starts beating and erythroblasts enter the circulation, where they subsequently mature into primitive erythrocytes. Zebrafish primitive macrophages are mainly generated from the rostral blood island (RBI) that is derived from the anterior lateral mesoderm (ALM). Specification of the myeloid fate is determined by the expression of the gene encoding the transcription factor PU.1 *(pu.1)* in a subset of precursor cells also expressing *scl*, *lmo2*, and *fli1a* in RBI (Herbomel et al. 1999).

Primitive hematopoiesis is transient and is subsequently replaced by the definitive wave of hematopoiesis that generates the hematopoietic stem cells (HSCs), which give rise to all adult blood lineages throughout life. As in mammals, zebrafish definitive HSCs are generated in the aorta–gonads–mesonephrons (AGM) region as detected by runx1 expression at 36 hpf (Fig. 3.1b). Recent studies using confocal imaging in live zebrafish embryos have provided direct evidence

Table 3.1 Stages of zebrafish developmental hematopoiesis

	Emergence time	Anatomic site	Cell type
Primitive hematopoiesis	11–12 hpf	ALM → RBI	Primitive macrophage
		PLM → ICM	Primitive erythrocyte
Definitive hematopoiesis	24–25 hpf	PBI	Erythro-myeloid progenitor
	26 hpf	AGM	HSC
Adult hematopoiesis	3 dpf	Thymus	T cell
	4 dpf	Kidney	HSC

hpf hours post-fertilization, *dpf* days post-fertilization, *ALM* anterior lateral mesoderm, *PLM* posterior lateral mesoderm, *RBI* rostral blood island, *ICM* intermediate cell mass, *PBI* posterior blood island, *AGM* aorta–gonad-mesonephros, *HSC* hematopoietic stem cell

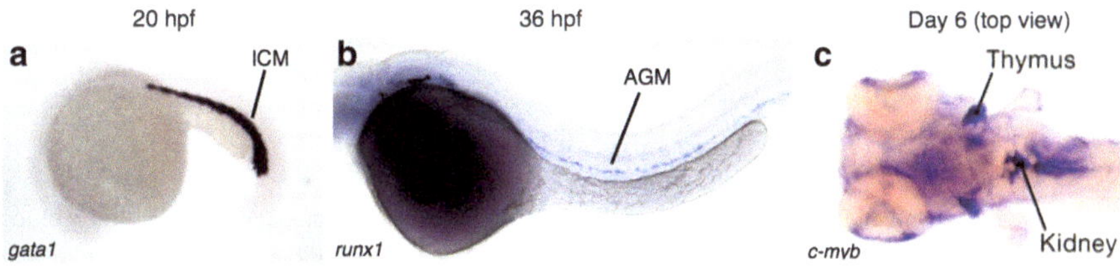

Fig. 3.1 Blood development in zebrafish. In situ hybridization for (**a**) *gata1* at 20 hpf (hours post-fertilization) in ICM, (**b**) for *runx1* at 36 hpf in AGM, and (**c**) for *c-myb* at day 6 (*top view*) to show expression in the kidney marrow and thymus

that HSCs are specified from the ventral endothelium in the dorsal aorta (Bertrand et al. 2010; Kissa and Herbomel 2010). Together with the results in mice (Boisset et al. 2010), these studies have proven the long-standing hypothesis about the existence of the "hemogenic endothelium." In addition, an intermediate wave was observed recently in both mammals and zebrafish that begins with committed erythromyeloid progenitors (EMP) (Bertrand et al. 2005, 2007; Yokota et al. 2006). In zebrafish, EMPs are found in the posterior blood island (PBI). They can only differentiate into erythroid and myeloid lineages but lack lymphoid potential and their differentiation is independent of Notch signaling (Bertrand et al. 2007).

After 48 hpf, definitive HSCs-derived progenitors are found to migrate to the caudal hematopoietic tissue (CHT) near the tail, where they undergo massive proliferation and further differentiation (Murayama et al. 2006). These cells eventually colonize the thymus to initiate lymphopoiesis, as well as the kidney that becomes the site for adult hematopoiesis and is the equivalent of mammalian bone marrow (Fig. 3.1c).

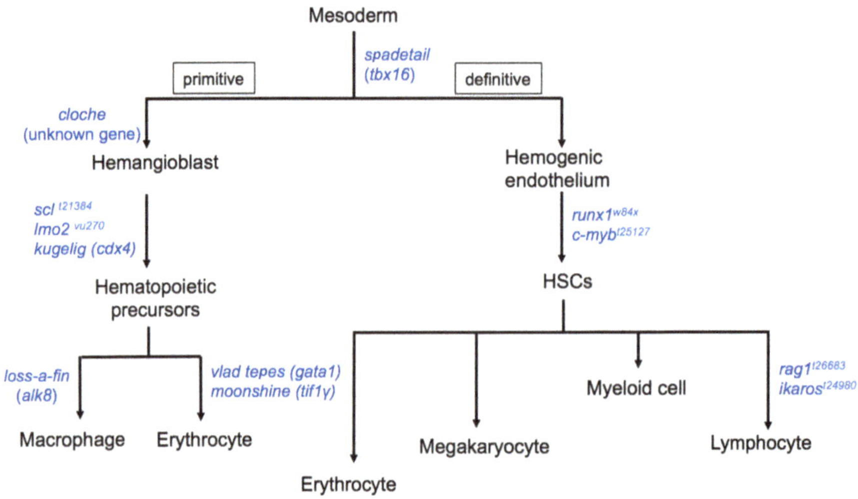

Fig. 3.2 Zebrafish genetic mutants of crucial genes regulating different stages of embryonic blood development. For mutants with specific names, the corresponding gene is shown in *parenthesis*

3.3 Genetic and Epigenetic Regulation of Zebrafish Hematopoiesis

The genetic program controlling hematopoiesis is highly conserved between zebrafish and mammals. This section will provide an overview of important transcription regulators and zebrafish mutants that affect distinct stages of embryonic hematopoiesis (Fig. 3.2).

3.3.1 Transcriptional Control of the Emergence of Hematopoietic Cells in Early Embryos

Like other vertebrates, blood cells in zebrafish originate from mesoderm. Following mesoderm patterning, ventral lateral mesoderm is further specified into blood, angioblast, and kidney progenitors. The close relationship between blood and angioblasts is revealed by cells coexpressing blood markers and endothelial markers during the early segmentation period, supporting the hypothesis that these two cell lineages share a common ancestor called "hemangioblast." A fate map study in zebrafish also supports the existence of hemangioblast by showing that fluorescent-labeled single cells during early gastrulation can give rise to both blood and vascular cells but not other mesodermal cell lineages (Vogeli et al. 2006). Additional genetic evidence comes from the zebrafish mutant *cloche* (*clo*), which completely lacks blood and vascular cells, as well as the heart endocardium but not other mesodermal organs (Stainier et al. 1995). Gene expression analyses have revealed a near complete absence of HSC markers and angioblast markers including *scl, lmo2, gata2, runx1, fli1*, and *flk* (Liao et al. 1998). Although the mutant gene in *clo* has not been identified due to the telomeric location of the gene, uncovering the mutation responsible for the *clo* mutant is expected to provide insight into the molecular events that direct commitment of mesoderm toward blood and/or endothelial fates.

The blood defects in *clo* mutant can be partially rescued by overexpression of *scl*, placing this basic helix–loop–helix (bHLH) transcription factor downstream of *clo*. The master functions of SCL and its associated partner the LIM-domain factor LMO2 in hematopoiesis have been extensively studied in mice. In zebrafish, these genes are expressed in both hematopoietic and endothelial progenitors and have been shown to act together to specify the hemangioblasts (Patterson et al. 2007). Mutants of *scl* and *lmo2* have been identified in zebrafish and both have severe defects in blood and vasculature with complete loss of primitive and definitive hematopoiesis (Bussmann et al. 2007; Weiss et al. 2012). Overexpression of *scl* by itself or together with *lmo2* expands the formation of blood and vascular cells at the expense of somatic tissue (Gering et al. 2003; Dooley et al. 2005). In *clo* mutants, forced expression of *scl* resulted in an expansion of hematopoietic but not endothelial tissue, suggesting that the function of *scl* in hemangioblasts is to specify blood rather than vascular fate (Dooley et al. 2005).

The specification of hematopoietic cell fate also relies on the *cdx-hox* pathway as elucidated by the zebrafish mutant *kugelig (kgg)*, which is characterized by severe anemia, shortened tail, and reduced yolk tube extension (Hammerschmidt et al. 1996). The defective gene *cdx4* belongs to the caudal-related homeobox transcription factor family that is implicated in anterior–posterior axis patterning through regulation of the *hox* genes (Davidson et al. 2003). Consistent with its role in *hox* regulation, *cdx4* mutants show an altered expression pattern of multiple *hox* genes. The number of ICM blood precursors is also reduced, indicating an early defect in hematopoiesis. Simultaneous knocking down of another cdx member *cdx1a* further interrupts the *hox* pattern and results in an almost complete loss of *scl* expression (Davidson and Zon 2006). The number of angioblasts expressing the *flk* marker remains normal and the adjacent pronephric tissue is specified correctly, indicating that the disruption of hematopoietic progenitors is not caused by a general posterior patterning defect due to perturbed *hox* gene expression, as the number of *flk* + angioblasts remains normal and the adjacent pronephric tissue is specified correctly. Overexpression of *hoxb7* and *hoxa9* rescues the blood defect without correcting the tail morphology in *kgg* (Davidson et al. 2003), suggesting an integral role of *hox* genes in hematopoiesis. Inspired by the study in zebrafish, Wang et al. found that in mouse ES cells, ectopic *Cdx4* expression promotes hematopoietic mesoderm specification, increases hematopoietic progenitor formation, and, together with *HoxB4*, enhances multilineage hematopoietic engraftment of lethally irradiated adult mice (Wang et al. 2005). Taken together, these studies demonstrate the specific function of the *cdx-hox* pathway in vertebrate blood development.

3.3.2 Regulation of Definitive HSC Formation

At 26 hpf, definitive HSCs start to form in the AGM region in zebrafish embryos. By 36 hpf, HSCs can be easily detected along the ventral floor of the dorsal aorta by in situ hybridization for HSC markers such as the transcription factor genes *runx1* and *c-myb*.

The ontogeny of HSCs requires a coordinated interaction between signal transduction pathways and transcription factors. In zebrafish embryos, inhibition of the Hedgehog (Hh) pathway by genetic mutants or chemical inhibitors was found to greatly reduce *runx1*$^{+}$ HSCs in the AGM while leaving primitive erythrocytes intact, suggesting a specific requirement of the Hh signaling in definitive hematopoiesis (Gering and Patient 2005). Impaired migration of dorsal aorta angioblasts in these embryos indicates that the loss of HSCs results from improper patterning of the aorta. More recently, BMP signaling has been found to work together with the Hh pathway to polarize the dorsal aorta for HSC emergence from the ventral arterial endothelium (Wilkinson et al. 2009). Consistent with the zebrafish studies, the positive role of Hh in AGM HSC induction has also been found in murine organ culture (Peeters et al. 2009). Another signaling pathway required for artery identity and HSC induction is Notch (Burns et al. 2005). In the zebrafish *mindbomb* mutant that lacks Notch signaling, the expression of both arterial markers and HSC markers are greatly reduced, which can be rescued by *runx1* overexpression, placing *runx1* downstream or in parallel with the Notch pathway. More recently, Kim et al. showed that in both mouse ES cells and zebrafish embryos, Hh acts upstream of Notch during the generation of hemogenic endothelium and *Scl* induction mediates the conversion of hemogenic endothelial cells to hematopoietic cells (Kim et al. 2013).

The master regulatory role of the transcription factors RUNX1 and MYB has been demonstrated using targeted knockout mouse models. *Runx1* and *c-myb* mutants were also identified in zebrafish. Both mutants do not affect primitive hematopoiesis but severely impair AGM HSCs. Surprisingly, despite a larval "bloodless" phase due to the failure of AGM HSC formation, ~20 % of *runx1* mutant embryos are able to recover and generate CD41+ precursors that contribute to multilineage hematopoiesis in adulthood (Sood et al. 2010). The mutation in the *runx1* gene generates a premature truncation in the runt domain thus removing most of the residues important in *runx1* activity, suggesting a loss of *runx1* function in these mutants. One possible explanation for the recover of hematopoiesis in the absence of functional *runx1* is that other *runx* family members (*runx2a, runx2b, and runx3*) may compensate for *runx1* function. As noted, *runx3* was found to regulate both primitive and definitive hematopoiesis in zebrafish (Kalev-Zylinska et al. 2003). Knocking down *runx3* decreases AGM HSCs, while overexpression of *runx3* leads to an increase. Whether or not there are any *runx1*-independent pathways existing in mammals is not clear but *runx1* knockout embryonic stem cells have been reported to be capable of contributing to adult hematopoiesis in chimeric mice at a very low level (Kundu et al. 2005).

In contrast to the *runx1* mutant, zebrafish *c-myb* mutants completely fail to initiate definitive hematopoiesis (Soza-Ried et al. 2010). Although they can survive for 2–3 months with stunted growth, no adult hematopoiesis is observed in the kidney marrow. The reason for their prolonged survival may be due to the fact that unaffected primitive erythrocytes are present long enough to provide oxygen during the critical development period. The lack of definitive hematopoiesis makes the *c-myb* mutant fish a suitable model for hematopoietic cell transplantation assays, which have been hampered by the lack of inbred strains and the complex nature of MHC genes in zebrafish. Using this mutant as the recipient, successful hematopoietic reconstitution was achieved even with donor cells that have unmatched MHC loci (Hess et al. 2013).

3.3.3 Regulation of Lineage Specification

Once hematopoietic precursors are specified, additional transcription factors come into effect to direct lineage-specific differentiation. The essential role of *gata1* in erythropoietic differentiation was demonstrated by the zebrafish mutant *vlad tepes (vlt)* (Weinstein et al. 1996). The mutation in the *gata1* gene in *vlt* embryos results in a truncated protein unable to bind DNA and mediate GATA-specific transactivation (Lyons et al. 2002), leading to a complete loss of red cells. Moreover, a cross-inhibitory action was found between *gata1* and *pu.1* in that loss of *gata1* function transforms erythroid precursors into myeloid cells, and conversely, *pu.1* knockdown switches myeloid cells to a red cell fate (Galloway et al. 2005; Rhodes et al. 2005). These in vivo findings correlate well with studies in mammalian cell cultures showing that GATA1 and Pu.1 physically bind and cross antagonize each other (Zhang et al. 1999, 2000; Stopka et al. 2005; Liew et al. 2006), providing a good example on how transcription factors interact to control lineage fate.

Studies on transcription regulators for lineage differentiation have mainly focused on factors that affect transcription initiation by polymerase II (Pol II). However, recent genome-wide studies have revealed a critical role for transcription elongation in gene regulation suggested by the observation that Pol II is commonly "paused" on a large number of developmentally regulated genes (Guenther et al. 2007; Muse et al. 2007; Zeitlinger et al. 2007). A study on the zebrafish mutant *moonshine (mon)* directly supports this view (Bai et al. 2010). The mutant gene in *mon* is *tif1γ* (transcriptional intermediary factor 1 gamma), which encodes a ubiquitously expressed transcription cofactor that is highly enriched in the ICM blood island (Ransom et al. 2004). Loss of *tif1γ* results in a profound anemia in zebrafish embryos due to the block in red cell differentiation. Through a genetic suppressor screen approach, Bai et al. found that erythroid gene expression in *mon* can be restored by removing the factors that induce Pol II pausing (Bai et al. 2010). The study suggests that by physically interacting with both *scl-gata1* transcription complexes and Pol II elongation factors, TIF1γ recruits Pol II elongation machinery to erythroid genes and releases Pol II from pausing to promote their expression. TIF1γ has also been implicated in the erythroid-myeloid fate switch as revealed by increased definitive myelopoiesis in *mon* embryos and the development of a MPD (myeloid proliferative disease)-like phenotype in conditional knockout mice (Aucagne et al. 2011; Kusy et al. 2011; Monteiro et al. 2011; Bai et al. 2013). The role of *tif1γ* in gene regulation may be broader than regulating Pol II elongation as a recent study has found that it can also function as a chromatin factor to mediate TGFβ signaling in mouse ES cells by recognizing specific histone modification markers (Xi et al. 2011).

3.3.4 Epigenetic Regulators in Zebrafish Hematopoiesis

Increasing evidence has suggested epigenetic regulators act in concert with transcription factors and signaling pathways to regulate the development and homeostasis of the hematopoietic system. Studies using the morpholino knockdown approach and available genetic mutants in zebrafish have facilitated the identification of novel chromatin players in hematopoiesis. Li et al. have found that the repressive Mta3–NuRD chromatin-remodeling complex is essential for the initiation of primitive hematopoiesis in zebrafish embryos (Li et al. 2009). Inhibition of NuRD activity through depletion of Mta3 or HDAC inhibitors abolishes primitive hematopoietic lineages and causes abnormal angiogenesis, whereas overexpression of NuRD components enhances the expression of *scl* and *lmo2* in zebrafish embryos. The requirement for HDAC activity is not restricted in primitive hematopoiesis. In an insertional mutagenesis screen for factors affecting definitive HSCs in the AGM region, Burns et al. have shown that *hdac1* acts downstream of Notch signaling but upstream or in parallel to *runx1* to promote AGM HSC formation (Burns et al. 2009).

Among other chromatin factors affecting zebrafish hematopoiesis, a mutant of the cohesin subunit *rad21* was particularly interesting as cohesin mutations are responsible for a number of human developmental disorders such as the Cornelia de Lange syndrome (CdLS). More recently, mutations in cohesin genes have also been found in patient samples with myeloid diseases (Rocquain et al. 2010; Welch et al. 2012). Although the canonical role of cohesin is to mediate sister chromatid cohesion during cell division, increasing evidence have suggested a mitotic-independent role of cohesin in transcription. *Rad21* mutation in zebrafish causes a loss of early *runx1* expression from primitive progenitors without affecting other progenitor markers such as *scl* and *gata2* (Horsfield et al. 2007). The effect was observed even before cell division was blocked, suggesting that cohesin-dependent *runx1* expression is transcription related and separable from the mitotic defect. This zebrafish cohesin mutant thus provided the first evidence of cohesin-dependent gene regulation in a vertebrate.

To systematically identify epigenetic regulators of zebrafish hematopoiesis, Huang et al. performed a genetic screen for chromatin factors using a large-scale morpholino knockdown approach (Huang 2012). By injecting morpholinos targeting over 400 chromatin factors into the zebrafish embryos, they were able to identify more than 70 factors that affect primitive erythropoiesis or definitive HSCs or both. Their hits include some of the chromatin factors known to regulate hematopoiesis from previous studies, such as components in SWI/SNF complexes and HDACs, but most are novel regulators that have not been studied in hematopoiesis. One of them is CHD7, a chromodomain chromatin remodeler that is mutated in patients with CHARGE syndrome (Vissers et al. 2004). Knocking down *chd7* significantly increases both primitive and definitive blood production, and this effect is cell autonomous as determined by blastula transplantation. While the mechanism by which CHD7 represses hematopoiesis is still the subject of the ongoing research, their study has provided great insights on the role of chromatin factors in regulating gene transcription in hematopoietic cells.

3.4 Blood Disease Modeling Using Zebrafish

Besides being an excellent model for studying developmental hematopoiesis, recent advances in the field have demonstrated zebrafish as a valuable system for defining disease pathways and for developing novel therapies. A few examples of using zebrafish for blood disease modeling are listed here.

The zebrafish model has been used to study a number of human hematological diseases, including anemia, polycythemia, and porphyria. The characterization of the hypochromic anemia mutant *weissherbst* (*weh*) led to the discovery of ferroportin 1 as a novel iron transporter (Donovan et al. 2000). The human mutation of ferroportin 1 was subsequently found in patients affected by hemochromatosis disease with iron absorption defects (Montosi et al. 2001), demonstrating the power of zebrafish genetics for identifying novel disease-related genes. VHL-associated congenital polycythemia is modeled by a zebrafish mutant carrying the *von Hippel-Lindau* tumor suppressor (VHL) mutation (van Rooijen et al. 2009). The mutants faithfully recapitulate most of the key aspects of VHL-associated pathologies observed in human patients, such as the Chuvash form of polycythemia, severe neovascularization defects, and pronephros abnormalities (van Rooijen et al. 2011). Because the mutants can survive to postembryonic stages, they offer an excellent platform for the identification of genetic regulators of VHL disease progression and potential therapeutic agents. Several zebrafish models for Diamond Blackfan Anemia (DBA) were generated by genetic mutants or knockdown of ribosomal protein genes (Uechi et al. 2008; Danilova et al. 2011; Taylor and Zon 2011; Torihara et al. 2011). Characterization of the ribosomal defects in these animals revealed both p53-dependent and independent pathways. The unique forward and reverse genetic approaches in fish make it feasible to do systematic analyses of the contribution of ribosomal genes to blood cell function and will help identify novel pathways involved in DBA pathogenesis.

Since the first zebrafish transgenic cancer model of T cell acute lymphoblastic leukemia (T-ALL) (Langenau et al. 2003), many cancer models have been established in zebrafish by using transgenes to express mammalian oncoproteins. In the original T-ALL model, a GFP-tagged mouse *c-Myc* oncogene was expressed from the fish *rag2* promoter. The leukemia induced in this model was highly efficient with early onset. Later, a Cre-loxP system was introduced into this model to induce leukemia from a controllable heat shock promoter, thereby allowing the genetic dissection of molecular pathways involved in the disease progression (Feng et al. 2007, 2010). These studies identified *bcl2* and *Akt* pathways as modulators of T-ALL progression. AML1-ETO induced myeloid leukemia has also been modeled in zebrafish (Yeh et al. 2008). Induced human AML1-ETO expression in zebrafish embryos redirects myeloerythroid progenitor cells to granulocytic cell fate at the expense of the erythroid cell fate by downregulating *scl* expression. Inhibition of HDAC activity restores *scl* expression and antagonizes the cell fate switch by AML1-ETO, thereby revealing an important role of *scl* regulation in AML1-ETO-mediated leukemogenesis.

3.5 Conclusions and Future Perspectives

The process of hematopoiesis is highly regulated by complex interactions among growth factors, cytokines, and transcription factors. The versatile and unique features of zebrafish have made it an ideal model for dissecting genetic networks regulating vertebrate hematopoiesis. Moreover, zebrafish is becoming a more clinically relevant model for translational medicine.

An exciting advance in the zebrafish filed is in vivo drug screening. Compared to traditional cell culture-based screens, chemical screens performed in a whole organism may identify drugs with more in vivo potency and less toxicity. Zebrafish is well suited for large-scale chemical screens as live embryos can be placed in a 96-well plate and directly absorb chemicals that are added in the water. Tools have been designed to increase the scale of embryo collection so that thousands of developmentally synchronized embryos can be collected in minutes (Adatto et al. 2011). Several large chemical screens have been successfully conduced and novel therapeutic agents have been identified. For example, a chemical screen for AGM HSC regulators led to the discovery of prostaglandin E2 (PGE2) as a positive regulator of HSC formation and regeneration through interaction with the Wnt-signaling pathway (North et al. 2007; Goessling et al. 2009). Preclinical analyses have demonstrated the therapeutic potential of PGE2 treatment using human and nonhuman primate HSCs (Goessling et al. 2011), and a clinical trial has been initiated to test if PGE2 can enhance human HSC engraftment in patients treated with cord blood transplants. The same screen also identified many other compounds that hold great potential for clinical investigation. Such chemical screens in zebrafish can be further designed to identify compounds that reverse a specific disease phenotype. Using the AML1-ETO transgenic fish discussed in the previous section, Yeh et al. performed a chemical suppressor screen and identified COX-2 inhibitors as potential therapeutic agents to suppress the oncogenic function of AML1-ETO (Yeh et al. 2009).

The high throughput sequencing technologies and genome-association studies in recent years have identified huge amount of candidate mutations and polymorphisms for human diseases. The in vivo function of these candidate loci can be easily tested in zebrafish through a combination of convenient microinjection of morpholinos, mRNA, or cDNA. Recently, Gieger et al. carried out a high-powered meta-analysis of genome-wide association studies (GWAS) in nearly 67,000 individuals to identify putative novel regulators of megakaryopoiesis and platelet formation (Gieger et al. 2011). Using zebrafish as one of the models for functional assessment, they were able to verify 11 out of the 68 associated genomic loci as novel regulators for blood cell formation.

With fast improving technologies and genetic tools, it is conceivable that zebrafish will continually contribute to the identification of novel regulators of vertebrate hematopoiesis. Increased use of zebrafish in human disease modeling and therapeutic discovery are highly anticipated in the future.

Acknowledgments I thank Teresa Bowman for critical reading of the manuscript.

References

Adatto I, Lawrence C, Thompson M, Zon LI (2011) A new system for the rapid collection of large numbers of developmentally staged zebrafish embryos. PLoS One 6:e21715. doi:10.1371/journal.pone.0021715

Aucagne R, Droin N, Paggetti J et al (2011) Transcription intermediary factor 1gamma is a tumor suppressor in mouse and human chronic myelomonocytic leukemia. J Clin Invest 121:2361–2370. doi:10.1172/JCI4521345213

Bai X, Kim J, Yang Z et al (2010) TIF1gamma controls erythroid cell fate by regulating transcription elongation. Cell 142:133–143. doi:10.1016/j.cell.2010.05.028

Bai X, Trowbridge JJ, Riley E et al (2013) TiF1-gamma plays an essential role in murine hematopoiesis and regulates transcriptional elongation of erythroid genes. Dev Biol 373:422–430. doi:10.1016/j.ydbio.2012.10.008

Bertrand JY, Giroux S, Golub R et al (2005) Characterization of purified intraembryonic hematopoietic stem cells as a tool to define their site of origin. Proc Natl Acad Sci U S A 102:134–139. doi:10.1073/pnas.0402270102

Bertrand JY, Kim AD, Violette EP, Stachura DL, Cisson JL, Traver D (2007) Definitive hematopoiesis initiates through a committed erythromyeloid progenitor in the zebrafish embryo. Development 134:4147–4156. doi:10.1242/dev.012385

Bertrand JY, Chi NC, Santoso B, Teng S, Stainier DY, Traver D (2010) Haematopoietic stem cells derive directly from aortic endothelium during development. Nature 464:108–111. doi:10.1038/nature08738

Boisset JC, van Cappellen W, Andrieu-Soler C, Galjart N, Dzierzak E, Robin C (2010) In vivo imaging of haematopoietic cells emerging from the mouse aortic endothelium. Nature 464:116–120. doi:10.1038/nature08764

Burns CE, Traver D, Mayhall E, Shepard JL, Zon LI (2005) Hematopoietic stem cell fate is established by the Notch-Runx pathway. Genes Dev 19:2331–2342. doi:10.1101/gad.1337005

Burns CE, Galloway JL, Smith AC et al (2009) A genetic screen in zebrafish defines a hierarchical network of pathways required for hematopoietic stem cell emergence. Blood 113:5776–5782. doi:10.1182/blood-2008-12-193607

Bussmann J, Bakkers J, Schulte-Merker S (2007) Early endocardial morphogenesis requires Scl/Tal1. PLoS Genet 3:e140. doi:10.1371/journal.pgen.0030140

Chang N, Sun C, Gao L et al (2013) Genome editing with RNA-guided Cas9 nuclease in zebrafish embryos. Cell Res 23:465–472. doi:10.1038/cr.2013.45

Danilova N, Sakamoto KM, Lin S (2011) Ribosomal protein L11 mutation in zebrafish leads to haematopoietic and metabolic defects. Br J Haematol 152:217–228. doi:10.1111/j.1365-2141.2010.08396.x

Davidson AJ, Zon LI (2006) The caudal-related homeobox genes cdx1a and cdx4 act redundantly to regulate hox gene expression and the formation of putative hematopoietic stem cells during zebrafish embryogenesis. Dev Biol 292:506–518. doi:10.1016/j.ydbio.2006.01.003

Davidson AJ, Ernst P, Wang Y et al (2003) cdx4 mutants fail to specify blood progenitors and can be rescued by multiple hox genes. Nature 425:300–306. doi:10.1038/nature01973

Detrich HW 3rd, Kieran MW, Chan FY et al (1995) Intraembryonic hematopoietic cell migration during vertebrate development. Proc Natl Acad Sci U S A 92:10713–10717

Donovan A, Brownlie A, Zhou Y et al (2000) Positional cloning of zebrafish ferroportin1 identifies a conserved vertebrate iron exporter. Nature 403:776–781. doi:10.1038/35001596

Dooley KA, Davidson AJ, Zon LI (2005) Zebrafish scl functions independently in hematopoietic and endothelial development. Dev Biol 277:522–536. doi:10.1016/j.ydbio.2004.09.004

Ekker SC (2008) Zinc finger-based knockout punches for zebrafish genes. Zebrafish 5:121–123. doi:10.1089/zeb.2008.9988

Feng H, Langenau DM, Madge JA et al (2007) Heat-shock induction of T-cell lymphoma/leukaemia in conditional Cre/lox-regulated transgenic zebrafish. Br J Haematol 138:169–175. doi:10.1111/j.1365-2141.2007.06625.x

Feng H, Stachura DL, White RM et al (2010) T-lymphoblastic lymphoma cells express high levels of BCL2, S1P1, and ICAM1, leading to a blockade of tumor cell intravasation. Cancer Cell 18:353–366. doi:10.1016/j.ccr.2010.09.009

Galloway JL, Wingert RA, Thisse C, Thisse B, Zon LI (2005) Loss of gata1 but not gata2 converts erythropoiesis to myelopoiesis in zebrafish embryos. Dev Cell 8:109–116. doi:10.1016/j.devcel.2004.12.001

Gering M, Patient R (2005) Hedgehog signaling is required for adult blood stem cell formation in zebrafish embryos. Dev Cell 8:389–400. doi:10.1016/j.devcel.2005.01.010

Gering M, Yamada Y, Rabbitts TH, Patient RK (2003) Lmo2 and Scl/Tal1 convert non-axial mesoderm into haemangioblasts which differentiate into endothelial cells in the absence of Gata1. Development 130:6187–6199. doi:10.1242/dev.00875

Gieger C, Radhakrishnan A, Cvejic A et al (2011) New gene functions in megakaryopoiesis and platelet formation. Nature 480:201–208. doi:10.1038/nature10659

Goessling W, North TE, Loewer S et al (2009) Genetic interaction of PGE2 and Wnt signaling regulates developmental specification of stem cells and regeneration. Cell 136:1136–1147. doi:10.1016/j.cell.2009.01.015

Goessling W, Allen RS, Guan X et al (2011) Prostaglandin E2 enhances human cord blood stem cell xenotransplants and shows long-term safety in preclinical nonhuman primate transplant models. Cell Stem Cell 8:445–458. doi:10.1016/j.stem.2011.02.003

Guenther MG, Levine SS, Boyer LA, Jaenisch R, Young RA (2007) A chromatin landmark and transcription initiation at most promoters in human cells. Cell 130:77–88. doi:10.1016/j.cell.2007.05.042

Hammerschmidt M, Pelegri F, Mullins MC et al (1996) Mutations affecting morphogenesis during gastrulation and tail formation in the zebrafish, Danio rerio. Development 123:143–151

Herbomel P, Thisse B, Thisse C (1999) Ontogeny and behaviour of early macrophages in the zebrafish embryo. Development 126:3735–3745

Hess I, Iwanami N, Schorpp M, Boehm T (2013) Zebrafish model for allogeneic hematopoietic cell transplantation not requiring preconditioning. Proc Natl Acad Sci U S A 110:4327–4332. doi:10.1073/pnas.1219847110

Horsfield JA, Anagnostou SH, Hu JK et al (2007) Cohesin-dependent regulation of Runx genes. Development 134:2639–2649. doi:10.1242/dev.002485

Huang H-T (2012) Epigenetic regulation of hematopoiesis in zebrafish. Ph.D. thesis

Huang P, Xiao A, Zhou M, Zhu Z, Lin S, Zhang B (2011) Heritable gene targeting in zebrafish using customized TALENs. Nat Biotechnol 29:699–700. doi:10.1038/nbt.1939

Kalev-Zylinska ML, Horsfield JA, Flores MV et al (2003) Runx3 is required for hematopoietic development in zebrafish. Dev Dyn 228:323–336. doi:10.1002/dvdy.10388

Kettleborough RN, Busch-Nentwich EM, Harvey SA et al (2013) A systematic genome-wide analysis of zebrafish protein-coding gene function. Nature 496:494–497. doi:10.1038/nature11992

Kim PG, Albacker CE, Lu YF et al (2013) Signaling axis involving Hedgehog, Notch, and Scl promotes the embryonic endothelial-to-hematopoietic transition. Proc Natl Acad Sci U S A 110:E141–E150. doi:10.1073/pnas.1214361110

Kissa K, Herbomel P (2010) Blood stem cells emerge from aortic endothelium by a novel type of cell transition. Nature 464:112–115. doi:10.1038/nature08761

Kundu M, Compton S, Garrett-Beal L et al (2005) Runx1 deficiency predisposes mice to T-lymphoblastic lymphoma. Blood 106:3621–3624. doi:10.1182/blood-2005-04-1447

Kusy S, Gault N, Ferri F et al (2011) Adult hematopoiesis is regulated by TIF1gamma, a repressor of TAL1 and PU.1 transcriptional activity. Cell Stem Cell 8:412–425. doi:10.1016/j.stem.2011.02.005

Langenau DM, Traver D, Ferrando AA et al (2003) Myc-induced T cell leukemia in transgenic zebrafish. Science 299:887–890. doi:10.1126/science.1080280

Li X, Jia S, Wang S, Wang Y, Meng A (2009) Mta3-NuRD complex is a master regulator for initiation of primitive hematopoiesis in vertebrate embryos. Blood 114:5464–5472. doi:10.1182/blood-2009-06-227777

Liao EC, Paw BH, Oates AC, Pratt SJ, Postlethwait JH, Zon LI (1998) SCL/Tal-1 transcription factor acts downstream of cloche to specify hematopoietic and vascular progenitors in zebrafish. Genes Dev 12:621–626

Liew CW, Rand KD, Simpson RJ et al (2006) Molecular analysis of the interaction between the hematopoietic master transcription factors GATA-1 and PU.1. J Biol Chem 281:28296–28306. doi:10.1074/jbc.M602830200

Lyons SE, Lawson ND, Lei L, Bennett PE, Weinstein BM, Liu PP (2002) A nonsense mutation in zebrafish gata1 causes the bloodless phenotype in vlad tepes. Proc Natl Acad Sci U S A 99:5454–5459. doi:10.1073/pnas.082695299

Monteiro R, Pouget C, Patient R (2011) The gata1/pu.1 lineage fate paradigm varies between blood populations and is modulated by tif1gamma. EMBO J 30:1093–1103. doi:10.1038/emboj.2011.34

Montosi G, Donovan A, Totaro A et al (2001) Autosomal-dominant hemochromatosis is associated with a mutation in the ferroportin (SLC11A3) gene. J Clin Invest 108:619–623. doi:10.1172/JCI13468

Murayama E, Kissa K, Zapata A et al (2006) Tracing hematopoietic precursor migration to successive hematopoietic organs during zebrafish development. Immunity 25:963–975. doi:10.1016/j.immuni.2006.10.015

Muse GW, Gilchrist DA, Nechaev S et al (2007) RNA polymerase is poised for activation across the genome. Nat Genet 39:1507–1511. doi:10.1038/ng.2007.21

North TE, Goessling W, Walkley CR et al (2007) Prostaglandin E2 regulates vertebrate haematopoietic stem cell homeostasis. Nature 447:1007–1011. doi:10.1038/nature05883

Patterson LJ, Gering M, Eckfeldt CE, Green AR, Verfaillie CM, Ekker SC, Patient R (2007) The transcription factors Scl and Lmo2 act together during development of the hemangioblast in zebrafish. Blood 109:2389–2398. doi:10.1182/blood-2006-02-003087

Peeters M, Ottersbach K, Bollerot K, Orelio C, de Bruijn M, Wijgerde M, Dzierzak E (2009) Ventral embryonic tissues and Hedgehog proteins induce early AGM hematopoietic stem cell development. Development 136:2613–2621. doi:10.1242/dev.034728

Ransom DG, Haffter P, Odenthal J et al (1996) Characterization of zebrafish mutants with defects in embryonic hematopoiesis. Development 123:311–319

Ransom DG, Bahary N, Niss K et al (2004) The zebrafish moonshine gene encodes transcriptional intermediary factor 1gamma, an essential regulator of hematopoiesis. PLoS Biol 2:E237. doi:10.1371/journal.pbio.0020237

Rhodes J, Hagen A, Hsu K, Deng M, Liu TX, Look AT, Kanki JP (2005) Interplay of pu.1 and gata1 determines myelo-erythroid progenitor cell fate in zebrafish. Dev Cell 8:97–108. doi:10.1016/j.devcel.2004.11.014

Rocquain J, Gelsi-Boyer V, Adelaide J et al (2010) Alteration of cohesin genes in myeloid diseases. Am J Hematol 85:717–719. doi:10.1002/ajh.21798

Sander JD, Cade L, Khayter C, Reyon D, Peterson RT, Joung JK, Yeh JR (2011) Targeted gene disruption in somatic zebrafish cells using engineered TALENs. Nat Biotechnol 29:697–698. doi:10.1038/nbt.1934

Sood R, English MA, Belele CL et al (2010) Development of multilineage adult hematopoiesis in the zebrafish with a runx1 truncation mutation. Blood 115:2806–2809. doi:10.1182/blood-2009-08-236729

Soza-Ried C, Hess I, Netuschil N, Schorpp M, Boehm T (2010) Essential role of c-myb in definitive hematopoiesis is evolutionarily conserved. Proc Natl Acad Sci U S A 107:17304–17308. doi:10.1073/pnas.1004640107

Stainier DY, Weinstein BM, Detrich HW 3rd, Zon LI, Fishman MC (1995) Cloche, an early acting zebrafish gene, is required by both the endothelial and hematopoietic lineages. Development 121:3141–3150

Stopka T, Amanatullah DF, Papetti M, Skoultchi AI (2005) PU.1 inhibits the erythroid program by binding to GATA-1 on DNA and creating a repressive chromatin structure. EMBO J 24:3712–3723. doi:10.1038/sj.emboj.7600834

Taylor AM, Zon LI (2011) Modeling Diamond Blackfan anemia in the zebrafish. Semin Hematol 48:81–88. doi:10.1053/j.seminhematol.2011.02.002

Torihara H, Uechi T, Chakraborty A, Shinya M, Sakai N, Kenmochi N (2011) Erythropoiesis failure due to RPS19 deficiency is independent of an activated Tp53 response in a zebrafish model of Diamond-Blackfan anaemia. Br J Haematol 152:648–654. doi:10.1111/j.1365-2141.2010.08535.x

Uechi T, Nakajima Y, Chakraborty A, Torihara H, Higa S, Kenmochi N (2008) Deficiency of ribosomal protein S19 during early embryogenesis leads to reduction of erythrocytes in a zebrafish model of Diamond-Blackfan anemia. Hum Mol Genet 17:3204–3211. doi:10.1093/hmg/ddn216

van Rooijen E, Voest EE, Logister I et al (2009) Zebrafish mutants in the von Hippel-Lindau tumor suppressor display a hypoxic response and recapitulate key aspects of Chuvash polycythemia. Blood 113:6449–6460. doi:10.1182/blood-2008-07-167890

van Rooijen E, Santhakumar K, Logister I, Voest E, Schulte-Merker S, Giles R, van Eeden F (2011) A zebrafish model for VHL and hypoxia signaling. Methods Cell Biol 105:163–190. doi:10.1016/B978-0-12-381320-6.00007-2

Vissers LE, van Ravenswaaij CM, Admiraal R et al (2004) Mutations in a new member of the chromodomain gene family cause CHARGE syndrome. Nat Genet 36:955–957. doi:10.1038/ng1407

Vogeli KM, Jin SW, Martin GR, Stainier DY (2006) A common progenitor for haematopoietic and endothelial lineages in the zebrafish gastrula. Nature 443:337–339. doi:10.1038/nature05045

Wang Y, Yates F, Naveiras O, Ernst P, Daley GQ (2005) Embryonic stem cell-derived hematopoietic stem cells. Proc Natl Acad Sci U S A 102:19081–19086. doi:10.1073/pnas.0506127102

Weinstein BM, Schier AF, Abdelilah S et al (1996) Hematopoietic mutations in the zebrafish. Development 123:303–309

Weiss O, Kaufman R, Michaeli N, Inbal A (2012) Abnormal vasculature interferes with optic fissure closure in lmo2 mutant zebrafish embryos. Dev Biol 369:191–198. doi:10.1016/j.ydbio.2012.06.029

Welch JS, Ley TJ, Link DC et al (2012) The origin and evolution of mutations in acute myeloid leukemia. Cell 150:264–278. doi:10.1016/j.cell.2012.06.023

Wilkinson RN, Pouget C, Gering M, Russell AJ, Davies SG, Kimelman D, Patient R (2009) Hedgehog and Bmp polarize hematopoietic stem cell emergence in the zebrafish dorsal aorta. Dev Cell 16:909–916. doi:10.1016/j.devcel.2009.04.014

Willett CE, Cortes A, Zuasti A, Zapata AG (1999) Early hematopoiesis and developing lymphoid organs in the zebrafish. Dev Dyn 214:323–336. doi:10.1002/(SICI)1097-0177(199904)214:4<323::AID-AJA5>3.0.CO;2–3

Xi Q, Wang Z, Zaromytidou AI et al (2011) A poised chromatin platform for TGF-beta access to master regulators. Cell 147:1511–1524. doi:10.1016/j.cell.2011.11.032

Yeh JR, Munson KM, Chao YL, Peterson QP, Macrae CA, Peterson RT (2008) AML1-ETO reprograms hematopoietic cell fate by downregulating scl expression. Development 135:401–410. doi:10.1242/dev.008904

Yeh JR, Munson KM, Elagib KE, Goldfarb AN, Sweetser DA, Peterson RT (2009) Discovering chemical modifiers of oncogene-regulated hematopoietic differentiation. Nat Chem Biol 5:236–243. doi:10.1038/nchembio.147

Yokota T, Huang J, Tavian M et al (2006) Tracing the first waves of lymphopoiesis in mice. Development 133:2041–2051. doi:10.1242/dev.02349

Zeitlinger J, Stark A, Kellis M et al (2007) RNA polymerase stalling at developmental control genes in the Drosophila melanogaster embryo. Nat Genet 39:1512–1516. doi:10.1038/ng.2007.26

Zhang P, Behre G, Pan J et al (1999) Negative cross-talk between hematopoietic regulators: GATA proteins repress PU.1. Proc Natl Acad Sci U S A 96:8705–8710

Zhang P, Zhang X, Iwama A et al (2000) PU.1 inhibits GATA-1 function and erythroid differentiation by blocking GATA-1 DNA binding. Blood 96:2641–2648

Part II
Epigenetic Mechanisms Regulating Mammalian Hematopoietic Stem Cell Development and Function

Chapter 4
Epigenetic and Transcriptional Mechanisms Regulating the Development of the Haematopoietic System in Mammals

Maud Fleury, Flor M. Perez-Campo, Guilherme Costa, Georges Lacaud, and Valerie Kouskoff

Abstract The haematopoietic system is established early during embryonic development and is maintained throughout adult life by haematopoietic stem cells. The cellular intermediates leading to the formation of mature blood cells are now fairly well characterised; from the mesoderm germ layer, successive steps of commitment give rise to haemangioblast, haemogenic endothelium, and haematopoietic stem and progenitor cells. Key transcription factors, such as ETV2, SCL, GATA2 or RUNX1, have been shown to specifically control some of these cell fate decisions. However, an integrated view of the transcriptional network controlling haematopoietic specification still remains to be established. Furthermore, it has become clear over the last decade that the transcriptional control of cell fate specification is globally regulated by epigenetic mechanisms. While the chromatin landscape is starting to be unravelled in adult haematopoiesis, virtually nothing is known about the epigenetic processes regulating the onset of haematopoiesis in the developing embryo. In this chapter, we describe the current state of our knowledge on the onset of mammalian haematopoiesis, focusing on murine development as it is by far the best characterised organism.

M. Fleury • V. Kouskoff (✉)
Stem Cell Haematopoiesis Group, Cancer Research UK Manchester Institute, The University of Manchester, Manchester, UK
e-mail: vKouskoff@PICR.man.ac.uk

F.M. Perez-Campo
Stem Cell Biology Group, Cancer Research UK Manchester Institute, The University of Manchester, Manchester, UK

Department of Internal Medicine, Hospital U.M. Valdecilla, IFIMAV, University of Cantabria, Santander, Spain

G. Costa • G. Lacaud
Stem Cell Biology Group, Cancer Research UK Manchester Institute, The University of Manchester, Manchester, UK

C. Bonifer and P.N. Cockerill (eds.), *Transcriptional and Epigenetic Mechanisms Regulating Normal and Aberrant Blood Cell Development*, Epigenetics and Human Health, DOI 10.1007/978-3-642-45198-0_4, © Springer-Verlag Berlin Heidelberg 2014

Keywords Embryonic Development • Haematopoietic Development • Gene Expression Regulation • Developmental Epigenetic modifiers • Developmental Haemangioblast / Haemogenic endothelium specification

4.1 Ontogeny of the Haematopoietic System in Mammals

Over the last century, detailed analyses of embryonic tissues have allowed to define a spatiotemporal map of haematopoiesis during development. These analyses revealed that haematopoiesis occurs at multiple sites during development, but that only some of these sites arc able to de novo generate haematopoietic cells. Despite numerous studies, the developmental origin of haematopoietic stem cells (HSCs) which maintain the haematopoietic system throughout adult life still remains unclear and controversial.

4.1.1 Anatomical Sites of Haematopoiesis During Development

In the mouse embryo, the first blood cells appear in the yolk sac (YS), an extra-embryonic supporting tissue, from embryonic day (E)7.5 before the vasculature connections are established with the embryo proper. Early haematopoietic production from the YS is referred to as the first wave of haematopoiesis or primitive haematopoiesis, as it mainly produces primitive erythrocytes (nucleated erythrocytes expressing embryonic globins), macrophages and megakaryocytes (Wong et al. 1986; Palis et al. 1999). It is believed that these cells will not contribute to the pool of haematopoietic cells found in the adult organism. A second wave of haematopoiesis arises shortly after and is called definitive haematopoiesis as it produces adult-type progenitor cells from E8.5 (Godin et al. 1995). During this second wave of haematopoiesis, HSCs which will give rise to the adult haematopoietic system are produced from E10.5. Based on milestone experiments performed with the avian model, it is now widely accepted that the first definitive HSCs originate from an intra-embryonic region (Dieterlen-Lievre 1975) and not from the YS as previously thought (Moore and Owen 1965). In the murine embryo, the AGM region (which encompasses the aorta together with the gonads and the mesonephros) was shown to contain multipotent progenitors between E9 and E12 (Medvinsky et al. 1993) and from E10 the first HSCs as defined by their ability to reconstitute the haematopoietic system of adult mouse recipients upon transplantation (Muller et al. 1994; Medvinsky and Dzierzak 1996). From E10.5 to E11, HSCs were also detected in the placenta (Gekas et al. 2005; Ottersbach and Dzierzak 2005) and the YS (Muller et al. 1994; Kumaravelu et al. 2002; Gekas et al. 2005), where they are thought to proliferate before reaching the foetal liver, the main site of HSC expansion from E12.5, then the bone marrow from E17.5 onward (reviewed by Mikkola and Orkin 2006).

The anatomical origin of HSCs has been the subject of intensive studies. Because the YS can produce adult-type progenitors just before the blood circulation is established (Palis et al. 1999), organ explants experiments have been used to confirm the intra-embryonic origin of HSCs previously suggested in the avian model. When explanted before the establishment of blood circulation, it has been shown that the presumptive AGM region contains developmental precursors of HSCs, while the YS is devoid of HSC activity (Cumano et al. 2001). Although these results confirm the theory of the intra-embryonic origin of HSCs, it has been suggested that the early YS might independently generate precursors requiring a specialised microenvironment such as the AGM to develop into definitive HSCs (Matsuoka et al. 2001). This hypothesis has recently gained further support through lineage tracing experiments demonstrating that the early YS contains precursors of adult HSCs (Samokhvalov et al. 2007; Tanaka et al. 2012a), although technical issues surrounding the experimental strategies undermine these studies. The allantois, which gives rise to the umbilical cord, has also been proposed as a site of de novo generation of multipotent haematopoietic progenitors, either using organ explants strategies (Zeigler et al. 2006; Corbel et al. 2007), or using mouse model devoid of heart beat (Rhodes et al. 2008).

4.1.2 Cellular Origin of Haematopoiesis

The cellular origin of haematopoietic precursors has also long been a subject of controversy. A first theory was proposed in the early twentieth century based on the anatomical studies of the first erythrocytes observed within the YS. These blood cells were surrounded by endothelial cells in histological structures called "blood islands" which seem to develop from a uniform mass of mesodermal cells. This observation led to the hypothesis that haematopoietic and endothelial cells from the YS shared a common developmental precursor (Maximov 1909; Sabin 1920), later termed the haemangioblast. This precursor was eventually identified using the embryonic stem (ES) cell differentiation model in which the blast colony-forming cell (BL-CFC), the in vitro equivalent of the putative haemangioblast, was identified as a tri-potential precursor which can differentiate into haematopoietic, endothelial and smooth muscle lineages and which can be enriched based on its expression of FLK1 (VEGF-receptor 2, KDR) and brachyury (Choi et al. 1998; Ema et al. 2003; Fehling et al. 2003). Using these markers, the haemangioblast was later isolated from the gastrulating embryo (Huber et al. 2004), where it was shown to localise within the primitive streak, and not within the mesodermal masses developing into blood islands, as first hypothesised. It is now believed that the haemangioblast is a transient mesodermal precursor which develops very quickly into the haematopoietic or endothelial lineages while migrating from the primitive streak to the extra-embryonic YS, and that as a result of both rapid commitment and migration, blood islands are polyclonal (Ferkowicz and Yoder 2005; Ueno and Weissman 2006).

Contemporary to the observation of the YS blood islands was the observation of blood cell clusters attaching to the endothelial wall of the dorsal aorta, which led to the theory of an endothelial origin of these haematopoietic cells (Jordan 1917). Phenotypic analysis of these clusters revealed a co-expression of both endothelial and haematopoietic markers (Garcia-Porrero et al. 1998; reviewed by Cumano and Godin 2007), suggesting a direct lineage relationship between blood and vasculature. It was later shown that sorted endothelial cells from E9.5 embryos can produce haematopoietic cells in vitro (Nishikawa et al. 1998). More convincing evidence demonstrating the production of haematopoietic cells from the endothelium in vivo came from lineage tracing experiments. The first strategy showed that Ac-LDL labelling of cells of the endothelial tree in living embryos resulted in production of labelled haematopoietic cells (Sugiyama et al. 2003, 2005). Another strategy involving inducible lineage-specific expression of a reporter gene demonstrated that part of the adult haematopoietic system was derived from an endothelial ancestor (Zovein et al. 2008). Recent technologies have now made possible the direct visualisation of "budding" haematopoietic cells from the endothelium in the AGM region (Boisset et al. 2010). This was also observed using the ES cell differentiation model (Eilken et al. 2009; Lancrin et al. 2009). The intra-aortic clusters are thought to contain the HSCs present within the AGM region, and it is hypothesised that these clusters are produced from the haemogenic endothelium although this remains to be formally demonstrated (Taoudi and Medvinsky 2007; Taoudi et al. 2008). Of interest, arterial haematopoietic activity is not restricted to the dorsal aorta as the vitelline and umbilical arteries also harbour HSCs at the same time as the aorta (de Bruijn et al. 2000) and contain intra-artery clusters (North et al. 2002). Haemogenic endothelial cells have also been isolated from the YS (Nishikawa et al. 1998; Li et al. 2005; Lancrin et al. 2009).

Although the theories of the haemangioblast and the haemogenic endothelium have been considered mutually exclusive, a unifying theory has been proposed where the haemangioblast produces haematopoietic cells through a haemogenic endothelial step (Fig. 4.1) [reviewed by Lancrin et al. (2010)]. This cellular hierarchy has been defined for the first wave of haematopoiesis; the cellular origin of the second wave of embryonic haematopoiesis is in contrast still not fully understood. It remains to be determined whether the haemogenic endothelium found within the AGM is directly derived from the first wave of already committed mesoderm or whether it is de novo specified from mesoderm via a haemangioblast intermediate. Defining the cellular origin of this second wave of haematopoiesis is critically important if one wants to understand how HSCs are generated.

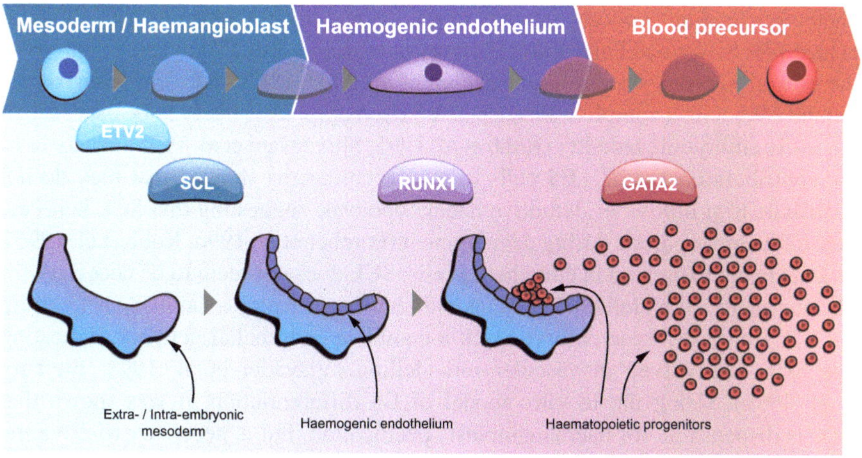

Fig. 4.1 The generation of embryonic blood cells. Mouse embryonic haematopoietic precursors are generated from extra- and intra-embryonic mesodermal cell populations through a transient haemogenic endothelium. Key transcription factors regulate the transitional steps involved in blood generation: ETV2 acts upstream of SCL and both proteins regulate the differentiation of mesoderm/haemangioblasts into haemogenic endothelial cells; RUNX1 is subsequently required for the emergence of blood precursors from the haemogenic endothelium; GATA2 is implicated in the maintenance and proliferation of the newly formed haematopoietic progenitor cells

4.2 Transcriptional Regulation of Haematopoietic Specification During Development

The spatiotemporal emergence of the haematopoietic system is a tightly controlled process orchestrated by multiple transcription factors. Several master regulators have been shown to coordinate the sequential steps leading to the production of mature blood cells. The complex transcriptional network controlling these cell fate decisions from mesoderm to haematopoietic cells is slowly been unravelled (Fig. 4.1).

4.2.1 From the Haemangioblast to the Haemogenic Endothelium

4.2.1.1 SCL

SCL (SCL/TAL1), a member of the bHLH (basic helix–loop–helix) transcription factor family, was initially identified as a target of chromosomal translocation in T-cell leukaemia (reviewed by Begley and Green 1999). In the developing embryo, SCL is expressed from E7.5 in the extra-embryonic mesoderm, and thereafter in the haematopoietic and endothelial cells of the YS and the embryo proper (Kallianpur et al. 1994; Silver and Palis 1997; Elefanty et al. 1999), but also in some neurons of the central nervous system (van Eekelen et al. 2003). SCL-deficient embryos die around E9.5 with a complete absence of YS haematopoiesis likely to be the cause of this early embryonic lethality (Robb et al. 1995; Shivdasani et al. 1995). Analysis of the contribution of $scl^{-/-}$ ES cells to mouse chimaeras showed that they do not contribute to primitive or definitive haematopoiesis, suggesting that SCL is necessary for haematopoiesis during development (Porcher et al. 1996; Robb et al. 1996). Despite being expressed in endothelial cells, SCL does not seem to be necessary for the emergence of endothelial cells, as null embryos display a capillary network in the YS. However, these embryos lack a complete endothelial network, indicating that SCL is involved in vascular remodelling (Visvader et al. 1998; Elefanty et al. 1999). Using the in vitro model of ES differentiation, it was shown that SCL is dispensable for haemangioblast specification, but is necessary to drive the haemangioblast toward haematopoietic fate (Ema et al. 2003; D'Souza et al. 2005). More recently, it was shown that SCL is specifically required for the transition from the haemangioblast to the haemogenic endothelium stage (Lancrin et al. 2009).

Recent studies have shown that SCL is also a regulator of mesoderm patterning: SCL can induce haematopoietic specification at the expense of other lineages such as cardiac or paraxial mesoderm (Ismailoglu et al. 2008), and YS endothelial cells are mis-specified toward cardiac fate in the absence of SCL (Van Handel et al. 2012).

4.2.1.2 LMO2

LMO2 (RTBN2) is a LIM domain transcription factor involved in chromosomal translocation in T-cell leukaemia (Boehm et al. 1991; Royer-Pokora et al. 1991). During development, LMO2 displays a similar pattern of expression as SCL, being detected in the mesoderm, the haemogenic sites of the cardiovascular system (Silver and Palis 1997; Manaia et al. 2000; Minko et al. 2003) and the nervous system (Hinks et al. 1997; Herberth et al. 2005). LMO2 is also expressed transiently in some somite derivatives and in the intra-embryonic endoderm (Manaia et al. 2000). Embryos deficient for LMO2 present a phenotype similar to SCL null embryos. They die around E10.5 from the absence of YS erythropoiesis (Warren et al. 1994), and null ES cells do not contribute to adult haematopoiesis cells in chimaera experiments (Yamada et al. 1998). Interestingly, LMO2 does not bind to DNA directly but has been shown to interact with transcriptional regulatory complexes involved in haematopoietic differentiation (Warren et al. 1994; Wadman et al. 1997; Rabbitts 1998; Xu et al. 2003). More specifically, a transcriptional complex involving LMO2, SCL, GATA-1, LDB1 and E2A was described in erythroid lineage. It is proposed that LMO2 acts as a scaffold to link SCL and GATA factors both of which are involved in DNA binding (Wadman et al. 1997). Since LMO2 is required for haematopoietic emergence, it is very likely that this factor is also involved in transcriptional complexes regulating haematopoietic specification, as it was recently suggested (Nottingham et al. 2007; Landry et al. 2008; Wilson et al. 2010).

4.2.1.3 ETV2

ETV2 (ETSRP71) is a member of the ETS family of transcription factors which was recently identified in zebrafish (Sumanas et al. 2005), mouse and human (Sumanas et al. 2008). ETV2 is expressed transiently in the developing embryo; it is first detected in the extra-embryonic mesoderm around E7.0, in the endothelial and haematopoietic cells of the blood island at E8.5, then in the endothelial cells of the developing vasculature of the embryo proper but is virtually absent after E11.5 (Lee et al. 2008; Kataoka et al. 2011; Koyano-Nakagawa et al. 2012; Wareing et al. 2012a).

Ectopic expression of ETV2 during ES cell differentiation promotes the formation of endothelial and haematopoietic progenitors (Lee et al. 2008; Koyano-Nakagawa et al. 2012; Liu et al. 2012). Embryos deficient for ETV2 die by E10.5 with a complete absence of blood progenitors and severe vascular defects (Lee et al. 2008; Kataoka et al. 2011; Koyano-Nakagawa et al. 2012; Wareing et al. 2012b), and null ES cells do not contribute to haematopoiesis or endothelial cells in chimaera experiments (Liu et al. 2012). Taken together, these data suggest a requirement of ETV2 in the establishment of haematopoietic and endothelial lineages. Supporting this theory, endogenous ETV2 expression is correlated with haematopoietic potential in the embryo and in the ES differentiation model (Kataoka et al. 2011; Koyano-Nakagawa et al. 2012; Wareing et al. 2012a). In particular, ETV2 expression marks the haemogenic endothelium population in vitro and in vivo and is required for its formation (Wareing et al. 2012a).

Interestingly, ETV2-deficient ES cells display increased cardiogenic potential (Liu et al. 2012) and ETV2 overexpression supresses development of the cardiac lineage (Rasmussen et al. 2011), a phenotype reminiscent of SCL activity. In line with these findings, it has been shown that SCL is a direct transcriptional target of ETV2 (Kataoka et al. 2011; Wareing et al. 2012b). Considering that ETV2 is expressed in primitive mesodermal precursors and only required until FLK1 expression as shown by conditional deletion experiments (Wareing et al. 2012a), it is proposed that ETV2 directs early mesoderm to differentiate toward haematopoietic and endothelial lineages and that its action is mediated by the induction of SCL.

4.2.2 *From the Haemogenic Endothelium to the Haematopoietic Progenitors*

4.2.2.1 RUNX1

RUNX1 (Acute Myeloid Leukaemia (AML-1), CBFα) is a member of the family of Runt-domain transcription factors and heterodimerizes with CBFβ to form the core-binding factor (CBF). Both subunits have been frequently found involved in chromosomal translocation in leukaemia and myelodysplasic syndromes (Look 1997). RUNX1 is expressed from E7.5 in the YS mesoderm and blood islands, and in haemogenic sites in the embryo later on. In the embryo proper, RUNX1 is expressed in endothelial cells of the aorta and vitelline artery as early as E8.5 and in the endothelial cells and haematopoietic clusters of the aorta and umbilical artery at E10.5 (North et al. 1999; Lacaud et al. 2002).

Embryos deficient for RUNX1 die around E12.5–E13.5, most likely due to haemorrhages in the central nervous system. Primitive erythropoiesis is observed in the YS, but no definitive haematopoiesis is present in the foetal liver (Okuda et al. 1996), suggesting a role for RUNX1 in the establishment of definitive but not primitive haematopoiesis. This hypothesis was confirmed using the ES cell differentiation model, where it was shown that *runx1*$^{-/-}$ ES cells generate primitive haematopoiesis in vitro (Lacaud et al. 2002), but do not contribute to definitive haematopoiesis in vivo in chimaeras embryos (Okuda et al. 1996). Deficient embryos are devoid of HSCs in the foetal liver and the YS but also in the AGM region (Cai et al. 2000; Mukouyama et al. 2000), suggesting a requirement of RUNX1 during the early steps of HSCs specification. This hypothesis was supported by the observation that intra-aortic clusters, believed to contain the emerging HSCs, are absent in RUNX1-deficient embryos (North et al. 1999). Further studies showed that RUNX1 is actually necessary for the generation of haematopoietic cells from the haemogenic endothelium. Indeed, endothelial cells sorted from E10.5 *runx1*$^{-/-}$ embryos do not generate haematopoietic cells in vitro (Yokomizo et al. 2001). Moreover, using a conditional KO mouse model, it was shown that RUNX1 expression is required in endothelial cells for HSCs emergence, but dispensable in cells already engaged in the haematopoietic lineage (Li et al. 2006; Chen et al. 2009). In vitro, it was shown that RUNX1 is required for the transition from the haemogenic endothelium population to haematopoietic cells (Lancrin et al. 2009).

Altogether, these studies revealed that RUNX1 is necessary for the emergence of definitive HSCs from the haemogenic endothelium during development. Furthermore, RUNX1 activity seems to be dose dependent since haploinsufficient embryos display HSC activity from E10, 1 day earlier than WT embryos. But at E11.5, although *runx1*$^{+/-}$ AGM contain the same number of HSCs, they are not able to amplify during ex vivo culture, suggesting that RUNX1 could also regulate the maintenance/expansion of HSCs in the embryo (Cai et al. 2000).

4.2.2.2 GATA2

GATA2 is a member of the zinc finger transcription factor and is part of the GATA family. During development, GATA2 is expressed at E7.5 in extra- and intra-embryonic mesoderm and in the YS blood islands (Minegishi et al. 1999; Kobayashi-Osaki et al. 2005). Later during development, GATA2 is expressed in the endothelial and haematopoietic cells of the AGM and foetal liver (Minegishi et al. 1999, 2003; Kobayashi-Osaki et al. 2005). GATA2 is also expressed in non-haematopoietic tissues such as the central nervous system and the uro-genital compartment (Nardelli et al. 1999; Siggers et al. 2002).

GATA2-deficient embryos die in utero around E10.5 with severe anaemia (Tsai et al. 1994). They display a markedly reduced primitive haematopoiesis. In chimaera experiments, null ES cells do participate in primitive haematopoiesis, although in a very low proportion, but are unable to contribute to definitive haematopoiesis (Tsai et al. 1994). Interestingly, deficient ES cells can produce definitive haematopoietic cells in vitro, but GATA2 is likely involved in the subsequent survival and/or proliferation of these haematopoietic progenitors (Tsai et al. 1994; Tsai and Orkin 1997). This hypothesis was further supported when it was shown that GATA2 haploinsufficiency leads to a decrease in HSCs number in the AGM, YS and foetal liver between E10 and E12 and that these HSCs display reduced proliferation during AGM explant culture. Compensatory mechanisms are likely to rescue GATA2 deficiency since older embryos and adult *gata2*$^{+/-}$ mice have normal numbers of HSCs. These HSCs, however, have a proliferative defect revealed in non-steady state physiology such as competitive engraftment experiments (Ling et al. 2004; Rodrigues et al. 2005).

4.2.3 Reconstruction of Gene Regulatory Networks

As described above, the emergence and specification of haematopoietic cells during development is regulated by transcription factors. Using loss and gain of function approaches, the systematic analysis of individual transcription factor has allowed to identify master regulators of this process and to understand at which specific time and stages they are acting. However, how these factors interact together within wider regulatory networks to orchestrate the tightly regulated process of haematopoietic development is still largely unknown. Gene regulatory networks are described as the functional interplay between transcription factor proteins and the '*cis*-regulatory modules' (CRM) associated with their target genes (reviewed by Pimanda and Gottgens 2010). Current strategies to decipher regulatory networks fall into two categories: the 'bottom-up' and the 'top-down' approaches.

4.2.3.1 Bottom-up Approaches

Bottom-up approaches start from one identified component of a specific network, such as a CRM or a transcription factor, and reconstruct the regulatory mechanisms upstream of this component within the network. An example of a bottom-up approach is the study of the transcriptional regulation of the *Scl* gene during haematopoietic development. Using in vivo transgenic reporter assays, several enhancers have been identified in the *Scl* locus, some of which have been shown to specifically direct the expression of a reporter gene in the haematopoietic and endothelial cells during development (Sanchez et al. 1999; Gottgens et al. 2002, 2004; Silberstein et al. 2005). Interestingly, the +19 enhancer, a 600-bp core sequence 19 kb downstream of the transcription start site, was shown to contain three binding motifs which were necessary for its activation (Gottgens et al. 2002). These motifs, two conserved ETS-binding sites and one conserved GATA-binding site, were bound by FLI1, ELF1 and GATA2 and it was shown that these three transcription factors formed a transcriptional complex regulating SCL expression. This *Ets/Ets/Gata* motif was the first CRM identified in the haematopoietic specification process. This CRM was then used in a genome-wide screening to identify new enhancers of other genes putatively involved in haematopoietic development, three of which (namely *Fli1*, *Hex* and *Smad6*) were found to display an expression pattern similar to the *Scl* +19 enhancer in vivo (Donaldson et al. 2005; Pimanda et al. 2007a).

Further studies of the *Fli1* enhancer lead to the identification of the first fully connected triad of HSC transcription factors, the SCL/FLI1/GATA2 triad (Pimanda et al. 2007b). This triad consists of genes encoding transcription factors regulating each other, thus forming a powerful feedforward loop stabilising the stem cell state. This activation is likely to take place early during haematopoietic specification as all three factors are co-expressed in presumptive HSCs of the AGM intra-aortic clusters. A network model was recently proposed where the triad is initiated by NOTCH and BMP4 signalling and modulated by RUNX1 in the haemogenic endothelium (Narula et al. 2010, 2013).

4.2.3.2 Top-Down Approaches

Top-down approaches are used to identify downstream targets of specific transcription factors. They are usually based on genome-wide expression profiling of cell populations following induction or inactivation of a transcription factor of interest. This approach generates a list of potential transcriptional targets based on functional relevance and the presence of predicted binding sites in the regulatory sequences of the candidates (as performed for ETV2 in Wareing et al. 2012b). Combining analyses for multiple transcription factors should then allow to highlight interconnections between transcriptional programmes and help reconstructing regulatory networks.

The main drawbacks of such approaches are the need to validate the direct binding of the transcription factor on its target and the risk of missing potential candidates because of lack of proper annotations of genomic sequences. These limitations are now being bypassed with the use of the ChIP-seq technology (chromatin immunoprecipitation combined with high-throughput sequencing). It is not yet possible to use this technology with low number of cells, thus precluding analysis of rare cell populations in the developing embryos. Nonetheless, this strategy was recently used to identify members of the transcriptional network controlled by SCL during haematopoietic development. A screen of direct transcriptional targets of SCL was conducted using a haematopoietic stem/progenitor cell line and 11 of these candidates were validated in vivo using ChIP and transgenic reporter assays for the corresponding enhancer region (Wilson et al. 2009). Using this strategy, the same group generated genome-wide binding profiles for ten major haematopoietic transcriptional regulators: SCL, LYL1, LMO2, GATA2, RUNX1, MEIS1, PU.1, ERG, FLI-1 and GFI1b (Wilson et al. 2010). This study suggested the existence a heptamer complex (containing SCL, LYL1, LMO2, GATA2, RUNX1, FLI-1 and ERG) acting to regulate the haematopoietic programme. Of interest, a collaboration involving RUNX1 with SCL, LYL1, LMO2 or GATA2 had not been described previously.

More recently, two studies have investigated the direct transcriptional targets of RUNX1 during haematopoietic development. In the first study, the combination of gene expression profiling and ChIP-seq analysis in differentiating ES cells identified a small number of direct transcriptional targets of RUNX1 during the haematopoietic specification. Among those, only 29 genes were shown to be affected by the loss of RUNX1 during haematopoietic emergence in the E7.5 YS (Tanaka et al. 2012b). Interestingly, 23 of these genes were previously described as direct transcriptional targets of the SCL/FLI1/GATA2 triad, thus reinforcing the hypothesis of transcriptional collaboration of these factors within an early haematopoietic regulatory network. In another study, the binding profile of SCL, FLI1 and RUNX1 were compared during the endothelial to haematopoietic transition in the ES cell differentiation model (Lichtinger et al. 2012). This study suggested that the binding profiles of SCL and FLI1 were altered during the transition, although most of the redistribution occurred within the same gene locus or in its vicinity; this shift in binding activity was shown to be in part a consequence of RUNX1 binding at the same locus. This study along with previous work from the same group demonstrated the implication of RUNX1 in modulating chromatin remodelling and epigenetic changes at the onset of haematopoiesis (Hoogenkamp et al. 2009; reviewed by Lichtinger et al. 2010).

While transcription factors implicated in haematopoietic development are further characterised, their role defined and their places within transcriptional networks uncovered, it becomes essential to integrate this wealth of knowledge into the higher levels of regulation encompassing the epigenetic control of genome-wide chromatin landscape.

4.3 Epigenetic Regulation of Transcription

Epigenetics describes the chromatin-based events regulating gene expression in a heritable manner. These events include DNA methylation and histone modifications. Changes in the DNA methylation pattern or alterations in the histone modifications landscape can modulate gene expression, either directly or by facilitating the recruitment of additional chromatin-modifying enzymes. The key role of chromatin modifying enzymes in haematopoiesis is highlighted by the fact that the function or expression of several of these enzymes is deregulated in leukaemia and other haematological malignancies (Fig. 4.2).

4.3.1 Post-translational Histone Modifications

The complexity of histone code is highlighted by the large range of possible modifications including acetylation, methylation, phosphorylation, ubiquitylation and SUMOylation, amongst others. Combinations of different modifications are thought to induce structural changes in the chromatin, and therefore modify the accessibility of transcription factors to regulatory sequences, allowing the regulation of gene expression in a time and tissue-specific manner. Our understanding of the specific functions of the proteins that direct chromatin modifications at different stages of the development or maintenance of the haematopoietic system still remains very limited.

4.3.1.1 Histone Acetyltransferases and Histone Deacetylases

The transfer of an acetyl group to specific lysine residues located at the histone tails is catalysed by histone acetyltransferases (HATs). This modification reduces the stability of the interaction between histone and DNA (Hong et al. 1993; Puig et al. 1998) relaxing the chromatin structure and thereby affecting gene expression (Shogren-Knaak et al. 2006; Campos and Reinberg 2009). In addition, this acetylated residue could also act as a docking site for bromodomain containing regulatory factors.

Among this group, the Monocytic Leukaemia Zinc Finger protein (MOZ, MYST3 or KAT6) is a member of the MYST family of HATs and transcription co-activators and was first identified as a translocation partner in various forms of AML (Borrow et al. 1996; Carapeti et al. 1998; Chaffanet et al. 2000; Esteyries et al. 2008). MOZ is the catalytic component of a large multi-subunit protein complex also harbouring ING5 (Inhibitor of Growth 5), the bromodomain PHD-finger protein (BRPF1) and EAF6 (Esa1-associated factor 6) (Doyon et al. 2006). Genetic deletion of MOZ results in severe defects in the development and maintenance of HSCs and the development of erythroid cells thus leading to

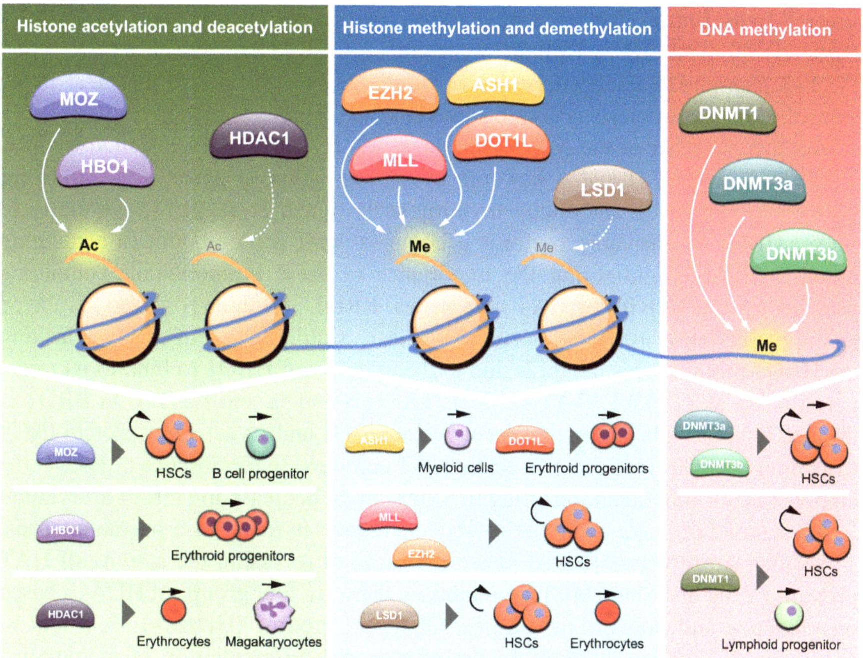

Fig. 4.2 Epigenetic modulators of haematopoietic development. Histone and DNA-modifying enzymes tightly control the proliferation and differentiation of haematopoietic cells. Histone acetylation: while MOZ acetylates lysine (K) residues of histone tails to promote haematopoietic stem cells (HSCs) proliferation and lymphoid differentiation, HBO1-mediated acetylation is required for erythroid progenitor differentiation; on the other hand, HDAC1 removes acetyl groups from histone tails and regulates erythroid and megakaryocyte differentiation. Histone methylation: ASH1 and DOT1L catalyse the methylation of histone 3 (H3) K36 and K79 and are involved in myeloid and erythroid progenitor differentiation, respectively; EZH2 and MLL methylate H3K27 and H3K4, respectively, and play crucial roles in HSCs proliferation; LSD1 removes methyl groups from H3K4 and H3K9 and it has been associated with HSCs proliferation and erythroid differentiation. DNA methylation: DNMT enzymes are responsible for the cytosine methylation in CpG islands of DNA; whilst DNMT3a and DNMT3b have been implicated in HSCs proliferation and differentiation, DNMT1 regulates HSCs proliferation and the differentiation of lymphoid cells. *White solid* and *dashed arrows* indicate addition and removal of modification groups, respectively; *black straight* and *circular arrows* indicate differentiation and proliferation of haematopoietic cells, respectively

embryonic lethality (Thomas et al. 2006; Katsumoto et al. 2008). A more specific abrogation of the HAT activity of MOZ ($MOZ^{\mathrm{HAT}-/-}$) indicated that the HAT activity of MOZ is critical for HSCs function (Perez-Campo et al. 2009). $MOZ^{\mathrm{HAT}-/-}$ embryos also display defects in the numbers and functionality of HSCs and committed progenitors as well as lower numbers of immature B cells (Perez-Campo et al. 2009). This phenotype is a direct consequence of the inability of HSCs and progenitor cells to proliferate, suggesting that MOZ-driven acetylation controls the appropriate balance between proliferation and differentiation in HSCs and progenitor cells. Beside this function, MOZ also acts as a co-activator for several transcription factors with haematopoietic specificity, such as RUNX1 (Kitabayashi et al. 2001; Holbert et al. 2007), MLL1 (Paggetti et al. 2010) and PU.1 (Katsumoto et al. 2006), further highlighting the relevance of this HAT in the control of haematopoiesis (Perez-Campo et al. 2013).

Another member of the MYST family of HATs, HBO1 (HAT Bound to ORC 1, MYST2 or KAT7), has a key role in the regulation of genes responsible for embryonic patterning and foetal erythropoiesis (Kueh et al. 2011). Similarly to MOZ, HBO1 is also able to form a complex with a bromodomain-containing protein, BRD1. Highlighting the importance of these bromodomain-containing proteins for the activity of HAT complexes, BRD1 null embryos, also displayed an important defect in foetal liver erythropoiesis, similar to that of embryos lacking the HBO1 protein. This defect is due to the inability of HBO1 to bind to its target genes (such as GATA1, GATA2, SCL/TAL1, STAT5a and ETO2), as BRD1 is thought to act as a bridging protein between HBO1 and its activator protein ING4 (Mishima et al. 2011) to form an active HAT complex. In the absence of BRD1, the complex is not functional, resulting in a substantial decrease in H3K14 acetylation levels at target promoters and the subsequent defects in foetal liver haematopoiesis.

Histone deacetylases (HDACs) are in charge of reversing the activity of HAT proteins (Yang and Seto 2008). In mammals, there are four groups of HDACs based on sequence and domain similarities. Of those, group I (HDACs1, 2, 3 and 8) encompasses proteins involved in the growth and differentiation of mammalian cells (Lagger et al. 2002). The expression of group I HDACs is very low in haematopoietic progenitor cells, but is induced in more differentiated progenitors and then either down-regulated during myeloid differentiation or retained during erythroid and megakaryocytic differentiation (Wada et al. 2009). Transcription of HDAC1 has been shown to be regulated by haematopoietic transcription factors. Indeed, GATA1 mediates the transcription of HDAC1 driving the differentiation of myeloid progenitors into erythroid–megakaryocytic lineages. In contrast, when HDAC1 transcription is down-regulated by members of the C/EBP transcription factors, myeloid progenitors differentiate into myeloid cells (Wada et al. 2009). Therefore, HDAC1 is implicated in early cell fate decisions during haematopoiesis. Due to the negative role of HDAC1 on myeloid differentiation, inhibitors of HDACs have been used, in conjunction with other agents, with good results to treat certain haematological disorders (Quintas-Cardama et al. 2011).

4.3.1.2 Histone Methyltransferases and Histone Demethylases

Histone methylation can take place on both lysine (K) and arginine (R) residues (Zhang and Reinberg 2001; Martin and Zhang 2005). Two groups of Histone Methyltransferases (HMTs) with opposing activities have a crucial role in the regulation of haematopoiesis and HSCs proliferation. One of them, the polycomb family (PcG) carries out the methylation at lysine 27 of histone H3 (H3K27), a mark linked to gene silencing, whereas the other group, the Trithorax (TrxG) methylates H3K4, a mark related to gene activation. These two groups of proteins regulate the expression of HOX genes among other targets.

EZH2 (Enhancer of Zeste homolog 2), a component of the Polycomb Repressor Complex 2 (PRC2), catalyses the di- and tri-methylation of H3K27. This modification is then bound by the Polycomb Repressor Complex 1 (PRC1), resulting in the transcriptional repression of the targeted genes. Ezh2 is essential for foetal liver HSCs proliferation and erythropoiesis, whereas its role in adult HSCs seems less important (Mochizuki-Kashio et al. 2011). Conversely, it has been shown that the forced expression of EZH2 enhances the number and proliferative potential of HSCs (Herrera-Merchan et al. 2012). A more detailed description of polycomb function in adult haematopoiesis and development can be found in the following Chaps. 5 and 6.

With the exception of MLL (Mixed Lineage Leukaemia) proteins, little is known about the role of TrxG genes in haematopoiesis. MLL plays a crucial role in the proliferation and differentiation of haematopoietic progenitors and maintains appropriate expression level of genes such as HOXA7 and HOXA9 during embryonic development. It has been proposed that the role of MLL in maintaining rather than in initiating HOX genes expression could be related to the fact that binding of MLL to specific clusters of CpG residues within HOXA9 locus protects this area from DNA methylation and subsequent silencing (Erfurth et al. 2008). Besides its role as a regulator of HOX genes expression, MLL can interact with other proteins with important roles in haematopoiesis, such as RUNX1 (Huang et al. 2011) or MOZ (Paggetti et al. 2010). Another member of the TrxG family, MLL5, has been shown to have an important role in haematopoiesis and HSC self-renewal (Heuser et al. 2009; Madan et al. 2009). In MLL5 null mice, HSCs display multiple haematopoietic defects, such as impairment in neutrophil function and in erythropoiesis, but more importantly in competitive repopulation capacity. Interestingly, HSCs from MLL5 null mice have a dramatically increased sensitivity to DNA demethylation-induced differentiation with 5 azadeoxycytidine indicating that MLL5 could be implicated in the regulation of HSC proliferation by a mechanism that involves DNA methylation (Heuser et al. 2009). For more details about MLL function in haematopoiesis, see the Chap. 7.

ASH1 (Absent small and homeotic disks protein 1 homologue) is a unique HMT that catalyses the methylation of H3K36 (Tanaka et al. 2007; An et al. 2011; Yuan et al. 2011). ASH1 is specifically expressed in HSCs in the bone marrow (Sung et al. 2006) and undifferentiated precursors of T cells in the thymus (Tanaka et al. 2008). ASH1 was recently shown to regulate HOX gene transcription synergizing with MLL, although this regulator activity seems to be independent of the HMTs catalytic domain (Tanaka et al. 2011). Knockdown of ASH1 in murine HSCs results in decreased number of macrophages and granulocytes, a phenotype similar to that induced by loss of MLL1 function indicating that this protein is a key epigenetic regulator of normal haematopoiesis.

DOT1L (disruptor of telomere silencing 1-like or KMT4) was shown to specifically methylate H3K79 as knockdown of this protein in mice results in a total loss of H3K79 methylation (Jones et al. 2008). Dot1L null mutant mice die between

E10.5 and E13.5 with severe and selective defects in erythroid, but not myeloid, differentiation (Feng et al. 2010). DOT1L methylation is a critical regulator of GATA2 and PU.1 transcription; mice lacking this HMT show an increase in PU.1 levels together with a decrease in GATA2 levels. Therefore, DOTL1 has a key role in early haematopoiesis, controlling the numbers of erythroid and myeloid cells (Feng et al. 2010). The role for DOT1L in erythropoiesis is also supported by a previous report showing that H3K79 methylation is enriched at the promoter of the β-globin locus (Im et al. 2003). DOT1L also seems to play a role in the development of leukaemia harbouring translocations of the MLL gene, as targets of the MLL-AF9 translocation display an alteration of their histone methylation pattern, specifically affecting the dimethylation of H3K79 (Bernt et al. 2011).

Histone lysine methylation is a dynamic process also regulated by the action of histone demethylases (HDMTs). LSD1 (Lysine-specific demethylase 1) which catalyses the demethylation of both H3K4 and H3K9, was the first HDMTs to be identified. Depletion of LSD1 in mice not only exacerbates the proliferation of HSCs and progenitor cells but also leads to severe defects in the differentiation of erythroid cells (Sprussel et al. 2012). LSD1 has also been shown to cooperate with the transcriptional repressor GFI1 to regulate differentiation of diverse haematopoietic lineages (Saleque et al. 2007).

4.3.2 DNA Methylation

DNA methylation consists in the addition of a methyl group to cytosines that precede guanines (CpGs). This process is catalysed by DNA methyltransferases (DNMTs) using S-adenosyl-methionine as donor of methyl groups (Hermann et al. 2004). A recent study showed that DNA methylation was extensively reprogrammed during early development in mammals with important changes taking place in the transition from the blastocyst to the post-implantation epiblast (Borgel et al. 2010). De novo methylation in the epiblast would be targeted to lineage-specific haematopoietic genes (such as Pou2af1, a gene that encodes a transcriptional co-activator involved in B-cell development, or Cytip, a gene expressed in leukocytes, among others). The promoters of these lineage-specific genes would be subsequently de-methylated during terminal differentiation. Both DNMT3a and DNMT3b are considered to be responsible for "de novo" methylation as they act preferably on un-methylated DNA substrates (Jaenisch and Bird 2003). DNMT3a and DNMT3b show different expression patterns and targets (Jaenisch and Bird 2003) and DNMT3b seems to be the main enzyme required for promoter methylation during implantation (Borgel et al. 2010). Initial studies on DNMT3a indicated that *Dnmt3a Dnmt3b* double-deficient HSCs, but not *Dnmt3a* or *Dnmt3b* single-deficient HSCs, were incapable of long-term reconstitution in transplantation assays (Tadokoro et al. 2007). However, in contrast to these studies where *Dnmt3a*-null HSCs contributed normally to haematopoiesis, the use of a conditional knock-out model to study the functions of DNMT3a (Challen et al. 2012) revealed that this protein is required to silence the HSCs self-renewal programme, thus permitting haematopoietic differentiation.

The importance of DNA methylation during haematopoiesis was initially suggested by gene deletion studies targeting *Dnmt1*. DNMT1 reproduces cytosine methylation patterns from a hemi-methylated substrate after DNA replication, and therefore is considered responsible for the maintenance and propagation of the methylation pattern (Jaenisch and Bird 2003). The ablation of DNMT1 expression in HSCs resulted in these cells undergoing apoptosis, whereas the expression of a hypomorphic allele lead to reduced repopulation capacity and decreased production of lymphoid progenitors while the development of myelo-erythroid progenitors remained normal (Broske et al. 2009). There is also evidence that DNMT1 interacts with GATA1, GFI1 and ZBP-89 in mouse erythro-leukemic cells (MEL).

Recent studies analysing DNA methylation during HSCs ontogeny have shown that, although overall DNA methylation landscape is essentially maintained during this process, the largest number of genes undergoing changes in their methylation pattern were highly expressed in downstream progenitors but not in HSCs. Indeed, only a few genes expressed in HSCs were differentially methylated (Beerman et al. 2013). These latest results suggest that DNA methylation in HSCs regulates the expression of genes that are activated during the differentiation of HSCs to defined lineages.

4.4 Concluding Remarks

The molecular and cellular control of haematopoietic development is an intricate process regulated at multiple levels. In this chapter, we have discussed the cellular complexity of haematopoietic specification, the transcriptional control regulating key steps of this differentiation process and how transcription factors may integrate into wider regulatory networks. To date, little is known about the epigenetic regulation of embryonic haematopoiesis and our understanding is mostly limited to the description of the phenotypes observed upon gene deletion of specific chromatin modifier proteins. Furthermore, the inaccessibility of the developing embryo and the low number of cells undergoing haematopoietic specification in these embryos hamper the biochemical study of the complexes of transcription factors and epigenetic regulators orchestrating haematopoietic development and maintenance. Devising novel technologies and experimental strategies will allow us to further push the boundaries to decipher the chromatin landscape during blood cell formation. By integrating together these multiple layers of regulation, we will further our understanding of haematopoietic specification. Ultimately, for a complete picture, this knowledge will have to be generated at the single cell level and linked to the extrinsic signals provided by the micro-environment which guides and instructs the developing blood cells.

References

An S, Yeo KJ, Jeon YH, Song JJ (2011) Crystal structure of the human histone methyltransferase ASH1L catalytic domain and its implications for the regulatory mechanism. J Biol Chem 286:8369–74. doi:10.1074/jbc.M110.203380

Beerman I, Bock C, Garrison BS, Smith ZD, Gu H, Meissner A, Rossi DJ (2013) Proliferation-dependent alterations of the DNA methylation landscape underlie hematopoietic stem cell aging. Cell Stem Cell 12:413–25. doi:10.1016/j.stem.2013.01.017

Begley CG, Green AR (1999) The SCL gene: from case report to critical hematopoietic regulator. Blood 93:2760–70

Bernt KM, Zhu N, Sinha AU, Vempati S, Faber J, Krivtsov AV, Feng Z, Punt N, Daigle A, Bullinger L, Pollock RM, Richon VM, Kung AL, Armstrong SA (2011) MLL-rearranged leukemia is dependent on aberrant H3K79 methylation by DOT1L. Cancer Cell 20:66–78. doi:10.1016/j.ccr.2011.06.010

Boehm T, Foroni L, Kaneko Y, Perutz MF, Rabbitts TH (1991) The rhombotin family of cysteine-rich LIM-domain oncogenes: distinct members are involved in T-cell translocations to human chromosomes 11p15 and 11p13. Proc Natl Acad Sci U S A 88:4367–71

Boisset JC, van Cappellen W, Andrieu-Soler C, Galjart N, Dzierzak E, Robin C (2010) In vivo imaging of haematopoietic cells emerging from the mouse aortic endothelium. Nature 464:116–20. doi:10.1038/nature08764

Borgel J, Guibert S, Li Y, Chiba H, Schubeler D, Sasaki H, Forne T, Weber M (2010) Targets and dynamics of promoter DNA methylation during early mouse development. Nat Genet 42:1093–100. doi:10.1038/ng.708

Borrow J, Stanton VP Jr, Andresen JM, Becher R, Behm FG, Chaganti RS, Civin CI, Disteche C, Dube I, Frischauf AM, Horsman D, Mitelman F, Volinia S, Watmore AE, Housman DE (1996) The translocation t(8;16)(p11;p13) of acute myeloid leukaemia fuses a putative acetyltransferase to the CREB-binding protein. Nat Genet 14:33–41. doi:10.1038/ng0996-33

Broske AM, Vockentanz L, Kharazi S, Huska MR, Mancini E, Scheller M, Kuhl C, Enns A, Prinz M, Jaenisch R, Nerlov C, Leutz A, Andrade-Navarro MA, Jacobsen SE, Rosenbauer F (2009) DNA methylation protects hematopoietic stem cell multipotency from myeloerythroid restriction. Nat Genet 41:1207–15. doi:10.1038/ng.463

Cai Z, de Bruijn M, Ma X, Dortland B, Luteijn T, Downing RJ, Dzierzak E (2000) Haploinsufficiency of AML1 affects the temporal and spatial generation of hematopoietic stem cells in the mouse embryo. Immunity 13:423–31

Campos EI, Reinberg D (2009) Histones: annotating chromatin. Annu Rev Genet 43:559–99. doi:10.1146/annurev.genet.032608.103928

Carapeti M, Aguiar RC, Goldman JM, Cross NC (1998) A novel fusion between MOZ and the nuclear receptor coactivator TIF2 in acute myeloid leukemia. Blood 91:3127–33

Chaffanet M, Gressin L, Preudhomme C, Soenen-Cornu V, Birnbaum D, Pebusque MJ (2000) MOZ is fused to p300 in an acute monocytic leukemia with t(8;22). Genes Chromosomes Cancer 28:138–44

Challen GA, Sun D, Jeong M, Luo M, Jelinek J, Berg JS, Bock C, Vasanthakumar A, Gu H, Xi Y, Liang S, Lu Y, Darlington GJ, Meissner A, Issa JP, Godley LA, Li W, Goodell MA (2012) Dnmt3a is essential for hematopoietic stem cell differentiation. Nat Genet 44:23–31. doi:10.1038/ng.1009

Chen MJ, Yokomizo T, Zeigler BM, Dzierzak E, Speck NA (2009) Runx1 is required for the endothelial to haematopoietic cell transition but not thereafter. Nature 457:887–91

Choi K, Kennedy M, Kazarov A, Papadimitriou JC, Keller G (1998) A common precursor for hematopoietic and endothelial cells. Development 125:725–32

Corbel C, Salaun J, Belo-Diabangouaya P, Dieterlen-Lievre F (2007) Hematopoietic potential of the pre-fusion allantois. Dev Biol 301:478–88. doi:10.1016/j.ydbio.2006.08.069

Cumano A, Godin I (2007) Ontogeny of the hematopoietic system. Annu Rev Immunol 25:745–85

Cumano A, Ferraz JC, Klaine M, Di Santo JP, Godin I (2001) Intraembryonic, but not yolk sac hematopoietic precursors, isolated before circulation, provide long-term multilineage reconstitution. Immunity 15:477–85

de Bruijn MF, Speck NA, Peeters MC, Dzierzak E (2000) Definitive hematopoietic stem cells first develop within the major arterial regions of the mouse embryo. EMBO J 19:2465–74

Dieterlen-Lievre F (1975) On the origin of haemopoietic stem cells in the avian embryo: an experimental approach. J Embryol Exp Morphol 33:607–19

Donaldson IJ, Chapman M, Kinston S, Landry JR, Knezevic K, Piltz S, Buckley N, Green AR, Gottgens B (2005) Genome-wide identification of cis-regulatory sequences controlling blood and endothelial development. Hum Mol Genet 14:595–601. doi:10.1093/hmg/ddi056

Doyon Y, Cayrou C, Ullah M, Landry A-J, Côté V, Selleck W, Lane WS, Tan S, Yang X-J, Côté J (2006) ING tumor suppressor proteins are critical regulators of chromatin acetylation required for genome expression and perpetuation. Mol Cell 21:51–64

D'Souza SL, Elefanty AG, Keller G (2005) SCL/Tal-1 is essential for hematopoietic commitment of the hemangioblast but not for its development. Blood 105:3862–70. doi:10.1182/blood-2004-09-3611

Eilken HM, Nishikawa S, Schroeder T (2009) Continuous single-cell imaging of blood generation from haemogenic endothelium. Nature 457:896–900. doi:10.1038/nature07760

Elefanty AG, Begley CG, Hartley L, Papaevangeliou B, Robb L (1999) SCL expression in the mouse embryo detected with a targeted lacZ reporter gene demonstrates its localization to hematopoietic, vascular, and neural tissues. Blood 94:3754–63

Ema M, Faloon P, Zhang WJ, Hirashima M, Reid T, Stanford WL, Orkin S, Choi K, Rossant J (2003) Combinatorial effects of Flk1 and Tal1 on vascular and hematopoietic development in the mouse. Genes Dev 17:380–93. doi:10.1101/gad.1049803

Erfurth FE, Popovic R, Grembecka J, Cierpicki T, Theisler C, Xia ZB, Stuart T, Diaz MO, Bushweller JH, Zeleznik-Le NJ (2008) MLL protects CpG clusters from methylation within the Hoxa9 gene, maintaining transcript expression. Proc Natl Acad Sci U S A 105:7517–22. doi:10.1073/pnas.0800090105

Esteyries S, Perot C, Adelaide J, Imbert M, Lagarde A, Pautas C, Olschwang S, Birnbaum D, Chaffanet M, Mozziconacci MJ (2008) NCOA3, a new fusion partner for MOZ/MYST3 in M5 acute myeloid leukemia. Leukemia 22:663–5. doi:10.1038/sj.leu.2404930

Fehling HJ, Lacaud G, Kubo A, Kennedy M, Robertson S, Keller G, Kouskoff V (2003) Tracking mesoderm induction and its specification to the hemangioblast during embryonic stem cell differentiation. Development 130:4217–27

Feng Y, Yang Y, Ortega MM, Copeland JN, Zhang M, Jacob JB, Fields TA, Vivian JL, Fields PE (2010) Early mammalian erythropoiesis requires the Dot1L methyltransferase. Blood 116:4483–91. doi:10.1182/blood-2010-03-276501

Ferkowicz MJ, Yoder MC (2005) Blood island formation: longstanding observations and modern interpretations. Exp Hematol 33:1041–7. doi:10.1016/j.exphem.2005.06.006

Garcia-Porrero JA, Manaia A, Jimeno J, Lasky LL, Dieterlen-Lievre F, Godin IE (1998) Antigenic profiles of endothelial and hemopoietic lineages in murine intraembryonic hemogenic sites. Dev Comp Immunol 22:303–19

Gekas C, Dieterlen-Lievre F, Orkin SH, Mikkola HK (2005) The placenta is a niche for hematopoietic stem cells. Dev Cell 8:365–75

Godin I, Dieterlen-Lievre F, Cumano A (1995) Emergence of multipotent hemopoietic cells in the yolk sac and paraaortic splanchnopleura in mouse embryos, beginning at 8.5 days postcoitus. Proc Natl Acad Sci U S A 92:773–7

Gottgens B, Nastos A, Kinston S, Piltz S, Delabesse EC, Stanley M, Sanchez MJ, Ciau-Uitz A, Patient R, Green AR (2002) Establishing the transcriptional programme for blood: the SCL stem cell enhancer is regulated by a multiprotein complex containing Ets and GATA factors. EMBO J 21:3039–50. doi:10.1093/emboj/cdf286

Gottgens B, Broccardo C, Sanchez MJ, Deveaux S, Murphy G, Gothert JR, Kotsopoulou E, Kinston S, Delaney L, Piltz S, Barton LM, Knezevic K, Erber WN, Begley CG, Frampton J,

Green AR (2004) The scl +18/19 stem cell enhancer is not required for hematopoiesis: identification of a 5′ bifunctional hematopoietic-endothelial enhancer bound by Fli-1 and Elf-1. Mol Cell Biol 24:1870–83

Herberth B, Minko K, Csillag A, Jaffredo T, Madarasz E (2005) SCL, GATA-2 and Lmo2 expression in neurogenesis. Int J Dev Neurosci 23:449–63. doi:10.1016/j.ijdevneu.2005.05.008

Hermann A, Gowher H, Jeltsch A (2004) Biochemistry and biology of mammalian DNA methyltransferases. Cell Mol Life Sci 61:2571–87. doi:10.1007/s00018-004-4201-1

Herrera-Merchan A, Arranz L, Ligos JM, de Molina A, Dominguez O, Gonzalez S (2012) Ectopic expression of the histone methyltransferase Ezh2 in haematopoietic stem cells causes myeloproliferative disease. Nat Commun 3:623. doi:10.1038/ncomms1623

Heuser M, Yap DB, Leung M, de Algara TR, Tafech A, McKinney S, Dixon J, Thresher R, Colledge B, Carlton M, Humphries RK, Aparicio SA (2009) Loss of MLL5 results in pleiotropic hematopoietic defects, reduced neutrophil immune function, and extreme sensitivity to DNA demethylation. Blood 113:1432–43. doi:10.1182/blood-2008-06-162263

Hinks GL, Shah B, French SJ, Campos LS, Staley K, Hughes J, Sofroniew MV (1997) Expression of LIM protein genes Lmo1, Lmo2, and Lmo3 in adult mouse hippocampus and other forebrain regions: differential regulation by seizure activity. J Neurosci 17:5549–59

Holbert MA, Sikorski T, Carten J, Snowflack D, Hodawadekar S, Marmorstein R (2007) The human monocytic leukemia zinc finger histone acetyltransferase domain contains DNA-binding activity implicated in chromatin targeting. J Biol Chem 282:36603–13. doi:10.1074/jbc.M705812200

Hong L, Schroth GP, Matthews HR, Yau P, Bradbury EM (1993) Studies of the DNA-binding properties of histone H4 amino terminus. Thermal-denaturation studies reveal that acetylation markedly reduces the binding constant of the H4 tail to DNA. J Biol Chem 268:305–314

Hoogenkamp M, Lichtinger M, Krysinska H, Lancrin C, Clarke D, Williamson A, Mazzarella L, Ingram R, Jorgensen H, Fisher A, Tenen DG, Kouskoff V, Lacaud G, Bonifer C (2009) Early chromatin unfolding by RUNX1: a molecular explanation for differential requirements during specification versus maintenance of the hematopoietic gene expression program. Blood 114:299–309. doi:10.1182/blood-2008-11-191890

Huang G, Zhao X, Wang L, Elf S, Xu H, Zhao X, Sashida G, Zhang Y, Liu Y, Lee J, Menendez S, Yang Y, Yan X, Zhang P, Tenen DG, Osato M, Hsieh JJ, Nimer SD (2011) The ability of MLL to bind RUNX1 and methylate H3K4 at PU.1 regulatory regions is impaired by MDS/AML-associated RUNX1/AML1 mutations. Blood 118:6544–52. doi:10.1182/blood-2010-11-317909

Huber TL, Kouskoff V, Fehling HJ, Palis J, Keller G (2004) Haemangioblast commitment is initiated in the primitive streak of the mouse embryo. Nature 432:625–30. doi:10.1038/nature03122

Im H, Park C, Feng Q, Johnson KD, Kiekhaefer CM, Choi K, Zhang Y, Bresnick EH (2003) Dynamic regulation of histone H3 methylated at lysine 79 within a tissue-specific chromatin domain. J Biol Chem 278:18346–52. doi:10.1074/jbc.M300890200

Ismailoglu I, Yeamans G, Daley GQ, Perlingeiro RC, Kyba M (2008) Mesodermal patterning activity of SCL. Exp Hematol 36:1593–603. doi:10.1016/j.exphem.2008.07.005

Jaenisch R, Bird A (2003) Epigenetic regulation of gene expression: how the genome integrates intrinsic and environmental signals. Nat Genet 33(Suppl):245–54. doi:10.1038/ng1089

Jones B, Su H, Bhat A, Lei H, Bajko J, Hevi S, Baltus GA, Kadam S, Zhai H, Valdez R, Gonzalo S, Zhang Y, Li E, Chen T (2008) The histone H3K79 methyltransferase Dot1L is essential for mammalian development and heterochromatin structure. PLoS Genet 4:e1000190. doi:10.1371/journal.pgen.1000190

Jordan HE (1917) Aortic cell clusters in vertebrate embryos. Proc Natl Acad Sci U S A 3:149–56

Kallianpur AR, Jordan JE, Brandt SJ (1994) The SCL/TAL-1 gene is expressed in progenitors of both the hematopoietic and vascular systems during embryogenesis. Blood 83:1200–8

Kataoka H, Hayashi M, Nakagawa R, Tanaka Y, Izumi N, Nishikawa S, Jakt ML, Tarui H (2011) Etv2/ER71 induces vascular mesoderm from Flk1+PDGFRalpha+ primitive mesoderm. Blood 118:6975–86. doi:10.1182/blood-2011-05-352658

Katsumoto T, Aikawa Y, Iwama A, Ueda S, Ichikawa H, Ochiya T, Kitabayashi I (2006) MOZ is essential for maintenance of hematopoietic stem cells. Genes Dev 20:1321–1330. doi:10.1101/gad.1393106

Katsumoto T, Yoshida N, Kitabayashi I (2008) Roles of the histone acetyltransferase monocytic leukemia zinc finger protein in normal and malignant hematopoiesis. Cancer Sci 99:1523–7. doi:10.1111/j.1349-7006.2008.00865.x

Kitabayashi I, Aikawa Y, Nguyen LA, Yokoyama A, Ohki M (2001) Activation of AML1-mediated transcription by MOZ and inhibition by the MOZ-CBP fusion protein. EMBO J 20:7184–96

Kobayashi-Osaki M, Ohneda O, Suzuki N, Minegishi N, Yokomizo T, Takahashi S, Lim KC, Engel JD, Yamamoto M (2005) GATA motifs regulate early hematopoietic lineage-specific expression of the Gata2 gene. Mol Cell Biol 25:7005–20. doi:10.1128/MCB.25.16.7005-7020.2005

Koyano-Nakagawa N, Kweon J, Iacovino M, Shi X, Rasmussen TL, Borges L, Zirbes KM, Li T, Perlingeiro RC, Kyba M, Garry DJ (2012) Etv2 is expressed in the yolk sac hematopoietic and endothelial progenitors and regulates Lmo2 gene expression. Stem Cells 30:1611–23. doi:10.1002/stem.1131

Kueh AJ, Dixon MP, Voss AK, Thomas T (2011) HBO1 is required for H3K14 acetylation and normal transcriptional activity during embryonic development. Mol Cell Biol 31:845–60. doi:10.1128/MCB.00159-10

Kumaravelu P, Hook L, Morrison AM, Ure J, Zhao S, Zuyev S, Ansell J, Medvinsky A (2002) Quantitative developmental anatomy of definitive haematopoietic stem cells/long-term repopulating units (HSC/RUs): role of the aorta-gonad-mesonephros (AGM) region and the yolk sac in colonisation of the mouse embryonic liver. Development 129:4891–9

Lacaud G, Gore L, Kennedy M, Kouskoff V, Kingsley P, Hogan C, Carlsson L, Speck N, Palis J, Keller G (2002) Runx1 is essential for hematopoietic commitment at the hemangioblast stage of development in vitro. Blood 100:458–66. doi:10.1182/blood-2001-12-0321

Lagger G, O'Carroll D, Rembold M, Khier H, Tischler J, Weitzer G, Schuettengruber B, Hauser C, Brunmeir R, Jenuwein T, Seiser C (2002) Essential function of histone deacetylase 1 in proliferation control and CDK inhibitor repression. EMBO J 21:2672–81. doi:10.1093/emboj/21.11.2672

Lancrin C, Sroczynska P, Stephenson C, Allen T, Kouskoff V, Lacaud G (2009) The haemangioblast generates haematopoietic cells through a haemogenic endothelium stage. Nature 457:892–5. doi:10.1038/nature07679

Lancrin C, Sroczynska P, Serrano AG, Gandillet A, Ferreras C, Kouskoff V, Lacaud G (2010) Blood cell generation from the hemangioblast. J Mol Med 88:167–72. doi:10.1007/s00109-009-0554-0

Landry JR, Kinston S, Knezevic K, de Bruijn MF, Wilson N, Nottingham WT, Peitz M, Edenhofer F, Pimanda JE, Ottersbach K, Gottgens B (2008) Runx genes are direct targets of Scl/Tal1 in the yolk sac and fetal liver. Blood 111:3005–14. doi:10.1182/blood-2007-07-098830

Lee D, Park C, Lee H, Lugus JJ, Kim SH, Arentson E, Chung YS, Gomez G, Kyba M, Lin S, Janknecht R, Lim DS, Choi K (2008) ER71 acts downstream of BMP, Notch, and Wnt signaling in blood and vessel progenitor specification. Cell Stem Cell 2:497–507. doi:10.1016/j.stem.2008.03.008

Li W, Ferkowicz MJ, Johnson SA, Shelley WC, Yoder MC (2005) Endothelial cells in the early murine yolk sac give rise to CD41-expressing hematopoietic cells. Stem Cells Dev 14:44–54. doi:10.1089/scd.2005.14.44

Li Z, Chen MJ, Stacy T, Speck NA (2006) Runx1 function in hematopoiesis is required in cells that express Tek. Blood 107:106–10. doi:10.1182/blood-2005-05-1955

Lichtinger M, Hoogenkamp M, Krysinska H, Ingram R, Bonifer C (2010) Chromatin regulation by RUNX1. Blood Cells Mol Dis 44:287–90. doi:10.1016/j.bcmd.2010.02.009
Lichtinger M, Ingram R, Hannah R, Muller D, Clarke D, Assi SA, Lie ALM, Noailles L, Vijayabaskar MS, Wu M, Tenen DG, Westhead DR, Kouskoff V, Lacaud G, Gottgens B, Bonifer C (2012) RUNX1 reshapes the epigenetic landscape at the onset of haematopoiesis. EMBO J 31:4318–33. doi:10.1038/emboj.2012.275
Ling KW, Ottersbach K, van Hamburg JP, Oziemlak A, Tsai FY, Orkin SH, Ploemacher R, Hendriks RW, Dzierzak E (2004) GATA-2 plays two functionally distinct roles during the ontogeny of hematopoietic stem cells. J Exp Med 200:871–82. doi:10.1084/jem.20031556
Liu F, Kang I, Park C, Chang LW, Wang W, Lee D, Lim DS, Vittet D, Nerbonne JM, Choi K (2012) ER71 specifies Flk-1+ hemangiogenic mesoderm by inhibiting cardiac mesoderm and Wnt signaling. Blood 119:3295–305. doi:10.1182/blood-2012-01-403766
Look AT (1997) Oncogenic transcription factors in the human acute leukemias. Science 278:1059–64
Madan V, Madan B, Brykczynska U, Zilbermann F, Hogeveen K, Dohner K, Dohner H, Weber O, Blum C, Rodewald HR, Sassone-Corsi P, Peters AH, Fehling HJ (2009) Impaired function of primitive hematopoietic cells in mice lacking the Mixed-Lineage-Leukemia homolog MLL5. Blood 113:1444–54. doi:10.1182/blood-2008-02-142638
Manaia A, Lemarchandel V, Klaine M, Max-Audit I, Romeo P, Dieterlen-Lievre F, Godin I (2000) Lmo2 and GATA-3 associated expression in intraembryonic hemogenic sites. Development 127:643–53
Martin C, Zhang Y (2005) The diverse functions of histone lysine methylation. Nat Rev Mol Cell Biol 6:838–49. doi:10.1038/nrm1761
Matsuoka S, Tsuji K, Hisakawa H, Xu M, Ebihara Y, Ishii T, Sugiyama D, Manabe A, Tanaka R, Ikeda Y, Asano S, Nakahata T (2001) Generation of definitive hematopoietic stem cells from murine early yolk sac and paraaortic splanchnopleures by aorta-gonad-mesonephros region-derived stromal cells. Blood 98:6–12
Maximov AA (1909) Investigations on blood and connective tissue: 1. The earliest developmental stages of blood and connective tissue cells in the mammalian embryo, until the beginning of blood formation in the liver. Arch Mikr Anat Entwicklungsgesch 73:444–561
Medvinsky A, Dzierzak E (1996) Definitive hematopoiesis is autonomously initiated by the AGM region. Cell 86:897–906
Medvinsky AL, Samoylina NL, Muller AM, Dzierzak EA (1993) An early pre-liver intraembryonic source of CFU-S in the developing mouse. Nature 364:64–7. doi:10.1038/364064a0
Mikkola HK, Orkin SH (2006) The journey of developing hematopoietic stem cells. Development 133:3733–44
Minegishi N, Ohta J, Yamagiwa H, Suzuki N, Kawauchi S, Zhou Y, Takahashi S, Hayashi N, Engel JD, Yamamoto M (1999) The mouse GATA-2 gene is expressed in the para-aortic splanchnopleura and aorta-gonads and mesonephros region. Blood 93:4196–207
Minegishi N, Suzuki N, Yokomizo T, Pan X, Fujimoto T, Takahashi S, Hara T, Miyajima A, Nishikawa S, Yamamoto M (2003) Expression and domain-specific function of GATA-2 during differentiation of the hematopoietic precursor cells in midgestation mouse embryos. Blood 102:896–905. doi:10.1182/blood-2002-12-3809
Minko K, Bollerot K, Drevon C, Hallais MF, Jaffredo T (2003) From mesoderm to blood islands: patterns of key molecules during yolk sac erythropoiesis. Gene Expr Patterns 3:261–72
Mishima Y, Miyagi S, Saraya A, Negishi M, Endoh M, Endo TA, Toyoda T, Shinga J, Katsumoto T, Chiba T, Yamaguchi N, Kitabayashi I, Koseki H, Iwama A (2011) The Hbo1-Brd1/Brpf2 complex is responsible for global acetylation of H3K14 and required for fetal liver erythropoiesis. Blood 118:2443–53. doi:10.1182/blood-2011-01-331892
Mochizuki-Kashio M, Mishima Y, Miyagi S, Negishi M, Saraya A, Konuma T, Shinga J, Koseki H, Iwama A (2011) Dependency on the polycomb gene Ezh2 distinguishes fetal from adult hematopoietic stem cells. Blood 118:6553–61. doi:10.1182/blood-2011-03-340554

Moore MA, Owen JJ (1965) Chromosome marker studies on the development of the haemopoietic system in the chick embryo. Nature 208:956–958

Mukouyama Y, Chiba N, Hara T, Okada H, Ito Y, Kanamaru R, Miyajima A, Satake M, Watanabe T (2000) The AML1 transcription factor functions to develop and maintain hematogenic precursor cells in the embryonic aorta-gonad-mesonephros region. Dev Biol 220:27–36. doi:10.1006/dbio.2000.9617

Muller AM, Medvinsky A, Strouboulis J, Grosveld F, Dzierzak E (1994) Development of hematopoietic stem cell activity in the mouse embryo. Immunity 1:291–301

Nardelli J, Thiesson D, Fujiwara Y, Tsai FY, Orkin SH (1999) Expression and genetic interaction of transcription factors GATA-2 and GATA-3 during development of the mouse central nervous system. Dev Biol 210:305–21. doi:10.1006/dbio.1999.9278

Narula J, Smith AM, Gottgens B, Igoshin OA (2010) Modeling reveals bistability and low-pass filtering in the network module determining blood stem cell fate. PLoS Comput Biol 6: e1000771. doi:10.1371/journal.pcbi.1000771

Narula J, Williams CJ, Tiwari A, Marks-Bluth J, Pimanda JE, Igoshin OA (2013) Mathematical model of a gene regulatory network reconciles effects of genetic perturbations on hematopoietic stem cell emergence. Dev Biol 379:258–69. doi:10.1016/j.ydbio.2013.04.016

Nishikawa SI, Nishikawa S, Kawamoto H, Yoshida H, Kizumoto M, Kataoka H, Katsura Y (1998) In vitro generation of lymphohematopoietic cells from endothelial cells purified from murine embryos. Immunity 8:761–9

North T, Gu TL, Stacy T, Wang Q, Howard L, Binder M, Marin-Padilla M, Speck NA (1999) Cbfa2 is required for the formation of intra-aortic hematopoietic clusters. Development 126:2563–75

North TE, de Bruijn MF, Stacy T, Talebian L, Lind E, Robin C, Binder M, Dzierzak E, Speck NA (2002) Runx1 expression marks long-term repopulating hematopoietic stem cells in the midgestation mouse embryo. Immunity 16:661–72

Nottingham WT, Jarratt A, Burgess M, Speck CL, Cheng JF, Prabhakar S, Rubin EM, Li PS, Sloane-Stanley J, Kong ASJ, de Bruijn MF (2007) Runx1-mediated hematopoietic stem-cell emergence is controlled by a Gata/Ets/SCL-regulated enhancer. Blood 110:4188–97. doi:10.1182/blood-2007-07-100883

Okuda T, van Deursen J, Hiebert SW, Grosveld G, Downing JR (1996) AML1, the target of multiple chromosomal translocations in human leukemia, is essential for normal fetal liver hematopoiesis. Cell 84:321–30

Ottersbach K, Dzierzak E (2005) The murine placenta contains hematopoietic stem cells within the vascular labyrinth region. Dev Cell 8:377–87. doi:10.1016/j.devcel.2005.02.001

Paggetti J, Largeot A, Aucagne R, Jacquel A, Lagrange B, Yang XJ, Solary E, Bastie JN, Delva L (2010) Crosstalk between leukemia-associated proteins MOZ and MLL regulates HOX gene expression in human cord blood CD34+ cells. Oncogene 29:5019–31. doi:10.1038/onc.2010.254

Palis J, Robertson S, Kennedy M, Wall C, Keller G (1999) Development of erythroid and myeloid progenitors in the yolk sac and embryo proper of the mouse. Development 126:5073–84

Perez-Campo FM, Borrow J, Kouskoff V, Lacaud G (2009) The histone acetyl transferase activity of monocytic leukemia zinc finger is critical for the proliferation of hematopoietic precursors. Blood 113:4866–74

Perez-Campo FM, Costa G, Lie-a-Ling M, Kouskoff V, Lacaud G (2013) The MYSTerious MOZ, a histone acetyltransferase with a key role in haematopoiesis. Immunology 139:161–5. doi:10.1111/imm.12072

Pimanda JE, Gottgens B (2010) Gene regulatory networks governing haematopoietic stem cell development and identity. Int J Dev Biol 54:1201–11. doi:10.1387/ijdb.093038jp

Pimanda JE, Donaldson IJ, de Bruijn MF, Kinston S, Knezevic K, Huckle L, Piltz S, Landry JR, Green AR, Tannahill D, Gottgens B (2007a) The SCL transcriptional network and BMP signaling pathway interact to regulate RUNX1 activity. Proc Natl Acad Sci U S A 104:840–5. doi:10.1073/pnas.0607196104

Pimanda JE, Ottersbach K, Knezevic K, Kinston S, Chan WY, Wilson NK, Landry JR, Wood AD, Kolb-Kokocinski A, Green AR, Tannahill D, Lacaud G, Kouskoff V, Gottgens B (2007b) Gata2, Fli1, and Scl form a recursively wired gene-regulatory circuit during early hematopoietic development. Proc Natl Acad Sci U S A 104:17692–7. doi:10.1073/pnas.0707045104

Porcher C, Swat W, Rockwell K, Fujiwara Y, Alt FW, Orkin SH (1996) The T cell leukemia oncoprotein SCL/tal-1 is essential for development of all hematopoietic lineages. Cell 86:47–57

Puig OM, Belles E, Lopez-Rodas G, Sendra R, Tordera V (1998) Interaction between N-terminal domain of H4 and DNA is regulated by the acetylation degree. Biochim Biophys Acta 1397:79–90

Quintas-Cardama A, Santos FP, Garcia-Manero G (2011) Histone deacetylase inhibitors for the treatment of myelodysplastic syndrome and acute myeloid leukemia. Leukemia 25:226–35. doi:10.1038/leu.2010.276

Rabbitts TH (1998) LMO T-cell translocation oncogenes typify genes activated by chromosomal translocations that alter transcription and developmental processes. Genes Dev 12:2651–7

Rasmussen TL, Kweon J, Diekmann MA, Belema-Bedada F, Song Q, Bowlin K, Shi X, Ferdous A, Li T, Kyba M, Metzger JM, Koyano-Nakagawa N, Garry DJ (2011) ER71 directs mesodermal fate decisions during embryogenesis. Development 138:4801–12. doi:10.1242/dev.070912

Rhodes KE, Gekas C, Wang Y, Lux CT, Francis CS, Chan DN, Conway S, Orkin SH, Yoder MC, Mikkola HK (2008) The emergence of hematopoietic stem cells is initiated in the placental vasculature in the absence of circulation. Cell Stem Cell 2:252–63

Robb L, Lyons I, Li R, Hartley L, Kontgen F, Harvey RP, Metcalf D, Begley CG (1995) Absence of yolk sac hematopoiesis from mice with a targeted disruption of the scl gene. Proc Natl Acad Sci U S A 92:7075–9

Robb L, Elwood NJ, Elefanty AG, Kontgen F, Li R, Barnett LD, Begley CG (1996) The scl gene product is required for the generation of all hematopoietic lineages in the adult mouse. EMBO J 15:4123–9

Rodrigues NP, Janzen V, Forkert R, Dombkowski DM, Boyd AS, Orkin SH, Enver T, Vyas P, Scadden DT (2005) Haploinsufficiency of GATA-2 perturbs adult hematopoietic stem-cell homeostasis. Blood 106:477–84. doi:10.1182/blood-2004-08-2989

Royer-Pokora B, Loos U, Ludwig WD (1991) TTG-2, a new gene encoding a cysteine-rich protein with the LIM motif, is overexpressed in acute T-cell leukaemia with the t(11;14)(p13;q11). Oncogene 6:1887–93

Sabin FR (1920) Studies on the origin of blood vessels and of red blood corpuscles as seen in the living blastoderm of the chick during the second day of incubation. Contrib Embryol 9:214–262

Saleque S, Kim J, Rooke HM, Orkin SH (2007) Epigenetic regulation of hematopoietic differentiation by Gfi-1 and Gfi-1b is mediated by the cofactors CoREST and LSD1. Mol Cell 27:562–72. doi:10.1016/j.molcel.2007.06.039

Samokhvalov IM, Samokhvalova NI, Nishikawa S (2007) Cell tracing shows the contribution of the yolk sac to adult haematopoiesis. Nature 446:1056–61

Sanchez M, Gottgens B, Sinclair AM, Stanley M, Begley CG, Hunter S, Green AR (1999) An SCL 3′ enhancer targets developing endothelium together with embryonic and adult haematopoietic progenitors. Development 126:3891–904

Shivdasani RA, Mayer EL, Orkin SH (1995) Absence of blood formation in mice lacking the T-cell leukaemia oncoprotein tal-1/SCL. Nature 373:432–4. doi:10.1038/373432a0

Shogren-Knaak M, Ishii H, Sun JM, Pazin MJ, Davie JR, Peterson CL (2006) Histone H4-K16 acetylation controls chromatin structure and protein interactions. Science 311:844–7. doi:10.1126/science.1124000

Siggers P, Smith L, Greenfield A (2002) Sexually dimorphic expression of Gata-2 during mouse gonad development. Mech Dev 111:159–62

Silberstein L, Sanchez MJ, Socolovsky M, Liu Y, Hoffman G, Kinston S, Piltz S, Bowen M, Gambardella L, Green AR, Gottgens B (2005) Transgenic analysis of the stem cell leukemia +19 stem cell enhancer in adult and embryonic hematopoietic and endothelial cells. Stem Cells 23:1378–88. doi:10.1634/stemcells.2005-0090

Silver L, Palis J (1997) Initiation of murine embryonic erythropoiesis: a spatial analysis. Blood 89:1154–64

Sprussel A, Schulte JH, Weber S, Necke M, Handschke K, Thor T, Pajtler KW, Schramm A, Konig K, Diehl L, Mestdagh P, Vandesompele J, Speleman F, Jastrow H, Heukamp LC, Schule R, Duhrsen U, Buettner R, Eggert A, Gothert JR (2012) Lysine-specific demethylase 1 restricts hematopoietic progenitor proliferation and is essential for terminal differentiation. Leukemia 26:2039–51. doi:10.1038/leu.2012.157

Sugiyama D, Ogawa M, Hirose I, Jaffredo T, Arai K, Tsuji K (2003) Erythropoiesis from acetyl LDL incorporating endothelial cells at the preliver stage. Blood 101:4733–8. doi:10.1182/blood-2002-09-2799

Sugiyama D, Arai K, Tsuji K (2005) Definitive hematopoiesis from acetyl LDL incorporating endothelial cells in the mouse embryo. Stem Cells Dev 14:687–96. doi:10.1089/scd.2005.14.687

Sumanas S, Jorniak T, Lin S (2005) Identification of novel vascular endothelial-specific genes by the microarray analysis of the zebrafish cloche mutants. Blood 106:534–41. doi:10.1182/blood-2004-12-4653

Sumanas S, Gomez G, Zhao Y, Park C, Choi K, Lin S (2008) Interplay among Etsrp/ER71, Scl, and Alk8 signaling controls endothelial and myeloid cell formation. Blood 111:4500–10. doi:10.1182/blood-2007-09-110569

Sung LY, Gao S, Shen H, Yu H, Song Y, Smith SL, Chang CC, Inoue K, Kuo L, Lian J, Li A, Tian XC, Tuck DP, Weissman SM, Yang X, Cheng T (2006) Differentiated cells are more efficient than adult stem cells for cloning by somatic cell nuclear transfer. Nat Genet 38:1323–8. doi:10.1038/ng1895

Tadokoro Y, Ema H, Okano M, Li E, Nakauchi H (2007) De novo DNA methyltransferase is essential for self-renewal, but not for differentiation, in hematopoietic stem cells. J Exp Med 204:715–22. doi:10.1084/jem.20060750

Tanaka Y, Katagiri Z, Kawahashi K, Kioussis D, Kitajima S (2007) Trithorax-group protein ASH1 methylates histone H3 lysine 36. Gene 397:161–8. doi:10.1016/j.gene.2007.04.027

Tanaka Y, Nakayama Y, Taniguchi M, Kioussis D (2008) Regulation of early T cell development by the PHD finger of histone lysine methyltransferase ASH1. Biochem Biophys Res Commun 365:589–594. doi:10.1016/J.Bbrc.2007.10.159

Tanaka Y, Kawahashi K, Katagiri Z, Nakayama Y, Mahajan M, Kioussis D (2011) Dual function of histone H3 lysine 36 methyltransferase ASH1 in regulation of Hox gene expression. PLoS One 6:e28171. doi:10.1371/journal.pone.0028171

Tanaka Y, Hayashi M, Kubota Y, Nagai H, Sheng G, Nishikawa S, Samokhvalov IM (2012a) Early ontogenic origin of the hematopoietic stem cell lineage. Proc Natl Acad Sci U S A 109:4515–20. doi:10.1073/pnas.1115828109

Tanaka Y, Joshi A, Wilson NK, Kinston S, Nishikawa S, Gottgens B (2012b) The transcriptional programme controlled by Runx1 during early embryonic blood development. Dev Biol 366:404–19. doi:10.1016/j.ydbio.2012.03.024

Taoudi S, Medvinsky A (2007) Functional identification of the hematopoietic stem cell niche in the ventral domain of the embryonic dorsal aorta. Proc Natl Acad Sci U S A 104:9399–403

Taoudi S, Gonneau C, Moore K, Sheridan JM, Blackburn CC, Taylor E, Medvinsky A (2008) Extensive hematopoietic stem cell generation in the AGM region via maturation of VE-cadherin+CD45+ pre-definitive HSCs. Cell Stem Cell 3:99–108

Thomas T, Corcoran LM, Gugasyan R, Dixon MP, Brodnicki T, Nutt SL, Metcalf D, Voss AK (2006) Monocytic leukemia zinc finger protein is essential for the development of long-term reconstituting hematopoietic stem cells. Genes Dev 20:1175–1186. doi:10.1101/gad.1382606

Tsai FY, Orkin SH (1997) Transcription factor GATA-2 is required for proliferation/survival of early hematopoietic cells and mast cell formation, but not for erythroid and myeloid terminal differentiation. Blood 89:3636–43

Tsai FY, Keller G, Kuo FC, Weiss M, Chen J, Rosenblatt M, Alt FW, Orkin SH (1994) An early haematopoietic defect in mice lacking the transcription factor GATA-2. Nature 371:221–6. doi:10.1038/371221a0

Ueno H, Weissman IL (2006) Clonal analysis of mouse development reveals a polyclonal origin for yolk sac blood islands. Dev Cell 11:519–33. doi:10.1016/j.devcel.2006.08.001

van Eekelen JA, Bradley CK, Gothert JR, Robb L, Elefanty AG, Begley CG, Harvey AR (2003) Expression pattern of the stem cell leukaemia gene in the CNS of the embryonic and adult mouse. Neuroscience 122:421–36

Van Handel B, Montel-Hagen A, Sasidharan R, Nakano H, Ferrari R, Boogerd CJ, Schredelseker J, Wang Y, Hunter S, Org T, Zhou J, Li X, Pellegrini M, Chen JN, Orkin SH, Kurdistani SK, Evans SM, Nakano A, Mikkola HK (2012) Scl represses cardiomyogenesis in prospective hemogenic endothelium and endocardium. Cell 150:590–605. doi:10.1016/j.cell.2012.06.026

Visvader JE, Fujiwara Y, Orkin SH (1998) Unsuspected role for the T-cell leukemia protein SCL/tal-1 in vascular development. Genes Dev 12:473–9

Wada T, Kikuchi J, Nishimura N, Shimizu R, Kitamura T, Furukawa Y (2009) Expression levels of histone deacetylases determine the cell fate of hematopoietic progenitors. J Biol Chem 284:30673–83. doi:10.1074/jbc.M109.042242

Wadman IA, Osada H, Grutz GG, Agulnick AD, Westphal H, Forster A, Rabbitts TH (1997) The LIM-only protein Lmo2 is a bridging molecule assembling an erythroid, DNA-binding complex which includes the TAL1, E47, GATA-1 and Ldb1/NLI proteins. EMBO J 16:3145–57. doi:10.1093/emboj/16.11.3145

Wareing S, Eliades A, Lacaud G, Kouskoff V (2012a) ETV2 expression marks blood and endothelium precursors, including hemogenic endothelium, at the onset of blood development. Dev Dyn 241:1454–64. doi:10.1002/dvdy.23825

Wareing S, Mazan A, Pearson S, Gottgens B, Lacaud G, Kouskoff V (2012b) The Flk1-Cre-mediated deletion of ETV2 defines its narrow temporal requirement during embryonic hematopoietic development. Stem Cells 30:1521–31. doi:10.1002/stem.1115

Warren AJ, Colledge WH, Carlton MB, Evans MJ, Smith AJ, Rabbitts TH (1994) The oncogenic cysteine-rich LIM domain protein rbtn2 is essential for erythroid development. Cell 78:45–57

Wilson NK, Miranda-Saavedra D, Kinston S, Bonadies N, Foster SD, Calero-Nieto F, Dawson MA, Donaldson IJ, Dumon S, Frampton J, Janky R, Sun XH, Teichmann SA, Bannister AJ, Gottgens B (2009) The transcriptional program controlled by the stem cell leukemia gene Scl/Tal1 during early embryonic hematopoietic development. Blood 113:5456–65. doi:10.1182/blood-2009-01-200048

Wilson NK, Foster SD, Wang X, Knezevic K, Schutte J, Kaimakis P, Chilarska PM, Kinston S, Ouwehand WH, Dzierzak E, Pimanda JE, de Bruijn MF, Gottgens B (2010) Combinatorial transcriptional control in blood stem/progenitor cells: genome-wide analysis of ten major transcriptional regulators. Cell Stem Cell 7:532–44. doi:10.1016/j.stem.2010.07.016

Wong PM, Chung SW, Chui DH, Eaves CJ (1986) Properties of the earliest clonogenic hemopoietic precursors to appear in the developing murine yolk sac. Proc Natl Acad Sci U S A 83:3851–4

Xu Z, Huang S, Chang LS, Agulnick AD, Brandt SJ (2003) Identification of a TAL1 target gene reveals a positive role for the LIM domain-binding protein Ldb1 in erythroid gene expression and differentiation. Mol Cell Biol 23:7585–99

Yamada Y, Warren AJ, Dobson C, Forster A, Pannell R, Rabbitts TH (1998) The T cell leukemia LIM protein Lmo2 is necessary for adult mouse hematopoiesis. Proc Natl Acad Sci U S A 95:3890–5

Yang XJ, Seto E (2008) The Rpd3/Hda1 family of lysine deacetylases: from bacteria and yeast to mice and men. Nat Rev Mol Cell Biol 9:206–18. doi:10.1038/nrm2346

Yokomizo T, Ogawa M, Osato M, Kanno T, Yoshida H, Fujimoto T, Fraser S, Nishikawa S, Okada H, Satake M, Noda T, Ito Y (2001) Requirement of Runx1/AML1/PEBP2alphaB for the generation of haematopoietic cells from endothelial cells. Genes Cells 6:13–23

Yuan W, Xu M, Huang C, Liu N, Chen S, Zhu B (2011) H3K36 methylation antagonizes PRC2-mediated H3K27 methylation. J Biol Chem 286:7983–9. doi:10.1074/jbc.M110.194027

Zeigler BM, Sugiyama D, Chen M, Guo Y, Downs KM, Speck NA (2006) The allantois and chorion, when isolated before circulation or chorio-allantoic fusion, have hematopoietic potential. Development 133:4183–92. doi:10.1242/dev.02596

Zhang Y, Reinberg D (2001) Transcription regulation by histone methylation: interplay between different covalent modifications of the core histone tails. Genes Dev 15:2343–60. doi:10.1101/gad.927301

Zovein AC, Hofmann JJ, Lynch M, French WJ, Turlo KA, Yang Y, Becker MS, Zanetta L, Dejana E, Gasson JC, Tallquist MD, Iruela-Arispe ML (2008) Fate tracing reveals the endothelial origin of hematopoietic stem cells. Cell Stem Cell 3:625–36

Chapter 5
Polycomb Complexes: Chromatin Regulators Required for Cell Diversity and Tissue Homeostasis

Miguel Vidal

Abstract The Polycomb group (PcG) products are a set of evolutionary conserved proteins that form chromatin regulator complexes that control expression of developmentally relevant genes. PcG activity is essential not only to maintain the developmental potential of pluripotent cells from which specialized cell types arise, but also to ensure the directionality of the differentiation process. In the adult, these PcG functions are essential for normal cell homeostasis and their deregulation is often associated with cell transformation events. PcG-dependent transcriptional control involves posttranslational modifications of histones, decreased DNA accessibility, and other mechanisms. While the stability of Polycomb-determined chromatin landscapes is rather stable in differentiated cells, in pluripotent cells it is characteristically dynamic in order to accommodate the execution of developmental genetic programs. Best known as repressors of gene expression, recent evidence points at roles during gene activation. Besides gene expression control, PcG products also participate in other essential functions such as DNA damage response, indicating that these proteins are involved in a wide spectrum of cellular and organismal functions in need of detailed characterization.

Keywords Polycomb • PRC1 • PRC2 • Chromatin regulators • Chromatin compaction • Histone modifiers • Developmental potential • CpG islands • Stem cell • Progenitors • Cell homeostasis

M. Vidal (✉)
Department of Cellular and Molecular Biology, Centro de Investigaciones Biológicas (CSIC), Ramiro de Maeztu, Madrid 28040, Spain
e-mail: mvidal@cib.csic.es

C. Bonifer and P.N. Cockerill (eds.), *Transcriptional and Epigenetic Mechanisms Regulating Normal and Aberrant Blood Cell Development*, Epigenetics and Human Health, DOI 10.1007/978-3-642-45198-0_5, © Springer-Verlag Berlin Heidelberg 2014

5.1 Introduction

The Polycomb group (PcG) of genes was first discovered during the genetic analysis of development in the fruit fly *Drosophila melanogaster*. A first mutant, named *extra sex combs* (Slifer 1942), referred to the presence of additional bristles in the legs that male flies use during mating. Mutations with similar phenotypes were isolated and the genes grouped under the denomination of one of them, Polycomb (Lewis 1978). These mutants showed homeotic transformations, i.e., a part of the body, for example, an anterior leg with no sex combs acquiring the identity of another part, as that of a posterior leg with sex combs (or, if considering embryos, anterior thoracic segments resembling posterior abdominal segments). The molecular nature of these defects lies on the ectopic expression of homeotic genes which are responsible for segment identity (Hox genes) (Riley et al. 1987). After molecular cloning of *Drosophila* Polycomb genes, mammalian homologs were identified and their inactivation in loss-of-function mouse models was also accompanied by homeotic transformations of the axial skeleton (Akasaka et al. 1996; del Mar Lorente et al. 2000; der Lugt et al. 1994).

For a long time, Polycomb products were considered exclusively as developmental regulators. Subsequent work showed their implication in a wide variety of functions that include parental imprinting (monoallelic expression), adult stem cell self-renewal, pluripotency, and, when deregulated, oncogenic transformation (Bracken and Helin 2009; Mills 2010; Sparmann and van Lohuizen 2006). Polycomb targets include genes associated with transitions within cell lineages on their way to full differentiation. Cell identity genes are Polycomb silenced just before their activation in the subsequent cell state and, at the same time, those genes defining the vanishing cell type are repressed in the new state (Bracken et al. 2006; Mohn et al. 2008; Xie et al. 2013). It is now absolutely clear that ordered differentiation of pluripotent cells cannot occur without the activity of the Polycomb system (Pasini et al. 2007; Shen et al. 2009). In turn, reprogramming from differentiated cells towards pluripotent states also requires Polycomb activity (Onder et al. 2012; Pereira et al. 2010). Importantly, Polycomb regulates self-renewal of pluripotent progenitors and proliferative of their differentiated progeny contributing to tissue homeostasis (Calés et al. 2008; Klauke et al. 2013; Lessard and Sauvageau 2003; Luis et al. 2011). Thus, Polycomb is a malleable regulatory system for selective use of the genome in the generation of cell diversity.

Polycomb functions depend, at least in part, on their activities as catalyzers of chromatin modifications. Polycomb products are a heterogeneous collection of proteins that act in complexes. Their best-known activity in transcriptional control is as negative regulators of gene expression, although reportedly they are also associated with gene activity. Polycomb complexes contain, in addition to PcG products, "non-Polycomb" subunits that were not identified in the original genetic screens. The Polycomb system is evolutionary ancient and conserved, from plants and fungi (not yeast) to mammals (Schuettengruber et al. 2007; Shaver et al. 2010; Whitcomb et al. 2007). Although thought specific for multicellular organisms, homologs are found in unicellular alga (Shaver et al. 2010) suggesting co-option for cell lineage functions.

Here, I will discuss recent advances in our understanding of the molecular aspects of Polycomb action and their role as chromatin regulators and architectural chromatin proteins. Recruitment to targets and their regulation, with a bias towards mammalian cells, is also examined [see some excellent recent reviews (Lanzuolo and Orlando 2012; Simon and Kingston 2013)]. I first present an overview of gene regulation, from DNA sequence and chromatin states to three-dimensional organization of the genome (Gibcus and Dekker 2012) as a framework to explain Polycomb action.

5.2 Chromatin Landscape, Topological Organization, and Selective Use of the Genome

The diversity of cell types in multicellular eukaryotes is the result of differential use of the coding potential of the genome. This is achieved through regulated access of genomic sites to DNA-binding proteins (transcription factors). Controlled localization determines the nature of contacts between sites in chromatin within a highly, topologically organized structure.

5.2.1 Chromatin States

Polycomb complexes are endowed with catalytic activities that can modify histones and other substrates. DNA access is influenced by nucleosomes, whose mobility, in turn, can be conditioned by posttranslational modifications in canonical histones and by the presence of histone variants (Cosgrove et al. 2004). These modifications also affect binding and activity of chromatin-associated proteins, confirming coevolution of regulated DNA accessibility with packaging mechanisms for large DNA molecules. The close relationship between chromatin regulators, histone modifications, and transcriptional activities is apparent in the predictive power of chromatin states to identify DNA regulatory elements (Zhou et al. 2010). Remarkably, out of the large collection of possible combinations of histone marks, just a small number of functionally meaningful sets, or chromatin states, can be distilled. Thus, thousands of promoters and enhancers can be categorized into three and four discrete chromatin state types, respectively, whereas all genomic regions depending on whether transcriptionally active or repressed fit into three and four states, respectively. For example, nucleosomes with histone H3 di- and tri-methylated at lysine 4 and acetylated at lysines 9 and 27 correlate with active promoters, while mono- and di-methylated K4 in histone H3 is found in weak/poised enhancers (Ernst et al. 2011). Characteristically, one of the silenced states is identified by nucleosomes enriched in histone H3 tri-methylated at lysine 27 (H3K27me3), a Polycomb-specific modification (Margueron and Reinberg 2011).

Similarly, combinations of chromatin regulators that add or remove covalent modifications, also known as "writers" and "erasers," respectively, as well as proteins that recognize these modifications, i.e., the "readers" (Musselman et al. 2012; Taverna et al. 2007), correlate with distinctive sets of chromatin states (Ram et al. 2011). Six major combinations of chromatin-associated modifiers and "readers," or regulatory modules, have been identified in pluripotent and hematopoietic cells. Four of these correspond to two types each of promoters and enhancers, another to transcribed regions, and a last one to repressed regions binding Polycomb proteins. Generally, these modules include modifiers of opposing activity, but modifiers at Polycomb-silenced promoters are all of repressive nature (Ram et al. 2011). Independently, Drosophila chromatin is partitioned to five states (Filion et al. 2010): two distinct classes of transcriptionally active euchromatic domains, two distinct transcriptionally inactive domains, heterochromatic states, of which one is enriched in heterochromatin protein 1 (HP1) while the other contains Polycomb proteins, and chromatin associated with the nuclear lamina; the latter (Lamin-Associated Domains, LADs) includes a large fraction of the genome and is transcriptionally inert (Filion et al. 2010).

5.2.2 *Topological Organization of Chromatin and Gene Control*

The definition of chromatin states does not take into account restrictions derived from the three-dimensional configuration resulting from chromatin fiber folding. How this actually occurs is still not known. However, it is clear that it is subjected to limitations imposed by the long polymeric nature of chromatin and the effects of associated proteins (Iyer et al. 2011). Computationally generated models have been tested for their ability to fit experimental observations (Dekker et al. 2013). In one of them, the Multi-Loop-Subcompartment model, chromatin segments of $\simeq$1 megabase (Mb) pairs are proposed to fold in small loops separated by short linkers, in a rosette-like configuration (Jhunjhunwala et al. 2008). Looping, as an organizing principle, is consistent with genome-wide chromatin contacts mapped using chromosome conformation capture techniques (de Wit and de Laat 2012). At high resolution—high DNA sequencing depth and comparisons of contacts between smaller DNA fragment, <100 kb—the analysis shows chromatin organized in domains termed Topologically Associating Domains (TADs) (Dixon et al. 2012; Hou et al. 2012; Nora et al. 2012; Sexton et al. 2012). TADs are defined by differences in the probabilities of contacts between sites, whereby sites contained within the domains contact more frequently than with sites outside. TADs across cell types and between mouse and humans are highly similar and independent from transcriptional status (Dixon et al. 2012), indicating a strong architectural underlying principle. TADs are separated by short genomic segments or domain boundaries, enriched in CCCTC-binding factor CTCF (Shen et al. 2012), one of the proteins bound to insulators. These are DNA segments defined in transgenic assays by their ability to "shelter" regulatory elements from each other. TAD boundaries are important for spatial partitioning in domains (Nora et al. 2012). Cell type-specific contacts imply promoters and regulatory elements within the domains (Dixon et al. 2012; Nora et al. 2012) at loop-attachment points (Lin et al. 2012). At a lower resolution, chromosome conformation capture studies partition spatially the genome in interspersed compartments A and B. Compartment A correlates with gene-rich, highly expressed, DNAse I-sensitive genomic regions and contains accessible "open" chromatin, in opposition to closed chromatin in compartment B. Regions in compartment A, when analyzed as 1 Mb segments, also correlate with histone H3K36me3 and H3K27me3 marks. However, considered as shorter 100 kb segments, all above correlations hold except that for H3K27me3 (Lieberman-Aiden et al. 2009). Smaller, independently defined TADs are contained within A or B compartments. Three-dimensional chromatin architecture studied at yet higher resolution in pluripotent ES cells and neural progenitors showed that invariant TADs contain cell type-specific subdomains determined by looping interactions between regulatory sequences (Philips-Cremins et al. 2013). Major determinants of these spatial arrangements are, in addition to CTCF, the Mediator complex and cohesins, whose previously known roles as transcriptional regulators possibly derive from their activities as architectural proteins. Smaller chromatin loops linking enhancers and promoters involve Mediator and cohesins while interactions between more distant regions involve CTCF and cohesins. Cell lineage commitment and further differentiation would thus be characterized by specific sub-TAD level of chromatin organization (Philips-Cremins et al. 2013). In summary, eukaryotic chromosomes are folded in a highly ordered fashion within the 3D space of the nucleus.

Examples of how transcriptional activity is reflected in three-dimensional domain structure are the α-globin gene and the HoxD cluster (Baù et al. 2010; Noordermeer et al. 2011). At a larger scale, differentiation events correlate with spatial reorganization of chromatin; examples are the variations in LADs during neural differentiation of embryonic stem (ES) cells (Peric-Hupkes et al. 2010) or the changes in chromatin contacts that accompany B-cell development (Lin et al. 2012). By segregating genes encoding regulators of developmental competence (Kohwi et al. 2013) or cell lineage commitment (Lin et al. 2010) to transcriptionally inert regions (as in compartments B), the stability and direction of developmental processes are insured. Then, upon differentiation signals, activating transcription factors confer transcriptional competency to a previously silent compartment. Contacts between enhancer–promoter and promoter–promoter (Li et al. 2012; Lin et al. 2012) within TADs as well as with those in adjacent TADs coalesce into spatially discrete RNA pol II-enriched sites, possibly coinciding with transcription factories (Chakalova et al. 2005; Cook 2010). Inactive genes in these TADS, however, would locate away from the factories, in a configuration

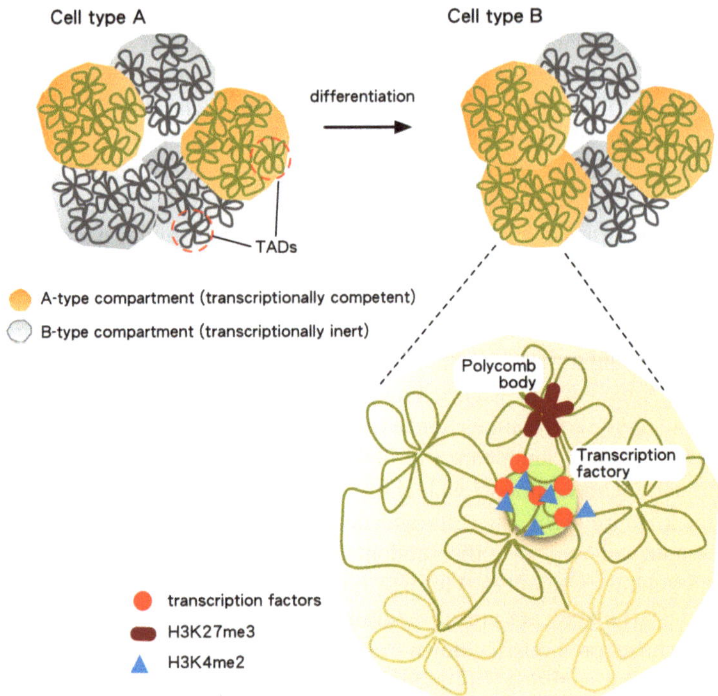

Fig. 5.1 Simplified overview of Polycomb repression and chromatin topology. Chromatin is segregated in large compartments depending on transcriptional activity. Within these compartments chromatin is folded in much smaller architectural units (Topological Associating Domains, or TADs) regardless of transcriptional status. CTCF and cohesins (not shown) delineate and sustain contacts at TADs boundaries. Differentiation cues resulting in differentiation of cell A into cell B concur with acquisition of transcriptional competence that allows coordinated activation of loci (organized in tissue-specific chromatin interactions) within a given TAD. Association of repressed genes is (reversibly) stabilized by Polycomb proteins, whereas transcription factor-dependent association between promoter/enhancer within TADs and with those in other TADs stabilizes association into regions of localized transcription (transcription factories). Only two of the associated histone marks, characteristic of repressed and active genes, are indicated. By stabilizing contacts between not activated loci, Polycomb contributes to decrease undesired fluctuations in gene expression. While robust, the silent state of Polycomb targets is responsive to developmental programs

characterized by H3K27me3 enrichment (Lin et al. 2012). Figure 5.1 depicts a simplified view of chromatin organization linking changes in transcription status and nuclear location during differentiation. Clustering of silent loci is often visualized as speckled areas enriched in Polycomb products known as Polycomb bodies (Mao et al. 2011). Contacts between Polycomb-repressed genes (Bantignies et al. 2011) in Polycomb bodies and their contribution to functional spatial segregation within the topological organization of chromatin are well documented in flies (Delest et al. 2012).

5.2.3 Control of Gene Expression by Regulation of RNA sPolymerase II Activity

Some correlative evidence links the presence of Polycomb products on promoters to an essential step in the regulation of RNA pol II activity: pausing transcriptionally engaged polymerase to prevent productive elongation (Core et al. 2012; Rahl et al. 2010). On a majority of promoters, RNA Pol II is stalled by the activity of the negative elongation factor (NELF) and DRB sensitivity inducing factor (DSIF) or pausing factors (Adelman and Lis 2012; Levine 2011; Zhou et al. 2012). Following binding in an initially hypophosphorylated state, cyclin-dependent kinase 7, a subunit of general transcription factor complex TFIIH, phosphorylates serine 5 (S5P) in the multicopy (52 times) heptapeptide YSTSPS located at RNA pol II C-terminal region. Along with this modification, a short nascent transcript is synthesized, 7-methyl-guanosine added to its 5′ end, and then pausing factors halt elongation. Release from the paused state into full elongation occurs when cyclin-dependent kinase 9 (Cdk9), a subunit of P-TEFb complex, phosphorylates (and inactivates) DSIF, NELF, and also serine 2 of RNA pol II (S2). In vivo imaging shows Cdk9 co-localization in transcription factories, with the paused (S5P) form of RNA pol II, but no so much with the form engaged in processive polymerization (S2P) (Ghamari et al. 2013). In mammalian pluripotent cells, developmental loci repressed by Polycomb bind the nonproductive form of RNA pol II phosphorylated at S5, but not at S2 (Brookes et al. 2012).

5.3 Polycomb-Mediated Posttranslational Modifications

5.3.1 Polycomb-Specific Histone Modifications

Catalytic activities in Polycomb subunits are essential for gene repression and other functions. Substrates of Polycomb-dependent posttranslational modifications include principally histones, but also a variety of other proteins.

In addition to histone H3 methylation (H3K27me3), Polycomb complexes mono-ubiquitylate the C-terminal region of histone H2A (H2AUb1), at lysine 119. The enzymes responsible for these modifications reside in separate biochemical entities or Polycomb-Repressive Complexes (PRCs). Histone ubiquitylation activity resides in PRC1 complexes, whereas histone methyltransferase (HMTase) belongs to PRC2 complexes (the number reflects that the complex was isolated after PRC1). The precise function of these and other histone modifications is intensely debated. A "histone code," as determined by specific combinations of histone modifications, would reflect instructions for transcription changes (Strahl and Allis 2000). Thus, some histone marks are considered as "activating" and other "repressing." The enrichment in both marks, H3K4me3 (activating) and

H3K27me3 (repressing), at Polycomb-silenced promoters underlies, in part, their naming as bivalent regions (Bernstein et al. 2006a). Beyond the semantic part of the argument, other authors propose that histone modifications are primarily determined by transcription and chromatin remodeling (Henikoff and Shilatifard 2011). Certainly, specificity can be appreciated in the binding of chromatin complexes to regions with particular combinations of histone marks (Musselman et al. 2012). However, the complexity of these combinations is rather limited, as stated by the small number of chromatin states observed. Therefore, more important than directing binding, histones modified in one or another way probably allosterically influence the activity of chromatin regulatory proteins (Rando 2012). Indeed, Polycomb HMTase is just one example (see below).

5.3.2 Polycomb Methyltransferases

In mammalian cells, a PRC2 complex containing Enhancer of Zeste homolog 2 (EZH2), Suppressor of Zeste 12 homolog (SUZ12), and Embryonic Ectoderm Development (EED) marks in vitro nucleosomes with H3K27me3 (Cao et al. 2002; Kuzmichev et al. 2002). Of all subunits in the complex, which also contained AE-binding protein 2 (AEBP2) and the retinoblastoma binding protein 4 (RBBP4/RbAP48), only EZH2 contains a SET domain, characteristic of most lysine methyltransferases. A similar complex, containing the ortholog E(Z), was identified in Drosophila (Czermin et al. 2002; Müller et al. 2002). Complexes in mammalian cells containing the paralog EZH1 also show H3K27-specific HMTase activity (Margueron et al. 2008; Shen et al. 2008) and, in some contexts, as in ES cells, EZH1 and EZH2 are functionally redundant (Shen et al. 2008). Additionally, mammalian EZH2 in a PRC2 variant has been shown to methylate in vitro lysine 26 of linker histone H1 (Kuzmichev et al. 2004).

H3K27me3 is the hallmark of Polycomb activity, although how mechanistically it is linked to transcriptional silencing actually is still unclear. SET domain deletion in EZH2 drastically decreases H3K27me3 levels (Shen et al. 2008). Important new evidence strongly supports that unmodified histone H3K27 is the in vivo substrate of Polycomb methyltransferase and that gene repression is linked to methylation: using Drosophila as a model, the deletion of the gene encoding histone H3 and subsequent complementation with unmethylatable K27R variant were found to phenocopy the E(Z) mutation (Pengelly et al. 2013). This demonstrated that Polycomb-dependent repression is inexorably linked to H3K27 methylation. For some targets at least, this function may be linked to PRC1 recruiting (Cao et al. 2002) (see below).

EZH2 HMTase activity depends on its association with subunits EED, SUZ12, RBBP4, and AEBP2. Some of these subunits sense chromatin structure through specific histone contacts so that H3K27me3 nucleosomes stimulate and H3K4me3 or H3K36me2,3 nucleosomes inhibit EZH2 activity (Ciferri et al. 2012; Margueron et al. 2009; Schmitges et al. 2011). HMTase substrate specificity is determined by the SET domain, as indicated by mutations Y641F or A677G, which make H3K27me2 a preferred substrate rather than H3K27me0 and H3k27me1 used by wild type EZH2 (McCabe et al. 2012b; Sneeringer et al. 2010). Interestingly, these mutations were identified in patients with B-cell lymphoma and correlate with augmented H3K27me3 levels (McCabe et al. 2012a; Sneeringer et al. 2010).

The recent modeling of the three-dimensional structure of PRC2 has helped to explain the contrasting effects of interactions with the chromatin landscape. Critical contacts between the SET motif and the SANT domains of EZH2 are thought to respond to conformational changes in EED and SUZ12, the samplers of histone H3 methylated at K27 or K4/K36, respectively (Ciferri et al. 2012). The model also explains why EZH2 catalytic activity is prevented only on the K27 that resides in the same histone tail with methylated K4 or K36 (Voigt et al. 2012). AEBP2 contacts all other PRC2 subunits assisting in its integrated responses. Thus, PRC2 appears to be a catalytic device with intrinsic ability for spreading repression-compatible histone modifications towards adjacent nucleosomes until it is confronted with inhibitory signals from transcriptionally active regions.

H3K27 methylation is reversed by the action of specific members of the family of Jumonji C (JMJC) demethylases [for more details, see a recent review (Kooistra and Helin 2012)]. KDM1 lysine (K)-specific demethylase 6B (KDM6b/JMJD3) and lysine (K)-specific demethylase 6A (KDM6a/UTX) remove methyl groups from H3K27me3 and H3K27me2 up to the mono-methylated form. Only the Jumonji C domain-containing histone demethylase 1 homolog D (JHDM1D/KDM7A) demethylates H3K27me1 (and other methylated histones too). H3K27 demethylases are recruited to Polycomb targets in pluripotent cells for differentiation-required gene activation (Agger et al. 2007; Lan et al. 2007; Lee et al. 2007). However, often they are associated with active sites, counteracting any EZH2 activity that could interfere with gene expression (Dahle et al. 2010; De Santa et al. 2009). These JMJC proteins, however, can also act independently of their activity as demethylases, for instance, localizing elongation factors to active genes (Chen et al. 2012).

5.3.3 Polycomb H2A Mono-ubiquitin Ligases

Polycomb-dependent histone mono-ubiquitylation of histone H2A, a modification found on 5–15 % of total H2A in mammalian cells (Goldknopf et al. 1975), was identified through biochemical fractionation and following the catalytic activity responsible for the modification (Wang et al. 2004). The addition of the 76 amino acid Ubiquitin (Ub) polypeptide is mediated by an activating enzyme (E1) that transfers Ub to one of several conjugating enzymes (E2); subsequently, E2-Ub associate with a third component, the so-called E3 ligase, that brings in proximity the substrate for ubiquitylation [recently reviewed (Komander and Rape 2012)]. H2A ubiquitylation copurified with a PRC1 complex and functional testing of individual PRC1 subunits found most activity on the RING-finger protein RING1B/RNF2. This was consistent with the known role of RING-finger proteins as E3 ligases. Other Polycomb RING-finger proteins were present in the complex, but only the RING1 paralogs (RING1A/RING1 and RING1B/RNF2; SCE in Drosophila) act as E3 mono-ubiquitin ligases. The other RING-finger subunits (members of the family of Polycomb group ring finger (PCGF) proteins) function as positive cofactors in the ubiquitylation reaction (Cao et al. 2005; Wang et al. 2004). Thus, Polycomb E3 ligases, as other RING-finger E3 ligases, act as dimers of RING-finger proteins. In vitro studies show that UBCH5C/UBE2D3 is the preferred E2 element in H2A mono-ubiquitylation (Buchwald et al. 2006). Structural studies show that UBCH5C/UBE2D3 associates with RING1B through an interface resulting from the folding of the RING finger, away from the region that binds PCGF subunits (Bentley et al. 2011; Buchwald et al. 2006; Li et al. 2006). Binding to the nucleosome substrate involves DNA and an acidic patch on histone H4 that contact a basic interface demarcated by a RING1B-BMI1/PCGF4 dimer (Bentley et al. 2011). Pairs of RING1-PCGF proteins are the defining unit PRC1 complexes (see below). It is generally assumed that the E3 ligase activity lies mostly with RING1B/RNF2; however, both in vitro (Buchwald et al. 2006) and in vivo evidence (de Napoles et al. 2004) demonstrates that RING1A/RING1 also acts as an E3 ligase.

Polycomb RING1 proteins are the major histone H2A ubiquitin ligases, as shown by the undetectable levels in cells depleted from these proteins (de Napoles et al. 2004). Likewise, SCE is the major H2A ubiquitin ligase in Drosophila (Gutierrez et al. 2011). However, in some contexts additional E3 ubiquitin ligases mono-ubiquitylate histone H2A. For instance, RNA-binding RING-dependent ubiquitin protein ligase (hRUL138/DZIP3) acts as part of a NCoR-HDAC complex that represses chemokine genes (Zhou et al. 2008) or ubiquitin protein ligase E3 component n-recognin 2 (UBR2) that modifies histone H2A during spermatogenesis (An et al. 2010). Also, the Cullin4B-Ring E3 ligase complex (CRL4B), a member of the family of cullin-RING E3 ligases (Jackson and Xiong 2009), has been shown to mono-ubiquitylate histone H2A in cancer cells (Hu et al. 2012), an unexpected observation given its inability to modify nucleosomal H2A in vitro (Wang et al. 2006). The histone variant H2A.Z (H2Av in Drosophila) is found at the silent X-chromosome but also in transcriptionally active regions and in Polycomb-regulated bivalent domains [not in stably Polycomb-silenced sites, though (Creyghton et al. 2008; Ku et al. 2012)]. It can also be mono-ubiquitylated in a RING1-dependent manner (Ku et al. 2012; Sarcinella et al. 2007). Interestingly, H2A.Z ubiquitylation occurs not only at lysine 120 (equivalent to H2A K119) but also at lysines 121 and, to a less extent, 125 (Ku et al. 2012).

What are the consequences of H2A mono-ubiquitylation on transcription? Correlative evidence shows a link between histone Polycomb-dependent H2AUb1 and gene repression in ES cells. Thus, upregulation of gene expression concurrent with H2AUb1 loss in RING1-deficient cells is rescued by wild type RING1B but not by catalytically inert forms (RING1B mutants I53S or I53A) (Endoh et al. 2012).

H2AUb1 dependent and independent Polycomb repression is also seen in Drosophila (Gutierrez et al. 2011). Mechanistically, the question remains to this day without clear answer. In vitro, H2AUb1 nucleosomes are not efficiently tri-methylated at histone H3K4, and this results in transcription initiation failure (Nakagawa et al. 2008).

Regardless of the silencing mechanism, the correlation between gene repression and histone H2AUb1 modification is generally consistent with activation associated with ubiquitin proteases that remove the Ub moiety from histone H2A (Joo et al. 2007; Zhu et al. 2007). Histone H2A deubiquitinating enzymes are a large and structurally diverse set, some acting on several substrates, in addition to H2A. They are members of the family of Ub-specific proteases [USP10 (Draker et al. 2011), USP12 (Joo et al. 2011), USP16 (Joo et al. 2007), USP21 (Nakagawa et al. 2008), USP22 (Zhao et al. 2008b), and USP46 (Joo et al. 2011)], of the Ub C-terminal hydrolases [Brca1-associated protein 1(BAP1) (Scheuermann et al. 2010)], and of the JAB1/MPN/Mov34 metalloenzyme (JAMM) metalloproteases [myb-like, SWIRM and MPN domains 1 (MYSM1) (Zhu et al. 2007)]. Of these, at least USP10 also deubiquitinates H2A.Z (Draker et al. 2011). Another protease, USP16/UBP-M, is responsible for the deubiquitination wave that accompanies mitosis (Joo et al. 2007). It appears that these proteases function in a local context. For instance, in prostate cancer cells, MYSM1, as part of a histone acetyltransferase (HAT)-containing complex, activates androgen receptor (AR)-regulated genes, in a process coupled to removal of linker histone H1 (Zhu et al. 2007). In hematopoietic cells, MYSM1 associates with BRAHMA/SMARCA2, an ATPase of the SWI/SNF type of chromatin remodelers, to activate the B-cell lineage transcription factor EBF1 (Jiang et al. 2011b). These results indicate that MYSM1 and perhaps other H2A deubiquitinases act as part of varied complexes involved in transcriptional activation. However, not every H2A deubiquitinase participates in gene activation. In Drosophila, inactivation of H2A ubiquitin protease Calypso (the homolog in mammals is BRCA1-associated protein 1, BAP1) results in loss of repression at a subset of Polycomb targets (Gutierrez et al. 2011; Scheuermann et al. 2010). Calypso, together with the Polycomb member Additional sex combx (ASX), is part of a Polycomb-repressive deubiquitinase complex (PR-DUB) complex that associates with Polycomb response elements [PREs, DNA sequences that recruit Polycomb complexes (see below)] (Scheuermann et al. 2010). In the absence of Calypso, ubiquitylation and deubiquitylation cycles, a process that has been proposed as necessary for repression, cannot take place. In mammalian cells, BAP1 may function independently of its in vitro H2A-deubiquitylating activity (Scheuermann et al. 2010). Its major impact may result from its ability to stabilize other regulators such as host cell factor-1 (HCF-1) and O-linked *N*-acetylglucosamine transferase (OGT) (Dey et al. 2012) (see below). In agreement with this, Polycomb-dependent repression of Hox genes is not affected by BAP1 inactivation (Abdel-Wahab et al. 2012).

5.3.4 Other Histone Modifying Activities

Some of the subunits in Polycomb complexes not identified genetically as Polycomb products are also histone modifiers. Among them is FBXL10/KDM2B, a DNA-binding protein involved in PRC1 recruiting (see below). FBXL10/KDM2B has a JMJC domain that can demethylate histone H3K36 (He et al. 2008) and H3K4 (Frescas et al. 2007), although how influential this activity is in gene control is not established.

5.3.5 Non-histone Substrates of Polycomb Enzymes

The catalytic activities of Polycomb complexes are not restricted to histones. Even the well-known histone modifiers EZH2 and RING1B/RNF2 have been shown to act on non-histone substrates. An example is the EZH2-dependent methylation of transcription factor GATA4, a modification that weakens its binding to HAT p300 and thus reduces its activating ability (He et al. 2012). Another substrate is transformation-related protein 53 (TRP53) poly-ubiquitylation by RING1B/RNF2 in some tumor cells (Su et al. 2013).

5.3.6 SUMO Modification

Small ubiquitin-like modifier (SUMO) family proteins alter the function of covalently bound substrates analogously to ubiquitylation. SUMO modifications also occur in a stepwise manner: an E1 activating enzyme transfers SUMO polypeptide to the E2 ligase (ubiquitin-conjugating enzyme E2I/UBC9) which upon binding to a substrate-bound E3 adaptor links the SUMO moiety to the substrate [reviewed in Geiss-Friedlander and Melchior (2007)]. The activity of PRC1 subunit chromobox 4 (CBX4/PC2) as a SUMO adaptor was found serendipitously in cotransfection assays with C-terminal-binding protein 2 (CTBP2), an interacting partner known to be SUMOylated (Kagey et al. 2003). Besides CTBP2, CBX4/PC2 SUMOylates a variety of substrates, including de novo DNA methyltransferase 3a (Dnmt3a) (Li et al. 2007), CTCF (MacPherson et al. 2009), or homeodomain interacting protein kinase 2 (HIPK2) (Roscic et al. 2006). CBX4/PC2 itself can be SUMOylated and together with UBC9 and other modified substrates localizes at nuclear bodies enriched in Polycomb products, or Polycomb bodies (Kagey et al. 2003). CBX4/Pc2 SUMOylation regulates PRC1 assembly on chromatin, as deduced from the increased association of complexes containing hyperSUMOylated CBX4/PC2 in tissues deficient in the SUMO-specific protease 2 (Senp2) (Kang et al. 2010). A similar positive effect on Polycomb association is seen upon SUMOylation of *C. elegans* Polycomb protein SOP-2 (Zhang et al. 2004). In contrast, as a puzzling observation, sumoylation of SOP-2 homolog in Drosophila, Sex Comb on midleg (SCM), is linked to decreased binding to PREs and repressing activity (Smith et al. 2011). These are examples of profound impact on Polycomb complexes mediated by reversible posttranslational modification of their subunits.

5.3.7 Protein Glycosylation

The addition of a single O-linked *N*-acetylgucosamine to serine or threonine residues is a posttranslational modification of functionally diverse proteins, including many important transcriptional regulators [reviewed in Hanover et al. (2012)], among them Drosophila Polyhomeotic (PH) (Gambetta et al. 2009). In fly embryonic tissues, the maintenance of Polycomb-dependent repression is lost in mutants lacking O-linked GlcNAcylation, explaining that the gene encoding the O-linked *N*-acetylglucosamine transferase (OGT), Super sex combs (SXC), is categorized as a Polycomb gene (Gambetta et al. 2009; Sinclair et al. 2009). O-GlcNacetylated proteins are found at Polycomb Regulatory Elements (PRE) DNA sequences. However, while global PH binding decreases in SXC mutant cells, neither H3K27me3 marks of E(Z) occupancy are affected (Gambetta et al. 2009). The full elucidation of OGT impact on Polycomb function needs further studies.

5.4 Polycomb Biochemical Entities

Polycomb complexes are conveniently categorized into PRC1 and PRC2 classes, that not only contain non-overlapping sets of subunits but are enzymatically characterized by their abilities to modify histones H2A (PRC1) or H3 (PRC2). Although biochemically heterogeneous, a minimum set of subunits or complex core is strictly required for their enzymatic activities and is shared among complexes within the same class. Other subunits add regulatory functionality to PRC1 and PRC2, although for many of them their roles have not been determined. A detailed description of known complexes is included in this book in Chap. 6. Here, I present a brief overview of PRC-specific complex cores and additional subunits, focusing on protein motifs related to their activities.

5.4.1 PRC2 Complexes

The organization and regulation of PRC2 has recently been reviewed (O'Meara and Simon 2012). A functional Polycomb HMTase consists of: the catalytic subunit (paralogs, EZH1 and EZH2), histone binding modules (RBBP4/RAbp48, EED), and regulator (SUZ12) and scaffold (AEBP2) components. EED and RBBP4 are proteins with propeller-like folded WD40 repeats, a structure found in other histone binding proteins. SUZ12 has a VEFS domain (an acidic cluster and a tryptophan/methionine-rich sequence named after its presence at the C-terminal region of proteins VRN2-EMF2-FIS2-Su(z)12) which is essential for HMTase inhibition. EZH paralogs contain, in addition to a lysine methyltransferase SET domain, two SANT domains. From the above described model for the core PRC2 complex between two nucleosomes (Ciferri et al. 2012) it appears that EED binding to histone H3K27me3 contacts a SANT domain to allosterically activate EZH2 (Margueron et al. 2009); conversely, RBBP4-bound histone H3K4me3 or H3K46me3 inhibits EZH2 (Schmitges et al. 2011) through contacts mediated by SUZ12. AEBP2 contacts all other subunits and its three zinc fingers hold potential for DNA binding (Kim et al. 2009). The model suggest that the presence of EED isoforms, differing at their N-terminal region (Kuzmichev et al. 2005), could be functionally relevant given its contact with EZH2 SANT domain. The PRC2 core is organized as a regulatory unit whose stability is crippled in the absence of some subunits, as seen after depletion of EED or SUZ12 (Montgomery et al. 2005; Pasini et al. 2004).

Non-core PRC2 subunits are mostly involved in PRC2 interaction with histones. These include the Plant homeodomain (PHD) proteins of the Polycomb-like (PCL) family: PHD finger protein 1 (PHF1/PCL1), metal response element binding transcription factor 2 (MTF2/PCL2) and PHD finger protein 19 (PHF19/PCL3) and jumonji, AT-rich interactive domain 2 (JARID2). One or another PCL subunit facilitates association with H3K36me3 regions through their PH domains and JARID2 plays important roles in PRC2 binding and modulation of its activity.

5.4.2 PRC1 Complexes

The core element of PRC1 complexes is a heterodimer of RING-finger proteins: a E3 ligase for histone H2A mono-ubiquitylation (either RING1A or its paralog RING1B) and a member of the Polycomb group of Ring-Finger (PCGF) family, which act as a positive cofactor. A variable number of additional subunits, in distinct sets, associate with core elements defined by each of the six PCGF proteins (Gao et al. 2012; Gearhart et al. 2006; Levine et al. 2002; Ogawa et al. 2002; Sánchez et al. 2007).

PRC1 complexes have been named after the PCGF member present. Thus, complexes with PCGF2/MEL18 or PCGF4/BMI1 were termed PRC1.2 and PRC1.4, respectively, and are considered the canonical PRC1 complex. Characteristically, these PRC1 complexes, but not others, contain Polyhomeotic-like paralogs (PHC1, PHC2, PHC3), proteins with a sterile alpha motif (SAM) widely used domain in protein–protein interactions (Qiao and Bowie 2005) which are instrumental in Polycomb repression (Isono et al. 2013); additional PRC1.2 and PRC1.4 subunits with SAM motifs are the Sex comb on midleg paralogs (SCML1,

SCML2), one of which (SCML2) also has a malignant brain tumor (MBT) motif, a binding domain for methylated histone H3K9 (Bonasio et al. 2010). Another feature of PRC1.2 and PRC2.4 is the presence of one or more paralogs of the CBX family of N-terminal chromodomain-containing proteins (CBX2/M33, CBX4/PC2, CBX6, CBX7 and CBX8), the homologs of Drosophyla Polycomb. Chromodomains, as MBT repeats, recognize histone methylated at lysines, and those in CBX proteins preferentially bind tri-methylated H3K27 (Bernstein et al. 2006b; Fischle et al. 2003).

While PCGF and RING1 proteins associate through their N-terminal RING-finger motifs, the C-terminal region of RING1 proteins interacts with a conserved Polycomb repressor box at the C-terminal region of CBX proteins (Satijn et al. 1997; Schoorlemmer et al. 1997). That same RING1 region binds the RING1 and YY1-binding protein (RYBP) (García et al. 1999) and its paralog YY1-associated factor 2 (YAF2) (Kalenik et al. 1997). RING1 proteins bind either CBX or RYBP exclusively (Wang et al. 2010). This probably explains why the other PRC1 complexes (PRC1.1, PRC1.3, PRC1.5, and PRC1.6) contain, instead of CBX subunits, RYBP or YAF2 subunits (Gao et al. 2012). The RING1-PCGF1/NSPC1 core is found with KDM2B (a DNA-binding protein) and BCOR paralogs (Gearhart et al. 2006; Sánchez et al. 2007); PRC1.6 contains RING1-PCGF6/MBLR; heterodimers DP1-E2F6 and MAX-MGA that bind DNA sequences for E2F sites and E2 boxes, respectively; the MBT-repeat protein l(3)mbt-like 2 (L3MBTL2) and other subunits (Ogawa et al. 2002); PRC1.3 and PCR1-5, finally, are defined by heterodimers RING1-PCGF3 and RING1-PCGF5 and contain, yet, additional subunits. Altogether, PRC1 complexes are far more heterogeneous than PRC2. PCGF subunits bind chromatin in partially overlapping patterns (Gao et al. 2012), suggesting distinctive activities for PRC1 complexes, although this remains largely unknown.

5.4.3 Other Complexes with Polycomb Subunits

While simplified PRC1 forms and PRC2 are recognizable in Drosophila, other complexes found in flies seem not to have corresponding homologs in mammals. A protein assembly recently isolated containing Sex comb on midleg with four MBT domains (SFMBT) homologs is proposed to be the counterpart of PHO-repressive complex (PHO-RC), a heterodimer of PHO and SFMBT proteins (Klymenko et al. 2006). The mammalian complex contains additional subunits, including well-known chromatin modifiers as LSD1 and COREST (Zhang et al. 2013). Analogously to PHO-RC, mammalian SMFBT complexes also interact with PRC1 (Zhang et al. 2013).

As mentioned earlier, Drosophila PR-DUB complex contains ubiquitin protease Calypso and ASX (Scheuermann et al. 2010). Calypso homolog in mammalian cells, BAP1, also associates with homologs ASXL1 and ASXL2, but unlike Drosophila PR-DUB, they form part of much diverse biochemical entities (Dey et al. 2012).

5.5 Targeting Polycomb Function

Transitions between cell states, from pluripotent to more differentiated cell types, are accompanied by changes in the genomic regions marked by Polycomb activity (Bracken et al. 2006; Mohn et al. 2008). In Drosophila cells, nucleosomes at Polycomb-targeted promoters are in a highly dynamic state (Mito et al. 2007) and steady-state histone modifications requires continued Polycomb recruitment. Indeed, Polycomb association with chromatin, as measured by live imaging (FRAP), shows very short residence times, within the same range as transcription factors (Steffen et al. 2012). Of note, exchange rates are highest at pluripotent cells and tend to slow down in more mature cells (Fonseca et al. 2012). During differentiation, Polycomb colonization of new sites is accompanied by eviction from sites destined to be derepressed, reflecting a different outcome of antagonic influences on Polycomb association at these sites. In contrast, at stably silenced regions, Polycomb presence probably is maintained by a lower rate of chromatin remodeling and the spreading of Polycomb-modified nucleosomes, thereby contributing to the developmental restriction that goes with cell differentiation (Zhu et al. 2013). In some cases, however, loci silenced by Polycomb progressively acquire a stably silent state maintained by Polycomb-independent means, generally involving DNA methylation (van Arensbergen et al. 2013).

How Polycomb complexes are directed to their targets is a subject of intense research. Seminal work with pluripotent mammalian cells has mapped PRC1 and PRC2 binding preferentially to promoters of loci encoding developmental regulators (Boyer et al. 2006; Lee et al. 2006). These promoters are located in a subset of specialized, methylation-free GC-rich sequences (CpG islands, CGI) (Ku et al. 2008; Mikkelsen et al. 2007). Nucleosomes at these sites are enriched in H3K4me3 and H3K27me3 marks, usually thought of as "activating" and "repressing" marks. In general, these loci show little or no expression in pluripotent cells. However, upon differentiation their status changes and promoters retain one or another mark depending on activation or silencing of the locus in the new cell state (Azuara et al. 2006; Bernstein et al. 2006a; Cui et al. 2009; Mikkelsen et al. 2007). Indeed, removal of H3K27 methylation through EED inactivation results in derepression of these promoters (Boyer et al. 2006); on the other hand, decreased H3K4 methylation at these promoters, upon downregulation of dpy-30 homolog (DPY30), a subunit of SET1/MLL complexes, interferes with transcriptional activation needed at genes induced during differentiation(Jiang et al. 2011a). It has been proposed that such a singular chromatin configuration (bivalent domains) (Bernstein et al. 2006a) allows genes encoding developmentally relevant transcription factors and signaling molecules to be silent while poised for activation. Polycomb regulation in Drosophila, however, occurs in the absence of CGIs or "bivalent domains." Instead, functionally similar regions are identified, bound by Polycomb and Trithorax (TrxG) products (some of which are MLL homologs). These regions are thought to be in a "balanced" state and—although enriched in H3K27me3—have no H3K4me3 marks (Gaertner et al. 2012; Schwartz et al. 2010). Recently, ChIP studies in *D. melanogaster* showed that in addition to transcriptionally silent loci, PRC1 subunits also bind transcriptionally active promoters co-occupied by cohesins, where they participate in promoting expression from these loci (Schaaf et al. 2013b).

The association of Polycomb complexes with chromatin is influenced by DNA-binding proteins, noncoding RNAs, and interactions with resident proteins such as histones. It is conceivable that the nature of these associations and the possibility of their mutual reinforcement determine the overall avidity of binding. Therefore, while recruiting has been usually considered to be instructed, for instance, by proteins or RNAs recognizing specific DNA sequences, it is becoming increasingly accepted that Polycomb association with targets is a consequence of chromatin sampling, thereby being responsive to transcriptional status (Klose et al. 2013). First, I will discuss mechanisms that influence binding of Polycomb complexes to its targets and then their maintenance or eviction.

5.5.1 Polycomb Recruiting Through DNA-Binding Proteins

With the exception of Drosophila Pleiohomeotic (PHO) and its paralog (PHO-L) genetically defined Polycomb products lack ability to bind DNA (PHO-L) (Brown et al. 1998, 2003). PHO, PHO-L and its vertebrate homolog YY1 transcription factor (YY1) bind DNA through four conserved zinc-finger motifs (Brown et al. 1998). In mammals, however, evidence for YY1-dependent association of Polycomb proteins to targets is limited (Woo et al. 2010) and it appears likely that YY1 cannot be considered as a general Polycomb recruiter in mammals (Mendenhall et al. 2010). In Drosophila, Polycomb-repressive elements (PREs), genomic regions with sites for PHO and other DNA-binding proteins recruit Polycomb complexes and mediate repression of transgenic constructs and endogenous targets (Müller and Kassis 2006). Other DNA-binding proteins functionally linked to Polycomb silencing are GAGA factor (GAF), Dorsal Switch Protein 1 (DSP1), Pipsqueak (PSQ), Grayny Head-like (GRH), Zeste, and SPPS (a member of the Sp1/KLF family of zinc-finger proteins) (Ringrose and Paro 2007). Polycomb recruiting to PREs is most likely indirect, through subunits that interact with DNA-binding proteins, as illustrated by Polycomb (PC) association with PSQ and GRH (Strübbe et al. 2011). PRE-like sequences are hardly known in mammalian cells (Sing et al. 2009; Woo et al. 2010). However, comparative mapping of H3K27me3-marked regions and RNA transcripts in a neural differentiation model identifies intergenic sequences (Transcribed Intergenic Polycomb sites, TIPs) which might be analogous to intergenic PREs in Drosophila (Hekimoglu-Balkan et al. 2012). At any rate in Drosophila cells, in addition to PREs, PRC1 proteins bind, facilitated by cohesins, many promoters (Enderle et al. 2011), although in this case not for silencing functions (Schaaf et al. 2013b).

5.5.2 *Proteins Binding GC-Rich DNA as Recruiters of Polycomb Complexes in Vertebrates*

In mammalian pluripotent cells, EZH2 and SUZ12 occupy CGI regions (Ku'08), unusual genomic domains which are unmethylated genomic domains interdispersed in a landscape of methylated DNA (Deaton and Bird 2011; Illingworth and Bird 2009; Stadler et al. 2011). About 70 % of mammalian promoters, including many at intergenic sites are contained within CGIs (Illingworth et al. 2010). Gene expression, divergent transcription, RNA pol II pausing, and nucleosome destabilization, all of them features of a permissive chromatin state concur at CGIs (Blackledge and Klose 2011; Core et al. 2008; Deaton and Bird 2011; Fenouil et al. 2012). Recent work shows that CGI-like, non-methylated Polycomb marked regions are present throughout vertebrates and, therefore, are not unique to warm-blood vertebrates as previously thought (Long et al. 2013b). PRC1 subunits also locate to CGI, although co-localization with PRC2 products is restricted to the subset of larger size CGIs (Ku et al. 2008). Gene bodies of Polycomb-repressed genes in ES cells are marked by H3K27me3 and H2AUb1, but enrichment peaks map close to the transcription initiation site (TSS) (Brookes et al. 2012).

To test whether the prevalent location of Polycomb complexes at CGI is mediated by DNA-binding proteins, computational searches for binding motifs recognized by transcription factors yielded a reduced number of sites for repressors, mostly expressed in differentiated cells, i.e., nonfunctional in ES cells. Moreover, Polycomb-bound regions showed a remarkable absence of binding motifs for transcriptional activators (Ku et al. 2008). Thus, the best predictor for Polycomb association is a high content in GC sequences (Mendenhall et al. 2010). In an alternative approach, searching in pluripotent cells for transcription factors contained in Cbx-containing PRC1 complexes, the RE1-silencing transcription factor (REST) was identified (Dietrich et al. 2012; Ren and Kerppola 2011). This DNA-binding protein that also interacts with PRC2 (Dietrich et al. 2012) was among the very few transcription factors identified during a computational search of TF motifs in Polycomb-bound CGIs in ES cells (Ku et al. 2008). However, whether REST is directly recruiting Polycomb to their targets is not clear, since even though RING1B or SUZ12 is enriched among REST binding sites, RING1B occupies only a very small subset of REST motifs (Dietrich et al. 2012) and genes derepressed upon inactivation of REST overlap only partially with those upregulated in RING1B-deficient cells (Dietrich et al. 2012). Despite this, independent experiments showed REST motifs appearing in a different computational search that combined the occurrence of predicted binding sites for transcription factors with the dynamic changes in H3K27me3 occurring during neural differentiation of pluripotent cells (Arnold et al. 2013). This study also revealed motifs for members of the SNAIL family of transcription factors that together with REST motifs were found predictive of transient H3K27me3 marks taking place during differentiation of neural progenitors. Furthermore, DNA fragments containing REST or SNAIL binding sites confer H3K27m3 enrichment to linked sequences in transgenes, demonstrating the ability of transcription factors to configure chromatin landscapes (Arnold et al. 2013).

The PRC1 subunit FBXL10/KDM2b has also been shown to be involved in recruiting Polycomb complexes. FBXL10/KDM2b has a CXXC zinc-finger motif similar to that of other proteins known to bind non-methylated CpG sequences (Long et al. 2013a). ChIP studies show that most PRC2- and PRC1-bound sites in ES cells are also enriched in FBXL10/KDM2b (Farcas et al. 2012; He et al. 2013; Wu et al. 2013) and that binding depends on the CXXC domain (He et al. 2013; Wu et al. 2013). FBXL10/KDM2b interacts directly with RING1B (Sánchez et al. 2007) and PCGF1/NSPC1 (Wu et al. 2013). Although RING1B enrichment at Polycomb targets decreases modestly when FBXL10/KDM2b is downregulated, overall levels of total H2AUb1 are clearly reduced (Wu et al. 2013). CBX7 association, in contrast, is not affected, consistent with its ability to bind H3K27me3. Thus, Polycomb proteins bound at their targets in the absence of FBXL10/KDM2b account for poor derepression (Farcas et al. 2012; He et al. 2013; Wu et al. 2013). FBXL10/KDM2b is bound to most CGIs (Wu et al. 2013), suggesting that only binding to DNA is not sufficient for recruitment of this PRC1 complex. Whether the extended contacts offered by large Polycomb-bound CGIs or the activity of additional players help locating FBXL10/KDM2b Polycomb partners to CGIs remains to be established. The use of DNA-binding proteins that recognize non-methylated DNA may explain why in hypo-methylated cells H3K27me3 marks appear at ectopic sites while their presence decreases at Polycomb targets, which concomitantly are upregulated (Reddington et al. 2013). Therefore, the activity of DNA methyltransferases and the selective recognition of methylated/unmethylated DNA may be important during the establishment of Polycomb domains after epigenetic reprogramming at the earliest stages of development.

5.5.3 Other DNA-Binding Proteins as Polycomb Recruiters

At least two PRC2 subunits with potential for DNA binding may play a role in recruiting the complex to their targets. One, JARID2, was not found in initial isolates of PRC2 complexes. JARID2 has JMJC domain related to that found in histone demethylases, although it is catalytically inactive (Tsukada et al. 2006). Several groups found that JARID2 and EZH2 or SUZ12 co-occupy a large number of genomic sites (Landeira et al. 2010; Li et al. 2010; Pasini et al. 2010a; Peng et al. 2009; Shen et al. 2009). JARID2 inactivation is accompanied by decreased PRC2 binding. However, the effects on H3K27me3 levels differ among studies and are interpreted proposing HMTase-inhibiting (Peng et al. 2009; Shen et al. 2009) or activating (Li et al. 2010; Pasini et al. 2010a) roles for JARID2. These discrepancies remain unresolved and the existence of distinct PRC2 complexes, one without JARID, responsible for most H3K27 methylation, and another with JARID2, strongly bound to DNA, has been suggested by way of explanation (Herz and

Shilatifard 2010). In vitro, JARID2 binds DNA through its AT-rich interaction domain (ARID) (Li et al. 2010), but the in vivo effect of this possible binding has not been determined.

PRC2 subunit AEBP2 is a three zinc-finger protein binds to an unusual, CTT(N) 15-23cagGCC sequence. A very small collection of genomic sites bound in brain tissue by AEBP2 was also bound by SUZ12 (Kim et al. 2009). It is not clear if all AEBP2 bound depends on its DNA-binding activity and if this capacity would serve to target PRC2 or if, on the contrary, most AEBP bound to chromatin is a part of PRC2 targeted by other means.

PRC2 recruitment is affected by loss-of-function mutations in ASXL1 (Abdel-Wahab et al. 2012). The levels of H3K27me3 and derepression of Polycomb targets also are associated with ASXL1 inactivation. ASXL1 belongs to complexes with ubiquitin protease BAP1 (Dey et al. 2012; Scheuermann et al. 2010), and although it is not found in PRC2 complexes, it co-immunoprecipitates with the PRC2 subunit SUZ12 (Abdel-Wahab et al. 2012). There is no evidence for ASXL proteins binding DNA directly, although bioinformatic analysis identifies a N-terminal domain compatible with a winged helix-turn-helix fold found in other DNA-binding proteins (Aravind and Iyer 2012).

PRC1 recruitment through DNA-binding proteins has been described in hematopoietic cells. Co-occupancy of genomic sites bound by RING1B and the runt-related transcription factor 1 (RUNX1), a heterodimeric DNA-binding protein found as a fused product in acute myeloid leukemia, has been observed in hematopoietic cells (Yu et al. 2012). Moreover, upon RUNX1 deletion, RING1B occupancy is reduced, consisting with a role for RUNX1 in Polycomb recruiting. Biochemical analysis shows that this can occur through direct interaction between PCFG4/BMI1 and RUNX1 (Yu et al. 2012).

Finally, the PRC1 subunit RYBP which binds non-specifically DNA in vitro (Neira et al. 2009) has been proposed as a mediator of PRC1 recruiting independent of binding to H2K27me3 (Tavares et al. 2012). RYBP binds many genomic sites occupied by RING1B (Gao et al. 2012; Hisada et al. 2012; Morey et al. 2013; Tavares et al. 2012) and its association with chromatin, in contrast to that of RING1B, is not affected by EED depletion (i.e., lack of H3K27me3) (Hisada et al. 2012; Tavares et al. 2012). In the absence of H3K27me3, RING1B binding is very much decreased (Leeb et al. 2010; Tavares et al. 2012), and therefore, it is difficult to evaluate the actual contribution of RYBP to PRC1 recruitment in the presence of H3K27me3. After RYBP inactivation, the extent of PcG targets occupancy by PRC1 is affected mildly (Hisada et al. 2012; Morey et al. 2013) or more substantially (Tavares et al. 2012) at the same time that H2AUb levels decrease (Gao et al. 2012; Morey et al. 2013; Tavares et al. 2012).

In summary, it is clear that the association of PRC complexes with chromatin can be facilitated by DNA-binding proteins. Of these, proteins recognizing generic DNA features (i.e., CpG-rich sequences) play a more prevalent role than conventional transcription factors. However, within specific cell lineage or developmental time contexts, these may contribute effectively to PRC recruitment to specific targets.

5.5.4 Polycomb Association with Chromatin Through Interaction with Histones

Some Polycomb subunits recognize and bind specific sites in histones. As for many other chromatin modifiers, this represents opportunities to promote binding and stabilization of its association or, on the contrary, to repel contact. These activities can be determined not only by covalent modifications at histone tails, but also by nucleosome density.

Chromobox-containing subunits of PRC1 complexes recognize and bind in vitro tri-methylated H3K27 (Bernstein et al. 2006b; Fischle et al. 2003). For a long time PRC2-dependent recruitment of PRC1 has been considered to be essential for PRC1 targeting. Chromatin binding of chromobox CBX7 PRC1 subunit is severely affected in EED-deficient (without H3K27me3 marks) cells (Tavares et al. 2012), just as it is the association of RING1B (Leeb et al. 2010; Tavares et al. 2012), presumably due to its CBX7-dependent binding, indirectly, through CBX7. However, PRC1 subunits (or H2AUb1 marks) only co-localize partially with PRC2-bound/H3K27me3-enriched sites (Ku et al. 2008). Moreover histone H2AUb1 or RING1B recruitment to the silenced X-chromosome is little affected in cells without H3K27me3 (Leeb et al. 2010; Schoeftner et al. 2006; Tavares et al. 2012). Together, these observations support the existence of alternative means for PRC1 targeting.

As described above, the methylation status of specific residues of histone H3 influences PRC2 association as well as the catalytic activity of EZH2 (Margueron et al. 2009; Schmitges et al. 2011). Thus, methylated H3K4 and H3K36 are refractory to PRC2 association, while methylated H3K27 stimulates binding and methyltransferase activity. Most likely this indicates that such contacts are mainly mechanisms by which alterations in histone modifications spread to adjacent nucleosomes. Chromatin modifiers that participate in propagation of chromatin states often act through binding to the product of the activity of the catalytic subunit, thereby enhancing the processivity of the modification (Hathaway et al. 2012). However, there is an apparent inconsistency of PRC2 HMTase inhibition by H3K4me3 (Schmitges et al. 2011) and the coexistence of bound Polycomb at nucleosomes with H3K4me3 and H3K27me3 in bivalent domains. An explanation for this finding is that these modifications are on separate H3 tails in vivo and that PRC2 inhibition only occurs when K4 and K27 marks are in a nucleosomal symmetric fashion, but not if asymmetric (Voigt et al. 2012).

5.5.5 Noncoding RNAs as Polycomb Recruiters

Noncoding RNAs (ncRNAs) are a large collection of nuclear and cytoplasmic RNAs synthesized similarly to mRNAs and that engage in a variety of regulatory functions (Batista and Chang 2013; Guttman and Rinn 2012; Mercer and Mattick

2013). ncRNAs fold in stable high-order structures which determine their function. Often they are the product of divergent transcription, a characteristic of RNA pol II promoters (Core et al. 2008; Seila et al. 2008), in which the paired transcript is a protein-coding mRNA (Sigova et al. 2013). Biochemical analysis shows molecular interactions between some ncRNAs and chromatin modifiers, including Polycomb products (Guttman et al. 2011; Khalil et al. 2009; Zhao et al. 2010).

The idea that ncRNAs may recruit Polycomb complexes to targets originated in studies about the function of a ncRNA expressed from the HOXC gene cluster, HOX Antisense Intergenic RNA (HOTAIR). Its inactivation correlates with upregulation of a segment of the HOXD cluster encoding the late-expressing genes HOXD8 to HOXD13 (Rinn et al. 2007). Moreover, this derepression is accompanied by loss of H3K27me3 and reduced SUZ12 occupancy. Since HOTAIR binds SUZ12 and EZH2, it was suggested that ncRNAS could target Polycomb-dependent repression in trans (Rinn et al. 2007). In agreement with this idea, ectopic HOTAIR expression in epithelial tumor cells results in altered distribution of H3K27me3 and PRC2 occupancy of new sites (Gupta et al. 2010). Other examples of Polycomb recruiting through ncRNAs are found at the silenced X-chromosome and some imprinted loci on mouse chromosomes 7 and 12. For example, RepA, a ncRNA encoded in the Xist locus (Zhao et al. 2008a), or ncRNAs from Kcnq1ot1 or Meg3 loci (Pandey et al. 2008; Zhao et al. 2010) also bind PRC2 products and are required for sustained H3K27me3 levels and locus silencing. In all cases, targeting occurs in cis, unlike HOTAIR which operates in trans. Binding of PRC1 complexes to other ncRNAs has also been described (Guttman et al. 2011; Yap et al. 2010). The best studied, ANRIL, an antisense transcript overlapping the Ink4 locus in human cells (encoding tumor suppressors), recruits CBX7 in cis (Yap et al. 2010). Polycomb binding to ncRNAs occurs through RNA sequences folded in complementary stem-loop structures (Zhao et al. 2008a). Rather than restricted to a few ncRNAS, a large number of them are found in pull-down assays with anti-SUZ12 and anti-EZH2 antibodies (Khalil et al. 2009; Zhao et al. 2010). In addition, many short ncRNAs, $\simeq$50–200 nt in length, associated with CGI regions, contain sequences with potential stem-loop folding that bind SUZ12 (Kanhere et al. 2010). These short ncRNAs use TSSs distinct from those of mRNAs, are expressed independently of Polycomb, and are lost from loci derepressed during differentiation (Kanhere et al. 2010). It is not known whether, as longer ncRNAs (Guttman et al. 2011; Tsai et al. 2010), they also bind other chromatin regulators.

Specific protein domains involved in ncRNA binding have not been defined, except for the chromobox of CBX7, which binds ANRIL although through residues not involved in H3K27me3 recognition (Yap et al. 2010). On the other hand, EZH2 affinity for HOTAIR is affected by cyclin-dependent kinase 1 (CDK1) phosphorylation (Kaneko et al. 2010).

Despite the known cases of ncRNA-mediated Polycomb targeting to specific genes, it is not clear whether this is a general mechanism for specific recruiting. HOTAIR activity, for instance, is not restricted to the HOXC cluster; instead many other sites are found to bind HOTAIR as identified by a Chromatin Isolation by RNA Purification (ChIRP) method (Chu et al. 2011). On the other hand, recruitment appears coordinated with other chromatin modifying activities, since a single ncRNA is able to bind at the same time Polycomb subunits and other chromatin regulators (Guttman et al. 2011; Tsai et al. 2010). It is likely that if short ncRNAs are going to act as Polycomb recruiters, they would function as a way to sense transcriptional state, rather than to identify specific targets.

5.5.6 Switching Transcriptional States at Polycomb-Regulated Targets

CGIs are genomic regions conducive to transcription initiation (Deaton and Bird 2011) and are focal points of the competition between Polycomb activity and effective transcription (Lynch et al. 2011). Histone modifications unfavorable to Polycomb residence or the recruitment of transcriptional activator complexes will switch a previously Polycomb-silenced promoter to an active state. Likewise, transcription cessation or active repression would set up a scenario for incoming Polycomb complexes to take over as silencing agents.

Polycomb function in Drosophila is antagonized by TrxG complexes (Schuettengruber et al. 2011). A TrxG subunit that provides a clue about how this may occur is the CREB-binding protein (CBP, CREBBP), a histone acetyltransferase which acetylates H3K27 (Tie et al. 2009). Its homolog in mammalian cells, CREBBP/KAT3A and the HAT E1A-binding protein p300 (Ep300) have been found to acetylate histone H3K27 (Pasini et al. 2010b). H3K27 acetylation prevents its methylation by EZH2, thus facilitating reversal of Polycomb-dependent repression. An indication of the effects caused by alterations in the relative levels of antagonic modifiers of H3K27 is the increase in H3K27ac in pluripotent cells lacking PRC2 subunit SUZ12 (Pasini et al. 2010b). Conversely, hyperactive mutant E(Z) results in reduced H3K27ac and inappropriate silencing in Drosophila embryos (Stepanik and Harte 2012). Interestingly, acetylation of histone H3K27 is a feature of active enhancers (Creyghton et al. 2010) possibly underlying Polycomb eviction associated with enhancer activation (Vernimmen et al. 2011).

Activation of Polycomb-repressed genes is often a response to developmental signals transduced through kinases (Sawarkar and Paro 2010). Some of these environmental cues are transmitted through histone phosphorylation events mediated by members of the mitogen- and stress-activated kinases (MSK), both in Drosophila and in mammalian cells. Under mitogenic stimulation, or retinoic acid-induced differentiation, MSK1 and 2 phosphorylate histone H3K27me3 at serine 28 (Gehani et al. 2010; Lau and Cheung 2011). Such a modification is accompanied by Polycomb eviction and acquisition of H3K27Ac marks. A similar activity is seen in Drosophila, where recruiting of JIL1, a MSK homolog, correlates with the establishment of H3K27acS28ph marks at promoters and enhancers (Kellner et al. 2012). Polycomb displacement resulting from H3S28 phosphorylation is effective not only in interphase, but also during prometaphase and mitosis, as seen by in vivo imaging of PC (Fonseca et al. 2012). However, the detailed mechanism of the reversion of a Polycomb-silenced state remains to be elucidated. In the likely sequence of events, early phosphorylation would promote Polycomb eviction. It is not certain that histone demethylases would play a role in this switch, at least in mammalian early development, because loss of methyltransferase and loss of demethylase correlate with phenotypes at distinct developmental times (Shpargel et al. 2012). Moreover, combined action of distinct demethylases would be required in order to fully demethylate H3K27 (Kooistra and Helin 2012) to an acetylation substrate.

For gene-specific switching to a Polycomb-repressed state, deacetylation of histone H3K27ac may be a first step. This has been documented in ES cells, where recruitment of the NuRD complex to its targets results in concurrent deacetylation and subsequent methylation of H3K27 (Reynolds et al. 2012). Alternatively, PRC2 complexes could also be recruited to transcriptionally active, H3K36me3-marked, sites, through binding of containing Polycomb-like PCL subunits via their TUDOR domains (Ballaré et al. 2012; Brien et al. 2012; Cai et al. 2013). At least in one case, H3K36me3 demethylase NO66 associated with PCL protein PHF19 (Brien et al. 2012) would initiate the transition of an active state to Polycomb-repressed state. In addition, chromatin compaction after transcription termination stimulates the HMTase activity of PRC2 (Yuan et al. 2012) and therefore assists in the establishment of a repressed state.

5.5.7 Maintenance of Histone Marks on Polycomb-Modified Nucleosomes

Specific gene expression and chromatin states are perpetuated throughout cell divisions, thereby ensuring the stability of differentiation stages. During DNA replication, the incorporation in nucleosomes of newly synthesized histones necessitates the deployment of mechanisms that propagate histone marks patterns to daughter cells (Zhu and Reinberg 2011). Maintenance processes are also demanded by nucleosome turnover that occurs at transcribed genes and active DNA regulatory elements during interphase (Henikoff 2008). Preserving histone modifications in relation with replication-independent turnover of nucleosomes could occur at least in two ways: deposition of pre-marked histones and residence of histone modifiers at the turnover site. Here, adjacent histone H3K27me3 could serve as an anchor (and catalytic activator) of Polycomb HMTase (Margueron et al. 2009; Yuan et al. 2012). Additional factors, in analogy with the ATRX helicase linking DAXX histone chaperon-dependent assembly of histone H3.3 nucleosomes (Eustermann et al. 2011), could also be involved.

In proliferating cells, H3–H4 tetramers do not dissociate during genome replication. Thus, daughter DNA strands contain both newly synthesized histones and those from parental origin (Xu et al. 2010). As parental histones contain specific modifications which are bound by complexes containing specific modification-recognition modules (i.e., EED for H3K27me3), the catalytic module (EZH1, EZH2) of such complexes would reinstate these modifications in the nucleosome. Alternatively, the association of a histone modifier with the replication machinery could ensure the modification of reformed nucleosomes. The interaction of chromatin modifiers with elements of the replicating machinery such as PCNA (Rowbotham et al. 2011) or the CAF1 chaperone (Loyola et al. 2009) has indeed been demonstrated. Similarly, EZH2 has been shown to co-localize with BrdU-labeled foci (Hansen et al. 2008), and, in Drosophila embryos, PRC2 and PRC1 subunits are in close proximity to replisome components (Petruk et al. 2012). Also, in assays in vitro, PRC1's subunits PSC, PC, and SCE are found stably associated with replicating DNA (Follmer et al. 2012; Francis et al. 2009). However, no tri-methylated H3K27 or H3K4 are found on nucleosomes repositioned some time after passage of the replication fork in Drosophila embryos (Petruk et al. 2012). This observation is consistent with those of studies in mammalian cells showing that H3K27 tri-methylation starts at S-phase and is completed only after mitosis, during G1 phase (Zee et al. 2012). Also in mammalian cells, approximately half of H3K27me3 on newly synthesized histone H3 is produced from unmodified K27 in S and G2 phases, whereas the remaining modification takes place in G1 from histones in di-methylated form (Zee et al. 2012). The stepwise nature of Polycomb HMTase action suggests that the maintenance of transcriptional states may be compatible with fluctuations at histone marks (Huang et al. 2012). PRC1-dependent modification of histone H2A also takes place during the G1 phase, after USP16-driven global deubiquitination wave in G2 and mitosis (Joo et al. 2007).

5.6 Mechanisms of Polycomb-Dependent Repression

How Polycomb impacts transcriptional activity is still an unresolved issue. Linking Polycomb abilities, i.e., catalytic activities and protein–protein interactions with gene control mechanisms has proven to be difficult. For some time, it was accepted that Polycomb repression was related to "chromatin compaction," analogous to the largely absent gene expression within "closed" heterochromatic regions. However, this turned out to be not true. Rather than being simple ON/OFF switches, Polycomb act in a dynamic fashion just as is being realized for other chromatin modifiers (Reynolds et al. 2013). In cells with a developmental potential, Polycomb complexes act on genes still capable of changing their expression state by fine-tuning their transcription status by a variety of mechanisms, while a less dynamic scenario may be at play on the large inactive Polycomb domains of differentiated cells. A summary of Polycomb complexes, biochemical activities of their subunits, and major functions is shown in Fig. 5.2.

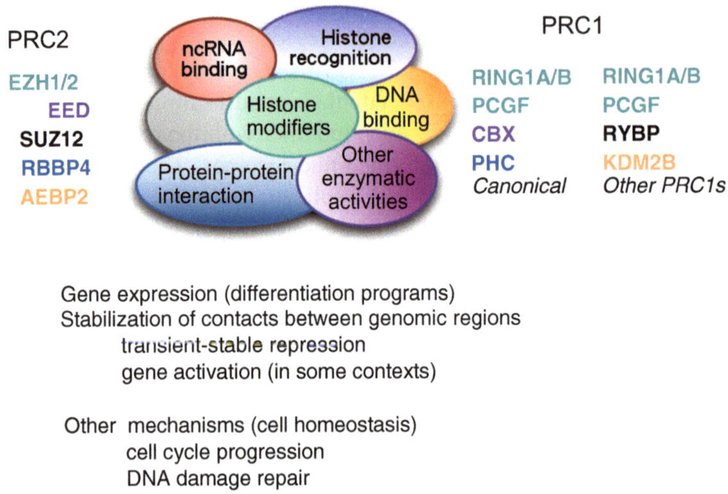

Fig. 5.2 Summary of Polycomb complexes and their activities. A hypothetical, unifying, Polycomb-repressive complex is shown, indicating a possible core of subunits and their biochemical activities. In association with targets (by histone recognition, DNA contacts, ncRNAs), reinforcing and maintenance (histone modifications) of their clustering (protein–protein interactions), and ability to become dissociated, Polycomb complexes regulate gene expression. Core subunits of PRC2 and PRC1 complexes are shown, together with their major associated activity. Generation of cell diversity and maintenance of cell homeostasis functions are categorized under transcriptional and non-transcriptional mechanisms

5.6.1 Polycomb Function and RNA Polymerase II Activity

A first hint about Polycomb action at transcription initiation was drawn from transgenic studies in Drosophila. Here, PRE repression of a heat shock promoter (known to bind paused RNA pol II prior to induction) was found to occur even in the presence of recruited RNA pol II and TFIID (general transcription factor essential for initiation). The repressed transgene was unable to produce mRNA (Dellino et al. 2004). Additional evidence, in Drosophila, pointing at a possible link between Polycomb function and RNA pol II pausing is PRC1 enrichment at stalled, proximal promoters that produce short sense transcripts in Drosophila cells (Enderle et al. 2011; Kharchenko et al. 2010; Muse et al. 2007; Nechaev et al. 2010; Zeitlinger et al. 2007). By studying muscle tissue during Drosophila embryogenesis, it was found that paused RNA pol II associates with muscle-specific promoters in a stage-specific, but not tissue-specific manner and that the repressed state correlated with tissue-specific Polycomb targeting (Gaertner et al. 2012). In this case, it would appear that polymerase release from pausing was restricted by Polycomb, although by unknown mechanisms. In *extra sex combs* embryos (mutation in the gene that encodes PRC2 subunit ESC), RNA pol II occupancy increases at many promoters, including those not bound by paused polymerase in wild type embryos (Chopra et al. 2011).

Loss of histone H2AUb1 in pluripotent ES cells after inactivation of RING1 proteins correlates with an increase in total RNA pol II bound at Polycomb-repressed promoters (Endoh et al. 2012; Stock et al. 2007). In a different experimental model, elongation inhibition was associated with the H2A ubiquitylating activity of hRUL138/DZIP3 (Zhou et al. 2008). Transcriptionally engaged RNA pol II in mammalian cells, as identified by genome-wide sequencing of run-on transcripts, peaks only at the TSS of PRC2-occupied promoters, whereas at PRC2 and PRC1 (bivalent) promoters the levels are very low (Min et al. 2011). Detailed studies of RNA pol II associated with Polycomb-repressed genes in ES cells finds a variant phosphorylated at S5 but not at S2 or S7. This RNA pol II species accumulates at TSSs, but it is also found throughout the entire transcriptional unit up to the transcription end site (Brookes et al. 2012). Loci with this unusual chromatin configuration lack H3K36me3 marks (a sign of active transcription elongation) and produce no mature mRNA. Unfortunately, molecular characterization of these promoters has not clarified yet how Polycomb would act through transcriptional pausing.

Unexpectedly, recent studies in Drosphila, however, support a role for PRC1 complexes assisting the pausing factors NELS and DSIF in polymerase modification at promoters for effective transcription (Schaaf et al. 2013b). In these studies, it was found that, in addition to the expected location on silent, H3K27me3-marked loci, PRC1 was found also on active, H3K27me3-free, genes which were also bound by cohesin (Schaaf et al. 2013b). Cohesins are known PRC1 interactors (Strübbe et al. 2011) and are required for PRC1 recruitment to active Drosophila promoters (Schaaf et al. 2013b). In addition, while cohesins associate with genes with promoter-proximal transcriptional pausing they do not, with a few exceptions, bind Polycomb-repressed loci (Schaaf et al. 2013a). For Polycomb-silenced genes, PRC1 down-regulation resulted in increases of the elongating form or RNA pol II (S2P) RNA pol II and of mRNA, in agreement with the release of a gene repression function. In contrast, active genes showed, upon PRC1 inactivation, decreased levels of total and S2 RNA pol II at gene bodies, with a concomitant reduction of mRNA levels, suggesting that PRC1 and pausing factors work together for effective transcription (Schaaf et al. 2013b).

5.6.2 Nucleosome Compaction by Polycomb

Reconstituted Polycomb complexes condense nucleosomal arrays in vitro, as determined by electronic microscopy (Francis et al. 2004). Thus, similar to HP1, high mobility proteins and others, Polycomb subunits could be categorized as chromatin architectural proteins (Luger et al. 2012; McBryant et al. 2006).

Evidence for chromatin compaction has been gathered for subunits of PRC1 complexes and also for the PRC2 subunit EZH1. In a first observation, a Drosophila PRC1 complex was shown to compact chromatin as assessed by a decrease in the average internucleosomal distances in preassembled arrays. This activity locates to the C-terminal region of PSC and is independent of DNA sequence and histone tails (Francis et al. 2004). However, the Drosophila C-terminal PSC region is not conserved in plants or metazoans. Nevertheless, a reconstituted mouse PRC1 complex, in a CBX2/M33-dependent manner, was shown to act similarly to PSC (Grau et al. 2011). Structural studies, however, determined that a conformationally disordered, highly charged region identified in chromo domain-containing and RING-finger-containing PRC1 subunits is sufficient for nucleosomal compaction (Beh et al. 2012; Grau et al. 2011). In vivo, PRC1 repression through DNA compaction has been shown for clustered Hox genes in ES cells. Here, fluorescent in situ hybridization shows that following RING1B/RNF2 depletion, Hox genes at the end of the cluster are activated and move away from the compact structure formed by the rest of silent genes (Eskeland et al. 2010). L3MBTL2, a MBT-domain PRC1 subunit, is also able to compact nucleosomal arrays in vitro. In contrast to its requirement for methylated H3 or H4 histone N-tails, chromatin compaction activity, just as that of PSC or CBX2/M33, does not require histone tails (Trojer et al. 2011).

Reconstituted PRC2 complexes containing EZH1, but not those containing its paralog EZH2, are highly active compacting chromatin in vitro but only as part of the complex (Margueron et al. 2008). Another difference with PRC1 compaction is that histone tails are needed. A single PRC2-EZH1 aggregate brings together three/four nucleosomes. In tissue culture cells, chromatin accessibility (measured as sensitivity to DNAse) at reporter constructs and endogenous genes decreased when bound by EZH1, in line with in vitro activity. Interestingly, transcriptional repression through PRC2-EZH1-mediated chromatin compaction maybe uncoupled from H3K27me3 (Margueron et al. 2008).

An in vitro effect of PRC1-dependent nucleosome compaction is the inhibition of ATP-dependent chromatin remodelers (Shao et al. 1999). Some in vivo evidence for this activity can be inferred from gain or loss of Polycomb occupancy at targets, depending on downregulation or ectopic expression of SNF5/SMARCB1, a core component of subunit of chromatin remodeler SWI/SNF (Kia et al. 2008; Wilson et al. 2010). However, in a different model (ES cells), no relationship could be found between SNF5 and Polycomb repression (You et al. 2013). Thus, the overall relevance of chromatin compaction in Polycomb function remains largely unknown. And yet, correlative evidence would suggest that the large increase in H3K27me3-marked nucleosomes observed in differentiated but not in pluripotent cells is due to diminished chromatin remodeling activity compared to that of cells with high developmental potential (Hawkins et al. 2010; Meshorer et al. 2006; Zhu et al. 2013).

5.6.3 In Polycomb Bodies, Away from Transcription Factories

Polycomb complexes form large macromolecular assemblies within the cell, so-called Polycomb bodies. In apparent contradiction with its chromatin compaction function, Polycomb bodies appear to localize to perichromatin, the interface between interchromatin regions and condensed chromatin (Cheutin and Cavalli 2012; Cmarko et al. 2002). Polycomb bodies in Drosophila include silent Polycomb targets, in particular large genomic regions enriched in H3K27me3-marked nucleosomes and characterized by high occupancy of Polycomb subunits (Cheutin and Cavalli 2012). In the microscope, these regions are seen as very large speckles. However, smaller Polycomb domains do not form stable bodies. The data are consistent with contacts between Polycomb-bound sites (Bantignies et al. 2011; Sexton et al. 2012) and suggest that these bodies form at sites of high Polycomb density rather than as coalescent points where genes locate for repression (Cheutin and Cavalli 2012). PRE-containing transgenes co-localize to Polycomb bodies when repressed (Bantignies et al. 2003; Grimaud et al. 2006) However, detailed studies with transgenes indicate that such co-localization depends on insulator elements rather than on PREs and Polycomb complexes (Li et al. 2011). Transgenes containing enhancers localize to different nuclear domains called transcription factories and this association is also dependent on insulator function (Li et al. 2013). Thus, for effective repression, Polycomb proteins seem to, in a reversible manner, stabilize gene location at transcriptionally silent sites.

Polycomb-related gene repositioning phenomena can also involve ncRNAs as exemplified by transcriptional units in human tissue culture cells controlled by the cell cycle regulator E2F1. Under proliferating conditions, these genes are transcribed and localize to interchromatin granules at nuclear bodies identified by the presence of splicing factors (Mao et al. 2011), whereas in quiescence, they are silent and localize to Polycomb bodies. A PRC1 chromobox protein, CBX4/PC2, co-localizes to these promoters through E2F1 association. Importantly, however, the residence of loci in transcriptionally inactive (Polycomb bodies) or active (interchromatin granules) environments depends on CBX4/PC2 associating with distinct ncRNAs, TUG1 and NEAT2, respectively (Yang et al. 2011). Selective affinity for one or the other is determined by posttranslational modification of CBX4/PC2, in this case methylation by the well-known HMTase SUV39H1. In the presence of mitogens, cell cycle kinases inactivate SUV39H1; CBX4/PC2 is demethylated by histone demethylase JARID1A/KDM4c; demethylated CBX4/PC2 loses affinity for TUG1 ncRNA and gains affinity for NEAT2 ncRNA at interchromatin granules. Relocation to a transcriptionally conducive environment is accompanied by recruitment of CDCA7L, a RING-class E3 ubiquitin ligase that mono-ubiquitylates H2B through binding to SUMOylated E2F1 (by CBX4/PC2) (Yang et al. 2011). This example demonstrates that we have barely scratched the surface of the complexities of how Polycomb is involved in regulating the balance between active and inactive gene expression states.

5.6.4 *Sometimes, Polycomb Subunits Participate in Gene Activation*

Although the best know functions of Polycomb are those concerning PRCs, activities as individual subunits are also reported, for instance in gene activation events. In prostate cancer cells, EZH2 HMTase activity is needed for gene expression (Xu et al. 2012). As no H3K27me3 is involved, it is suggested that other regulators, probably the androgen receptor in this case, may be a substrate for EZH2 catalytic activity. Likewise, EZH1 inactivation in a tissue culture model of skeletal differentiation results in defective RNA pol II occupancy and activation of myogenic genes. In this case, EZH1 interacts with RNA pol II and acts as a positive regulator of transcriptional elongation (Mousavi et al. 2012). Finally, Cbx8, in a complex with HAT TIP60/KAT5 and MLL-AF9, is necessary for transcriptional activation associated with a MLL-AF9-triggered leukemogenic program (Tan et al. 2011).

5.7 Non-transcriptional Functions of Polycomb

Besides its role in transcriptional regulation, Polycomb directly influences also other important cellular functions such as DNA damage repair (Gieni et al. 2011; Vissers et al. 2012) and cell cycle progression. The latter does not include repression of proliferation inhibitors such as well-known Polycomb targets Cdk2nb/p15, Cdkn2a/p16, that encode cyclin-dependent kinase inhibitors that halt the cell cycle by impeding entrance in S-phase.

Roles for Polycomb in DNA damage have been inferred from the higher sensitivity of mutant cells to agents that induce DNA breaks (Chagraoui et al. 2011; Ginjala et al. 2011; Ismail et al. 2010; Pan et al. 2011; Wu et al. 2011). PRC1 and PRC2 subunits are rapidly recruited to sites of induced DNA damage after laser or ultraviolet irradiation (Chou et al. 2010; Hong et al. 2008). How this occurs exactly is obscured by contradictory evidence: for instance, BMI1/PCGF4 recruitment was found to be dependent and independent of poly(ADP-ribosyl) polymerase (PARP) activity (Chagraoui et al. 2011; Ginjala et al. 2011). Distinct contributions by several mechanisms acting, in a context-dependent manner, may be at the basis of these discrepancies. Co-localization of BMI1/PCGF4 with DNA-damage foci occurs before full H2AX phosphorylation (γH2AX), which is an early event occurring at sites of DNA damage that acts as a docking element for recruitment of the repair machinery (Papamichos-Chronakis and Peterson 2012; Soria et al. 2012). In fact, BMI1/PCGF4 and RING1B/RNF2 have been found to mono-ubiquitylate γH2AX as a step prior to the assembly of DNA repair proteins (Ginjala et al. 2011; Pan et al. 2011; Wu et al. 2011). Perhaps as a consequence of impaired DNA repair by homologous recombination, BMI1/PCGF4-deficient cells accumulate at G2/M (Ginjala et al. 2011). In the case of nucleotide excision repair, histone H2A ubiquitylation occurring upon ultraviolet

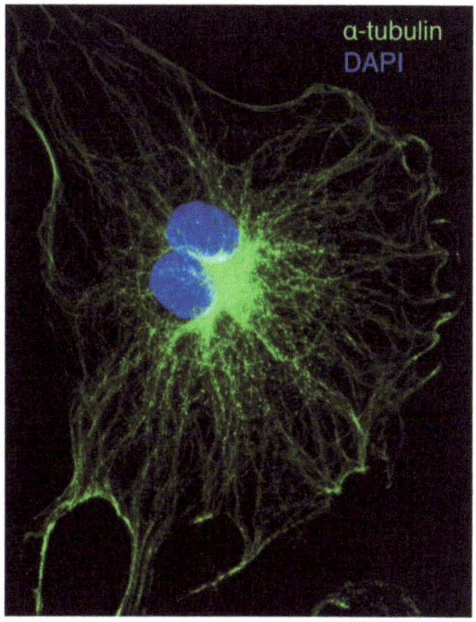

Fig. 5.3 Non-transcriptional functions of Polycomb proteins. Mitotic defects in cells lacking RING1A and RING1B. Example of binucleated cell, appearing in a culture of primary fibroblasts after RING1 protein inactivation, probably a consequence of failed cytokinesis

irradiation is also RING1B/RNF2 dependent (Bergink et al. 2006). However, the precise mechanism by which Polycomb complexes influence DNA repair still remains to be elucidated.

In addition to the contribution to DNA damage repair, Polycomb influences cell cycle progression via posttranslational modifications of cell proliferation regulators. For instance, loss of Drosophila PRC1 subunit PSC results in cells that accumulate at the G2/M phase. In contrast, inactivation of other PRC1 products, such as PC or SCE has no effect (Mohd-Sarip et al. 2012). PSC is found in complexes other than PRC1 and is associated with cell cycle regulators such as CDK1/CDC2, cyclin B (CCNB) and subunits of the Anaphase Promoting Complex (APC). CDK1-CCNB phosphorylates a collection of proteins involved in the transition form interphase to mitosis, including nuclear membrane breakdown and mitotic spindle assembly. Mitotic segregation defects seen in PSC-deficient cells correlate with decreased levels of poly-ubiquitylated CCNB, which appear to depend on PSC (Mohd-Sarip et al. 2012). The observation is surprising, considering that APC activity is directed to destroy CCNB towards the end of mitosis. It is possible that PSC modification of CCNB may therefore not be related to its proteasomal degradation. Mutations in other Drosophila PRC1 subunits showed no proliferative defects but inactivation of RING1 paralogs in mammalian fibroblasts results in mitotic aberrations as indicated by the presence of micronuclei and binucleate cells (Fig. 5.3). Another proliferative defect associated with Polycomb-dependent posttranslational modifications is the accumulation of geminin, a negative regulator of replication through inhibition of licensing factor CDT1. It is thought that defective poly-ubiquitylation in cells deficient in PRC1 subunit PHC1 results in unscheduled geminin stabilization and quiescence (Ohtsubo et al. 2008).

5.8 Concluding Remarks

The Polycomb field has exploded in the last few years and while we still tend to talk of two "types" of complexes (PRC1 and PRC2) the real situation is far more complicated. While their main functions are as transcriptional repressors, this article shows that Polycomb proteins are part of a dynamic and extensive protein network that performs diverse tasks in a number of different contexts and is also regulated by external signals. The different subunits of Polycomb complexes can be modified, exchanged, and associated with diverse types of other proteins and bind even to noncoding RNA and all of this in a cell type- and cell stage-specific fashion. System-wide studies are now urgently needed to link the epigenetic function of Polycomb complexes with the proteome. At the mechanistic level, as for other chromatin modifiers, there are still many gaps in our understanding of the molecular mechanisms by which Polycomb represses transcription. However, without such mechanistic insights we will not be able to counteract situations where Polycomb function is aberrant, as outlined in Chap. 6 about the role of Polycomb in leukemia. Much recent work has examined the location of genomic sites bound by Polycomb products and the associated histone marks. Future efforts should now attempt to put these linear maps of chromatin states into three-dimensional regulatory spaces and investigate the impact of Polycomb-dependent changes in nuclear architecture on transcription regulation. Single-cell approaches need to be established that provide access to details that are lost in cell population analyses and inform of the dynamic, rather than static, nature of the system. A great interest exists in translating new knowledge on Polycomb function into therapeutic/diagnostic possibilities, be in harnessing the power of these complexes in regulating self-renewal of stem cells for regenerative medicine or in taming/suppressing transformed cells. At any rate, we still have a long way to go until we understand the workings of such an evolutionary successful system in the generation of cell diversity and tissue homeostasis. There is still much scope for exciting and satisfying research.

Acknowledgments Work in the lab is supported by grants BFU2010-18146 (MINECO), Fundación Areces, Oncocycle S2010/BMD2470 (CAM), and FP7-People-2011-ITN.

References

Abdel-Wahab O, Adli M, LaFave LM et al (2012) ASXL1 mutations promote myeloid transformation through loss of PRC2-mediated gene repression. Cancer Cell 22:180–193

Adelman K, Lis JT (2012) Promoter-proximal pausing of RNA polymerase II: emerging roles in metazoans. Nat Rev Genet 13:720–731

Agger K, Cloos PA, Christensen J et al (2007) UTX and JMJD3 are histone H3K27 demethylases involved in HOX gene regulation and development. Nature 449:731–734

Akasaka T, Kanno M, Balling R et al (1996) A role for mel-18, a Polycomb group-related vertebrate gene, during the anteroposterior specification of the axial skeleton. Development 122:1513–1522

An JY, Kim E-A, Jiang Y et al (2010) UBR2 mediates transcriptional silencing during spermatogenesis via histone ubiquitination. Proc Natl Acad Sci USA 107:1912–1917
Aravind L, Iyer LM (2012) The HARE-HTH and associated domains: novel modules in the coordination of epigenetic DNA and protein modifications. Cell Cycle 11:119–131
Arnold P, Scholer A, Pachkov M et al (2013) Modeling of epigenome dynamics identifies transcription factors that mediate Polycomb targeting. Genome Res 23:60–73
Azuara V, Perry P, Sauer S et al (2006) Chromatin signatures of pluripotent cell lines. Nat Cell Biol 8:532–538
Ballaré C, Lange M, Lapinaite A et al (2012) Phf19 links methylated Lys36 of histone H3 to regulation of Polycomb activity. Nat Struct Mol Biol 19:1257–1265
Bantignies F, Grimaud C, Lavrov S et al (2003) Inheritance of Polycomb-dependent chromosomal interactions in Drosophila. Genes Dev 17:2406–2420
Bantignies F, Roure V, Comet I et al (2011) Polycomb-dependent regulatory contacts between distant Hox Loci in Drosophila. Cell 144:214–226
Batista PJ, Chang HY (2013) Long noncoding RNAs: cellular address codes in development and disease. Cell 152:1298–1307
Baù D, Sanyal A, Lajoie BR et al (2010) The three-dimensional folding of the α-globin gene domain reveals formation of chromatin globules. Nat Struct Mol Biol 18:107–114
Beh LY, Colwell LJ, Francis NJ (2012) A core subunit of Polycomb repressive complex 1 is broadly conserved in function but not primary sequence. Proc Natl Acad Sci USA 109:E1063–E1071
Bentley ML, Corn JE, Dong KC et al (2011) Recognition of UbcH5c and the nucleosome by the Bmi1/Ring1b ubiquitin ligase complex. EMBO J 30:3285–3297
Bergink S, Salomons FA, Hoogstraten D et al (2006) DNA damage triggers nucleotide excision repair-dependent monoubiquitylation of histone H2A. Genes Dev 20:1343–1352
Bernstein BE, Mikkelsen TS, Xie X et al (2006a) A bivalent chromatin structure marks key developmental genes in embryonic stem cells. Cell 125:315–326
Bernstein E, Duncan EM, Masui O et al (2006b) Mouse Polycomb proteins bind differentially to methylated histone H3 and RNA and are enriched in facultative heterochromatin. Mol Cell Biol 26:2560–2569
Blackledge NP, Klose R (2011) CpG island chromatin: a platform for gene regulation. Epigenetics 6:147–152
Bonasio R, Lecona E, Reinberg D (2010) MBT domain proteins in development and disease. Semin Cell Dev Biol 21:221–230
Boyer LA, Plath K, Zeitlinger J et al (2006) Polycomb complexes repress developmental regulators in murine embryonic stem cells. Nature 441:349–353
Bracken AP, Helin K (2009) Polycomb group proteins: navigators of lineage pathways led astray in cancer. Nat Rev Cancer 9:773–784
Bracken AP, Dietrich N, Pasini D et al (2006) Genome-wide mapping of Polycomb target genes unravels their roles in cell fate transitions. Genes Dev 20:1123–1136
Brien GL, Gambero G, O'Connell DJ et al (2012) Polycomb PHF19 binds H3K36me3 and recruits PRC2 and demethylase NO66 to embryonic stem cell genes during differentiation. Nat Struct Mol Biol 19:1273–1281
Brookes E, de Santiago I, Hebenstreit D et al (2012) Polycomb associates genome-wide with a specific RNA polymerase II variant, and regulates metabolic genes in ESCs. Cell Stem Cell 10:157–170
Brown JL, Mucci D, Whiteley M et al (1998) The Drosophila Polycomb group gene pleiohomeotic encodes a DNA binding protein with homology to the transcription factor YY1. Mol Cell 1:1057–1064
Brown JL, Fritsch C, Mueller J, Kassis JA (2003) The Drosophila pho-like gene encodes a YY1-related DNA binding protein that is redundant with pleiohomeotic in homeotic gene silencing. Development 130:285–294

Buchwald G, der Stoop van P, Weichenrieder O et al (2006) Structure and E3-ligase activity of the Ring–Ring complex of Polycomb proteins Bmi1 and Ring1b. EMBO J 25:2465–2474
Cai L, Rothbart SB, Lu R et al (2013) An H3K36 methylation-engaging Tudor motif of Polycomb-like proteins mediates PRC2 complex targeting. Mol Cell 49:571–582
Calés C, Román-Trufero M, Pavón L et al (2008) Inactivation of the Polycomb group protein Ring1B unveils an antiproliferative role in hematopoietic cell expansion and cooperation with tumorigenesis associated to Ink4a deletion. Mol Cell Biol 28:1018–1028
Cao R, Wang L, Wang H et al (2002) Role of histone H3 lysine 27 methylation in Polycomb-group silencing. Science 298:1039–1043
Cao R, Tsukada Y-I, Zhang Y (2005) Role of Bmi-1 and Ring1A in H2A ubiquitylation and Hox gene silencing. Mol Cell 20:845–854
Chagraoui J, Hebert J, Girard S, Sauvageau G (2011) An anticlastogenic function for the Polycomb group gene Bmi1. Proc Natl Acad Sci USA 108:5284–5289
Chakalova L, Debrand E, Mitchell JA et al (2005) Replication and transcription: shaping the landscape of the genome. Nat Rev Genet 6:669–677
Chen S, Ma J, Wu F et al (2012) The histone H3 Lys 27 demethylase JMJD3 regulates gene expression by impacting transcriptional elongation. Genes Dev 26:1364–1375
Cheutin T, Cavalli G (2012) Progressive Polycomb assembly on H3K27me3 compartments generates Polycomb bodies with developmentally regulated motion. PLoS Genet 8:e1002465
Chopra VS, Hendrix DA, Core LJ et al (2011) The Polycomb group mutant esc leads to augmented levels of paused Pol II in the Drosophila embryo. Mol Cell 42:837–844
Chou DM, Adamson B, Dephoure NE et al (2010) A chromatin localization screen reveals poly (ADP ribose)-regulated recruitment of the repressive Polycomb and NuRD complexes to sites of DNA damage. Proc Natl Acad Sci USA 107:18475–18480
Chu C, Qu K, Zhong FL et al (2011) Genomic maps of long noncoding RNA occupancy reveal principles of RNA-chromatin interactions. Mol Cell 44:667–678
Ciferri C, Lander GC, Maiolica A et al (2012) Molecular architecture of human Polycomb repressive complex 2. Elife 1:e00005
Cmarko D, Verschure PJ, Otte AP et al (2002) Polycomb group gene silencing proteins are concentrated in the perichromatin compartment of the mammalian nucleus. J Cell Sci 116: 335–343
Cook PR (2010) A model for all genomes: the role of transcription factories. J Mol Biol 395:1–10
Core LJ, Waterfall JJ, Lis JT (2008) Nascent RNA sequencing reveals widespread pausing and divergent initiation at human promoters. Science 322:1845–1848
Core LJ, Waterfall JJ, Gilchrist DA et al (2012) Defining the status of RNA polymerase at promoters. Cell Rep 2:1025–1035
Cosgrove MS, Boeke JD, Wolberger C (2004) Regulated nucleosome mobility and the histone code. Nat Struct Mol Biol 11:1037–1043
Creyghton MP, Markoulaki S, Levine SS et al (2008) H2AZ is enriched at Polycomb complex target genes in ES cells and is necessary for lineage commitment. Cell 135:649–661
Creyghton MP, Cheng AW, Welstead GG et al (2010) Histone H3K27ac separates active from poised enhancers and predicts developmental state. Proc Natl Acad Sci USA 107:21931–21936
Cui K, Zang C, Roh T-Y et al (2009) Chromatin signatures in multipotent human hematopoietic stem cells indicate the fate of bivalent genes during differentiation. Cell Stem Cell 4:80–93
Czermin B, Melfi R, McCabe D et al (2002) Drosophila enhancer of Zeste/ESC complexes have a histone H3 methyltransferase activity that marks chromosomal Polycomb sites. Cell 111: 185–196
Dahle Ø, Kumar A, Kuehn MR (2010) Nodal signaling recruits the histone demethylase Jmjd3 to counteract Polycomb-mediated repression at target genes. Sci Signal 3:ra48
de Napoles M, Mermoud JE, Wakao R et al (2004) Polycomb group proteins Ring1A/B link ubiquitylation of histone H2A to heritable gene silencing and X inactivation. Dev Cell 7: 663–676

De Santa F, Narang V, Yap ZH et al (2009) Jmjd3 contributes to the control of gene expression in LPS-activated macrophages. EMBO J 28:3341–3352
de Wit E, de Laat W (2012) A decade of 3C technologies: insights into nuclear organization. Genes Dev 26:11–24
Deaton AM, Bird A (2011) CpG islands and the regulation of transcription. Genes Dev 25: 1010–1022
Dekker J, Marti-Renom MA, Mirny LA (2013) Exploring the three-dimensional organization of genomes: interpreting chromatin interaction data. Nat Rev Genet 14:390–403
del Mar LM, Marcos-Gutiérrez C, Pérez C et al (2000) Loss- and gain-of-function mutations show a Polycomb group function for Ring1A in mice. Development 127:5093–5100
Delest A, Sexton T, Cavalli G (2012) Polycomb: a paradigm for genome organization from one to three dimensions. Curr Opin Cell Biol 24:405–414
Dellino GI, Schwartz YB, Farkas G et al (2004) Polycomb silencing blocks transcription initiation. Mol Cell 13:887–893
der Lugt van NM, Domen J, Linders K et al (1994) Posterior transformation, neurological abnormalities, and severe hematopoietic defects in mice with a targeted deletion of the bmi-1 proto-oncogene. Genes Dev 8:757–769
Dey A, Seshasayee D, Noubade R et al (2012) Loss of the tumor suppressor BAP1 causes myeloid transformation. Science 337:1541–1546
Dietrich N, Lerdrup M, Landt E et al (2012) REST–mediated recruitment of Polycomb repressor complexes in mammalian cells. PLoS Genet 8:e1002494
Dixon JR, Selvaraj S, Yue F et al (2012) Topological domains in mammalian genomes identified by analysis of chromatin interactions. Nature 485:376–380
Draker R, Sarcinella E, Cheung P (2011) USP10 deubiquitylates the histone variant H2A.Z and both are required for androgen receptor-mediated gene activation. Nucleic Acids Res 39: 3529–3542
Enderle D, Beisel C, Stadler MB et al (2011) Polycomb preferentially targets stalled promoters of coding and noncoding transcripts. Genome Res 21:216–226
Endoh M, Endo TA, Endoh T et al (2012) Histone H2A mono-ubiquitination is a crucial step to mediate PRC1-dependent repression of developmental genes to maintain ES cell identity. PLoS Genet 8:e1002774
Ernst J, Kheradpour P, Mikkelsen TS et al (2011) Mapping and analysis of chromatin state dynamics in nine human cell types. Nature 473:43–49
Eskeland R, Leeb M, Grimes GR et al (2010) Ring1B compacts chromatin structure and represses gene expression independent of histone ubiquitination. Mol Cell 38:452–464
Eustermann S, Yang J-C, Law MJ et al (2011) Combinatorial readout of histone H3 modifications specifies localization of ATRX to heterochromatin. Nat Struct Mol Biol 18:777–782
Farcas AM, Blackledge NP, Sudbery I et al (2012) KDM2B links the Polycomb repressive complex 1 (PRC1) to recognition of CpG islands. Elife 1:e00205
Fenouil R, Cauchy P, Koch F et al (2012) CpG islands and GC content dictate nucleosome depletion in a transcription-independent manner at mammalian promoters. Genome Res 22: 2399–2408
Filion GJ, van Bemmel JG, Braunschweig U et al (2010) Systematic protein location mapping reveals five principal chromatin types in Drosophila cells. Cell 143:212–224. doi:10.1016/j.cell.2010.09.009
Fischle W, Wang Y, Jacobs SA et al (2003) Molecular basis for the discrimination of repressive methyl-lysine marks in histone H3 by Polycomb and HP1 chromodomains. Genes Dev 17: 1870–1881
Follmer NE, Wani AH, Francis NJ (2012) A Polycomb group protein is retained at specific sites on chromatin in mitosis. PLoS Genet 8:e1003135
Fonseca JP, Steffen PA, Muller S et al (2012) In vivo Polycomb kinetics and mitotic chromatin binding distinguish stem cells from differentiated cells. Genes Dev 26:857–871

Francis NJ, Kingston RE, Woodcock CL (2004) Chromatin compaction by a Polycomb group protein complex. Science 306:1574–1577

Francis NJ, Follmer NE, Simon MD et al (2009) Polycomb proteins remain bound to chromatin and DNA during DNA replication in vitro. Cell 137:110–122

Frescas D, Guardavaccaro D, Bassermann F et al (2007) JHDM1B/FBXL10 is a nucleolar protein that represses transcription of ribosomal RNA genes. Nature 450:309–313

Gaertner B, Johnston J, Chen K et al (2012) Poised RNA polymerase II changes over developmental time and prepares genes for future expression. Cell Rep 2:1670–1683

Gambetta MC, Oktaba K, Müller J (2009) Essential role of the glycosyltransferase sxc/Ogt in Polycomb repression. Science 325:93–96. doi:10.1126/science.1169727

Gao Z, Zhang J, Bonasio R et al (2012) PCGF homologs, CBX proteins, and RYBP define functionally distinct PRC1 family complexes. Mol Cell 45:344–356

García E, Marcos-Gutiérrez C, del Mar LM et al (1999) RYBP, a new repressor protein that interacts with components of the mammalian Polycomb complex, and with the transcription factor YY1. EMBO J 18:3404–3418

Gearhart MD, Corcoran CM, Wamstad JA, Bardwell VJ (2006) Polycomb group and SCF ubiquitin ligases are found in a novel BCOR complex that is recruited to BCL6 targets. Mol Cell Biol 26:6880–6889

Gehani SS, Agrawal-Singh S, Dietrich N et al (2010) Polycomb group protein displacement and gene activation through MSK-dependent H3K27me3S28 phosphorylation. Mol Cell 39: 886–900

Geiss-Friedlander R, Melchior F (2007) Concepts in sumoylation: a decade on. Nat Rev Mol Cell Biol 8:947–956

Ghamari A, van de Corput MPC, Thongjuea S et al (2013) In vivo live imaging of RNA polymerase II transcription factories in primary cells. Genes Dev 27:767–777

Gibcus JH, Dekker J (2012) The context of gene expression regulation. F1000 Biol Rep 4:8

Gieni RS, Ismail IH, Campbell S, Hendzel MJ (2011) Polycomb group proteins in the DNA damage response: a link between radiation resistance and "stemness". Cell Cycle 10:883–894

Ginjala V, Nacerddine K, Kulkarni A et al (2011) BMI1 is recruited to DNA breaks and contributes to DNA damage-induced H2A ubiquitination and repair. Mol Cell Biol 31: 1972–1982

Goldknopf IL, Taylor CW, Baum RM et al (1975) Isolation and characterization of protein A24, a "histone-like" non-histone chromosomal protein. J Biol Chem 250:7182–7187

Grau DJ, Chapman BA, Garlick JD et al (2011) Compaction of chromatin by diverse Polycomb group proteins requires localized regions of high charge. Genes Dev 25:2210–2221

Grimaud C, Bantignies F, Pal-Bhadra M et al (2006) RNAi components are required for nuclear clustering of Polycomb group response elements. Cell 124:957–971

Gupta RA, Shah N, Wang KC et al (2010) Long non-coding RNA HOTAIR reprograms chromatin state to promote cancer metastasis. Nature 464:1071–1076

Gutierrez L, Oktaba K, Scheuermann JC et al (2011) The role of the histone H2A ubiquitinase Sce in Polycomb repression. Development 139:117–127

Guttman M, Rinn JL (2012) Modular regulatory principles of large non-coding RNAs. Nature 482:339–346

Guttman M, Donaghey J, Carey BW et al (2011) lincRNAs act in the circuitry controlling pluripotency and differentiation. Nature 477:295–300

Hanover JA, Krause MW, Love DC (2012) Bittersweet memories: linking metabolism to epigenetics through O-GlcNAcylation. Nat Rev Mol Cell Biol 13:312–321

Hansen K, Bracken A, Pasini D et al (2008) A model for transmission of the H3K27me3 epigenetic mark. Nat Cell Biol 10:1291–1300

Hathaway NA, Bell O, Hodges C et al (2012) Dynamics and memory of heterochromatin in living cells. Cell 149:1447–1460

Hawkins RD, Hon GC, Lee LK et al (2010) Distinct epigenomic landscapes of pluripotent and lineage-committed human cells. Cell Stem Cell 6:479–491

He J, Kallin EM, Tsukada Y-I, Zhang Y (2008) The H3K36 demethylase Jhdm1b/Kdm2b regulates cell proliferation and senescence through p15Ink4b. Nat Struct Mol Biol 15: 1169–1175
He A, Shen X, Ma Q et al (2012) PRC2 directly methylates GATA4 and represses its transcriptional activity. Genes Dev 26:37–42
He J, Shen L, Wan M et al (2013) Kdm2b maintains murine embryonic stem cell status by recruiting PRC1 complex to CpG islands of developmental genes. Nat Cell Biol 15:373–384
Hekimoglu-Balkan B, Aszodi A, Heinen R et al (2012) Intergenic Polycomb target sites are dynamically marked by non-coding transcription during lineage commitment. RNA Biol 9:314–325
Henikoff S (2008) Nucleosome destabilization in the epigenetic regulation of gene expression. Nat Rev Genet 9:15–26
Henikoff S, Shilatifard A (2011) Histone modification: cause or cog? Trends Genet 27:389–396
Herz H-M, Shilatifard A (2010) The JARID2-PRC2 duality. Genes Dev 24:857–861
Hisada K, Sanchez C, Endo TA et al (2012) RYBP represses endogenous retroviruses and preimplantation- and germ line-specific genes in mouse embryonic stem cells. Mol Cell Biol 32:1139–1149
Hong Z, Jiang J, Lan L et al (2008) A Polycomb group protein, PHF1, is involved in the response to DNA double-strand breaks in human cell. Nucleic Acids Res 36:2939–2947
Hou C, Li L, Qin ZS, Corces VG (2012) Gene density, transcription, and insulators contribute to the partition of the Drosophila genome into physical domains. Mol Cell 48:471–484
Hu H, Yang Y, Ji Q et al (2012) CRL4B catalyzes H2AK119 monoubiquitination and coordinates with PRC2 to promote tumorigenesis. Cancer Cell 22:781–795
Huang C, Xu M, Zhu B (2012) Epigenetic inheritance mediated by histone lysine methylation: maintaining transcriptional states without the precise restoration of marks? Philos Trans R Soc Lond B Biol Sci 368:1471–1475
Illingworth RS, Bird AP (2009) CpG islands–"a rough guide". FEBS Lett 583:1713–1720
Illingworth RS, Gruenewald-Schneider U, Webb S et al (2010) Orphan CpG islands identify numerous conserved promoters in the mammalian genome. PLoS Genet 6:e1001134
Ismail IH, Andrin C, McDonald D, Hendzel MJ (2010) BMI1-mediated histone ubiquitylation promotes DNA double-strand break repair. J Cell Biol 191:45–60
Isono K, Endo TA, Ku M et al (2013) SAM domain polymerization links subnuclear clustering of PRC1 to gene silencing. Dev Cell 26:565–577
Iyer BV, Kenward M, Arya G (2011) Hierarchies in eukaryotic genome organization: insights from polymer theory and simulations. BMC Biophys 4:8
Jackson S, Xiong Y (2009) CRL4s: the CUL4-RING E3 ubiquitin ligases. Trends Biochem Sci 34: 562–570
Jhunjhunwala S, van Zelm MC, Peak MM et al (2008) The 3D structure of the immunoglobulin heavy-chain locus: implications for long-range genomic interactions. Cell 133:265–279
Jiang H, Shukla A, Wang X et al (2011a) Role for Dpy-30 in ES cell-fate specification by regulation of H3K4 methylation within bivalent domains. Cell 144:513–525
Jiang X-X, Nguyen Q, Chou Y et al (2011b) Control of B cell development by the histone H2A deubiquitinase MYSM1. Immunity 35:883–896
Joo H-Y, Zhai L, Yang C et al (2007) Regulation of cell cycle progression and gene expression by H2A deubiquitination. Nature 449:1068–1072
Joo H-Y, Jones A, Yang C et al (2011) Regulation of histone H2A and H2B deubiquitination and Xenopus development by USP12 and USP46. J Biol Chem 286:7190–7201
Kagey MH, Melhuish TA, Wotton D (2003) The Polycomb protein Pc2 is a SUMO E3. Cell 113:127–137
Kalenik JL, Chen D, Bradley ME et al (1997) Yeast two-hybrid cloning of a novel zinc finger protein that interacts with the multifunctional transcription factor YY1. Nucleic Acids Res 25: 843–849

Kaneko S, Li G, Son J et al (2010) Phosphorylation of the PRC2 component Ezh2 is cell cycle-regulated and up-regulates its binding to ncRNA. Genes Dev 24:2615–2620
Kang X, Qi Y, Zuo Y et al (2010) SUMO-specific protease 2 is essential for suppression of Polycomb group protein-mediated gene silencing during embryonic development. Mol Cell 38:191–201
Kanhere A, Viiri K, Araújo CC et al (2010) Short RNAs are transcribed from repressed Polycomb target genes and interact with Polycomb repressive complex-2. Mol Cell 38:675–688
Kellner WA, Ramos E, Van Bortle K et al (2012) Genome-wide phosphoacetylation of histone H3 at Drosophila enhancers and promoters. Genome Res 22:1081–1088
Khalil AM, Guttman M, Huarte M et al (2009) Many human large intergenic noncoding RNAs associate with chromatin-modifying complexes and affect gene expression. Proc Natl Acad Sci USA 106:11667–11672
Kharchenko PV, Alekseyenko AA, Schwartz YB et al (2010) Comprehensive analysis of the chromatin landscape in Drosophila melanogaster. Nature 471:480–485
Kia SK, Gorski MM, Giannakopoulos S, Verrijzer CP (2008) SWI/SNF mediates Polycomb eviction and epigenetic reprogramming of the INK4b-ARF-INK4a locus. Mol Cell Biol 28: 3457–3464
Kim H, Kang K, Kim J (2009) AEBP2 as a potential targeting protein for Polycomb Repression Complex PRC2. Nucleic Acids Res 37:2940–2950
Klauke K, Radulović V, Broekhuis M et al (2013) Polycomb Cbx family members mediate the balance between haematopoietic stem cell self-renewal and differentiation. Nat Cell Biol 15: 353–362
Klose RJ, Cooper S, Farcas AM et al (2013) Chromatin sampling—an emerging perspective on targeting Polycomb repressor proteins. PLoS Genet 9:e1003717–e1003718
Klymenko T, Papp B, Fischle W et al (2006) A Polycomb group protein complex with sequence-specific DNA-binding and selective methyl-lysine-binding activities. Genes Dev 20: 1110–1122
Kohwi M, Lupton JR, Lai S-L et al (2013) Developmentally regulated subnuclear genome reorganization restricts neural progenitor competence in Drosophila. Cell 152:97–108
Komander D, Rape M (2012) The ubiquitin code. Annu Rev Biochem 81:203–229
Kooistra SM, Helin K (2012) Molecular mechanisms and potential functions of histone demethylases. Nat Rev Mol Cell Biol 13:297–311
Ku M, Koche RP, Rheinbay E et al (2008) Genomewide analysis of PRC1 and PRC2 occupancy identifies two classes of bivalent domains. PLoS Genet 4:e1000242
Ku M, Jaffe JD, Koche RP et al (2012) H2A.Z landscapes and dual modifications in pluripotent and multipotent stem cells underlie complex genome regulatory functions. Genome Biol 13: R85
Kuzmichev A, Nishioka K, Erdjument-Bromage H et al (2002) Histone methyltransferase activity associated with a human multiprotein complex containing the Enhancer of Zeste protein. Genes Dev 16:2893–2905
Kuzmichev A, Jenuwein T, Tempst P, Reinberg D (2004) Different EZH2-containing complexes target methylation of histone H1 or nucleosomal histone H3. Mol Cell 14:183–193
Kuzmichev A, Margueron R, Vaquero A et al (2005) Composition and histone substrates of Polycomb repressive group complexes change during cellular differentiation. Proc Natl Acad Sci USA 102:1859–1864
Lan F, Bayliss P, Rinn JL et al (2007) A histone H3 lysine 27 demethylase regulates animal posterior development. Nature 449:689–694
Landeira D, Sauer S, Poot R et al (2010) Jarid2 is a PRC2 component in embryonic stem cells required for multi-lineage differentiation and recruitment of PRC1 and RNA Polymerase II to developmental regulators. Nat Cell Biol 12:618–624
Lanzuolo C, Orlando V (2012) Memories from the Polycomb group proteins. Annu Rev Genet 46: 561–589

Lau PNI, Cheung P (2011) Histone code pathway involving H3 S28 phosphorylation and K27 acetylation activates transcription and antagonizes Polycomb silencing. Proc Natl Acad Sci USA 108:2801–2806
Lee TI, Jenner RG, Boyer LA et al (2006) Control of developmental regulators by Polycomb in human embryonic stem cells. Cell 125:301–313
Lee MG, Villa R, Trojer P et al (2007) Demethylation of H3K27 regulates Polycomb recruitment and H2A ubiquitination. Science 318:447–450
Leeb M, Pasini D, Novatchkova M et al (2010) Polycomb complexes act redundantly to repress genomic repeats and genes. Genes Dev 24:265–276
Lessard J, Sauvageau G (2003) Bmi-1 determines the proliferative capacity of normal and leukaemic stem cells. Nature 423:255–260
Levine M (2011) Paused RNA polymerase II as a developmental checkpoint. Cell 145:502–511
Levine SS, Weiss A, Erdjument-Bromage H et al (2002) The core of the Polycomb repressive complex is compositionally and functionally conserved in flies and humans. Mol Cell Biol 22:6070–6078
Lewis EB (1978) A gene complex controlling segmentation in Drosophila. Nature 276:565–570
Li Z, Cao R, Wang M et al (2006) Structure of a Bmi-1-Ring1B Polycomb group ubiquitin ligase complex. J Biol Chem 281:20643–20649
Li B, Zhou J, Liu P et al (2007) Polycomb protein Cbx4 promotes SUMO modification of de novo DNA methyltransferase Dnmt3a. Biochem J 405:369–378
Li G, Margueron R, Ku M et al (2010) Jarid2 and PRC2, partners in regulating gene expression. Genes Dev 24:368–380
Li HB, Müller M, Bahechar IA et al (2011) Insulators, not Polycomb response elements, are required for long-range interactions between Polycomb targets in Drosophila melanogaster. Mol Cell Biol 31:616–625
Li G, Ruan X, Auerbach RK et al (2012) Extensive promoter-centered chromatin interactions provide a topological basis for transcription regulation. Cell 148:84–98
Li H-B, Ohno K, Gui H, Pirrotta V (2013) Insulators target active genes to transcription factories and Polycomb-repressed genes to Polycomb bodies. PLoS Genet 9:e1003436
Lieberman-Aiden E, van Berkum NL, Williams L et al (2009) Comprehensive mapping of long-range interactions reveals folding principles of the human genome. Science 326:289–293
Lin YC, Jhunjhunwala S, Benner C et al (2010) A global network of transcription factors, involving E2A, EBF1 and Foxo1, that orchestrates B cell fate. Nat Immunol 11:635–643
Lin YC, Benner C, Månsson R et al (2012) Global changes in the nuclear positioning of genes and intra- and interdomain genomic interactions that orchestrate B cell fate. Nat Immunol 13:1196–1204
Long HK, Blackledge NP, Klose RJ (2013a) ZF-CxxC domain-containing proteins, CpG islands and the chromatin connection. Biochem Soc Trans 41:727–740
Long HK, Sims D, Heger A et al (2013b) Epigenetic conservation at gene regulatory elements revealed by non-methylated DNA profiling in seven vertebrates. Elife 2:e00348
Loyola A, Tagami H, Bonaldi T et al (2009) The HP1alpha-CAF1-SetDB1-containing complex provides H3K9me1 for Suv39-mediated K9me3 in pericentric heterochromatin. EMBO Rep 10:769–775
Luger K, Dechassa ML, Tremethick DJ (2012) New insights into nucleosome and chromatin structure: an ordered state or a disordered affair? Nat Rev Mol Cell Biol 13:436–447
Luis NM, Morey L, Mejetta S et al (2011) Regulation of human epidermal stem cell proliferation and senescence requires Polycomb-dependent and -independent functions of Cbx4. Cell Stem Cell 9:233–246
Lynch MD, Smith AJH, De Gobbi M et al (2011) An interspecies analysis reveals a key role for unmethylated CpG dinucleotides in vertebrate Polycomb complex recruitment. EMBO J 31: 317–329
MacPherson MJ, Beatty LG, Zhou W et al (2009) The CTCF insulator protein is posttranslationally modified by SUMO. Mol Cell Biol 29:714–725

Mao YS, Zhang B, Spector DL (2011) Biogenesis and function of nuclear bodies. Trends Genet 27:295–306
Margueron R, Reinberg D (2011) The Polycomb complex PRC2 and its mark in life. Nature 469: 343–349
Margueron R, Li G, Sarma K et al (2008) Ezh1 and Ezh2 maintain repressive chromatin through different mechanisms. Mol Cell 32:503–518
Margueron R, Justin N, Ohno K et al (2009) Role of the Polycomb protein EED in the propagation of repressive histone marks. Nature 461:762–767
McBryant SJ, Adams VH, Hansen JC (2006) Chromatin architectural proteins. Chromosome Res 14:39–51
McCabe MT, Graves AP, Ganji G et al (2012a) Mutation of A677 in histone methyltransferase EZH2 in human B-cell lymphoma promotes hypertrimethylation of histone H3 on lysine 27 (H3K27). Proc Natl Acad Sci USA 109:2989–2994
McCabe MT, Ott HM, Ganji G et al (2012b) EZH2 inhibition as a therapeutic strategy for lymphoma with EZH2-activating mutations. Nature 492:108–112
Mendenhall EM, Koche RP, Truong T et al (2010) GC-rich sequence elements recruit PRC2 in mammalian ES cells. PLoS Genet 6:e1001244
Mercer TR, Mattick JS (2013) Structure and function of long noncoding RNAs in epigenetic regulation. Nat Struct Mol Biol 20:300–307
Meshorer E, Yellajoshula D, George E et al (2006) Hyperdynamic plasticity of chromatin proteins in pluripotent embryonic stem cells. Dev Cell 10:105–116
Mikkelsen TS, Ku M, Jaffe DB et al (2007) Genome-wide maps of chromatin state in pluripotent and lineage-committed cells. Nature 448:553–561
Mills AA (2010) Throwing the cancer switch: reciprocal roles of Polycomb and trithorax proteins. Nat Rev Cancer 10:669–682
Min IM, Waterfall JJ, Core LJ et al (2011) Regulating RNA polymerase pausing and transcription elongation in embryonic stem cells. Genes Dev 25:742–754
Mito Y, Henikoff JG, Henikoff S (2007) Histone replacement marks the boundaries of cis-regulatory domains. Science 315:1408–1411
Mohd-Sarip A, Lagarou A, Doyen CM et al (2012) Transcription-independent function of Polycomb group protein PSC in cell cycle control. Science 336:744–747
Mohn F, Weber M, Rebhan M et al (2008) Lineage-specific Polycomb targets and de novo DNA methylation define restriction and potential of neuronal progenitors. Mol Cell 30:755–766
Montgomery ND, Yee D, Chen A et al (2005) The murine Polycomb group protein Eed is required for global histone H3 lysine-27 methylation. Curr Biol 15:942–947
Morey L, Aloia L, Cozzuto L et al (2013) RYBP and Cbx7 define specific biological functions of Polycomb complexes in mouse embryonic stem cells. Cell Rep 3:60–69
Mousavi K, Zare H, Wang AH, Sartorelli V (2012) Polycomb protein Ezh1 promotes RNA polymerase II elongation. Mol Cell 45:255–262
Müller J, Kassis JA (2006) Polycomb response elements and targeting of Polycomb group proteins in Drosophila. Curr Opin Genet Dev 16:476–484
Müller J, Hart CM, Francis NJ et al (2002) Histone methyltransferase activity of a Drosophila Polycomb group repressor complex. Cell 111:197–208
Muse GW, Gilchrist DA, Nechaev S et al (2007) RNA polymerase is poised for activation across the genome. Nat Genet 39:1507–1511
Musselman CA, Lalonde M-E, Côté J, Kutateladze TG (2012) Perceiving the epigenetic landscape through histone readers. Nat Struct Mol Biol 19:1218–1227
Nakagawa T, Kajitani T, Togo S et al (2008) Deubiquitylation of histone H2A activates transcriptional initiation via trans-histone cross-talk with H3K4 di- and trimethylation. Genes Dev 22:37–49
Nechaev S, Fargo DC, dos Santos G et al (2010) Global analysis of short RNAs reveals widespread promoter-proximal stalling and arrest of Pol II in Drosophila. Science 327:335–338

Neira J, Román-Trufero M, Contreras L et al (2009) The transcriptional repressor RYBP is a natively unfolded protein which folds upon binding to DNA. Biochemistry 48:1348–1360
Noordermeer D, Leleu M, Splinter E et al (2011) The dynamic architecture of Hox gene clusters. Science 334:222–225
Nora EP, Lajoie BR, Schulz EG et al (2012) Spatial partitioning of the regulatory landscape of the X-inactivation centre. Nature 485:381–385
O'Meara MM, Simon JA (2012) Inner workings and regulatory inputs that control Polycomb repressive complex 2. Chromosoma 121:221–234
Ogawa H, Ishiguro K-I, Gaubatz S et al (2002) A complex with chromatin modifiers that occupies E2F- and Myc-responsive genes in G0 cells. Science 296:1132–1136
Ohtsubo M, Yasunaga S, Ohno Y et al (2008) Polycomb-group complex 1 acts as an E3 ubiquitin ligase for Geminin to sustain hematopoietic stem cell activity. Proc Natl Acad Sci USA 105:10396–10401
Onder TT, Kara N, Cherry A et al (2012) Chromatin-modifying enzymes as modulators of reprogramming. Nature 483:598–602
Pan M-R, Peng G, Hung W-C, Lin S-Y (2011) Monoubiquitination of H2AX protein regulates DNA damage response signaling. J Biol Chem 286:28599–28607
Pandey RR, Mondal T, Mohammad F et al (2008) Kcnq1ot1 antisense noncoding RNA mediates lineage-specific transcriptional silencing through chromatin-level regulation. Mol Cell 32: 232–246
Papamichos-Chronakis M, Peterson CL (2012) Chromatin and the genome integrity network. Nat Rev Genet 14:62–75
Pasini D, Bracken AP, Jensen MR et al (2004) Suz12 is essential for mouse development and for EZH2 histone methyltransferase activity. EMBO J 23:4061–4071
Pasini D, Bracken AP, Hansen JB et al (2007) The Polycomb group protein Suz12 is required for embryonic stem cell differentiation. Mol Cell Biol 27:3769–3779
Pasini D, Cloos PAC, Walfridsson J et al (2010a) JARID2 regulates binding of the Polycomb repressive complex 2 to target genes in ES cells. Nature 464:306–310
Pasini D, Malatesta M, Jung HR et al (2010b) Characterization of an antagonistic switch between histone H3 lysine 27 methylation and acetylation in the transcriptional regulation of Polycomb group target genes. Nucleic Acids Res 38:4958–4969
Peng JC, Valouev A, Swigut T et al (2009) Jarid2/Jumonji coordinates control of PRC2 enzymatic activity and target gene occupancy in pluripotent cells. Cell 139:1290–1302
Pengelly AR, Copur Ö, Jäckle H et al (2013) A histone mutant reproduces the phenotype caused by loss of histone-modifying factor Polycomb. Science 339:698–699
Pereira CF, Piccolo FM, Tsubouchi T et al (2010) ESCs require PRC2 to direct the successful reprogramming of differentiated cells toward pluripotency. Cell Stem Cell 6:547–556
Peric-Hupkes D, Meuleman W, Pagie L et al (2010) Molecular maps of the reorganization of genome-nuclear lamina interactions during differentiation. Mol Cell 38:603–613
Petruk S, Sedkov Y, Johnston DM et al (2012) TrxG and PcG proteins but not methylated histones remain associated with DNA through replication. Cell 150:922–933
Philips-Cremins JE, Sauria MEG, Sanyal A et al (2013) Architectural protein subclasses shape 3D organization of genomes during lineage commitment. Cell 153:1281–1295
Qiao F, Bowie JU (2005) The many faces of SAM. Sci STKE 2005:re7
Rahl PB, Lin CY, Seila AC et al (2010) c-Myc regulates transcriptional pause release. Cell 141: 432–445
Ram O, Goren A, Amit I et al (2011) Combinatorial patterning of chromatin regulators uncovered by genome-wide location analysis in human cells. Cell 147:1628–1639
Rando OJ (2012) Combinatorial complexity in chromatin structure and function: revisiting the histone code. Curr Opin Genet Dev 22:148–155
Reddington JP, Perricone SM, Nestor CE et al (2013) Redistribution of H3K27me3 upon DNA hypomethylation results in de-repression of Polycomb-target genes. Genome Biol 14:R25

Ren X, Kerppola TK (2011) REST interacts with Cbx proteins and regulates Polycomb repressive complex 1 occupancy at RE1 elements. Mol Cell Biol 31:2100–2110
Reynolds N, Salmon-Divon M, Dvinge H et al (2012) NuRD-mediated deacetylation of H3K27 facilitates recruitment of Polycomb Repressive Complex 2 to direct gene repression. EMBO J 31:593–605
Reynolds N, O'Shaughnessy A, Hendrich B (2013) Transcriptional repressors: multifaceted regulators of gene expression. Development 140:505–512
Riley PD, Carroll SB, Scott MP (1987) The expression and regulation of Sex combs reduced protein in Drosophila embryos. Genes Dev 1:716–730
Ringrose L, Paro R (2007) Polycomb/Trithorax response elements and epigenetic memory of cell identity. Development 134:223–232
Rinn JL, Kertesz M, Wang JK et al (2007) Functional demarcation of active and silent chromatin domains in human HOX loci by noncoding RNAs. Cell 129:1311–1323
Roscic A, Möller A, Calzado MA et al (2006) Phosphorylation-dependent control of Pc2 SUMO E3 ligase activity by its substrate protein HIPK2. Mol Cell 24:77–89
Rowbotham SP, Barki L, Neves-Costa A et al (2011) Maintenance of silent chromatin through replication requires SWI/SNF-like chromatin remodeler SMARCAD1. Mol Cell 42:285–296
Sánchez C, Sánchez I, Demmers JAA et al (2007) Proteomics analysis of Ring1B/Rnf2 interactors identifies a novel complex with the Fbxl10/Jhdm1B histone demethylase and the Bcl6 interacting corepressor. Mol Cell Proteomics 6:820–834
Sarcinella E, Zuzarte PC, Lau PNI et al (2007) Monoubiquitylation of H2A.Z distinguishes its association with euchromatin or facultative heterochromatin. Mol Cell Biol 27:6457–6468
Satijn DP, Gunster MJ, der Vlag van J et al (1997) RING1 is associated with the Polycomb group protein complex and acts as a transcriptional repressor. Mol Cell Biol 17:4105–4113
Sawarkar R, Paro R (2010) Interpretation of developmental signaling at chromatin: the Polycomb perspective. Dev Cell 19:651–661
Schaaf CA, Kwak H, Koenig A et al (2013a) Genome-wide control of RNA polymerase II activity by cohesin. PLoS Genet 9:e1003382
Schaaf CA, Misulovin Z, Gause M et al (2013b) Cohesin and Polycomb proteins functionally interact to control transcription at silenced and active genes. PLoS Genet 9:e1003560
Scheuermann JC, De Ayala Alonso AG, Oktaba K et al (2010) Histone H2A deubiquitinase activity of the Polycomb repressive complex PR-DUB. Nature 465:243–249
Schmitges FW, Prusty AB, Faty M et al (2011) Histone methylation by PRC2 is inhibited by active chromatin marks. Mol Cell 42:330–341
Schoeftner S, Sengupta AK, Kubicek S et al (2006) Recruitment of PRC1 function at the initiation of X inactivation independent of PRC2 and silencing. EMBO J 25:3110–3122
Schoorlemmer J, Marcos-Gutiérrez C, Were F et al (1997) Ring1A is a transcriptional repressor that interacts with the Polycomb-M33 protein and is expressed at rhombomere boundaries in the mouse hindbrain. EMBO J 16:5930–5942
Schuettengruber B, Chourrout D, Vervoort M et al (2007) Genome regulation by Polycomb and trithorax proteins. Cell 128:735–745
Schuettengruber B, Martinez A-M, Iovino N, Cavalli G (2011) Trithorax group proteins: switching genes on and keeping them active. Nat Rev Mol Cell Biol 12:799–814
Schwartz YB, Kahn TG, Stenberg P et al (2010) Alternative epigenetic chromatin states of Polycomb target genes. PLoS Genet 6:e1000805
Seila AC, Calabrese JM, Levine SS et al (2008) Divergent transcription from active promoters. Science 322:1849–1851
Sexton T, Yaffe E, Kenigsberg E et al (2012) Three-dimensional folding and functional organization principles of the drosophila genome. Cell 148:458–472
Shao Z, Raible F, Mollaaghababa R et al (1999) Stabilization of chromatin structure by PRC1, a Polycomb complex. Cell 98:37–46

Shaver S, Casas-Mollano JA, Cerny RL, Cerutti H (2010) Origin of the Polycomb repressive complex 2 and gene silencing by an E(z) homolog in the unicellular alga Chlamydomonas. Epigenetics 5:301–312
Shen X, Liu Y, Hsu Y-J et al (2008) EZH1 mediates methylation on histone H3 lysine 27 and complements EZH2 in maintaining stem cell identity and executing pluripotency. Mol Cell 32:491–502
Shen X, Kim W, Fujiwara Y et al (2009) Jumonji modulates Polycomb activity and self-renewal versus differentiation of stem cells. Cell 139:1303–1314
Shen Y, Yue F, McCleary DF et al (2012) A map of the cis-regulatory sequences in the mouse genome. Nature 488:116–120
Shpargel KB, Sengoku T, Yokoyama S, Magnuson T (2012) UTX and UTY demonstrate histone demethylase-independent function in mouse embryonic development. PLoS Genet 8:e1002964
Sigova AA, Mullen AC, Molinie B et al (2013) Divergent transcription of long noncoding RNA/mRNA gene pairs in embryonic stem cells. Proc Natl Acad Sci USA 110:2876–2881
Simon JA, Kingston RE (2013) Occupying chromatin: Polycomb mechanisms for getting to genomic targets, stopping transcriptional traffic, and staying put. Mol Cell 49:808–824
Sinclair DAR, Syrzycka M, Macauley MS et al (2009) Drosophila O-GlcNAc transferase (OGT) is encoded by the Polycomb group (PcG) gene, super sex combs (sxc). Proc Natl Acad Sci USA 106:13427–13432
Sing A, Pannell D, Karaiskakis A et al (2009) A vertebrate Polycomb response element governs segmentation of the posterior hindbrain. Cell 138:885–897
Slifer EH (1942) A mutant stock of Drosophila with extra sex-combs. J Exp Zool 90:31–40
Smith M, Mallin DR, Simon JA, Courey AJ (2011) Small ubiquitin-like modifier (SUMO) conjugation impedes transcriptional silencing by the Polycomb group repressor sex comb on midleg. J Biol Chem 286:11391–11400
Sneeringer CJ, Scott MP, Kuntz KW et al (2010) Coordinated activities of wild-type plus mutant EZH2 drive tumor-associated hypertrimethylation of lysine 27 on histone H3 (H3K27) in human B-cell lymphomas. Proc Natl Acad Sci USA 107:20980–20985
Soria G, Polo SE, Almouzni G (2012) Prime, repair, restore: the active role of chromatin in the DNA damage response. Mol Cell 46:722–734
Sparmann A, van Lohuizen M (2006) Polycomb silencers control cell fate, development and cancer. Nat Rev Cancer 6:846–856
Stadler MB, Murr R, Burger L et al (2011) DNA-binding factors shape the mouse methylome at distal regulatory regions. Nature 480:490–495
Steffen PA, Fonseca JP, Ringrose L (2012) Epigenetics meets mathematics: towards a quantitative understanding of chromatin biology. Bioessays 34:901–913
Stepanik VA, Harte PJ (2012) A mutation in the E(Z) methyltransferase that increases trimethylation of histone H3 lysine 27 and causes inappropriate silencing of active Polycomb target genes. Dev Biol 364:249–258
Stock J, Giadrossi S, Casanova M et al (2007) Ring1-mediated ubiquitination of H2A restrains poised RNA polymerase II at bivalent genes in mouse ES cells. Nat Cell Biol 9:1428–1435
Strahl BD, Allis CD (2000) The language of covalent histone modifications. Nature 403:41–45
Strübbe G, Popp C, Schmidt A et al (2011) Polycomb purification by in vivo biotinylation tagging reveals cohesin and Trithorax group proteins as interaction partners. Proc Natl Acad Sci USA 108:5572–5577
Su W-J, Fang J-S, Cheng F et al (2013) RNF2/Ring1b negatively regulates p53 expression in selective cancer cell types to promote tumor development. Proc Natl Acad Sci USA 110: 1720–1725
Tan J, Jones M, Koseki H et al (2011) CBX8, a Polycomb group protein, is essential for MLL-AF9-induced leukemogenesis. Cancer Cell 20:563–575
Tavares L, Dimitrova E, Oxley D et al (2012) RYBP-PRC1 complexes mediate H2A ubiquitylation at Polycomb target sites independently of PRC2 and H3K27me3. Cell 148: 664–678

Taverna SD, Li H, Ruthenburg AJ et al (2007) How chromatin-binding modules interpret histone modifications: lessons from professional pocket pickers. Nat Struct Mol Biol 14:1025–1040
Tie F, Banerjee R, Stratton CA et al (2009) CBP-mediated acetylation of histone H3 lysine 27 antagonizes Drosophila Polycomb silencing. Development 136:3131–3141
Trojer P, Cao AR, Gao Z et al (2011) L3MBTL2 protein acts in concert with PcG protein-mediated monoubiquitination of H2A to establish a repressive chromatin structure. Mol Cell 42:438–450
Tsai M-C, Manor O, Wan Y et al (2010) Long noncoding RNA as modular scaffold of histone modification complexes. Science 329:689–693
Tsukada Y-I, Fang J, Erdjument-Bromage H et al (2006) Histone demethylation by a family of JmjC domain-containing proteins. Nature 439:811–816
van Arensbergen J, García-Hurtado J, Maestro MA et al (2013) Ring1b bookmarks genes in pancreatic embryonic progenitors for repression in adult β cells. Genes Dev 27:52–63
Vernimmen D, Lynch MD, De Gobbi M et al (2011) Polycomb eviction as a new distant enhancer function. Genes Dev 25:1583–1588
Vissers JHA, van Lohuizen M, Citterio E (2012) The emerging role of Polycomb repressors in the response to DNA damage. J Cell Sci 125:3939–3948
Voigt P, Leroy G, Drury WJ III et al (2012) Asymmetrically modified nucleosomes. Cell 151: 181–193
Wang H, Wang L, Erdjument-Bromage H et al (2004) Role of histone H2A ubiquitination in Polycomb silencing. Nature 431:873–878
Wang H, Zhai L, Xu J et al (2006) Histone H3 and H4 ubiquitylation by the CUL4-DDB-ROC1 ubiquitin ligase facilitates cellular response to DNA damage. Mol Cell 22:383–394
Wang R, Taylor AB, Leal BZ et al (2010) Polycomb group targeting through different binding partners of RING1B C-terminal domain. Structure 18:966–975
Whitcomb SJ, Basu A, Allis CD, Bernstein E (2007) Polycomb group proteins: an evolutionary perspective. Trends Genet 23:494–502
Wilson BG, Wang X, Shen X et al (2010) Epigenetic antagonism between Polycomb and SWI/SNF complexes during oncogenic transformation. Cancer Cell 18:316–328
Woo CJ, Kharchenko PV, Daheron L et al (2010) A region of the human HOXD cluster that confers Polycomb-group responsiveness. Cell 140:99–110
Wu C-Y, Kang H-Y, Yang W-L et al (2011) Critical role of monoubiquitination of histone H2AX protein in histone H2AX phosphorylation and DNA damage response. J Biol Chem 286: 30806–30815
Wu X, Johansen JV, Helin K (2013) Fbxl10/Kdm2b recruits Polycomb repressive complex 1 to CpG islands and regulates H2A ubiquitylation. Mol Cell 49:1134–1146
Xie R, Everett LJ, Lim H-W et al (2013) Dynamic chromatin remodeling mediated by Polycomb proteins orchestrates pancreatic differentiation of human embryonic stem cells. Cell Stem Cell 12:224–237
Xu M, Long C, Chen X et al (2010) Partitioning of histone H3-H4 tetramers during DNA replication-dependent chromatin assembly. Science 328:94–98
Xu K, Wu ZJ, Groner AC et al (2012) Ezh2 oncogenic activity in castration-resistant prostate cancer cells is Polycomb-independent. Science 338:1465–1469
Yang L, Lin C, Liu W et al (2011) ncRNA- and Pc2 methylation-dependent gene relocation between nuclear structures mediates gene activation programs. Cell 147:773–788
Yap KL, Li S, Muñoz-Cabello AM et al (2010) Molecular interplay of the noncoding RNA ANRIL and methylated histone H3 lysine 27 by Polycomb CBX7 in transcriptional silencing of INK4a. Mol Cell 38:662–674
You JS, De Carvalho DD, Dai C et al (2013) SNF5 is an essential executor of epigenetic regulation during differentiation. PLoS Genet 9:e1003459
Yu M, Mazor T, Huang H et al (2012) Direct recruitment of Polycomb repressive complex 1 to chromatin by core binding transcription factors. Mol Cell 45:330–343
Yuan W, Wu T, Fu H et al (2012) Dense chromatin activates Polycomb repressive complex 2 to regulate H3 lysine 27 methylation. Science 337:971–975

Zee BM, Britton LMP, Wolle D et al (2012) Origins and formation of histone methylation across the human cell cycle. Mol Cell Biol 32:2503–2514
Zeitlinger J, Stark A, Kellis M et al (2007) RNA polymerase stalling at developmental control genes in the Drosophila melanogaster embryo. Nat Genet 39:1512–1516
Zhang H, Smolen GA, Palmer R et al (2004) SUMO modification is required for in vivo Hox gene regulation by the Caenorhabditis elegans Polycomb group protein SOP-2. Nat Genet 36: 507–511
Zhang J, Bonasio R, Strino F et al (2013) SFMBT1 functions with LSD1 to regulate expression of canonical histone genes and chromatin-related factors. Genes Dev 27:749–766
Zhao J, Sun BK, Erwin JA et al (2008a) Polycomb proteins targeted by a short repeat RNA to the mouse X chromosome. Science 322:750–756
Zhao Y, Lang G, Ito S et al (2008b) A TFTC/STAGA module mediates histone H2A and H2B deubiquitination, coactivates nuclear receptors, and counteracts heterochromatin silencing. Mol Cell 29:92–101
Zhao J, Ohsumi TK, Kung JT et al (2010) Genome-wide identification of Polycomb-associated RNAs by RIP-seq. Mol Cell 40:939–953
Zhou W, Zhu P, Wang J et al (2008) Histone H2A monoubiquitination represses transcription by inhibiting RNA polymerase II transcriptional elongation. Mol Cell 29:69–80
Zhou VW, Goren A, Bernstein BE (2010) Charting histone modifications and the functional organization of mammalian genomes. Nat Rev Genet 12:7–18
Zhou Q, Li T, Price DH (2012) RNA polymerase II elongation control. Annu Rev Biochem 81: 119–143
Zhu B, Reinberg D (2011) Epigenetic inheritance: uncontested? Cell Res 21:435–441
Zhu P, Zhou W, Wang J et al (2007) A histone H2A deubiquitinase complex coordinating histone acetylation and H1 dissociation in transcriptional regulation. Mol Cell 27:609–621
Zhu J, Adli M, Zou JY et al (2013) Genome-wide chromatin state transitions associated with developmental and environmental cues. Cell 152:642–654

Chapter 6
The Role of Polycomb Group Proteins in Hematopoietic Stem Cell (HSC) Self-Renewal and Leukemogenesis

Vincent van den Boom, Hein Schepers, Annet Z. Brouwers-Vos, and Jan Jacob Schuringa

Abstract Throughout embryonic development as well as during adult hematopoiesis Polycomb group (PcG) proteins fulfill important functions. Stem cell self-renewal but also lineage fate decisions are controlled by PcGs. Besides a role in normal hematopoiesis, PcGs are often deregulated in various types of cancer, including human leukemias. Within this chapter we will discuss the current understanding of complex composition of canonical and noncanonical Polycomb repressive complexes, how these can contribute to normal hematopoiesis, and how PcG proteins can participate in leukemic transformation.

Keywords Polycomb repressive complex • Human hematopoietic stem cells • Leukemia • Noncanonical • Stem cell self-renewal • Differentiation • AML • ALL

6.1 Introduction

Polycomb group (PcG) proteins are involved in epigenetic repression of gene transcription and generally reside in two distinct complexes: Polycomb repressive complex 1 (PRC1) and 2 (PRC2) (Fig. 6.1) (Simon and Kingston 2009). According to the classical model for PcG-mediated repression, the PRC2 complex, containing the methyltransferase EZH2, EED, and SUZ12, first trimethylates histone H3 at lysine 27 (H3K27me3) (Cao et al. 2002; Kirmizis et al. 2004; Kuzmichev et al. 2002). This epigenetic modification recruits the five-subunit PRC1 complex, most likely via the chromobox domain of the CBX subunit of the PRC1 complex (Bernstein et al. 2006; Levine et al. 2002). Subsequently, the PRC1 complex, via its RING1 subunit, can ubiquitinate histone H2A at Lysine 119 (H2AK119ub)

V. van den Boom • H. Schepers • A.Z. Brouwers-Vos • J.J. Schuringa (✉)
Department of Experimental Hematology, University Medical Center Groningen, University of Groningen, Hanzeplein 1, DA13, 9700, RB Groningen, The Netherlands
e-mail: j.j.schuringa@umcg.nl

C. Bonifer and P.N. Cockerill (eds.), *Transcriptional and Epigenetic Mechanisms Regulating Normal and Aberrant Blood Cell Development*, Epigenetics and Human Health, DOI 10.1007/978-3-642-45198-0_6,

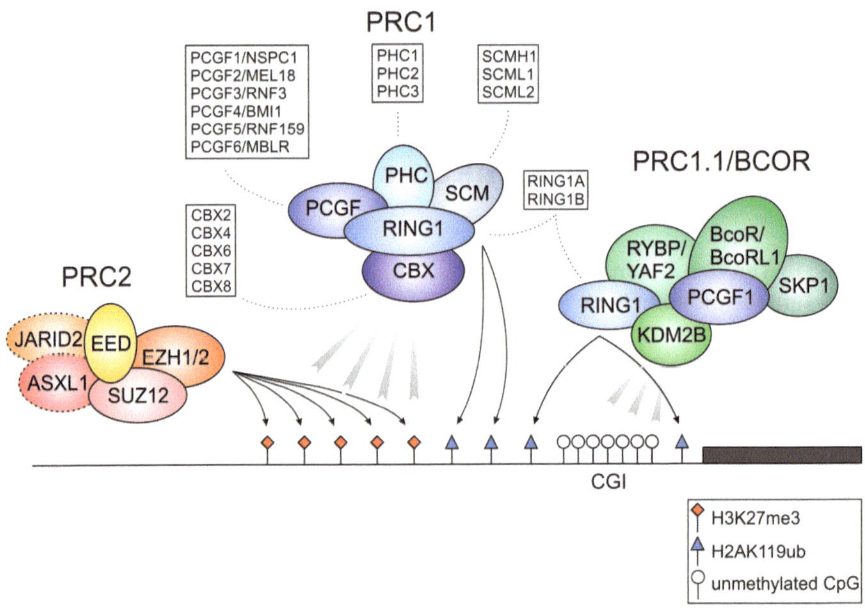

Fig. 6.1 Overview of canonical and noncanonical PcG complexes

(de Napoles et al. 2004; Wang et al. 2004). In humans, each of the PRC1 components has multiple paralog family members: six PCGF members (PCGF1/NSPC1, PCGF2/MEL18, PCGF3, PCGF4/BMI1, PCGF5, and PCGF6/MBLR), three PHC members (PHC1, PHC2, and PHC3), five CBX members (CBX2, CBX4, CBX6, CBX7, and CBX8), three Sex combs on midleg (SCM) members (SCML1, SCML2, and SMLH1), and two RING1 members (RING1A and RING1B). These paralogs allow a large diversity of distinct PRC1 complexes involved in PcG-mediated silencing (Whitcomb et al. 2007). This idea was supported by the identification of BMI1- and MEL18-containing PRC1 complexes and similarly PRC1 complexes with mutually exclusivity of CBX paralogs (Elderkin et al. 2007; Maertens et al. 2009; Vandamme et al. 2011). Indeed, PRC1 complexes were identified, which lack a CBX paralog family member but do contain RYBP or YAF2 and are targeted to PcG target genes independently of H3K27me3 (Gao et al. 2012; Tavares et al. 2012). Furthermore, some paralog family members like RING1A/B, PCGF1, and PCGF6 also reside in noncanonical PRC1 complexes such as the BCOR and E2F6 complexes (Gearhart et al. 2006; Ogawa et al. 2002; Sanchez et al. 2007; Trimarchi et al. 2001; Gao et al. 2012). Recently, the PRC1.1/BCOR complex was shown to target PcG genes by means of the KDM2B H3K36 demethylase (FBXL10), which specifically binds to unmethylated CpG islands (Fig. 6.1) (Farcas et al. 2012; He et al. 2013; Wu et al. 2013). These data show that noncanonical PRC1 complexes are involved in maintaining H2AK119ub levels as well. Also components of the core PRC2 complex display substoichiometric interactions with other (PcG) proteins. One such example is ASXL1, which is thought to mediate

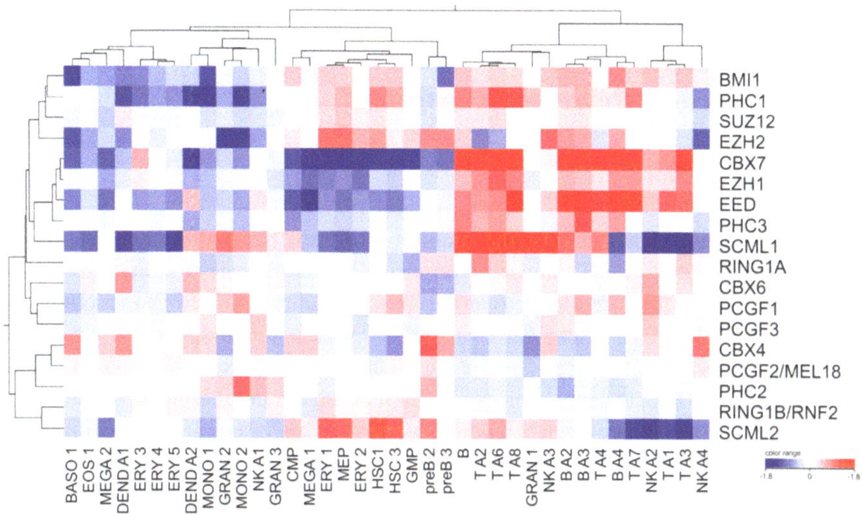

Fig. 6.2 Expression of PcG proteins in human CB cells throughout the hematopoietic compartment [based on data published by (Novershtern et al. 2011)]

recruitment and/or stabilization of the PRC2 complex to specific loci in the genome (Abdel-Wahab and Dey 2013).

The expression of PcG proteins throughout the hematopoietic compartment appears to be quite specific, with some PcGs being predominantly expressed in the immature stem/progenitor compartments, while other PcGs display a much more lineage-specific expression profile (Fig. 6.2). Taken together, these data suggest that many distinct PcG complexes exist that are likely to fulfill different functions. Yet, very little is known about possible differences in complex composition in, for instance, self-renewing hematopoietic stem cells (HSCs) versus non-self-renewing progenitors, or in normal versus leukemic stem cells (LSCs). Moreover, it is currently unclear whether specific loci would be preferentially occupied by certain PcG complexes across these different cell types. Lastly, we do not know exactly how repression of these different loci would be controlled by the various PRC complexes. Accumulating recent data indicates that molecular pathways regulating cell cycle, apoptosis, senescence, reactive oxygen species (ROS) metabolism and DNA repair are at least in part under control of PcG proteins. Whether PcG proteins fulfill similar roles in leukemia is currently unclear.

Understanding the molecular mechanisms by which PcG proteins affect stem cell fate will increase our insights into the biology of HSCs and will also aid in understanding the process of leukemic transformation, and ultimately in the identification of novel drug targets that might facilitate the eradication of LSCs. Here, we will provide an overview of the current understanding of the role of PcG proteins in HSC self-renewal and leukemogenesis.

6.2 Polycomb Function in Normal Hematopoiesis

6.2.1 Polycomb Repressive Complex 2

6.2.1.1 EZH1 and EZH2

Of all PRC2 complex subunits, the methyltransferase EZH2 has been studied in most detail. To study EZH2 function in murine hematopoiesis, conditional knockout models were required due to embryonic lethality of *Ezh2* knockout mice (O'Carroll et al. 2001). Using *Mx-Cre;Ezh2$^{fl/fl}$* mice it was first shown that B cell development was impaired in an *Ezh2* null background (Su et al. 2003). More specifically, pre-B cell generation was affected and accumulation of immature B cells in the bone marrow (BM) was observed. This was caused by impaired rearrangement of the immunoglobin heavy chain gene leading to reduced μ-chain expression. Transplantation experiments emphasized that this phenotype was cell autonomous and niche independent. In contrast, pro-B cell development was unaffected indicating that EZH2 has stage-specific functions in B-lymphopoiesis. EZH2 also plays a role in T-lymphocyte generation as *Ezh2* null thymocytes fail to develop further than early CD4/CD8 double negative CD44^{int}/CD25^{hi} stage in the thymus in a transplant setting (Su et al. 2005). In contrast to adult BM cells, fetal liver cells are highly sensitive to deletion of *Ezh2* (Mochizuki-Kashio et al. 2011). Using a *Tie2-Cre;Ezh2$^{fl/fl}$* conditional model it was shown that levels of LSK, CMP, GMP, and MEP cells were reduced in conditional *Ezh2$^{-/-}$* whole fetal livers at E12.5, indicating an important role for EZH2 during fetal hematopoiesis. Strikingly, *Ezh2* null cells could efficiently reconstitute hematopoiesis in adult BM in a competitive transplant setting. In addition, H3K27me3 levels, which are very low in fetal liver *Ezh2* KO cells, were largely restored in *Ezh2$^{-/-}$* fetal liver cells that reconstituted the BM of recipient mice. The authors suggest a model where *Ezh1*, which is strongly induced in adult versus fetal LSK cells, may rescue *Ezh2* null cells in an adult BM environment by inducing and/or maintaining proper H3K27me3 levels in the context of an EZH1–PRC2 complex (Mochizuki-Kashio et al. 2011). Murine EZH2 overexpression studies showed enhanced long-term repopulation potential in a serial transplant setting (Kamminga et al. 2006).

It was previously shown that in *Ezh2* null embryonic stem (ES) cells H3K27me3 levels are reduced but not completely lost (Shen et al. 2008). Furthermore, EZH1 was shown to interact with PRC2 components and locates to EZH2 target genes whereby most likely (lower) levels of the H3K27me3 mark can be maintained at these loci (Margueron et al. 2008; Shen et al. 2008). Conditional deletion of both *Ezh1* and *Ezh2* in the mouse skin showed a complete loss of H3K27me3 levels further supporting a model where *Ezh1* can at least partially compensate for loss of *Ezh2* (Ezhkova et al. 2011). Recently, the specific role of *Ezh1* in hematopoiesis was addressed in a hematopoietic conditional knockout model (Hidalgo et al. 2012). Conditional deletion of *Ezh1* in the hematopoietic system using a *Vav-Cre;Ezh1$^{fl/fl}$* mouse model showed development of BM hypoplasia most likely due to reduced numbers of LT-HSC, ST-HSC, MPP, and CLP cells, whereas CMP, GMP, and MEP cells were unaffected. Competitive transplantation experiments using a tamoxifen-inducible *Ezh1* conditional knockout model showed a loss of repopulating ability, which was attributed to increased levels of senescence in the primitive stem cell compartment. Increased senescence in this model is likely driven by de-repression of p16^{INK4A} since a *Vav-Cre;Ezh1*$^{fl/fl}$*;Ink4a-Arf*$^{fl/fl}$ showed restoration of the senescence phenotype. Similar to *Ezh2* knockout mice, B-cell lymphopoiesis was also affected in *Ezh1* conditional knockout mice (Su et al. 2003; Hidalgo et al. 2012). In contrast to the *Ezh2* conditional knockout model, deletion of *Ezh1* in the hematopoietic system also led to a reduction in pro-B cell numbers. Furthermore, fetal liver hematopoiesis was not affected by *Ezh1* deletion, suggesting that *Ezh2* is indeed the dominant H3K27 methyltransferase during fetal embryogenesis (Mochizuki-Kashio et al. 2011; Hidalgo et al. 2012).

Taken together, EZH1 and EZH2 seem to have both separate as well as overlapping functions in the hematopoietic system, whereby the loss of one methyltransferase can be partially compensated for by the other depending on the hematopoietic cell lineage and developmental stage.

6.2.1.2 EED

Apart from the methyltransferases EZH1 and EZH2, the PRC2 complex is also composed of the core subunits EED and SUZ12. The role of EED in murine hematopoiesis was first described by Lessard and colleagues using mice heterozygous for an *Eed* null allele or homozygous for an *Eed* hypomorphic allele (Lessard et al. 1999). Previously, homozygous *Eed* null embryos were shown to exert severe gastrulation defects (Schumacher et al. 1996; Faust et al. 1995). Biochemical analysis showed that EED interacts with H3K27me3 via its carboxy-terminal domain and that this interaction stimulates PRC2 methyltransferase activity suggesting a role for EED in propagation of H3K27me3 through DNA replication (Margueron et al. 2009). In murine hematopoiesis, EED likely acts as a negative regulator of cell proliferation and *Eed*$^{3354/+}$ mice showed increased levels of myeloid progenitors and pre-B cells. In addition, at older age these mice displayed hyperproliferation of both lymphoid and myeloid cells (Lessard et al. 1999). However, classical target genes like p16^{INK4A} and p19^{ARF} and Hox genes were not upregulated in this mouse model (Lessard et al. 1999).

6.2.1.3 SUZ12

Similar to *Ezh2* null and *Eed* null mice, also homozygous deletion of the *Suz12* gene resulted in early embryonic lethality (Pasini et al. 2004). This was accompanied by a complete loss of H3K27me2/3 suggesting that PRC2 activity is completely lost in these embryos. In murine hematopoiesis, an ENU-induced loss-of-function mutation in the *Suz12* gene was identified as a suppressor of thrombocytopenia and HSC defects in *cMpl*$^{-/-}$ mice (Majewski et al. 2008). More recently, it was shown that crossing heterozygous *Eed* null or *Ezh2* null mutations in a *cMpl*$^{-/-}$ background also resulted in an improvement in the thrombocytopenic phenotype, whereby increased white blood cell counts were observed (Majewski et al. 2010).

6.2.2 Polycomb Repressive Complex 1

As described above, the PRC1 complex consists of five subunits (PCGF, CBX, PHC, RING1, and SCM), and each of these subunits has a family of paralogs both in mice and humans. Below we will describe the involvement of the various paralog family members in hematopoiesis.

6.2.2.1 PCGF Paralog Family

The PCGF paralog family is composed of PCGF1 (NSPC1), PCGF2 (MEL18), PCGF3, PCGF4 (BMI1), PCGF5, and PCGF6 (MBLR). Of these, BMI1 has been studied most extensively in hematopoiesis. BMI1 was initially observed to be oncogenic in a retroviral integration site screen where it was identified as a collaborating hit in MMLV-induced B-cell lymphomas in Eμ-*myc* transgenic mice (Haupt et al. 1991; van Lohuizen et al. 1991). Homozygous deletion of *Bmi1* in mice resulted in reduced numbers of hematopoietic progenitors and more differentiated cells, eventually leading to hematopoietic failure (van der Lugt et al. 1994). More detailed analysis showed that BMI1 has a central regulatory role in self-renewal of HSCs by inducing symmetrical cell division both in mouse and human model systems (Park et al. 2003; Iwama et al. 2004; Lessard and Sauvageau 2003; Rizo et al. 2008). Accordingly, *Bmi1*$^{-/-}$ mice displayed dramatically reduced HSC frequencies. Mechanistically, the role of BMI1 in regulating HSC self-renewal is partially explained by its ability to repress the *Ink4a-Arf* locus (Jacobs et al. 1999; Park et al. 2003). Expression of p16$^{\mathrm{INK4A}}$ and p19^{ARF} in HSCs induces cell cycle arrest and p53-mediated cell death. Loss of BMI1 expression most likely results in a decreased H2AK119 ubiquitinating activity of the PRC1 complex at the *Ink4a-Arf* locus, inducing expression of p16$^{\mathrm{INK4A}}$ and p19^{ARF}. Importantly, the hematopoietic phenotype of *Bmi1*$^{-/-}$ mice is not only dependent on the induction of Ink4a/Arf. *Bmi1*$^{-/-}$;*Ink4a-Arf*$^{-/-}$ double knockout mice showed a partial recovery of hematopoietic cell counts but did not show a complete reversal of the *Bmi1* null phenotype, suggesting that other pathways are also involved (Bruggeman et al. 2005). In a separate study, competitive transplant experiments showed that *Bmi1*$^{-/-}$;*Ink4a-Arf*$^{-/-}$ cells showed peripheral blood chimerism levels comparable to wild-type cells, whereas *Bmi1*$^{-/-}$ cells did not contribute at all (Oguro et al. 2006). Although HSC self-renewal clearly correlated with p16$^{\mathrm{INK4A}}$ and p19^{ARF} expression, the typical hypoplastic BM phenotype of *Bmi1* null mice was not completely rescued by deletion of the *Ink4a-Arf* locus.

Gene expression analysis of the HSC/MPP fraction in *Bmi1* null mice surprisingly showed premature transcriptional activation of Ebf1 and Pax5, two regulators of B cell lymphopoiesis (Oguro et al. 2010). BMI1 is directly recruited to the promoters of these genes and *Bmi1*$^{-/-}$;*Ink4a-Arf*$^{-/-}$ cells are biased toward the B cell lineage at the expense of T cell lymphopoiesis.

A candidate *Ink4a-Arf*-independent pathway for BMI1 in controlling HSCs is regulation of ROS in the cell. *Bmi1*$^{-/-}$ mice displayed impaired mitochondrial function due to increased expression of PcG target genes involved in ROS metabolism (Liu et al. 2009). Long-term HSCs from *Bmi1*$^{-/-}$ mice showed increased levels of ROS and treatment of mice with the antioxidant *N*-acetylcysteine (NAC) resulted in a rescue of thymocyte cell numbers compared to non-treated *Bmi1*$^{-/-}$ mice. Furthermore, de-regulated ROS metabolism induced activation of the DNA damage response pathway. Knockout of *Chk2*, a component of the DNA damage response pathway, in a *Bmi1*$^{-/-}$ background resulted in partial reversal of the thymocyte phenotype observed in *Bmi1*$^{-/-}$ mice and increased LSK cell numbers in the BM. However, *Bmi1*$^{-/-}$;*Chk2*$^{-/-}$ cells failed to give long-term repopulation in a competitive transplant setting similar to *Bmi1*$^{-/-}$ mice, indicating that the HSC self-renewal phenotype of *Bmi1* null mice is not rescued by *Chk2* deletion. Interestingly, knocking down BMI1 in CD34^{+} human cord blood cells also induced increased ROS levels and apoptosis (Rizo et al. 2009). These data clearly show that apart from the role of BMI1 in regulating cell cycle and senescence through the *Ink4a-Arf* locus, BMI1 is also implicated in other pathways regulating oxygen metabolism.

PCGF1 has previously been identified as a member of the BCOR complex, containing RING1A, RING1B, BCOR, SKP1, and KDM2B (FBXL10) (Gearhart et al. 2006). Recently, the H3K36-specific demethylase KDM2B was shown to target this noncanonical PRC1 complex to unmethylated CpG islands in the promoters of lineage-specific genes in ES cells (Fig. 6.1) (Farcas et al. 2012; He et al. 2013; Wu et al. 2013). Depletion of KDM2B resulted in derepression, a loss of RING1B binding and decreased H2AK119ub levels at these target genes. In mouse hematopoietic cells, PCGF1 was picked up as a factor negative regulating self-renewal of lineage negative cells in a *Runx1* conditional knockout setting (Ross et al. 2012). PCGF1 knockdown was shown to induce expression of HOXA cluster genes and led to a loss of H2AK119ub at the promoters of these genes. Knockdown of PCGF1 in human CD34^{+} CB cells was recently shown to give a mild growth reduction in in vitro cultures and a loss of CFC frequency (van den Boom et al. 2013).

Mel18$^{-/-}$ mice display severe posterior transformations of the axial skeleton in a manner similar to *Bmi1* null mice (Akasaka et al. 1996; van der Lugt et al. 1994). Similarly, analysis of the hematopoietic compartments showed a strong reduction in cellularity in the thymus and spleens in both knockout models. Both models do not completely overlap in terms of phenotype since *Mel18* null mice develop abnormalities of the lower intestine, whereas *Bmi1*-deficient mice specifically display a cerebellum defect. However, generation of *Mel18*$^{-/-}$;*Bmi1*$^{-/-}$ mice showed that both proteins synergistically regulate *Hox* cluster expression (Akasaka et al. 2001). Competitive transplants using E14 fetal liver cells from *Mel18*$^{-/-}$ mice showed a only a mild reduction in repopulating activity compared to a severe loss of repopulating activity in *Bmi1*$^{-/-}$ cells (Iwama et al. 2004). Knockdown of MEL18 in human cord blood (CB) cells resulted in a dramatic reduction in proliferation in both BM stromal cocultures and liquid cultures and a strong reduction in CFC plating efficiency (van den Boom et al. 2013). However, while BMI1 depletion led to an upregulation of p14^{ARF} and p16^{INK4A}, MEL18 knockdown did not induce the expression of these genes. These data show that, although both involved in the regulation of hematopoiesis, BMI1 and MEL18 have distinct functions and BMI1 clearly has a much more prominent role in maintaining HSC self-renewal.

Concerning PCGF3, PCGF5, and PCGF6 no information is currently available on their potential function in murine hematopoiesis. PCGF6 knockdown in human CD34^{+} CB cells did not or only mildly affect cell growth in hematopoietic growth assays and did not alter CFC plating efficiency, suggesting that this gene is not involved in human hematopoiesis (van den Boom et al. 2013).

6.2.2.2 CBX Paralog Family

The CBX paralog family encompasses the CBX2, CBX4, CBX6, CBX7, and CBX8 proteins, which are homologues of the *Drosophila melanogaster* Polycomb protein and target the PRC1 complex to the chromatin through their interaction with H3K27me3. However, the chromodomains of the various CBX proteins show significant differences in substrate specificity in in vitro assays, since some interact with both H3K27me3 and H3K9me3, or selectively with H3K9me3 (Bernstein et al. 2006). Recently, two independent studies reported different functional modalities between CBX paralog family members in mouse ES cells (Morey et al. 2012; O'Loghlen et al. 2012). These data showed a key role for CBX7 in undifferentiated ES cells, whereas CBX2 and CBX4 were essential for lineage commitment. Gene expression data from human hematopoietic cell subsets also showed that the various CBX proteins are differentially expressed in HSCs, progenitors, and differentiated cells (Fig. 6.2) (van den Boom et al. 2013; Novershtern et al. 2011).

Two *Cbx2* (*M33*) knockout models were generated both of which showed skeletal transformations (Katoh-Fukui et al. 1998; Core et al. 1997). Interestingly, Coré and colleagues observed a reduction of cell numbers in the spleen, thymus, and BM of *Cbx2*$^{-/-}$ mice and found both B and T cell lymphopoiesis to be affected (Core et al. 2004; Core et al. 1997). However, competitive transplant experiments using fetal liver cells from *Cbx2*$^{-/-}$ mice did not show any change in repopulation capability compared to wild-type cells (Iwama et al. 2004). In contrast, knockdown of CBX2 in human CD34^{+} CB cells resulted in a severe phenotype in both the HSC and progenitor compartment (van den Boom et al. 2013). CBX2 knockdown cells displayed a reduced proliferation and increased apoptosis. Furthermore, the *CDKN1A* gene (*p21*) was directly targeted by CBX2 and concurrent knockdown of CBX2 and p21 partially rescued the CBX2 knockdown phenotype. It is possible that these differences in CBX2 function between human and mouse hematopoiesis are a consequence of species-specific differences in CBX2 function. For example, where p21 is a bona fide target of CBX2/PRC1 and PRC2 complexes in human $CD34^+$ CB cells, it has not been annotated as a PcG target gene in mouse cells.

CBX4 has the unique biochemical property among other CBXs that it is a SUMO E3 ligase stimulating UBC9-dependent sumoylation of the transcriptional repressor CtBP by tethering both to Polycomb bodies (Kagey et al. 2003). Interestingly, in human epidermal stem cells CBX4 preserved a slow-cycling and undifferentiated state and also prevents from senescence induction (Luis et al. 2011). Recent generation of a conditional *Cbx4* knockout mouse model showed that thymic development was impaired. However, this phenotype was caused by dysfunction of the thymic epithelial cells rather than a consequence of intrinsic defects of the developing thymocytes (Liu et al. 2013). CBX4 overexpression in murine hematopoietic cells mildly suppressed proliferation, inhibited replating potential in CFC assays, and reduced CAFC activity (Klauke et al. 2013). In contrast, knockdown of CBX4 in human $CD34^+$ CB cells strongly reduced cell proliferation in BM stromal cocultures and liquid cultures and diminished CFC plating efficiency (van den Boom et al. 2013).

Cbx7 knockout mice are born in Mendelian ratios and display a slight increase in body length (Forzati et al. 2012). MEFs from these mice showed increased proliferation and decreased induction of senescence and adult $Cbx7^{-/-}$ mice develop tumors in the liver and the lungs. However, effects of *Cbx7* knockout on the hematopoietic system were not studied in this report. Contrasting the tumor-suppressor role of CBX7 in other tissues, overexpression of CBX7 in murine hematopoietic cells increases proliferation in liquid cultures and enhances the in vitro proliferative capacity of LT-HSCs and ST-HSCs (Klauke et al. 2013). CBX7 overexpression in LT-HSCs and subsequent transplantation showed enhanced numbers of ST-HSCs and MPPs but not LT-HSCs at late stages after transplantation. The authors suggest a model where CBX7 preserves a HSC self-renewing state by specifically repressing genes involved in differentiation. CBX7 knockdown in human CB $CD34^+$ cells led to a mild proliferative disadvantage and reduction in CFC frequencies (van den Boom et al. 2013).

To study the role of CBX8 in normal hematopoiesis, both constitutive and conditional CBX8 knockout models were used (Tan et al. 2011). CBX8 deletion did not lead to changed peripheral blood cells numbers and bone marrow cellularity. Furthermore, LT-HSC functionality was not affected as shown by competitive transplant assays, suggesting that CBX8 is not involved in normal hematopoiesis. In contrast, in a separate study, overexpression of CBX8 in murine hematopoietic cells showed a phenotype opposite of CBX7, where frequencies of LT-HSCs, ST-HSCs, and MPP are all decreased (Klauke et al. 2013). Knockdown of CBX8 in human $CD34^+$ CB cells resulted in a mild negative phenotype in terms of cell proliferation, similar to CBX7 knockdown (van den Boom et al. 2013). Myeloid differentiation was not affected by knockdown of either CBX7 or CBX8.

6.2.2.3 RING1 Paralog Family

The RING1 paralog family consists of RING1A and RING1B. Both proteins possess H2AK119 ubiquitinating activity (Buchwald et al. 2006; de Napoles et al. 2004; Wang et al. 2004). Deletion of both *Ring1a* and *Ring1b* was necessary to achieve genome-wide depletion of H2AK119ub (Endoh et al. 2008; Stock et al. 2007). Furthermore, H2A ubiquitination is essential for repression of developmental genes and preserving ES cell identity (Endoh et al. 2012). Although H2AK119 ubiquitination is in part BMI1 dependent, recent studies showed RING1B-dependent H2AK119 ubiquitination in the context of noncanonical PRC1 complexes (*i.e.,* PRC1-RYBP, BCOR complexes) that are targeted to the chromatin independently of H3K27me3 (Farcas et al. 2012; Tlc et al. 2013; Wu et al. 2013; Kallin et al. 2009). *Ring1a*$^{-/-}$ mice showed skeletal abnormalities and slight deregulation of Hox gene expression (del Mar Lorente et al. 2000). In contrast, where *Ring1a* null mice are viable, deletion of *Ring1b* led to a severe gastrulation defects and embryonic lethality (Voncken et al. 2003). The generation of a conditional *Mx-Cre;Ring1b*$^{fl/fl}$ mouse model allowed the investigation of Ring1B function in hematopoiesis (Cales et al. 2008). This study showed that *Ring1b* depletion led to a mild increase in the primitive stem/progenitor compartment (LSK, Lin^{-} cells), whereas the total BM compartment was slightly decreased. In line with this phenotype, p16$^{\mathrm{INK4A}}$ was selectively activated in more differentiated cells but not in the Lin^{-} compartment, whereas the positive cell cycle regulator CycD2 was upregulated in most compartments. Simultaneous deletion of *p16*^{INK4A}, *p19*^{ARF}, and *Ring1b* resulted in a rescue of the negative effects on proliferation of the mature compartments, although the hyperproliferative phenotype of the primitive compartment was not corrected. Knockdown of RING1B in human CB CD34^{+} cells showed severe defects in long-term expansion and progenitor frequencies (van den Boom et al. 2013).

6.2.2.4 PHC Paralog Family

Both in mice and humans the Polyhomeotic paralog family encompasses three members: PHC1, PHC2, and PHC3. *Phc1*$^{-/-}$ (*Rae28*) mice showed skeletal abnormalities and a reduced spleen size (Takihara et al. 1997). Further studies showed involvement of *Phc1* in early B cell lymphopoiesis (Tokimasa et al. 2001). Furthermore, *Phc1* null mice displayed reduced hematopoietic progenitor activity in the fetal liver and loss of long-term repopulating activity of fetal liver cells in competitive transplant experiments (Ohta et al. 2002; Kim et al. 2004). *Phc2*$^{-/-}$ mice also showed skeletal abnormalities and deregulated expression of Hox genes and p16$^{\mathrm{INK4A}}$ and p19^{ARF} (Isono et al. 2005). In contrast to *Phc1*, null mice deletion of *Phc2* did not have an apparent phenotype in hematopoiesis.

6.2.2.5 SCM Paralog Family

The sex combs on midleg (SCM) paralog family consists of SCMH1, SCML1 and SCML2. SCMH1 was first identified as a substoichiometric subunit of the PRC1 complex (Levine et al. 2002). Protein interaction studies have showed that also SCML1 and SCML2 of this paralog family can interact with other PRC1 subunits in human cells (van den Boom et al. 2013; Gao et al. 2012). *Scmh1*$^{-/-}$ mice display characteristic features resembling other PcG knockout mouse models like skeletal abnormalities; however, studies concerning the hematopoietic system have not been reported (Takada et al. 2007).

6.3 Polycomb Function in Leukemia

6.3.1 Polycomb Repressive Complex 2

6.3.1.1 EZH1 and EZH2

Unlike its related family member EZH1, a potential role for the PRC2 member EZH2 in hematological malignancies has been heavily investigated. Both tumor suppressor as well as oncogenic functions have been described, which will be discussed here. Besides being overexpressed in a variety of malignancies (Bachmann et al. 2006), EZH2 has shown aberrant expression in mantle cell lymphoma (Visser et al. 2001), Hodgkin lymphoma (Raaphorst et al. 2000), and non-Hodgkin lymphoma (Table 6.1) (van Kemenade et al. 2001). In complex karyotype AML, overexpression of EZH2 has also been observed (Grubach et al. 2008), although in a panel of 60 AMLs separated in $CD34^+$ and $CD34^-$ fractions, higher expression was only observed within the $CD34^-$ fraction, while the expression was significantly reduced within the $CD34^+$ compartment compared to normal bone marrow $CD34^+$ cells (Fig. 6.3) (Bonardi et al. 2013; de Jonge et al. 2011). In contrast, overexpression of EZH1 was observed within the AML $CD34^+$ compartment compared to NBM $CD34^+$ cells (Bonardi et al. 2013; de Jonge et al. 2011). More recently, mutations in the *EZH2* gene were discovered in ~22 % of follicular and diffuse large B cell lymphomas. Mutation of Tyrosine residue 641 (Y641) in the SET domain led to a severe decrease in enzymatic H3K27 trimethylation activity (Morin et al. 2010). EZH2 mutations were also reported for various myeloid malignancies like CMML, MDS, and AML (Abdel-Wahab et al. 2011; Ernst et al. 2012; Makishima et al. 2010; Cancer Genome Atlas Research Network. 2013). Although Y641 mutations were not found, these myeloid malignancies carried other *EZH2* mutations (R690, N693 and H694) that also affected the SET domain and thereby H3K27 trimethylation activity (Abdel-Wahab et al. 2011; Ernst et al. 2012; Makishima et al. 2010). In T-ALL ~18 % of the patients displayed truncating or missense mutations before the SET domain in

Table 6.1 Involvement of PcG proteins in human hematological malignancies

Canonical PRC complex	Protein	Disease	Aberrancy	ref
PRC2	EZH1	AML	Overexpressed (AML $CD34^+$)	(de Jonge et al. 2011)
	EZH2	AML	Overexpressed (AML CK)	(Grubach et al. 2008)
		AML	Reduced expression (AML $CD34^+$)	(de Jonge et al. 2011)
		AML	Reduced expression in CBF-mutated AML	(Grubach et al. 2008)
		AML, MDS, MPN, CMML	Loss-of-function mutations SET domain (R690, N693, H694)	(Abdel-Wahab et al. 2011;Ernst et al. 2012;Makishima et al. 2010;Ernst et al. 2010;Nikoloski et al. 2010)
		MCL, NHL, HL	Overexpressed	(Visser et al. 2001;Raaphorst et al. 2000;van Kemenade et al. 2001)
		FL,DLBCL	Loss-of-function mutations SET domain (Y641)	(Morin et al. 2010)
		T-ALL	Missense mutations	(Ntziachristos et al. 2012)
		T-ALL	Reduced expression	(Simon et al. 2012)
		ETP-ALL	Loss-of-function deletion/mutations SET domain (R684)	(Zhang et al. 2012)
	EED	MDS/MPN	Loss-of-function deletion/mutations (protein stability, EZH2 interaction, H3K27 me3 binding)	(Ueda et al. 2012)
		ETP-ALL	Loss-of-function deletion/mutations (R684)	(Zhang et al. 2012;Ueda et al. 2012)
	SUZ12	T-ALL	Loss-of-function deletion/mutations (S369fs, others)	(Ntziachristos et al. 2012;Zhang et al. 2012)
		MPN	Loss-of-function deletion/mutations	(Brecqueville et al. 2012;Score et al. 2012)
PRC2 interactors	ASXL1	MDS, MPN, AML, CML	Mutations (decrease protein stability)	(Abdel-Wahab et al. 2011;Shih et al. 2012;Abdel-Wahab et al. 2012;Schnittger et al. 2013)
	JARID2	MDS/MPN, AML	Deletion	(Puda et al. 2012)
PRC1	PCGF2/MEL18	AML	Overexpressed (AML CK)	(Grubach et al. 2008)
		AML	Reduced expression in CBF-mutated AML	(Grubach et al. 2008)
	PCGF4/BMI1	AML	Overexpressed (AML $CD34^+$)	(de Jonge et al. 2011;Chowdhury et al. 2007;van Gosliga et al. 2007)
		AML	Overexpressed, predicts prognosis	(Chowdhury et al. 2007)
		MDS	Overexpressed, predicts prognosis	(Mihara et al. 2006;Xu et al. 2011)
		CML	Overexpressed, predicts prognosis	(Mohty et al. 2007)
		HL, NHL	Overexpressed	(Raaphorst et al. 2000;van Kemenade et al. 2001;Dukers et al. 2004)
	CBX7	FL	Overexpressed	(Scott et al. 2007)
		AML	Reduced expression (AML $CD34^+$)	(de Jonge et al. 2011)
	RING1A	AML	Overexpressed (AML $CD34^+$)	(de Jonge et al. 2011)
		AML/MDS	Overexpressed	(Xu et al. 2011)
	PHC1/RAE28	B-ALL	Reduced expression	(Tokimasa et al. 2001)
		AML	Overexpressed (AML $CD34^+$)	(de Jonge et al. 2011)
Non-canonical PRC complex	**Protein**	**Disease**	**Aberrancy**	
PRC1.1/BCOR complex	BCOR	AML	Mutated	(Grossmann et al. 2011)
		AML	Overexpressed (AML $CD34^+$)	(Bonardi et al. 2013;de Jonge et al. 2011)
	BCORL1	AML	Mutated	(Li et al. 2011)
Role of PcGs proteins in hematological malignancy model systems				
gene	**model**		**phenotype**	
EZH2	Conditional $Ezh2^{-/-}$ mouse model		T-ALL	(Simon et al. 2012)
	mBM MLL-AF9 Tx model		Leukemia development in 2nd mice impaired in $Ezh2^{-/-}$ cells	(Neff et al. 2012;Tanaka et al. 2012)
	mBM Ezh2 overexpression		Myeloproliferative disease,	(Herrera-Merchan et al. 2012)

(continued)

Table 6.1 (continued)

EED	mBM MLL-AF9/NRAS(G12D) Tx model	Reduced leukemic growth upon EED knockdown	(Shi et al. 2012)
SUZ12	mBM MLL-AF9/NRAS(G12D) Tx model	Reduced leukemic growth upon SUZ12 knockdown	(Shi et al. 2012)
PCGF4/BMI1	Primary AML	Reduced long-term growth upon BM1 knockdown	(Rizo et al. 2009)
	huCB CD34$^+$ BCR-ABL/BMI1	Myeloid/lymphoid transformation in vitro, B-ALL in vivo	(Rizo et al. 2010)
	mBM BCR-ABL/Bmi1 Tx model	B-ALL	(Sengupta et al. 2012;Waldron et al. 2011)
	mBM MYC/Bmil1 Tx model	B-ALL	(van Lohuizen M. et al. 1991)
	mBM HoxA9/Meis1 Tx model	Loss of serial transplantation in BMI1$^{-/-}$ BM	(Lessard and Sauvageau 2003)
	mBM AML1-ETO Tx model	Loss of leukemic CFC replating in BMI1$^{-/-}$ BM	(Smith et al. 2011)
	mBM PLZF-RARα Tx model	Loss of leukemic CFC replating in BMI1$^{-/-}$ BM	(Smith et al. 2011)
	mBM MLL-AF9 Tx model	Loss of transformation in *HoxA9*$^{-/-}$/*Bmi1*$^{-/-}$ background	(Smith et al. 2003)
CBX8	mBM MLL-AF9/ENL Tx model	Loss of transformation in *Cbx8*$^{-/-}$ background	(Tan et al. 2011b)
ASXL1	mBM NRAS(G12D) Tx model	loss of ASXL1 accelerated onset of leukemia	(Abdel-Wahab et al. 2012)

AML acute myeloid leukemia, *MDS* myelodysplastic syndrome, *MPN* myeloproliferative neoplasms, *CMML* chronic myelomonocytic leukemia, *MCL* mantle cell lymphoma, *NHL* non-hodgkin lymphoma, *HL* hodgkin lymphoma, *DLCLB* diffuse large B-cell lymphoma, *T-ALL* T cell acute lymphoblastic leukemia, *ETP-ALL* Early T-cell precursor acute lymphoblastic leukemia, *FL* follicular lymphoma, *CML* chronic myeloid leukemia, *mBM* mouse bone marrow; *Tx* transplantation; *CB* human cord blood; *CK* complex karyotype

EZH2 (Ntziachristos et al. 2012). Furthermore, 7 % of the T-ALLs demonstrated alterations in another PRC2 member, SUZ12 (Ntziachristos et al. 2012). Although no deletions or mutations of *EZH2* could be found in the remaining human T-ALL cases, transcriptome analysis indicated that EZH2 expression levels were strongly reduced (Simon et al. 2012). Conditional deletion of *Ezh2* in a mouse model led to the occurrence of T-ALL leukemias (Simon et al. 2012). H3K27 di- and trimethylation were reduced, but not absent, indicating that EZH1 may partially compensate for the loss of EZH2 (Simon et al. 2012). These data suggest that EZH2 functions as a tumor suppressor in T-ALL and that loss of EZH2 contributes to this malignancy. In a large proportion of the T-ALLs with *EZH2* or other PRC2 mutations, also oncogenic mutations in *NOTCH1* were observed (65 %) (Ntziachristos et al. 2012). This suggests that *NOTCH1* mutations and *EZH2* mutations collaborate in the induction of T-ALL.

In contrast, in an MLL-AF9 model of leukemia, deletion of *Ezh2* did not affect leukemic initiation, but maintenance of the leukemia was impaired. This is most likely the result of a reduction in the frequency of LSCs (Neff et al. 2012; Tanaka et al. 2012). H3K27me3 levels were strongly reduced, but not absent in these cells (Neff et al. 2012; Tanaka et al. 2012). As leukemic growth was not completely absent and H3K27 trimethylation persisted on a subset of genes, also here EZH1 may partially compensate for the loss of EZH2 (Neff et al. 2012; Shi et al. 2013). These results imply that EZH2 is required for the proper maintenance of MLL-AF9-dependent LSCs, which is opposite from its tumor suppressor function in T-ALL. As also knockdown of Eed and Suz12 led to a reduction in MLL-AF9/NRASG12D-induced leukemic growth, this further strengthens the dependence of these

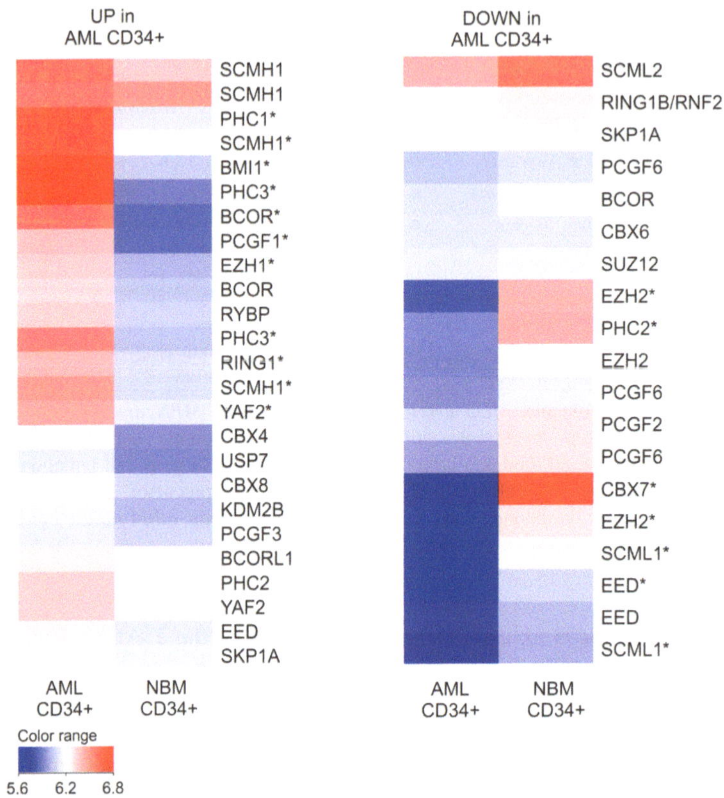

Fig. 6.3 Expression of PcG proteins in AML CD34⁺ and NBM CD34⁺ cells. Transcriptome profiling was performed on AML CD34⁺ (n = 60) and normal BM CD34⁺ cells (n = 40) (Bonardi et al. 2013; de Jonge et al. 2011). Differentially expressed genes were identified using an unpaired t-test with multiple testing correction (Benjamini-Hochberg, $p < 0.01$) and are marked with an *asterisk*

leukemias on PRC2 function (Shi et al. 2013). MLL-AF9-induced gene expression has been suggested to resemble an ES-like gene expression signature, which also comprises a PRC2 and MYC module (Kim et al. 2010; Somervaille et al. 2009). This MYC module is strongly reduced upon deletion of *Ezh2*, but only in secondary and not in primary leukemias. This suggests that PRC complexes play a role in the MLL-AF9-induced MYC module expression necessary for disease progression (Neff et al. 2012). In contrast, *EZH2*-deficient T-ALLs show an enhanced expression of MYC (Simon et al. 2012), suggesting that MLL-AF9 changes the gene regulatory functions of EZH2. Recent data showed that Menin, a partner of Mixed Lineage Leukemia (MLL), binds to the *Ezh2* promoter and enhances its expression (Thiel et al. 2013). Enhanced expression of EZH2 results in a myeloproliferative disease (Herrera-Merchan et al. 2012), but together with Menin causes a block in myeloid differentiation (Thiel et al. 2013). Since mutations that affect EZH2 methyltransferase activity have also been shown to induce MPD/MDS (Abdel-Wahab et al. 2011; Makishima et al. 2010), this suggests that this activity needs to be carefully regulated, as both hypo- and hyperactivity can result in myeloid malignancies. Therefore, the contrasting role of EZH2 as a tumor suppressor or oncogene can most likely be explained by the different genetic context in which EZH2 plays a role. The fact that PRC2 complexes are recruited by PML-RARα and PLZF-RARα to RAREs while PRC1 is only recruited by PLZF-RARα further adds to this notion (Boukarabila et al. 2009; Villa et al. 2007).

6.3.1.2 EED

Unlike *Ezh2*, deletion of *Eed*, another PRC2 family member, completely abrogated MLL-AF9-mediated leukemia initiation (Neff et al. 2012). But although primary MLL-AF9 targets were not affected by *Ezh2* deletion, *Eed* deletion did affect the expression of these primary targets. These discrepancies can be reconciled with the following assumptions: 1; EZH1 partially can compensate for EZH2 activity and 2; both EZH1 and EZH2 activity is dependent upon EED. The first assumption has been confirmed in an MLL-AF9/NRASG12D leukemic model, where reduction in both EZH1 and EZH2 were necessary to reduce leukemic growth in vitro (Shi et al. 2013). Furthermore, EZH2 activity is dependent on EED (Denisenko et al. 1998) and EED mutants have been observed in patients with myelodysplastic syndrome and early T-cell precursor ALL (Ueda et al. 2012; Zhang et al. 2012). These mutations in EED will affect EZH1 and EZH2 activities and thereby contribute to aberrant myelopoiesis and lymphopoiesis. Recent evidence for this hypothesis stems from the fact that besides *EZH2* mutations in ALL, also *EED* and *SUZ12* mutations have been discovered in early T-cell precursor ALL (Zhang et al. 2012), indicating that it is the overall PRC2 activity that is inactivated. Although EED and SUZ12 knockdown reduced in vitro leukemic growth of MLL-AF9/NRASG12D cells, the reduction in in vivo leukemia development was rather modest (Shi et al. 2013). Together this suggests that PRC2 activity is required for leukemic expansion, rather than leukemic engraftment.

6.3.1.3 SUZ12

A role for SUZ12 in T-ALL development can be extrapolated from the fact that in ~7–12 % of the T-ALL cases, loss of function mutations was detected (Ntziachristos et al. 2012; Zhang et al. 2012). Downregulation of SUZ12 resulted in the expression of NOTCH1 target genes, comparable to EZH2 downregulation (Ntziachristos et al. 2012). The observation that ~33 % of early T-cell precursor ALLs has mutations in *SUZ12*, *EED*, and *EZH2* suggests that PRC2 has a tumor suppressor function in NOTCH1-mediated T-ALL.

In myeloid malignancies such as CML or other myeloproliferative neoplasms (MPNs), the role of SUZ12 is less clear. In non-CML MPNs, like polycythemia vera (PV), essential thrombocytopenia, (ET) and myelofibrosis (MF), rare mutations in *SUZ12* have been observed (Brecqueville et al. 2012; Score et al. 2012), but a higher incidence (31 %) of *ASXL1* mutations was identified (Brecqueville et al. 2012). As *ASXL1* mutations have been shown to result in loss of PRC2 activity (Abdel-Wahab et al. 2012), it is likely that these are mutually exclusive, since they all point to a tumor suppressive mechanism of PRC2 in these non-CML MPNs. This is consistent with deletion of another substoichiometric PRC2 member JARID2 in MPNs (Puda et al. 2012) and these data are in line with a tumor suppressive function of PRC2.

However, progression of chronic phase CML has been linked to the increased expression of SUZ12 in a WNT-dependent manner (Pizzatti et al. 2010). Knockdown of SUZ12 induced differentiation of these chronic phase blasts, indicating that PRC2 activity blocks differentiation. Similar, in an MLL-AF9/NRASG12D model of myeloid leukemia, reduction of SUZ12 severely hampered leukemic expansion, as did reduction in EED and EZH2 (Shi et al. 2013). These data are more in line with an oncogenic role of PRC2 activity, although it affects leukemic expansion more than leukemic engraftment/initiation. It is currently unclear if this PRC2 activity is only oncogenic in myeloid malignancies and tumor suppressive in lymphoid malignancies or whether this is imposed by the collaborating oncogenic insults like MLL-AF9 or NOTCH1.

6.3.1.4 ASXL1

ASXL1 is putative member of the PcG protein family and was recently identified as an important factor involved in regulating PRC2 activity (Abdel-Wahab et al. 2012). Mutations in ASXL1 have been detected in a variety of hematologic malignancies, like CML, MPN, AML, and MDS (Abdel-Wahab et al. 2011; Shih et al. 2012; Cancer Genome Atlas Research Network. 2013), with a worse overall survival in MDS (Bejar et al. 2011) and AML (Metzeler et al. 2011). These mutations were shown to decrease the stability of the protein, as ASXL1 was undetectable in cells with *ASXL1* mutations (Abdel-Wahab et al. 2012). This suggests that ASXL1 has a tumor suppressive function in these myeloid malignancies. Downregulation of ASXL1 led to upregulation of a gene signature that resembled the gene signature observed when MLL-AF9 is expressed in cells (Abdel-Wahab et al. 2012). The upregulated genes included genes in the posterior HOXA cluster (i.e., HOXA9) that are both classical MLL-AF9 and PcG target genes. These data suggested that ASXL1 might modulate the activity of PcG complexes on the *HOXA* locus. Indeed, knockdown of ASXL1 led to a decrease in H3K27me3 levels on the HOXA cluster, which was due to impaired recruitment of the PRC2 complex, and hence loss of EZH2 activity at the HOXA cluster (Abdel-Wahab et al. 2012). ASXL1 was found to directly interact with members of the PRC2 complex suggesting a direct role for ASXL in PRC2 targeting. In an NRAS

(G12D) model of leukemia, it was subsequently shown that loss of ASXL1 not only increased the self-renewal of leukemic cells but also accelerated the onset of leukemia (Abdel-Wahab et al. 2012). This data is consistent with leukemias induced by MLL-AF9/NRAS(G12D), where the HOX-mediated self-renewal component was supplied by the MLL-AF9 fusion (Shi et al. 2013).

ASXL1 has also been shown to form a complex with the deubiquitinase BAP1, which removes the monoubiquitin from histone H2A at Lysine 119 (Scheuermann et al. 2010). This complex (PR-DUB) which was originally identified in *Drosophila melanogaster* binds PcG target genes and is essential for repression of *Hox* genes in *Drosophila* (Scheuermann et al. 2010). This suggested a model where *ASXL1* mutations interfere with both PRC2 and PRC1 function, coordinately contributing to upregulation of the *HOX* cluster. However, both in mouse and human hematopoietic cells, knockout/knockdown of BAP1 did not induce expression of the *HOXA* genes, and loss of ASXL1 did not result in a decrease in H2AK119Ub (Abdel-Wahab et al. 2012; Abdel-Wahab and Dey 2013; Dey et al. 2012). Using a *Bap1*$^{fl/fl}$*;creERT2* conditional knockout model it was shown that loss of *Bap1* induces a MDS/CMML-like disease (Dey et al. 2012). Interestingly, the authors show that apart from influencing PcG-mediated repression, the ASXL1/BAP1 complex also mediates ubiquitination, and thereby the stability of the epigenetic regulators such as OGT (Dey et al. 2012). OGT mediates O-GlcNAcylation of Ser-112 of histone H2B and can be recruited to CpG-rich transcription start sites of active genes via its interaction with TET proteins. It will be interesting to see how abrogation of these distinct functional pathways of ASXL1 contributes to leukemic transformation in ASXL1 mutant cells.

6.3.2 Polycomb Repressive Complex 1

6.3.2.1 PCGF Paralog Family

From the 6 PCGF family members, the best-described role in leukemic transformation is for PCGF4/BMI1. Although PCGF2/MEL18 expression has been shown to decrease upon differentiation of the leukemic cell line HL60, no causal role has been described yet (Jo et al. 2011). The role of BMI1 has been better documented. The first evidence for a possible involvement of BMI1 in the development of hematological malignancies came from murine models in which BMI1 was identified as a cooperating factor with MYC in the induction of B cell lymphomagenesis (van Lohuizen et al. 1991). Next, Sauvageau and colleagues demonstrated that BMI1 not only determines the proliferative capacity of normal stem cells but also of LSCs (Lessard and Sauvageau 2003). In a mouse model in which coexpression of the oncogenes HOXA9 and MEIS1 resulted in a quick onset of myeloid leukemia, no disease was observed in secondary recipients in a *Bmi1*-deficient background (Lessard and Sauvageau 2003). These data indicated that BMI1 is essential for the maintenance of HOXA9-MEIS1 LSCs in vivo. Similar, serial replating of leukemic colony-forming units after AML-ETO or PLZF-RARα-mediated transformation was severely hampered by deletion of *Bmi1* (Smith et al. 2011). However, lower proliferation and serial replating of leukemic stem and progenitor cells deficient for *Bmi1* were not observed in leukemias with an MLL-AF9 background (Smith et al. 2011; Tan et al. 2011). In that setting, HOXA9 was strongly upregulated by MLL-AF9 and was able to bypass oncogene-induced senescence in the absence of BMI1, as MLL-AF9 transformation was lost in the $Bmi1^{-/-}$;$Hoxa9^{-/-}$ background (Smith et al. 2011).

While BMI1 is required for maintenance and self-renewal of HSCs, no data has been reported indicating that overexpression of BMI1 is sufficient to induce leukemia as a single event. However, a role for BMI1 during myeloid and lymphoid leukemic transformation has been inferred from studies indicating a correlation between high expression of BMI1 and disease progression in various leukemias and MDS/MPD (Chowdhury et al. 2007; Grubach et al. 2008; Mihara et al. 2006; Mohty et al. 2007; Xu et al. 2011; Yong et al. 2011; de Jonge et al. 2011; van Gosliga et al. 2007; Raaphorst et al. 2000; van Kemenade et al. 2001; Dukers et al. 2004). In primary human acute myeloid leukemia patient samples, BMI1 was among the highest upregulated PcG genes in AML $CD34^+$ cells compared to normal BM $CD34^+$ cells (Fig. 6.3) and downregulation of BMI1 impaired long-term expansion and self-renewal properties of LSCs (Rizo et al. 2009). Upon aging, BMI1 expression goes down in lymphoid progenitors, resulting in an upregulation of $p16^{INK4A}$ and $p19^{ARF}$ (Signer et al. 2008). However, reintroduction of Bmi1 was sufficient to render aged lymphoid progenitor cells susceptible for BCR-ABL-induced transformation (Signer et al. 2008). While one paper indicated that a C18Y polymorphism exists in BMI1 that resulted in an increase in proteasome-mediated degradation (Zhang and Sarge 2009), activating mutations in BMI1 have not been described. Although BMI1 function appears to be predominantly regulated at the expression level in tumor cells, posttranslational modifications have also been reported that alter the activity of BMI1. For instance, AKT-induced serine phosphorylation has been shown to inhibit BMI1-mediated HSC self-renewal, INK4A-ARF repression, and its ability to promote tumor growth (Liu et al. 2012). In contrast, in prostate cancer it was demonstrated that AKT-mediated phosphorylation can enhance the oncogenic potential of BMI1, independent of INK4A-ARF repression (Nacerddine et al. 2012). In AML1-ETO-positive leukemias, it was shown that aberrant signaling via mutated cKit can cause loss of Polycomb-mediated repression (Ray et al. 2013). Together, these data indicate that cytokine/growth factor signaling can directly influence PcG proteins, and it will be very interesting to further delineate the role of posttranslational modifications of PcG proteins in the future.

Over the past decades, a concept has emerged in which leukemia is regarded as a multistep process in which a number of (epi)genetic events are required in order to induce overt disease. As discussed above, to date it has not been demonstrated that expression of BMI1 alone is sufficient to induce leukemia. However, various lines of evidence suggest that BMI1 might act as an important collaborating factor in the transformation process. In a tumor model in which the oncogene TLS-ERG was introduced into human hematopoietic progenitors, in a limited number of cases the transduced cells underwent a stepwise transformation and immortalization in which upregulation of BMI1 was identified as one of the cooperating hits (Warner et al. 2005). BMI1 can cooperate with H-RAS to induce aggressive breast cancer with brain metastases (Datta et al. 2007). Primary human epithelial cells could efficiently be immortalized by co-expressing hTERT and BMI1 (Haga et al. 2007). One report indicated that some of the X-linked SCID patients transplanted with retrovirally transduced $CD34^+$ BM cells in order to re-express the IL2γ receptor developed T cell leukemias. Integration site analysis revealed that in one patient the vectors had integrated near the *BMI1* gene (Hacein-Bey-Abina et al. 2008), leaving open the possibility that an upregulation of BMI1 might have contributed to the development of leukemia in this patient as well. Together with BCR-ABL, BMI1 is able to induce myeloid and lymphoid transformation in vitro and a serially transplantable $CD19^+$ B-lymphoid leukemia in vivo (Rizo et al. 2010; Waldron et al. 2011). This collaboration between BCR-ABL and BMI1 was recently confirmed in a mouse model where B-lymphoid progenitors were transformed to B-ALL blasts upon overexpression of BMI1 (Sengupta et al. 2012). Where BMI1 induced self-renewal, the collaborating BCR-ABL oncogene prevented apoptosis and maintained proliferation (Sengupta et al. 2012).

Our understanding of the mechanisms by which BMI1 exerts its phenotypes is steadily increasing. While gain of BMI1 function might be involved in extending the lifespan of normal and LSCs by bypassing senescence, more direct control over the fate of HSC divisions appears to exist also. Although the molecular mechanisms remain to be elucidated, the symmetry of cell division of HSCs is directed toward a more symmetric mode of cell division upon overexpression of BMI1 (Iwama et al. 2004). While under normal homeostasis HSCs might divide asymmetrically, resulting in one new HSC and one daughter cell that has lost stem cell integrity and will differentiate, high BMI1 levels might dictate a more symmetric distribution of specific proteins, mRNAs, or other metabolites during mitosis whereby stem cell integrity is maintained in both daughter cells. How BMI1 would be involved in such processes remains unclear.

Protection against oxidative stress and apoptosis emerges as an important BMI1-downstream pathway as well, either by reducing p53 levels via BMI1-mediated repression of the INK4A/ARF locus or via modulation of the oxidative stress response in an INK4A/ARF-independent manner. Downmodulation of BMI1 resulted in an accumulation of ROS levels, both in knockout mouse models as well as in human $CD34^+$ cells transduced with lentiviral BMI1 RNAi vectors (Liu et al. 2009; Rizo et al. 2009). In other non-hematopoietic model systems, it was also shown that downmodulation of BMI1 results in p53-mediated apoptosis, whereby ROS levels were increased (Alajez et al. 2009; Chatoo et al. 2009). The induction of ROS in the absence of BMI1 could be counteracted by treatment with antioxidants such as NAC, but appeared to be independent of INK4A/ARF in hematopoietic cells (Liu et al. 2009). However, in *Atm*-deficient astrocytes, oxidative stress resulted in an increase in ROS levels, which inhibited cell growth via a MEK-ERK1-BMI1-$p16^{INK4A}$-dependent pathway (Kim and Wong 2009). In *Bmi1*$^{-/-}$ mice, the increase in ROS coincided with an increase in DNA damage and an activation of the DNA damage repair pathways, and treatment with NAC or removal of CHK2 at least partially restored some the phenotypes (Liu et al. 2009). A number of genes that have been described to regulate intracellular redox homeostasis were found to be derepressed in *Bmi1*$^{-/-}$ mice (Liu et al. 2009). In human CD34^{+} cells, downmodulation of BMI1 coincided with decreased expression of FOXO3 (Rizo et al. 2009). Foxo3a$^{-/-}$ HSCs were defective in their competitive repopulation capacity, lost their quiescence, and displayed elevated ROS levels (Miyamoto et al. 2007). Thus, BMI1 might be required to protect hematopoietic stem/progenitor cells from apoptosis or loss of quiescence induced by oxidative stress conditions. In human leukemias, besides facilitating symmetric stem cell divisions, the LSC might utilize enhanced expression of BMI1 as a mode to protect itself from oxidative stress.

6.3.2.2 CBX Paralog Family

From the five CBX family members, CBX2,4,6,7, and 8, only CBX7 and CBX8 have been described to play a role in leukemia or lymphoma. CBX7 is a PcG member that has been shown to extend cellular lifespan in similar fashion as BMI1 (Gil et al. 2004). CBX7 was demonstrated to bypass senescence through repression of the *Ink4a/Arf* and the *Cdkn1a* locus (Gil et al. 2004). Enhanced expression of CBX7 was observed in human follicular lymphomas, and T cell lymphomas also appeared when CBX7 was ectopically expressed in murine lymphoid cells (Scott et al. 2007). In these lymphomas a decrease in p16^{INK4A} and p14^{ARF} expression was observed, suggesting that the CBX7-mediated bypass of cellular senescence contributes to the malignant phenotype. However, the long latency and incomplete penetrance of these lymphomas suggested that increased expression of CBX7 is not sufficient to drive lymphomagenesis on its own and that additional collaborating events are necessary (Scott et al. 2007). Overexpression of CBX7 in 5FU-treated BM cells recently confirmed the contribution of CBX7 to T-cell malignancies, although in 30 % of the malignancies also erythroid and immature leukemias were observed (Klauke et al. 2013).

The CBX family member CBX8 has been shown to interact with AF9 and ENL (Hemenway et al. 2001; Monroe et al. 2011; Garcia-Cuellar et al. 2001; Mueller et al. 2007). These proteins are common fusion partners for the *Mixed Lineage Leukemia* (*MLL*) gene in juvenile and adult leukemias. Although CBX family members are usually viewed as transcriptional repressors, in the context of MLL-AF9, MLL-ENL, and potentially also other MLL fusion proteins, CBX8 is actually required to induce gene expression (Tan et al. 2011). CBX8 was shown to be essential for MLL-AF9-induced HoxA9 expression (Tan et al. 2011), which had been demonstrated to be required for MLL-AF9-induced leukemic transformation (Faber et al. 2009; Zeisig et al. 2004). Indeed, deletion of *Cbx8* completely abolished MLL-AF9- and MLL-ENL-induced leukemogenesis (Tan et al. 2011). Depletion of CBX8 did not affect the binding of MLL fusion proteins to their target promoters, but rather affected the binding of RNA polymerase II and subsequent promoter activation (Tan et al. 2011). As CBX family members were originally discovered to have repressive functions, this furthermore highlights that oncogenic MLL-AF9 and MLL-ENL fusions alter the function of CBX8.

6.3.2.3 RING1 Paralog Family

For the PRC1 complex member RING1, not much is known regarding its role in leukemia. RING1 expression is generally higher in MDS and AML as compared to normal bone marrow cells (Xu et al. 2011), and particularly within the AML $CD34^+$ compartment, RING1A is significantly upregulated compared to normal BM $CD34^+$ cells with no significant differences in RING1B expression (Fig. 6.3) (Bonardi et al. 2013; de Jonge et al. 2011). By means of protein pulldown assays, RING1 was found to interact with the MLL-fusion partners AF9 and ENL (Monroe et al. 2011; Mueller et al. 2007). This suggested that MLL-AF9 and MLL-ENL could potentially bind PRC1 through RING1 as well. However, for AF9 it was shown that AF9 does not directly bind RING1B, but uses CBX8 as an intermediate that binds both (Hemenway et al. 2001). As also ENL has been shown to bind to CBX8 (Mueller et al. 2007), it is therefore conceivable that CBX8 also acts as an intermediate between MLL-ENL and RING1. Very little is currently known about the different or overlapping roles of RING1A and RING1B in leukemogenesis which will need to be further investigated in detail. Intriguingly, *Ring1b* depletion had no effect on MLL-AF9-mediated transformation (Tan et al. 2011). Since depletion of Cbx8 did severely affect MLL-AF9-induced leukemia (Tan et al. 2011), this suggests that MLL-AF9-induced leukemic transformation might be independent from canonical PRC1 signaling.

6.3.2.4 PHC Paralog Family

Expression analysis of the PHC family member PHC1 in BM mononuclear cells from patients with B-ALL indicated a complete loss of PHC1 expression (Tokimasa et al. 2001). Deletion of *Phc1* in mice leads to a complete block in B cell maturation between the pro-B and pre-B cell stages in neonatal splenocytes (Tokimasa et al. 2001), whereas T cell development appeared normal. This is consistent with its constitutive expression during B cell development and suggests that loss of PHC1 is underlying the B cell developmental arrest in ALL. How PHC1 expression is lost during B-ALL development is unclear, as Southern blot analysis has shown that both alleles were present in the B-ALL samples (Tokimasa et al. 2001). In AML $CD34^+$ cells, PHC1 and PHC3 expression was found to be significantly upregulated in the AML $CD34^+$ compartment (Fig. 6.3) (Bonardi et al. 2013; de Jonge et al. 2011).

6.3.2.5 SCM Paralog Family

Not much is known about a role for the SCML family members SCMH1, SCML1, and SCML2 in leukemic transformation. Grubach and colleagues have investigated the expression of various PcG genes in a panel of 126 AML patients (Grubach et al. 2008). SCML2 appeared to be significantly higher expressed in AML compared with normal bone marrow cells, especially in AML patients with an underlying t(8;21) or inv(16) translocation. We investigated gene expression in a panel of AML patients subdivided in $CD34^+$ and $CD34^-$ fractions and observed that SCMH1 was significantly upregulated and SCML1 was significantly downregulated in AML $CD34^+$ cells compared to normal BM $CD34^+$ cells (Bonardi et al. 2013; de Jonge et al. 2011).

6.4 Noncanonical PRC1 Complexes

Apart from the canonical five-subunit PRC1 complex, various other noncanonical PRC1 complexes have been described. One category of noncanonical PRC1 complexes contains RYBP or YAF2 instead of a CBX subunit and are targeted to chromatin in a manner independent of H3K27me3 (Gao et al. 2012; Tavares et al. 2012). In addition, the noncanonical BCOR and E2F6 complexes contain PCGF1/NSPC1 and PCGF6/MBLR, respectively (Gao et al. 2012; Gearhart et al. 2006; Ogawa et al. 2002; Sanchez et al. 2007; Trimarchi et al. 2001; Qin et al. 2012; Trojer et al. 2011). Recently, the E2F6 complex was shown to have H2AK119 ubiquitination activity through its RING1 subunits and to induce a repressive chromatin structure (Trojer et al. 2011; Gao et al. 2012). Whereas the complex is essential for mouse development, recent knockdown studies in human CB $CD34^+$ cells did not show a dramatic phenotype upon PCGF6 shRNA expression and suggests that PCGF6 does not play an important role in normal hematopoiesis (Qin et al. 2012; van den Boom et al. 2013).

Interestingly, recent whole-exome sequencing approaches in AML have identified recurrent mutations in the BCOR complex subunits *BCOR* and *BCORL1* (Tiacci et al. 2012; Grossmann et al. 2011; Li et al. 2011). BCOR and BCORL1 are large nuclear proteins that act as corepressor of BCL6 or other transcriptional regulators. Translocations have also been described like t(X;17)(p11;q12) resulting in the formation of a BCOR-RARα fusion in a patient with acute promyelocytic leukemia (APL) (Yamamoto et al. 2010). Most of the identified alterations are nonsense mutations, out-of-frame insertions/deletions, or splice site mutations that most likely result in truncated proteins that lack the C-terminal nuclear receptor recruitment motif. This would suggest that BCOR/BCORL1 would act as a tumor-suppressor gene that is inactivated by mutations in a subset of AML (Tiacci et al. 2012; Li et al. 2011). On the other hand, it has been observed that BCOR together with its binding partners PCGF1 and RING1A in the noncanonical PRC1 complex are among the most highly upregulated PcG genes in AML $CD34^+$ cells compared to normal BM CD34$^+$ cells (Fig. 6.3) (de Jonge et al. 2011), suggesting that these increased expression levels might participate in the process of leukemic transformation. In line with this notion, it was recently shown that ectopic expression of KDM2B, which targets the PRC1.1/BCOR complex to chromatin, is sufficient to transform hematopoietic progenitors (He et al. 2011). Reversely, depletion of *Kdm2b* significantly impaired Hoxa9/Meis1-induced leukemic transformation by mediating silencing of p15^{INK4B} expression through active demethylation of H3K36me2, suggesting that KDM2B functions as an oncogene (He et al. 2011). Recently, KDM2B was shown to tether the BCOR complex to non-methylated CpG islands in developmental genes, enforcing gene repression by RING1-dependent H2AK119 ubiquitination (Farcas et al. 2012; He et al. 2013; Wu et al. 2013). Further mechanistic studies are required to gain further insight into the possible mechanisms by which the noncanonical BCOR complex might contribute to leukemia development.

6.5 Summarizing Remarks

Challenges for the future lie in the further unraveling of gene networks that are under the control of PcG proteins and how regulation of these genes affects the fate of normal hematopoietic and LSCs. It is becoming clear that multiple distinct PRC1 complexes can be composed of (most likely) specific functions, and it will be interesting to determine whether differences in complex composition exist between HSCs and progenitors, or between leukemic and normal stem cells, and ultimately how complex composition might relate to specific target gene regulation. Also, PcG proteins can act independently of the canonical PRC1 complexes, but whether and how these noncanonical PRC1 complexes may participate in leukemia development still needs to be unraveled. Posttranslational modifications of PcG proteins are beginning to be identified, although our understanding of how these affect Polycomb signaling are far from complete. Future studies will help to further delineate the role of PcG proteins in the normal hematopoietic system as well as in the process of leukemic transformation.

References

Abdel-Wahab O, Dey A (2013) The ASXL-BAP1 axis: new factors in myelopoiesis, cancer and epigenetics. Leukemia 27:10–15

Abdel-Wahab O, Pardanani A, Patel J, Wadleigh M, Lasho T, Heguy A, Beran M, Gilliland DG, Levine RL, Tefferi A (2011) Concomitant analysis of EZH2 and ASXL1 mutations in myelofibrosis, chronic myelomonocytic leukemia and blast-phase myeloproliferative neoplasms. Leukemia 25:1200–1202

Abdel-Wahab O, Adli M, Lafave LM, Gao J, Hricik T, Shih AH, Pandey S, Patel JP, Chung YR, Koche R, Perna F, Zhao X, Taylor JE, Park CY, Carroll M, Melnick A, Nimer SD, Jaffe JD,

Aifantis I, Bernstein BE, Levine RL (2012) ASXL1 mutations promote myeloid transformation through loss of PRC2-mediated gene repression. Cancer Cell 22:180–193
Akasaka T, Kanno M, Balling R, Mieza MA, Taniguchi M, Koseki H (1996) A role for mel-18, a Polycomb group-related vertebrate gene, during theanteroposterior specification of the axial skeleton. Development 122:1513–1522
Akasaka T, van Lohuizen M, van der Lugt N, Mizutani-Koseki Y, Kanno M, Taniguchi M, Vidal M, Alkema M, Berns A, Koseki H (2001) Mice doubly deficient for the Polycomb Group genes Mel18 and Bmi1 reveal synergy and requirement for maintenance but not initiation of Hox gene expression. Development 128:1587–1597
Alajez NM, Shi W, Hui AB, Yue S, Ng R, Lo KW, Bastianutto C, O'Sullivan B, Gullane P, Liu FF (2009) Targeted depletion of BMI1 sensitizes tumor cells to P53-mediated apoptosis in response to radiation therapy. Cell Death Differ 16:1469–1479
Bachmann IM, Halvorsen OJ, Collett K, Stefansson IM, Straume O, Haukaas SA, Salvesen HB, Otte AP, Akslen LA (2006) EZH2 expression is associated with high proliferation rate and aggressive tumor subgroups in cutaneous melanoma and cancers of the endometrium, prostate, and breast. J Clin Oncol 24:268–273
Bejar R, Stevenson K, Abdel-Wahab O, Galili N, Nilsson B, Garcia-Manero G, Kantarjian H, Raza A, Levine RL, Neuberg D, Ebert BL (2011) Clinical effect of point mutations in myelodysplastic syndromes. N Engl J Med 364:2496–2506
Bernstein E, Duncan EM, Masui O, Gil J, Heard E, Allis CD (2006) Mouse polycomb proteins bind differentially to methylated histone H3 and RNA and are enriched in facultative heterochromatin. Mol Cell Biol 26:2560–2569
Bonardi F, Fusetti F, Deelen P, van Gosliga D, Vellenga E, Schuringa JJ (2013) A proteomics and transcriptomics approach to identify leukemic stem cell (LSC) markers. Mol Cell Proteomics 12:626–637
Boukarabila H, Saurin AJ, Batsche E, Mossadegh N, van Lohuizen M, Otte AP, Pradel J, Muchardt C, Sieweke M, Duprez E (2009) The PRC1 Polycomb group complex interacts with PLZF/RARA to mediate leukemic transformation. Genes Dev 23:1195–1206
Brecqueville M, Rey J, Bertucci F, Coppin E, Finetti P, Carbuccia N, Cervera N, Gelsi-Boyer V, Arnoulet C, Gisserot O, Verrot D, Slama B, Vey N, Mozziconacci MJ, Birnbaum D, Murati A (2012) Mutation analysis of ASXL1, CBL, DNMT3A, IDH1, IDH2, JAK2, MPL, NF1, SF3B1, SUZ12, and TET2 in myeloproliferative neoplasms. Genes Chromosomes Cancer 51:743–755
Bruggeman SW, Valk-Lingbeek ME, van der Stoop PP, Jacobs JJ, Kieboom K, Tanger E, Hulsman D, Leung C, Arsenijevic Y, Marino S, van Lohuizen M (2005) Ink4a and Arf differentially affect cell proliferation and neural stem cell self-renewal in Bmi1-deficient mice. Genes Dev 19:1438–1443
Buchwald G, van der Stoop P, Weichenrieder O, Perrakis A, van Lohuizen M, Sixma TK (2006) Structure and E3-ligase activity of the Ring-Ring complex of polycomb proteins Bmi1 and Ring1b. EMBO J 25:2465–2474
Cales C, Roman-Trufero M, Pavon L, Serrano I, Melgar T, Endoh M, Perez C, Koseki H, Vidal M (2008) Inactivation of the polycomb group protein Ring1B unveils an antiproliferative role in hematopoietic cell expansion and cooperation with tumorigenesis associated with Ink4a deletion. Mol Cell Biol 28:1018–1028
Cancer Genome Atlas Research Network (2013) Genomic and epigenomic landscapes of adult de novo acute myeloid leukemia. N Engl J Med 368:2059–2074
Cao R, Wang L, Wang H, Xia L, Erdjument-Bromage H, Tempst P, Jones RS, Zhang Y (2002) Role of histone H3 lysine 27 methylation in Polycomb-group silencing. Science 298:1039–1043
Chatoo W, Abdouh M, David J, Champagne MP, Ferreira J, Rodier F, Bernier G (2009) The polycomb group gene Bmi1 regulates antioxidant defenses in neurons by repressing p53 pro-oxidant activity. J Neurosci 29:529–542

Chowdhury M, Mihara K, Yasunaga S, Ohtaki M, Takihara Y, Kimura A (2007) Expression of Polycomb-group (PcG) protein BMI-1 predicts prognosis in patients with acute myeloid leukemia. Leukemia 21:1116–1122

Core N, Bel S, Gaunt SJ, Aurrand-Lions M, Pearce J, Fisher A, Djabali M (1997) Altered cellular proliferation and mesoderm patterning in Polycomb-M33-deficient mice. Development 124:721–729

Core N, Joly F, Boned A, Djabali M (2004) Disruption of E2F signaling suppresses the INK4a-induced proliferative defect in M33-deficient mice. Oncogene 23:7660–7668

Datta S, Hoenerhoff MJ, Bommi P, Sainger R, Guo WJ, Dimri M, Band H, Band V, Green JE, Dimri GP (2007) Bmi-1 cooperates with H-Ras to transform human mammary epithelial cells via dysregulation of multiple growth-regulatory pathways. Cancer Res 67:10286–10295

de Jonge HJ, Woolthuis CM, Vos AZ, Mulder A, van den Berg E, Kluin PM, van der Weide K, de Bont ES, Huls G, Vellenga E, Schuringa JJ (2011) Gene expression profiling in the leukemic stem cell-enriched CD34+ fraction identifies target genes that predict prognosis in normal karyotype AML. Leukemia 25:1825–1833

de Napoles M, Mermoud JE, Wakao R, Tang YA, Endoh M, Appanah R, Nesterova TB, Silva J, Otte AP, Vidal M, Koseki H, Brockdorff N (2004) Polycomb group proteins Ring1A/B link ubiquitylation of histone H2A to heritable gene silencing and X inactivation. Dev Cell 7:663–676

del Mar Lorente M, Marcos-Gutierrez C, Perez C, Schoorlemmer J, Ramirez A, Magin T, Vidal M (2000) Loss- and gain-of-function mutations show a polycomb group function for Ring1A in mice. Development 127:5093–5100

Denisenko O, Shnyreva M, Suzuki H, Bomsztyk K (1998) Point mutations in the WD40 domain of Eed block its interaction with Ezh2. Mol Cell Biol 18:5634–5642

Dey A, Seshasayee D, Noubade R, French DM, Liu J, Chaurushiya MS, Kirkpatrick DS, Pham VC, Lill JR, Bakalarski CE, Wu J, Phu L, Katavolos P, Lafave LM, Abdel-Wahab O, Modrusan Z, Seshagiri S, Dong K, Lin Z, Balazs M, Suriben R, Newton K, Hymowitz S, Garcia-Manero G, Martin F, Levine RL, Dixit VM (2012) Loss of the tumor suppressor BAP1 causes myeloid transformation. Science 337(6101):1541–1546

Dukers DF, van Galen JC, Giroth C, Jansen P, Sewalt RG, Otte AP, Kluin-Nelemans HC, Meijer CJ, Raaphorst FM (2004) Unique polycomb gene expression pattern in Hodgkin's lymphoma and Hodgkin's lymphoma-derived cell lines. Am J Pathol 164:873–881

Elderkin S, Maertens GN, Endoh M, Mallery DL, Morrice N, Koseki H, Peters G, Brockdorff N, Hiom K (2007) A phosphorylated form of Mel-18 targets the Ring1B histone H2A ubiquitin ligase to chromatin. Mol Cell 28:107–120

Endoh M, Endo TA, Endoh T, Fujimura Y, Ohara O, Toyoda T, Otte AP, Okano M, Brockdorff N, Vidal M, Koseki H (2008) Polycomb group proteins Ring1A/B are functionally linked to the core transcriptional regulatory circuitry to maintain ES cell identity. Development 135:1513–1524

Endoh M, Endo TA, Endoh T, Isono K, Sharif J, Ohara O, Toyoda T, Ito T, Eskeland R, Bickmore WA, Vidal M, Bernstein BE, Koseki H (2012) Histone H2A mono-ubiquitination is a crucial step to mediate PRC1-dependent repression of developmental genes to maintain ES cell identity. PLoS Genet 8:e1002774

Ernst T, Chase AJ, Score J, Hidalgo-Curtis CE, Bryant C, Jones AV, Waghorn K, Zoi K, Ross FM, Reiter A, Hochhaus A, Drexler HG, Duncombe A, Cervantes F, Oscier D, Boultwood J, Grand FH, Cross NC (2010) Inactivating mutations of the histone methyltransferase gene EZH2 in myeloid disorders. Nat Genet 42:722–726

Ernst T, Pflug A, Rinke J, Ernst J, Bierbach U, Beck JF, Hochhaus A, Gruhn B (2012) A somatic EZH2 mutation in childhood acute myeloid leukemia. Leukemia 26:1701–1703

Ezhkova E, Lien WH, Stokes N, Pasolli HA, Silva JM, Fuchs E (2011) EZH1 and EZH2 cogovern histone H3K27 trimethylation and are essential for hair follicle homeostasis and wound repair. Genes Dev 25:485–498

Faber J, Krivtsov AV, Stubbs MC, Wright R, Davis TN, Heuvel-Eibrink M, Zwaan CM, Kung AL, Armstrong SA (2009) HOXA9 is required for survival in human MLL-rearranged acute leukemias. Blood 113:2375–2385

Farcas AM, Blackledge NP, Sudbery I, Long HK, McGouran JF, Rose NR, Lee S, Sims D, Cerase A, Sheahan TW, Koseki H, Brockdorff N, Ponting CP, Kessler BM, Klose RJ (2012) KDM2B links the Polycomb Repressive Complex 1 (PRC1) to recognition of CpG islands. Elife 1:e00205

Faust C, Schumacher A, Holdener B, Magnuson T (1995) The eed mutation disrupts anterior mesoderm production in mice. Development 121:273–285

Forzati F, Federico A, Pallante P, Abbate A, Esposito F, Malapelle U, Sepe R, Palma G, Troncone G, Scarfo M, Arra C, Fedele M, Fusco A (2012) CBX7 is a tumor suppressor in mice and humans. J Clin Invest 122:612–623

Gao Z, Zhang J, Bonasio R, Strino F, Sawai A, Parisi F, Kluger Y, Reinberg D (2012) PCGF homologs, CBX proteins, and RYBP define functionally distinct PRC1 family complexes. Mol Cell 45:344–356

Garcia-Cuellar MP, Zilles O, Schreiner SA, Birke M, Winkler TH, Slany RK (2001) The ENL moiety of the childhood leukemia-associated MLL-ENL oncoprotein recruits human Polycomb 3. Oncogene 20:411–419

Gearhart MD, Corcoran CM, Wamstad JA, Bardwell VJ (2006) Polycomb group and SCF ubiquitin ligases are found in a novel BCOR complex that is recruited to BCL6 targets. Mol Cell Biol 26:6880–6889

Gil J, Bernard D, Martinez D, Beach D (2004) Polycomb CBX7 has a unifying role in cellular lifespan. Nat Cell Biol 6:67–72

Grossmann V, Tiacci E, Holmes AB, Kohlmann A, Martelli MP, Kern W, Spanhol-Rosseto A, Klein HU, Dugas M, Schindela S, Trifonov V, Schnittger S, Haferlach C, Bassan R, Wells VA, Spinelli O, Chan J, Rossi R, Baldoni S, De CL, Goetze K, Serve H, Peceny R, Kreuzer KA, Oruzio D, Specchia G, Di RF, Fabbiano F, Sborgia M, Liso A, Farinelli L, Rambaldi A, Pasqualucci L, Rabadan R, Haferlach T, Falini B (2011) Whole-exome sequencing identifies somatic mutations of BCOR in acute myeloid leukemia with normal karyotype. Blood 118:6153–6163

Grubach L, Juhl-Christensen C, Rethmeier A, Olesen LH, Aggerholm A, Hokland P, Ostergaard M (2008) Gene expression profiling of Polycomb, Hox and Meis genes in patients with acute myeloid leukaemia. Eur J Haematol 81:112–122

Hacein-Bey-Abina S, Garrigue A, Wang GP, Soulier J, Lim A, Morillon E, Clappier E, Caccavelli L, Delabesse E, Beldjord K, Asnafi V, MacIntyre E, Dal CL, Radford I, Brousse N, Sigaux F, Moshous D, Hauer J, Borkhardt A, Belohradsky BH, Wintergerst U, Velez MC, Leiva L, Sorensen R, Wulffraat N, Blanche S, Bushman FD, Fischer A, Cavazzana-Calvo M (2008) Insertional oncogenesis in 4 patients after retrovirus-mediated gene therapy of SCID-X1. J Clin Invest 118:3132–3142

Haga K, Ohno S, Yugawa T, Narisawa-Saito M, Fujita M, Sakamoto M, Galloway DA, Kiyono T (2007) Efficient immortalization of primary human cells by p16INK4a-specific short hairpin RNA or Bmi-1, combined with introduction of hTERT. Cancer Sci 98:147–154

Haupt Y, Alexander WS, Barri G, Klinken SP, Adams JM (1991) Novel zinc finger gene implicated as myc collaborator by retrovirally accelerated lymphomagenesis in E mu-myc transgenic mice. Cell 65:753–763

He J, Nguyen AT, Zhang Y (2011) KDM2b/JHDM1b, an H3K36me2-specific demethylase, is required for initiation and maintenance of acute myeloid leukemia. Blood 117:3869–3880

He J, Shen L, Wan M, Taranova O, Wu H, Zhang Y (2013) Kdm2b maintains murine embryonic stem cell status by recruiting PRC1 complex to CpG islands of developmental genes. Nat Cell Biol 15:373–384

Hemenway CS, de Erkenez AC, Gould GC (2001) The polycomb protein MPc3 interacts with AF9, an MLL fusion partner in t(9;11)(p22;q23) acute leukemias. Oncogene 20:3798–3805

Herrera-Merchan A, Arranz L, Ligos JM, de Molina A, Dominguez O, Gonzalez S (2012) Ectopic expression of the histone methyltransferase Ezh2 in haematopoietic stem cells causes myeloproliferative disease. Nat Commun 3:623

Hidalgo I, Herrera-Merchan A, Ligos JM, Carramolino L, Nunez J, Martinez F, Dominguez O, Torres M, Gonzalez S (2012) Ezh1 Is required for hematopoietic stem cell maintenance and prevents senescence-like cell cycle arrest. Cell Stem Cell 11:649–662

Isono K, Fujimura Y, Shinga J, Yamaki M, Wang J, Takihara Y, Murahashi Y, Takada Y, Mizutani-Koseki Y, Koseki H (2005) Mammalian polyhomeotic homologues Phc2 and Phc1 act in synergy to mediate polycomb repression of Hox genes. Mol Cell Biol 25:6694–6706

Iwama A, Oguro H, Negishi M, Kato Y, Morita Y, Tsukui H, Ema H, Kamijo T, Katoh-Fukui Y, Koseki H, van Lohuizen M, Nakauchi H (2004) Enhanced self-renewal of hematopoietic stem cells mediated by the polycomb gene product Bmi-1. Immunity 21:843–851

Jacobs JJ, Kieboom K, Marino S, DePinho RA, van Lohuizen M (1999) The oncogene and Polycomb-group gene bmi-1 regulates cell proliferation and senescence through the ink4a locus. Nature 397:164–168

Jo S, Lee H, Kim S, Hwang EM, Park JY, Kang SS, Chung H (2011) Inhibition of PCGF2 enhances granulocytic differentiation of acute promyelocytic leukemia cell line HL-60 via induction of HOXA7. Biochem Biophys Res Commun 416:86–91

Kagey MH, Melhuish TA, Wotton D (2003) The polycomb protein Pc2 is a SUMO E3. Cell 113:127–137

Kallin EM, Cao R, Jothi R, Xia K, Cui K, Zhao K, Zhang Y (2009) Genome-wide uH2A localization analysis highlights Bmi1-dependent deposition of the mark at repressed genes. PLoS Genet 5:e1000506

Kamminga LM, Bystrykh LV, de Boer A, Houwer S, Douma J, Weersing E, Dontje B, de Haan G (2006) The Polycomb group gene Ezh2 prevents hematopoietic stem cell exhaustion. Blood 107:2170–2179

Katoh-Fukui Y, Tsuchiya R, Shiroishi T, Nakahara Y, Hashimoto N, Noguchi K, Higashinakagawa T (1998) Male-to-female sex reversal in M33 mutant mice. Nature 393:688–692

Kim J, Wong PK (2009) Oxidative stress is linked to ERK1/2-p16 signaling-mediated growth defect in ATM-deficient astrocytes. J Biol Chem 284:14396–14404

Kim JY, Sawada A, Tokimasa S, Endo H, Ozono K, Hara J, Takihara Y (2004) Defective long-term repopulating ability in hematopoietic stem cells lacking the Polycomb-group gene rae28. Eur J Haematol 73:75–84

Kim J, Woo AJ, Chu J, Snow JW, Fujiwara Y, Kim CG, Cantor AB, Orkin SH (2010) A Myc network accounts for similarities between embryonic stem and cancer cell transcription programs. Cell 143:313–324

Kirmizis A, Bartley SM, Kuzmichev A, Margueron R, Reinberg D, Green R, Farnham PJ (2004) Silencing of human polycomb target genes is associated with methylation of histone H3 Lys 27. Genes Dev 18:1592–1605

Klauke K, Radulovic V, Broekhuis M, Weersing E, Zwart E, Olthof S, Ritsema M, Bruggeman S, Wu X, Helin K, Bystrykh L, de Haan G (2013) Polycomb Cbx family members mediate the balance between haematopoietic stem cell self-renewal and differentiation. Nat Cell Biol 15:353–362

Kuzmichev A, Nishioka K, Erdjument-Bromage H, Tempst P, Reinberg D (2002) Histone methyltransferase activity associated with a human multiprotein complex containing the Enhancer of Zeste protein. Genes Dev 16:2893–2905

Lessard J, Sauvageau G (2003) Bmi-1 determines the proliferative capacity of normal and leukaemic stem cells. Nature 423:255–260

Lessard J, Schumacher A, Thorsteinsdottir U, van Lohuizen M, Magnuson T, Sauvageau G (1999) Functional antagonism of the Polycomb-Group genes eed and Bmi1 in hemopoietic cell proliferation. Genes Dev 13:2691–2703

Levine SS, Weiss A, Erdjument-Bromage H, Shao Z, Tempst P, Kingston RE (2002) The core of the polycomb repressive complex is compositionally and functionally conserved in flies and humans. Mol Cell Biol 22:6070–6078

Li M, Collins R, Jiao Y, Ouillette P, Bixby D, Erba H, Vogelstein B, Kinzler KW, Papadopoulos N, Malek SN (2011) Somatic mutations in the transcriptional corepressor gene BCORL1 in adult acute myelogenous leukemia. Blood 118:5914–5917

Liu J, Cao L, Chen J, Song S, Lee IH, Quijano C, Liu H, Keyvanfar K, Chen H, Cao LY, Ahn BH, Kumar NG, Rovira II, Xu XL, van Lohuizen M, Motoyama N, Deng CX, Finkel T (2009) Bmi1 regulates mitochondrial function and the DNA damage response pathway. Nature 459:387–392

Liu Y, Liu F, Yu H, Zhao X, Sashida G, Deblasio A, Harr M, She QB, Chen Z, Lin HK, Di GS, Elf SE, Yang Y, Miyata Y, Huang G, Menendez S, Mellinghoff IK, Rosen N, Pandolfi PP, Hedvat CV, Nimer SD (2012) Akt phosphorylates the transcriptional repressor bmi1 to block its effects on the tumor-suppressing ink4a-arf locus. Sci Signal 5:ra77

Liu B, Liu YF, Du YR, Mardaryev AN, Yang W, Chen H, Xu ZM, Xu CQ, Zhang XR, Botchkarev VA, Zhang Y, Xu GL (2013) Cbx4 regulates the proliferation of thymic epithelial cells and thymus function. Development 140:780–788

Luis NM, Morey L, Mejetta S, Pascual G, Janich P, Kuebler B, Cozutto L, Roma G, Nascimento E, Frye M, Di CL, Benitah SA (2011) Regulation of human epidermal stem cell proliferation and senescence requires polycomb- dependent and -independent functions of Cbx4. Cell Stem Cell 9:233–246

Maertens GN, El Messaoudi-Aubert S, Racek T, Stock JK, Nicholls J, Rodriguez-Niedenfuhr M, Gil J, Peters G (2009) Several distinct polycomb complexes regulate and co-localize on the INK4a tumor suppressor locus. PLoS One 4:e6380

Majewski IJ, Blewitt ME, de Graaf CA, McManus EJ, Bahlo M, Hilton AA, Hyland CD, Smyth GK, Corbin JE, Metcalf D, Alexander WS, Hilton DJ (2008) Polycomb repressive complex 2 (PRC2) restricts hematopoietic stem cell activity. PLoS Biol 6:e93

Majewski IJ, Ritchie ME, Phipson B, Corbin J, Pakusch M, Ebert A, Busslinger M, Koseki H, Hu Y, Smyth GK, Alexander WS, Hilton DJ, Blewitt ME (2010) Opposing roles of polycomb repressive complexes in hematopoietic stem and progenitor cells. Blood 116:731–739

Makishima H, Jankowska AM, Tiu RV, Szpurka H, Sugimoto Y, Hu Z, Saunthararajah Y, Guinta K, Keddache MA, Putnam P, Sekeres MA, Moliterno AR, List AF, McDevitt MA, Maciejewski JP (2010) Novel homo- and hemizygous mutations in EZH2 in myeloid malignancies. Leukemia 24:1799–1804

Margueron R, Li G, Sarma K, Blais A, Zavadil J, Woodcock CL, Dynlacht BD, Reinberg D (2008) Ezh1 and Ezh2 maintain repressive chromatin through different mechanisms. Mol Cell 32:503–518

Margueron R, Justin N, Ohno K, Sharpe ML, Son J, Drury WJ III, Voigt P, Martin SR, Taylor WR, De Marco V, Pirrotta V, Reinberg D, Gamblin SJ (2009) Role of the polycomb protein EED in the propagation of repressive histone marks. Nature 461:762–767

Metzeler KH, Becker H, Maharry K, Radmacher MD, Kohlschmidt J, Mrozek K, Nicolet D, Whitman SP, Wu YZ, Schwind S, Powell BL, Carter TH, Wetzler M, Moore JO, Kolitz JE, Baer MR, Carroll AJ, Larson RA, Caligiuri MA, Marcucci G, Bloomfield CD (2011) ASXL1 mutations identify a high-risk subgroup of older patients with primary cytogenetically normal AML within the ELN Favorable genetic category. Blood 118:6920–6929

Mihara K, Chowdhury M, Nakaju N, Hidani S, Ihara A, Hyodo H, Yasunaga S, Takihara Y, Kimura A (2006) Bmi-1 is useful as a novel molecular marker for predicting progression of myelodysplastic syndrome and patient prognosis. Blood 107:305–308

Miyamoto K, Araki KY, Naka K, Arai F, Takubo K, Yamazaki S, Matsuoka S, Miyamoto T, Ito K, Ohmura M, Chen C, Hosokawa K, Nakauchi H, Nakayama K, Nakayama KI, Harada M, Motoyama N, Suda T, Hirao A (2007) Foxo3a is essential for maintenance of the hematopoietic stem cell pool. Cell Stem Cell 1:101–112

Mochizuki-Kashio M, Mishima Y, Miyagi S, Negishi M, Saraya A, Konuma T, Shinga J, Koseki H, Iwama A (2011) Dependency on the polycomb gene Ezh2 distinguishes fetal from adult hematopoietic stem cells. Blood 118:6553–6561

Mohty M, Yong AS, Szydlo RM, Apperley JF, Melo JV (2007) The polycomb group BMI1 gene is a molecular marker for predicting prognosis of chronic myeloid leukemia. Blood 110:380–383

Monroe SC, Jo SY, Sanders DS, Basrur V, Elenitoba-Johnson KS, Slany RK, Hess JL (2011) MLL-AF9 and MLL-ENL alter the dynamic association of transcriptional regulators with genes critical for leukemia. Exp Hematol 39(1):77–86

Morey L, Pascual G, Cozzuto L, Roma G, Wutz A, Benitah SA, Di CL (2012) Nonoverlapping functions of the Polycomb group Cbx family of proteins in embryonic stem cells. Cell Stem Cell 10:47–62

Morin RD, Johnson NA, Severson TM, Mungall AJ, An J, Goya R, Paul JE, Boyle M, Woolcock BW, Kuchenbauer F, Yap D, Humphries RK, Griffith OL, Shah S, Zhu H, Kimbara M, Shashkin P, Charlot JF, Tcherpakov M, Corbett R, Tam A, Varhol R, Smailus D, Moksa M, Zhao Y, Delaney A, Qian H, Birol I, Schein J, Moore R, Holt R, Horsman DE, Connors JM, Jones S, Aparicio S, Hirst M, Gascoyne RD, Marra MA (2010) Somatic mutations altering EZH2 (Tyr641) in follicular and diffuse large B-cell lymphomas of germinal-center origin. Nat Genet 42:181–185

Mueller D, Bach C, Zeisig D, Garcia-Cuellar MP, Monroe S, Sreekumar A, Zhou R, Nesvizhskii A, Chinnaiyan A, Hess JL, Slany RK (2007) A role for the MLL fusion partner ENL in transcriptional elongation and chromatin modification. Blood 110:4445–4454

Nacerddine K, Beaudry JB, Ginjala V, Westerman B, Mattiroli F, Song JY, van der Poel H, Ponz OB, Pritchard C, Cornelissen-Steijger P, Zevenhoven J, Tanger E, Sixma TK, Ganesan S, van Lohuizen M (2012) Akt-mediated phosphorylation of Bmi1 modulates its oncogenic potential, E3 ligase activity, and DNA damage repair activity in mouse prostate cancer. J Clin Invest 122:1920–1932

Neff T, Sinha AU, Kluk MJ, Zhu N, Khattab MH, Stein L, Xie H, Orkin SH, Armstrong SA (2012) Polycomb repressive complex 2 is required for MLL-AF9 leukemia. Proc Natl Acad Sci USA 109:5028–5033

Nikoloski G, Langemeijer SM, Kuiper RP, Knops R, Massop M, Tonnissen ER, van der Heijden A, Scheele TN, Vandenberghe P, de Witte T, van der Reijden BA, Jansen JH (2010) Somatic mutations of the histone methyltransferase gene EZH2 in myelodysplastic syndromes. Nat Genet 42:665–667

Novershtern N, Subramanian A, Lawton LN, Mak RH, Haining WN, McConkey ME, Habib N, Yosef N, Chang CY, Shay T, Frampton GM, Drake AC, Leskov I, Nilsson B, Preffer F, Dombkowski D, Evans JW, Liefeld T, Smutko JS, Chen J, Friedman N, Young RA, Golub TR, Regev A, Ebert BL (2011) Densely interconnected transcriptional circuits control cell states in human hematopoiesis. Cell 144:296–309

Ntziachristos P, Tsirigos A, Vlierberghe PV, Nedjic J, Trimarchi T, Flaherty MS, Ferres-Marco D, da Ros V, Tang Z, Siegle J, Asp P, Hadler M, Rigo I, Keersmaecker KD, Patel J, Huynh T, Utro F, Poglio S, Samon JB, Paietta E, Racevskis J, Rowe JM, Rabadan R, Levine RL, Brown S, Pflumio F, Dominguez M, Ferrando A, Aifantis I (2012) Genetic inactivation of the polycomb repressive complex 2 in T cell acute lymphoblastic leukemia. Nat Med 18:298–303

O'Carroll D, Erhardt S, Pagani M, Barton SC, Surani MA, Jenuwein T (2001) The polycomb-group gene Ezh2 is required for early mouse development. Mol Cell Biol 21:4330–4336

Ogawa H, Ishiguro K, Gaubatz S, Livingston DM, Nakatani Y (2002) A complex with chromatin modifiers that occupies E2F- and Myc-responsive genes in G0 cells. Science 296:1132–1136

Oguro H, Iwama A, Morita Y, Kamijo T, Kamijo T, van Lohuizen M, van Nakauchi H (2006) Differential impact of Ink4a and Arf on hematopoietic stem cells and their bone marrow microenvironment in Bmi1-deficient mice. J Exp Med 203:2247–2253

Oguro H, Yuan J, Ichikawa H, Ikawa T, Yamazaki S, Kawamoto H, Nakauchi H, Iwama A (2010) Poised lineage specification in multipotential hematopoietic stem and progenitor cells by the polycomb protein Bmi1. Cell Stem Cell 6:279–286

Ohta H, Sawada A, Kim JY, Tokimasa S, Nishiguchi S, Humphries RK, Hara J, Takihara Y (2002) Polycomb group gene rae28 is required for sustaining activity of hematopoietic stem cells. J Exp Med 195:759–770

O'Loghlen A, Munoz-Cabello AM, Gaspar-Maia A, Wu HA, Banito A, Kunowska N, Racek T, Pemberton HN, Beolchi P, Lavial F, Masui O, Vermeulen M, Carroll T, Graumann J, Heard E, Dillon N, Azuara V, Snijders AP, Peters G, Bernstein E, Gil J (2012) MicroRNA regulation of Cbx7 mediates a switch of Polycomb orthologs during ESC differentiation. Cell Stem Cell 10:33–46

Park IK, Qian D, Kiel M, Becker MW, Pihalja M, Weissman IL, Morrison SJ, Clarke MF (2003) Bmi-1 is required for maintenance of adult self-renewing haematopoietic stem cells. Nature 423:302–305

Pasini D, Bracken AP, Jensen MR, Lazzerini DE, Helin K (2004) Suz12 is essential for mouse development and for EZH2 histone methyltransferase activity. EMBO J 23:4061–4071

Pizzatti L, Binato R, Cofre J, Gomes BE, Dobbin J, Haussmann ME, D'Azambuja D, Bouzas LF, Abdelhay E (2010) SUZ12 is a candidate target of the non-canonical WNT pathway in the progression of chronic myeloid leukemia. Genes Chromosomes Cancer 49:107–118

Puda A, Milosevic JD, Berg T, Klampfl T, Harutyunyan AS, Gisslinger B, Rumi E, Pietra D, Malcovati L, Elena C, Doubek M, Steurer M, Tosic N, Pavlovic S, Guglielmelli P, Pieri L, Vannucchi AM, Gisslinger H, Cazzola M, Kralovics R (2012) Frequent deletions of JARID2 in leukemic transformation of chronic myeloid malignancies. Am J Hematol 87:245–250

Qin J, Whyte WA, Anderssen E, Apostolou E, Chen HH, Akbarian S, Bronson RT, Hochedlinger K, Ramaswamy S, Young RA, Hock H (2012) The polycomb group protein L3mbtl2 assembles an atypical PRC1-family complex that is essential in pluripotent stem cells and early development. Cell Stem Cell 11:319–332

Raaphorst FM, van Kemenade FJ, Blokzijl T, Fieret E, Hamer KM, Satijn DP, Otte AP, Meijer CJ (2000) Coexpression of BMI-1 and EZH2 polycomb group genes in Reed-Sternberg cells of Hodgkin's disease. Am J Pathol 157:709–715

Ray D, Kwon SY, Tagoh H, Heidenreich O, Ptasinska A, Bonifer C (2013) Lineage-inappropriate PAX5 expression in t(8;21) acute myeloid leukemia requires signaling-mediated abrogation of polycomb repression. Blood 122:759–769

Rizo A, Dontje B, Vellenga E, de Haan G, Schuringa JJ (2008) Long-term maintenance of human hematopoietic stem/progenitor cells by expression of BMI1. Blood 111:2621–2630

Rizo A, Olthof S, Han L, Vellenga E, de Haan G, Schuringa JJ (2009) Repression of BMI1 in normal and leukemic human CD34(+) cells impairs self-renewal and induces apoptosis. Blood 114:1498–1505

Rizo A, Horton SJ, Olthof S, Dontje B, Ausema A, van Os R, van den Boom V, Vellenga E, de Haan G, Schuringa JJ (2010) BMI1 collaborates with BCR-ABL in leukemic transformation of human CD34+ cells. Blood 116:4621–4630

Ross K, Sedello AK, Todd GP, Paszkowski-Rogacz M, Bird AW, Ding L, Grinenko T, Behrens K, Hubner N, Mann M, Waskow C, Stocking C, Buchholz F (2012) Polycomb group ring finger 1 cooperates with Runx1 in regulating differentiation and self-renewal of hematopoietic cells. Blood 119:4152–4161

Sanchez C, Sanchez I, Demmers JA, Rodriguez P, Strouboulis J, Vidal M (2007) Proteomics analysis of Ring1B/Rnf2 interactors identifies a novel complex with the Fbxl10/Jhdm1B histone demethylase and the Bcl6 interacting corepressor. Mol Cell Proteomics 6:820–834

Scheuermann JC, de Ayala Alonso AG, Oktaba K, Ly-Hartig N, McGinty RK, Fraterman S, Wilm M, Muir TW, Muller J (2010) Histone H2A deubiquitinase activity of the Polycomb repressive complex PR-DUB. Nature 465:243–247

Schnittger S, Eder C, Jeromin S, Alpermann T, Fasan A, Grossmann V, Kohlmann A, Illig T, Klopp N, Wichmann HE, Kreuzer KA, Schmid C, Staib P, Peceny R, Schmitz N, Kern W,

Haferlach C, Haferlach T (2013) ASXL1 exon 12 mutations are frequent in AML with intermediate risk karyotype and are independently associated with an adverse outcome. Leukemia 27:82–91

Schumacher A, Faust C, Magnuson T (1996) Positional cloning of a global regulator of anterior-posterior patterning in mice. Nature 384:648

Score J, Hidalgo-Curtis C, Jones AV, Winkelmann N, Skinner A, Ward D, Zoi K, Ernst T, Stegelmann F, Dohner K, Chase A, Cross NC (2012) Inactivation of polycomb repressive complex 2 components in myeloproliferative and myelodysplastic/myeloproliferative neoplasms. Blood 119:1208–1213

Scott CL, Gil J, Hernando E, Teruya-Feldstein J, Narita M, Martinez D, Visakorpi T, Mu D, Cordon-Cardo C, Peters G, Beach D, Lowe SW (2007) Role of the chromobox protein CBX7 in lymphomagenesis. Proc Natl Acad Sci USA 104:5389–5394

Sengupta A, Ficker AM, Dunn SK, Madhu M, Cancelas JA (2012) Bmi1 reprograms CML B-lymphoid progenitors to become B-ALL-initiating cells. Blood 119:494–502

Shen X, Liu Y, Hsu YJ, Fujiwara Y, Kim J, Mao X, Yuan GC, Orkin SH (2008) EZH1 mediates methylation on histone H3 lysine 27 and complements EZH2 in maintaining stem cell identity and executing pluripotency. Mol Cell 32:491–502

Shi J, Wang E, Zuber J, Rappaport A, Taylor M, Johns C, Lowe SW, Vakoc CR (2013) The Polycomb complex PRC2 supports aberrant self-renewal in a mouse model of MLL-AF9;Nras (G12D) acute myeloid leukemia. Oncogene 32(7):930–938

Shih AH, Abdel-Wahab O, Patel JP, Levine RL (2012) The role of mutations in epigenetic regulators in myeloid malignancies. Nat Rev Cancer 12:599–612

Signer R, Montecino-Rodriguez E, Witte O, Dorshkind K (2008) Aging and cancer resistance in lymphoid progenitors are linked processes conferred by p16Ink4a and Arf. Genes Dev 22:3115–3120

Simon JA, Kingston RE (2009) Mechanisms of polycomb gene silencing: knowns and unknowns. Nat Rev Mol Cell Biol 10:697–708

Simon C, Chagraoui J, Krosl J, Gendron P, Wilhelm B, Lemieux S, Boucher G, Chagnon P, Drouin S, Lambert R, Rondeau C, Bilodeau A, Lavallee S, Sauvageau M, Hebert J, Sauvageau G (2012) A key role for EZH2 and associated genes in mouse and human adult T-cell acute leukemia. Genes Dev 26:651–656

Smith KS, Chanda SK, Lingbeek M, Ross DT, Botstein D, van Lohuizen M, Cleary ML (2003) Bmi-1 regulation of INK4A-ARF is a downstream requirement for transformation of hematopoietic progenitors by E2a-Pbx1. Mol Cell 12:393–400

Smith LL, Yeung J, Zeisig BB, Popov N, Huijbers I, Barnes J, Wilson AJ, Taskesen E, Delwel R, Gil J, Van Lohuizen M, So CW (2011) Functional crosstalk between Bmi1 and MLL/Hoxa9 axis in establishment of normal hematopoietic and leukemic stem cells. Cell Stem Cell 8:649–662

Somervaille TC, Matheny CJ, Spencer GJ, Iwasaki M, Rinn JL, Witten DM, Chang HY, Shurtleff SA, Downing JR, Cleary ML (2009) Hierarchical maintenance of MLL myeloid leukemia stem cells employs a transcriptional program shared with embryonic rather than adult stem cells. Cell Stem Cell 4:129–140

Stock JK, Giadrossi S, Casanova M, Brookes E, Vidal M, Koseki H, Brockdorff N, Fisher AG, Pombo A (2007) Ring1-mediated ubiquitination of H2A restrains poised RNA polymerase II at bivalent genes in mouse ES cells. Nat Cell Biol 9:1428–1435

Su IH, Basavaraj A, Krutchinsky AN, Hobert O, Ullrich A, Chait BT, Tarakhovsky A (2003) Ezh2 controls B cell development through histone H3 methylation and Igh rearrangement. Nat Immunol 4:124–131

Su IH, Dobenecker MW, Dickinson E, Oser M, Basavaraj A, Marqueron R, Viale A, Reinberg D, Wulfing C, Tarakhovsky A (2005) Polycomb group protein ezh2 controls actin polymerization and cell signaling. Cell 121:425–436

Takada Y, Isono K, Shinga J, Turner JM, Kitamura H, Ohara O, Watanabe G, Singh PB, Kamijo T, Jenuwein T, Burgoyne PS, Koseki H (2007) Mammalian Polycomb Scmh1 mediates exclusion of Polycomb complexes from the XY body in the pachytene spermatocytes. Development 134:579–590

Takihara Y, Tomotsune D, Shirai M, Katoh-Fukui Y, Nishii K, Motaleb MA, Nomura M, Tsuchiya R, Fujita Y, Shibata Y, Higashinakagawa T, Shimada K (1997) Targeted disruption of the mouse homologue of the Drosophila polyhomeotic gene leads to altered anteroposterior patterning and neural crest defects. Development 124:3673–3682

Tan J, Jones M, Koseki H, Nakayama M, Muntean AG, Maillard I, Hess JL (2011) CBX8, a Polycomb Group Protein, Is Essential for MLL-AF9-Induced Leukemogenesis. Cancer Cell 20:563–575

Tanaka S, Miyagi S, Sashida G, Chiba T, Yuan J, Mochizuki-Kashio M, Suzuki Y, Sugano S, Nakaseko C, Yokote K, Koseki H, Iwama A (2012) Ezh2 augments leukemogenicity by reinforcing differentiation blockage in acute myeloid leukemia. Blood 120:1107–1117

Tavares L, Dimitrova E, Oxley D, Webster J, Poot R, Demmers J, Bezstarosti K, Taylor S, Ura H, Koide H, Wutz A, Vidal M, Elderkin S, Brockdorff N (2012) RYBP-PRC1 complexes mediate H2A ubiquitylation at polycomb target sites independently of PRC2 and H3K27me3. Cell 148:664–678

Thiel AT, Feng Z, Pant DK, Chodosh LA, Hua X (2013) The trithorax protein partner menin acts in tandem with EZH2 to suppress C/EBPalpha and differentiation in MLL-AF9 leukemia. Haematologica 98:918–927

Tiacci E, Grossmann V, Martelli MP, Kohlmann A, Haferlach T, Falini B (2012) The corepressors BCOR and BCORL1: two novel players in acute myeloid leukemia. Haematologica 97:3–5

Tokimasa S, Ohta H, Sawada A, Matsuda Y, Kim JY, Nishiguchi S, Hara J, Takihara Y (2001) Lack of the Polycomb-group gene rae28 causes maturation arrest at the early B-cell developmental stage. Exp Hematol 29:93–103

Trimarchi JM, Fairchild B, Wen J, Lees JA (2001) The E2F6 transcription factor is a component of the mammalian Bmi1-containing polycomb complex. Proc Natl Acad Sci USA 98:1519–1524

Trojer P, Cao AR, Gao Z, Li Y, Zhang J, Xu X, Li G, Losson R, Erdjument-Bromage H, Tempst P, Farnham PJ, Reinberg D (2011) L3MBTL2 protein acts in concert with PcG protein-mediated monoubiquitination of H2A to establish a repressive chromatin structure. Mol Cell 42:438–450

Ueda T, Sanada M, Matsui H, Yamasaki N, Honda ZI, Shih LY, Mori H, Inaba T, Ogawa S, Honda H (2012) EED mutants impair polycomb repressive complex 2 in myelodysplastic syndrome and related neoplasms. Leukemia 26:2557–2560

van den Boom V, Rozenveld-Geugien M, Bonardi F, Malanga D, van Gosliga D, Heijink AM, Viglietto G, Morrone G, Fusetti F, Vellenga E, Schuringa JJ (2013) Nonredundant and locus-specific gene repression functions of PRC1 paralog family members in human hematopoietic stem/progenitor cells. Blood 121:2452–2461

van der Lugt NM, Domen J, Linders K, van Roon M, Robanus-Maandag E, te Riele H, van der Valk M, Deschamps J, Sofroniew M, van Lohuizen M (1994) Posterior transformation, neurological abnormalities, and severe hematopoietic defects in mice with a targeted deletion of the bmi-1 proto-oncogene. Genes Dev 8:757–769

van Gosliga D, Schepers H, Rizo A, van der Kolk D, Vellenga E, Schuringa JJ (2007) Establishing long-term cultures with self-renewing acute myeloid leukemia stem/progenitor cells. Exp Hematol 35:1538–1549

van Kemenade FJ, Raaphorst FM, Blokzijl T, Fieret E, Hamer KM, Satijn DP, Otte AP, Meijer CJ (2001) Coexpression of BMI-1 and EZH2 polycomb-group proteins is associated with cycling cells and degree of malignancy in B-cell non-Hodgkin lymphoma. Blood 97:3896–3901

van Lohuizen M, Verbeek S, Scheijen B, Wientjens E, van der Gulden H, Berns A (1991) Identification of cooperating oncogenes in E mu-myc transgenic mice by provirus tagging. Cell 65:737–752

Vandamme J, Volkel P, Rosnoblet C, Le Faou P, Angrand PO (2011) Interaction proteomics analysis of polycomb proteins defines distinct PRC1 complexes in mammalian cells. Mol Cell Proteomics 10:M110

Villa R, Pasini D, Gutierrez A, Morey L, Occhionorelli M, Vire E, Nomdedeu JF, Jenuwein T, Pelicci PG, Minucci S, Fuks F, Helin K, Di CL (2007) Role of the polycomb repressive complex 2 in acute promyelocytic leukemia. Cancer Cell 11:513–525

Visser HP, Gunster MJ, Kluin-Nelemans HC, Manders EM, Raaphorst FM, Meijer CJ, Willemze R, Otte AP (2001) The Polycomb group protein EZH2 is upregulated in proliferating, cultured human mantle cell lymphoma. Br J Haematol 112:950–958

Voncken JW, Roelen BA, Roefs M, de Vries S, Verhoeven E, Marino S, Deschamps J, van Lohuizen M (2003) Rnf2 (Ring1b) deficiency causes gastrulation arrest and cell cycle inhibition. Proc Natl Acad Sci USA 100:2468–2473

Waldron T, De Dominici M, Soliera AR, Audia A, Iacobucci I, Lonetti A, Martinelli G, Zhang Y, Martinez R, Hyslop T, Bender TP, Calabretta B (2011) c-Myb and its target Bmi1 are required for p190BCR/ABL leukemogenesis in mouse and human cells. Leukemia 26:644–653

Wang H, Wang L, Erdjument-Bromage H, Vidal M, Tempst P, Jones RS, Zhang Y (2004) Role of histone H2A ubiquitination in Polycomb silencing. Nature 431:873–878

Warner JK, Wang JC, Takenaka K, Doulatov S, McKenzie JL, Harrington L, Dick JE (2005) Direct evidence for cooperating genetic events in the leukemic transformation of normal human hematopoietic cells. Leukemia 19:1794–1805

Whitcomb SJ, Basu A, Allis CD, Bernstein E (2007) Polycomb Group proteins: an evolutionary perspective. Trends Genet 23:494–502

Wu X, Johansen JV, Helin K (2013) Fbxl10/Kdm2b recruits polycomb repressive complex 1 to CpG islands and regulates H2A ubiquitylation. Mol Cell 49:1134–1146

Xu F, Li X, Wu L, Zhang Q, Yang R, Yang Y, Zhang Z, He Q, Chang C (2011) Overexpression of the EZH2, RING1 and BMI1 genes is common in myelodysplastic syndromes: relation to adverse epigenetic alteration and poor prognostic scoring. Ann Hematol 90:643–653

Yamamoto Y, Tsuzuki S, Tsuzuki M, Handa K, Inaguma Y, Emi N (2010) BCOR as a novel fusion partner of retinoic acid receptor alpha in a t(X;17)(p11;q12) variant of acute promyelocytic leukemia. Blood 116:4274–4283

Yong AS, Stephens N, Weber G, Li Y, Savani BN, Eniafe R, Keyvanfar K, Kurlander R, Rezvani K, Barrett AJ (2011) Improved outcome following allogeneic stem cell transplantation in chronic myeloid leukemia is associated with higher expression of BMI-1 and immune responses to BMI-1 protein. Leukemia 25:629–637

Zeisig BB, Milne T, Garcia-Cuellar MP, Schreiner S, Martin ME, Fuchs U, Borkhardt A, Chanda SK, Walker J, Soden R, Hess JL, Slany RK (2004) Hoxa9 and Meis1 are key targets for MLL-ENL-mediated cellular immortalization. Mol Cell Biol 24:617–628

Zhang J, Sarge KD (2009) Identification of a polymorphism in the RING finger of human Bmi-1 that causes its degradation by the ubiquitin-proteasome system. FEBS Lett 583:960–964

Zhang J, Ding L, Holmfeldt L, Wu G, Heatley SL, Payne-Turner D, Easton J, Chen X, Wang J, Rusch M, Lu C, Chen SC, Wei L, Collins-Underwood JR, Ma J, Roberts KG, Pounds SB, Ulyanov A, Becksfort J, Gupta P, Huether R, Kriwacki RW, Parker M, McGoldrick DJ, Zhao D, Alford D, Espy S, Bobba KC, Song G, Pei D, Cheng C, Roberts S, Barbato MI, Campana D, Coustan-Smith E, Shurtleff SA, Raimondi SC, Kleppe M, Cools J, Shimano KA, Hermiston ML, Doulatov S, Eppert K, Laurenti E, Notta F, Dick JE, Basso G, Hunger SP, Loh ML, Devidas M, Wood B, Winter S, Dunsmore KP, Fulton RS, Fulton LL, Hong X, Harris CC, Dooling DJ, Ochoa K, Johnson KJ, Obenauer JC, Evans WE, Pui CH, Naeve CW, Ley TJ, Mardis ER, Wilson RK, Downing JR, Mullighan CG (2012) The genetic basis of early T-cell precursor acute lymphoblastic leukaemia. Nature 481:157–163

Chapter 7
Role of the Trithorax (MLL): HOX Axis in HSC Development, Function, and Leukemia

Robert K. Slany

Abstract The discovery of mixed lineage leukemia (MLL) fusions as causal event for a very aggressive subtype of acute leukemia has spurred a large body of research investigating the importance of this protein and its fusion derivatives for normal and malignant hematopoiesis. Here, recent advances examining the chromatin-based functions of the histone methyltransferase MLL as epigenetic factor and cell cycle regulator are summarized. This is complemented by a review of the current knowledge describing the oncogenic mechanism of MLL fusion proteins that coordinately impact chromatin modification and transcriptional control. Finally, to complete the synopsis of MLL governed processes in blood cell development, this chapter ends with a short overview of the role of HOX-homeobox proteins as major MLL downstream effectors in the hematopoietic system.

Keywords MLL • HOX • Leukemia

7.1 A Historical and Clinical Primer: 11q23 Translocations and Mixed Lineage Leukemia

Undoubtedly, the development of modern, high-resolution karyotyping was one of the greatest breakthroughs that enabled identification of the major genetic players in malignant and normal hematopoiesis. This technique allowed the seminal discovery that the odd small speck of genetic material characterizing patients with chronic myeloid leukemia, the so-called philadelphia chromosome, actually consisted of a hybrid juxtaposing chromosomes 9 and 22 (Rowley 1973). This finding expedited the unveiling of the participating molecular players as the cellular counterpart of the

R.K. Slany (✉)
Department of Genetics, Friedrich Alexander Universitaet Erlangen, Erwin-Rommel Street 3, 91058 Erlangen, Germany
e-mail: robert.slany@fau.de

C. Bonifer and P.N. Cockerill (eds.), *Transcriptional and Epigenetic Mechanisms Regulating Normal and Aberrant Blood Cell Development*, Epigenetics and Human Health, DOI 10.1007/978-3-642-45198-0_7, © Springer-Verlag Berlin Heidelberg 2014

Abelson leukemia virus oncogene (c-ABL) and an unknown gene termed BCR for "breakpoint cluster region" (Heisterkamp et al. 1983; Shtivelman et al. 1985). In the wake of this discovery, a run started to identify the molecular constituents of other recurrent translocations that had been prominently identified and associated with certain subgroups of leukemia. In stark contrast to the often completely mangled genomes of solid tumors, leukemic cells very often carry only one or a few clearly recognizable genomic aberrations that can serve as a beacon within the sea of genetic information to pinpoint the genes involved in malignant transformation.

Already in the early 1980s the long arm of chromosome 11 was singled out as a unique hot spot for genomic aberrations in leukemia (Berger et al. 1982; Vermaelen et al. 1983). Soon a significant association of the "11q22–25" abnormality with infant leukemia (<1 year of age) was discovered. Closer examination revealed that almost all cases harbored reciprocal translocations where material of a variety of different partner chromosomes was joined to sequences at 11q23 and vice versa. By far the most frequent translocations involved chromosomes 4, 9, and 19 that donated material for fusion with the ominous locus at 11q23. The presence of these translocations was habitually correlated with "lineage heterogeneity" implying that affected blasts simultaneously displayed surface markers of lymphoid and myeloid lineage. Correspondingly, this special type of disease was designated mixed lineage leukemia (Childs et al. 1988; Kaneko et al. 1986; Stark et al. 1986).

As we know today, 11q23 translocations are also a hallmark of many cases of therapy-related acute myeloid leukemia (t-AML). This secondary disease is an unfortunate outcome in up to 10 % of patients that have been treated previously for an unrelated malignant disease. T-AML is closely associated with administration of topoisomerase inhibitors like etoposide (Rowley and Olney 2002). It has been suggested that certain DNAseI hypersensitive sites at 11q23 and the respective partner chromosomes are preferential breakpoints after poisoning of topoisomerase II. The resulting double strand breaks initiate the translocation process (Cowell et al. 2012). Besides translocations the 11q23 locus is also plagued by small tandem duplications (see below), particularly in adult acute myeloid leukemia.

From a clinical perspective the presence of 11q23 translocations is mostly associated with a dismal prognosis. Despite the improvements in supportive care, high-intensity chemotherapy, and the inclusion of bone marrow transplants in treatment schedules, the 5 year survival in infants with 11q23 leukemia hovers well below 50 % (Pui et al. 2011). The situation in adults (AML and ALL) is equally bleak with the possible exception of translocations involving 11q23 and chromosome 9. Overall, 5-year survival chances are again well below 50 % regardless of the actual partner chromosome involved (Krauter et al. 2009).

7.2 A Drosophila Embryonic Regulator in Mammals: Do Flies Suffer from Leukemia?

No fewer than four different research groups finished the race to identify the genes involved in 11q23 leukemia at more or less the same time (Djabali et al. 1992; Gu et al. 1992; Tkachuk et al. 1992; Ziemin-van der Poel et al. 1991). Initially, the affected gene on chromosome 11 was labeled dependent on the discovering groups: either *ALL1* (for acute lymphatic leukemia in analogy to AML1), *HRX* (human trithorax, for homology with a fly gene), or *MLL* (mixed lineage leukemia) with the latter prevailing (the actual technical nomenclature based on protein function is *KMT2a*, but this name has never been widely adopted in literature and for the remainder *MLL* will be used here).

The *MLL* gene is transcribed into an approximately 12 kb long cDNA that codes for a 3,972 amino acid protein with a relative molecular weight of roughly 430 kDa. From sequence comparison it became immediately clear that MLL was a mammalian homolog of *Drosophila* Trithorax (TRX), an important regulator of embryonic development. TRX mutations in flies perturb segmental identity and produce homeotic transformations (homeo = Greek: omoios = like; mutants that transform body parts to look like other body parts). Instead of three thoracic segments with different properties, TRX-mutants developed three uniform segments with the same appearance. Because similar phenotypes were also observed for animals with mutant homeobox (*HOX*) genes, the relationship of TRX and *HOX* was investigated in detail. In situ hybridizations revealed that TRX was absolutely necessary to maintain a preestablished *HOX* expression pattern. In the absence of TRX, correct *HOX* expression domains could be initially detected, but transcription was extinguished during further development [for a review see Schuettengruber et al. (2011)]. Therefore, TRX was somehow involved in endowing the individual cells with a "memory," thus maintaining a characteristic transcriptional state. Clearly this peculiar mode of action was different from other positively acting transcription factors known at that time. However, this behavior resembled a mirror image of the action of polycomb-group proteins which act as repressors of homeotic gene expression (for more detailed information on polycomb function see chapters by Vidal and van den Boom, in this book). Knockout studies in mice soon revealed a similar function also for Mll (Yu et al. 1995). Homozygous deletion of *Mll* was embryonic lethal around day 10 of embryonic development with *Hox* expression being largely absent. Heterozygous animals displayed shifts in *Hox* expression boundaries accompanied by homeotic phenotypes. *Mll* was also necessary for embryonic and adult hematopoiesis with reduced numbers of long- and short-term repopulating cells in fetal liver hematopoietic cells of $Mll^{-/-}$ animals and a perturbed adult hematopoietic development after inducible knockout of the *Mll* gene in hematopoietic cells (Ernst et al. 2004).

7.2.1 MLL Is a Histone Methyltransferase

Initially, the underlying mechanisms of MLL function remained enigmatic. A breakthrough discovery was made by identifying the highly conserved SET domain at the C terminus of MLL as the active site of a protein methyltransferase (Milne et al. 2002). The name SET is an acronym for "suppressor of variegation 3–9, enhancer of zeste, and trithorax," three proteins that share the same domain. These proteins are "epigenetic writers" that deposit methyl groups onto different lysines of histone proteins (hence the systematic name KMT2A for MLL indicating a lysin (K) methyltransferase). MLL is a histone H3, lysine 4-specific enzyme that catalyzes mono-, di-, and trimethylation. While monomethylation is a hallmark of enhancer sequences, H3K4m2/3 is found around the transcriptional start sites of nearly all actively transcribed genes (Herz et al. 2012; Pekowska et al. 2011). This modification also marks a particular subclass of promoters that are in a "bivalent" state where activation-associated chromatin modifications like H3K4 methylation and "repressive" marks like methylated H3K27 coexist (Voigt et al. 2012). These promoters frequently drive genes with important roles in differentiation, development, and immediate response where a rapid induction is a key requirement. It is estimated that up to 30 % of all genes in ESCs belong to this class (Vastenhouw and Schier 2012).

It is not well understood how methylation of H3K4 supports transcription. In general most of the histone marks likely serve as docking sites for chromatin "readers." These proteins are responsible for the biochemical effects of chromatin modification. Readers can act either by themselves or as members of protein complexes. Several proteins use plant homeodomain (PHD) domains to specifically recognize methylated H3K4. Prominent examples are CHD1, a nucleosome-stimulated ATPase that is also able to facilitate post-initiation processes like splicing (Sims et al. 2007), and the known nucleosome remodeling complex NURF (Wysocka et al. 2006). NURF is believed to create a nucleosome free region to expedite transcriptional initiation. Besides it has also been proposed that an internal PHD finger of MLL itself and WDR5, a protein associated with MLL in a multiprotein complex, also recognize methylated H3K4, thus "spreading" the modification across chromatin domains (Wysocka et al. 2005). Later this was partially contested (Schuetz et al. 2006), however, retaining the modifying enzymes during mitosis on the template strand would be a tempting mechanism to propagate modifications during replication (Petruk et al. 2012).

7.2.2 MLL Domains and Their Function

Next to the SET domain, MLL contains numerous functional units that have been investigated in detail (Fig. 7.1).

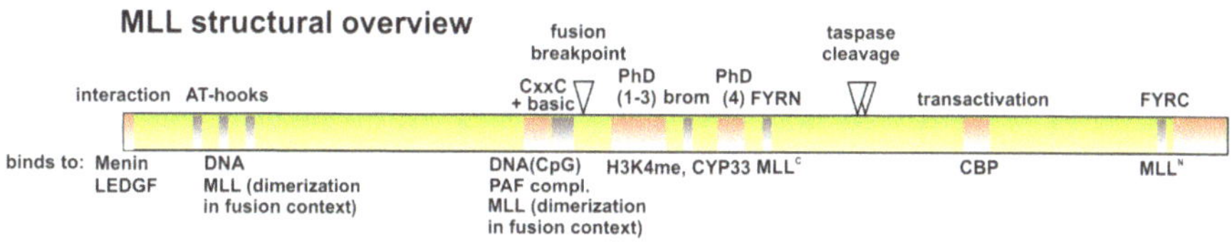

Fig. 7.1 Important MLL domains and their associated function

At the very N terminus MLL features an interaction domain that directly contacts the protein MENIN. Originally, MENIN (multiple endocrine neoplasia type I) was discovered as a tumor suppressor connected to a familiar cancer syndrome that is characterized by the occurrence of various tumors of the endocrine system like insulinomas, pituitary adenomas, and parathyroid tumors (Lakhani et al. 2007). Surprisingly, MENIN interaction is absolutely required for the oncogenic activity of MLL fusions and also for wt-MLL function (Chen et al. 2006; Grembecka et al. 2012; Huang et al. 2012a; Milne et al. 2005a; Yokoyama et al. 2005). MENIN ablation in the hematopoietic lineage reduces H3K4 methylation at *HOX* loci presumably because MLL recruitment is disturbed (Chen et al. 2006). Therefore, this protein is one of the rare examples where a protein, depending on the context, can be simultaneously tumor suppressor and oncoprotein. The MENIN-MLL interaction forms a scaffold that recruits another protein, the lens-epithelium-derived-growth factor (LEDGF) (Yokoyama and Cleary 2008). LEDGF contains a PWWP domain that mediates association with chromatin, and it has been also found interacting with various transcription factors as well as with RNA Pol II. Physiologically, LEDGF is involved in stress response. In the context of MLL it is believed that the MENIN/LEDGF interaction helps MLL to recognize specific chromatin regions; however, it is not clear if MLL requires MENIN for activation of all of its target genes.

Three so-called AT-hook motifs are present further downstream in MLL. AT-hooks are minor groove DNA binders with a preference for AT-rich sequences. This domain occurs frequently in chromatin-associated proteins like the HMG (high mobility group) proteins (Reeves 2001). Although the MLL AT-hooks have been shown to bind to distorted or cruciform DNA, their significance for MLL function is unclear (Zeleznik-Le et al. 1994). Deletion of all three AT-hooks alone does not affect transformation capability of MLL fusion proteins (Ayton et al. 2004), but no analogous experiment has been done for wt-MLL because there is no biological assay for MLL function besides knockout studies in mice. Latest results from our own laboratory indicate that the AT-hooks can also act as protein–protein interaction surface that mediates self-association of MLL fusion proteins (Maethner et al. 2013).

Another multifunctional MLL domain is located just upstream of the common breakpoint in 11q23 translocations. This area is characterized by a repeated CxxC consensus sequence that is part of an unusual zinc finger structure also present in DNA methyltransferase I and several CpG binding proteins. The MLL CxxC domain binds specifically non-methylated CpG dinucleotides which goes along well with the activating nature of MLL (Allen et al. 2006; Birke et al. 2002; Cierpicki et al. 2010). In this way access of MLL to methylated and therefore silenced DNA would be blocked. Next to DNA binding the CxxC region and a closely adjacent very basic region have been also implicated as protein–protein interaction domains. This domain recruits various repressor proteins like Bmi1 and histone deacetylases. Concomitantly, it can repress transcription in classical reporter assays (Xia et al. 2003). However, there was no evidence for the presence of repressor proteins after purification of the holo-MLL complex from cells (see below). Consistent with the proven role of MLL as transcriptional activator, the CxxC portion also mediates contact with the PAF complex (Milne et al. 2010; Muntean et al. 2010) that coordinates H3K4 methylation with transcriptional elongation, other transcription-associated chromatin modifications like ubiquitination of histone H2B, and also posttranscriptional events like splicing. Finally, next to these heterotypic interactions, the CxxC motif constitutes the second domain that allows specific self-association of MLL fusion proteins (Maethner et al. 2013). It will be interesting to see if all these various interactions can occur simultaneously of if they are an outcome of postranslational regulation representing different functional states of MLL. Because of this multipurpose nature, it is not surprising that the CxxC motif is absolutely essential for MLL fusion protein function (Slany et al. 1998) and that the design of this domain is a crucial determinant that discriminates MLL from its closest homolog MLL2 (Bach et al. 2009).

The first motif downstream of the breakpoint region and therefore consistently deleted from MLL fusion proteins is a series of four PHD fingers with an interspersed bromodomain between finger 3 and 4. PHD fingers are protein interaction domains with varying specificity. For MLL it has been shown that PHD finger number 3 binds to tri-methylated H3K4, thus endowing the histone-methyltransferase MLL to read its own mark. This feature seems to be essential for transactivation of MLL targets in vivo (Chang et al. 2010). Unrelated to this activity the PHD fingers appear to be the targets for regulation of normal MLL function by controlled proteolysis and/or conformational alterations. They are recognized by the ABS2 ubiquitin ligase (Wang et al. 2012) and mediate degradation of MLL through the proteasome system. In addition, the third PHD finger also has ubiquitin ligase activity on its own and binds cyclophilin Cyp33 (Anderson et al. 2002; Chen et al. 2008). It was suggested that the proline isomerase function of this protein elicits a conformational switch inducing corepressor binding. At present it is unknown if and how these various functions are interconnected. Yet, it is clear that this domain is incompatible with the activity of MLL **fusion** proteins. Artificial inclusion of the PHD fingers destroys the transforming potential of a test fusion (Muntean et al. 2008). This suggests that this domain has an important "negative control function" that limits activity of MLL proteins, and therefore it is incompatible with the "run-away" properties of the oncogenic counterparts.

Two closely spaced cleavage sites for posttranscriptional processing of the primary MLL protein can be found downstream of the PHD fingers (Hsieh et al. 2003b;

Yokoyama et al. 2002). Characterized by the amino acid sequence QV(L)D'G^{A}/$_{V}$DD (x = hydrophobic amino acid) that is also conserved in fly Trithorax and in the MLL homolog MLL2, these motifs are the target of a specialized protease called TASPASE1 (Hsieh et al. 2003a). This enzyme processes MLL into two fragments with apparent molecular weights of 320 kDa (MLL^{N}) and 180 kDa (MLL^{C}). Both parts remain associated by non-covalent interactions that have been mapped to two regions rich in phenylalanine and tyrosine that were labeled FRYN and FRYC because of their localization within the MLL^{N} and the MLL^{C} fragments, respectively. Posttranslational processing is important for MLL function and stability. In the absence of MLL^{C}, MLL^{N} becomes unstable and without MLL^{N} MLL^{C} shows aberrant subnuclear localization. In addition, knock down of TASPASE 1 in cell lines and a *taspase1* knockout in mice affects *HOX* gene expression as an indirect indicator of perturbed Mll function. In contrast to *MLL* deletion, TASPASE1-deficient animals are surviving. The reduced body size of taspase-negative animals uncovered an interesting link of MLL and cell cycle regulation (Takeda et al. 2006). Next to ensuring proper *HOX* expression, MLL is also involved in E2F-mediated transcription of *Cyclin* genes that are essential for cell cycle initiation and progression. Concomitant with the periodic activity of these genes, MLL shows a parallel cyclic expression pattern throughout the cell cycle. This biphasic presence of MLL is enforced by periodic degradation through the cell cycle proteolytic machinery containing SCF^{Skp2} and APC^{Cdc20}. This process is perturbed after genotoxic damage leading to phosphorylation of MLL by the checkpoint kinase ATR (Liu et al. 2010) thus contributing to stop the cell cycle for the necessary repair. Interestingly, MLL fusion proteins are not subject to degradation during cell cycle, and they are dominant negative mutants that abrogate MLL function in S-phase checkpoint. Therefore, it has been supposed that the ensuing overexpression of cyclins and the perturbation of proper DNA repair contribute to their oncogenic activity. Because of the cell cycle supporting role of MLL, inhibitors of taspase1 are in early studies as possible anticancer agents (Chen et al. 2010, 2012) .

Finally, a region with activator properties has been identified in the C-terminal portion of MLL. This motif interacts directly with the histone acetyltransferase CBP (Ernst et al. 2001). MLL contacts the KIX domain of CBP and this interaction promotes the association of CBP with various transcription factors that need to recruit histone acetyltransferases to prepare the adjacent chromatin for efficient transcription. The ternary interaction of MLL/CBP and transcription factors is likely an alternative possibility to direct MLL to specific target genes as suggested by the presence of MLL and CBP in a complex with the transcription factor MYB (Goto et al. 2002).

7.2.3 MLL Higher Order Complexes During Evolution and Their Role in Cellular Physiology

MLL does not act alone, but it is embedded into a macromolecular complex that shows a high degree of conservation from yeast to man. First isolated as the sole activity with H3K4 methyltransferase activity in *Saccharomyces cerevisiae*, this complex has been named COMPASS (complex associated with Set1) according to its core component Set1 (Miller et al. 2001). In yeast Set1 is responsible for all H3K4 methylation. During evolution this function diversified, and in mammals there are six confirmed H3K4 methylating enzymes that can be grouped in pairs (Shilatifard 2012). The most ancient enzymes are SET1A and SET1B that show the highest homology to the yeast ancestor. Knock down of the SET enzymes affects global H3K4 methylation indicating that these enzymes perform most of the methylation work (Ardehali et al. 2011; Mohan et al. 2011; Wu et al. 2008). In contrast, MLL and the close homolog MLL2 form a separate entity with a more specialized purpose (Wang et al. 2009). MLL-deficient cells still keep the majority of H3K4 methyl marks, and it has been proposed that MLL1 and MLL2 are selective for a subgroup of genes like (but not exclusively) the *HOX* loci and E2F-activated transcripts (Kerimoglu et al. 2013; Ladopoulos et al. 2013; Wang et al. 2009). Likewise MLL3 and MLL4 appear to be associated with activation of genes under control of nuclear hormone receptors (Mo et al. 2006). A somewhat enigmatic fifth homolog of MLL (MLL5) seems to have lost the catalytic activity despite the presence of a SET domain. Nevertheless, MLL5 regulates H3K4 methylation by indirect means, and MLL5 knockout mice have impairments in cell cycle control and defects in hematopoietic differentiation (Sebastian et al. 2009).

Despite this functional diversity SET1A/B and MLL1-4 are found in similar high molecular weight complexes that contain a common set of cofactors next to proteins specific for each H3K4 methyltransferase. At least four proteins are universally present in these assemblies. ASH2L, RBBP5, DPY30, and WDR5 are all required for optimal H3K4 methylation activity either by stabilizing the complex, enabling substrate contact, or augmenting the formation of an optimal active site. This topic has been reviewed recently, (Schuettengruber et al. 2011; Shilatifard 2012) and therefore it will not be covered any further here. Despite many unsolved questions, it becomes clear that MLL and related proteins as well as H3K4 methylation are very general features that are very likely required for all transcription in every cell. In addition, it has been shown that MLL is important for "bookmarking" during mitosis. MLL remains associated with chromatin during the cell cycle and facilitates early transcription after M-phase (Blobel et al. 2009). Consequently, the loss of function phenotypes is very pleiotropic. The effects of MLL on hematopoietic development are just one small facet of a more global role and much of it can be explained by the necessity of MLL for proper *HOX* expression (see below). This is a "caveat" for the development of therapeutics that targets these enzymes either directly or indirectly, for example, by inhibiting TASPASEI (Chen et al. 2010; Stauber et al. 2012). It is an easy prediction that these treatments will induce many side effects.

7.3 The Ugly Face of MLL; MLL Fusion Proteins in Leukemia

As indicated above, 11q23 translocations mutilate most of the MLL protein and conserve only approximately 1,400 amino acids of the N terminus that are fused in frame to the respective fusion partners. The most likely reason for the initiation of the prototypical reciprocal translocation is a double strand break. This link was suggested by the association of MLL translocations with prior treatment with topoisomerase inhibitors. The simultaneous presence of *MLL* and some of its common translocation partners in so-called transcription factories is an additional feature that promotes the generation of transforming MLL fusions (Cowell et al. 2012). Further corroboration for double strand breaks as initiating event comes from a large meta-study that uncovered a statistically significant correlation between background radiation and childhood leukemia in general (Kendall et al. 2013). A sizable portion of juvenile leukemia cases carries chromosomal translocations albeit not always within the MLL gene. Ionizing radiation is the major natural source of double stand breaks. For a review covering the mechanisms generating translocations see (Novo and Vizmanos 2006).

Regardless of the reason for the chromosomal aberration, one can deduce from the MLL motifs that are always in- or excluded, that MLL fusions should be still able to home in on MLL target sequences via interactions with MENIN/LEDGF/PAF, and the MLL-intrinsic DNA binding activities that are invariably retained in the chimeric proteins. In contrast the MLL derivatives lose H3K4 methyltransferase as well as histone acetyltransferase activity, and they are not cleaved by taspase or subject to control mechanisms exerted through the PHD fingers. Hence, their mechanism of action was a long-standing enigma. The most stunning feature of these proteins is the incredible variety of completely unrelated proteins that have been found to be joined to MLL by chromosomal translocations. Over 60 partners have been characterized, and new ones are still discovered (Meyer et al. 2009). Under closer scrutiny, however, this confusing collection can be ordered into several groups whose detailed investigations have yielded important hypotheses explaining most of the extraordinary transformation capability of these chimeric proteins.

7.3.1 *The Dirty Pack: ENL, AF9, AFF1-4 and Other Proteins Controlling Transcriptional Elongation*

From a numerical perspective a subgroup of MLL fusions clearly stands out. Only six different partners are involved in over 80 % of all clinically manifest cases. By far the most frequent of these is AF4 (ALL1 fused to on chromosome 4; also known as AFF1 = AF4-FMR2 family member 1) followed by ENL (eleven nineteen leukemia; alias MLLT1 = mixed-lineage-leukemia translocated 1) and AF9 (ALL1 fused to on chromosome 9; alias MLLT3). The runners up are AF10 (ALL1 fused to on chromosome 10, alias MLLT10), ELL (eleven-nineteen lysine rich leukemia gene), and AF6 (alias MLLT4).

Clearly, frequency of occurrence alone would not be an indication for a shared function because these genes might simply reside at loci most susceptible for DNA breaks. Yet, several lines of investigation revealed that there is indeed a common feature assigned to these proteins. All are involved in transcriptional elongation. The first fusion partner with an identified function in this process was ELL (Shilatifard et al. 1996). It was shown that ELL supports transcription by enabling RNA Polymerase II to skip pause sites after initiation. Control of elongation is widely used as transcriptional checkpoint with particular importance for "rapid response" genes, like those involved in differentiation, development, and cell cycle [for a review see Adelman and Lis (2012)]. The connection of transcriptional elongation and MLL fusion activity went unnoticed for a long time because in a structure function study ELL domains necessary for the elongation activity were dispensable for the transforming potential of MLL-ELL (DiMartino et al. 2000). This was determined by an in vitro CFC (colony forming cell) assay after retroviral transduction of primary hematopoietic precursors with the fusion construct. This test was initially developed to investigate MLL-ENL and relies on the fact that transformed cells become immortalized, and therefore they can be replated indefinitely in methylcellulose. In contrast normal cells start to differentiate and eventually arrest cell cycle (Lavau et al. 1997).

The link to elongation was revived when AF5q31 (alias AFF4, henceforth called AF5), a close homolog of AF4, was shown to co-purify with positive transcription elongation factor b (P-TEFb) (Estable et al. 2002). P-TEFb is a dimer of a cyclin-dependent kinase (CDK9) and a cyclin, mostly either cyclin T1 or T2. P-TEFb phosphorylates RNA Polymerase II within the so-called C-terminal repeat domain. This domain consists of a heptapeptide containing several serine residues that is reiterated 54 times in mammals. The serine at position 2 within the heptad serves as substrate for P-TEFb. This kinase reaction is a prerequisite for efficient elongation by RNA Pol II serving as a "landing pad" for several other transcription-associated proteins. In addition P-TEFb also phosphorylates and inactivates substrates with a negative influence on elongation like NELF (negative elongation factor) and DSIF (DRB sensitivity inducing factor). A review about this topic can be found at Zhou et al. (2012).

Another important step forward was the demonstration that AF10 interacts with the histone H3K79 methyltransferase DOT1L through a domain that is essential for the leukemogenic activity of the MLL-AF10 fusion (Okada et al. 2005). DOT1L is a highly conserved enzyme that catalyzes H3K79 mono-, di-, and tri-methylation. H3K79 me2/3 is distributed across actively transcribed chromatin and in yeast it has an "anti-silencing" function. Indeed, target genes of MLL fusion proteins are characterized by an unusually high level of this modification (Milne et al. 2005b). Despite the fact that transcribed chromatin seems to be universally characterized by H3K79 methylation, *Dot1l* knockout animals survive till mid-gestation and die with clear cardiovascular defects (Jones et al. 2008; Nguyen et al. 2011b) arguing against a general depression of transcription in the absence of this enzyme. Indeed DOT1L has been specifically found in complexes containing TCF, a transcriptional effector of Wnt signaling, and therefore DOT1L has been suggested to work preferentially within this particular signal pathway (Mohan et al. 2010). On the other hand a very recent publication challenges this view as there is no major phenotype after specific ablation of DOT1L in intestinal crypt cells, the cell type most dependent on Wnt-signaling (Ho et al. 2013). In summary, the role of DOT1L in general transcription is still enigmatic, but it would seem an unjustifiable expenditure of energy for a cell if it would methylate thousands of transcription units without any important biological function.

Independently, our lab has shown in two-hybrid studies that the frequent MLL fusion partners ENL, AF4, and AF10 can interact directly with each other and with chromatin (Zeisig et al. 2005). Moreover, we found that also ENL was able to recruit DOT1L. Taking an educated guess, (Bitoun et al. 2007) used this information to demonstrate by overexpression that AF4, ENL, and the ENL homolog AF9 can influence transcriptional elongation and DOT1L activity. Complementary to this information, we purified the natural, endogenous ENL complex, demonstrating that it constitutes a large, elongation-associated machinery (Mueller et al. 2007). This complex was initially called EAP (*e*longation *a*ssisting *p*roteins or *E*NL-*a*ssociated *p*roteins). EAP essentially contains four classes of proteins: (1) known MLL fusion partners including a member of the ENL family (ENL or AF9) and a representative of the AFF proteins (AF4, AF5, or FMR); (2) P-TEFb; (3) DOT1L, and (4) surprisingly and consistently, also parts of polycomb repressive complex I (see Fig. 7.2). Corroborating its presumed function, knockdown of essential EAP components affected global elongation rates. In a follow-up study (Mueller et al. 2009) we uncovered an intricate network of protein–protein interactions that stabilizes the higher order structure of EAP. It was shown that elongation promoting activity is firmly associated with the MLL-ENL fusion and that destabilizing EAP by mutations also destroys the transforming activity of MLL-ENL.

Subsequently, EAP-like complexes were isolated by several other laboratories (Lin et al. 2010; Monroe et al. 2011; Yokoyama et al. 2010). Interestingly, not all possible interactions within EAP seem to exist at the same time and therefore EAP can be subdivided in a complex with elongation promoting properties named "super elongation complex" (SEC) (Lin et al. 2010) and an assembly with chromatin modifying capacity containing DOT1L (DotCom) (Mohan et al. 2010). These

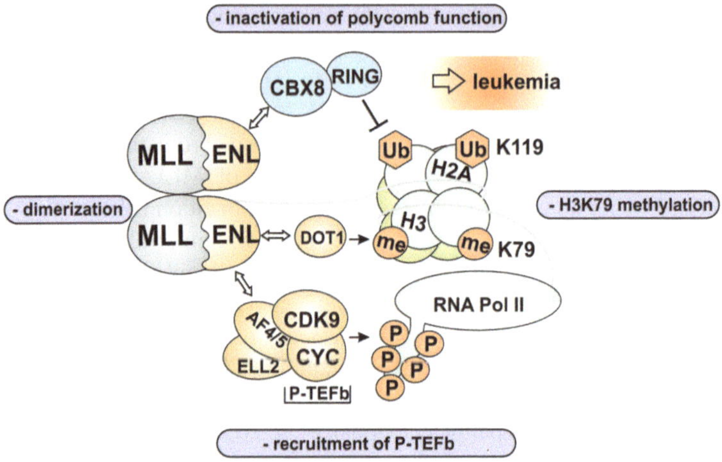

Fig. 7.2 The MLL fusion cancer machine. Fusions of MLL with the frequent partner proteins ENL/AF9 or AF4/AF5 recruit a multifunctional machinery that activates elongation and represses polycomb activity. For a detailed explanation see main text

alternative compositions of EAP assuming either a "SEC" or a "DotCom" configuration are controlled by ENL or AF9, respectively. These proteins serve as a "selector switch" that use an intrinsically disordered protein domain to bind exclusively to one of a variety of different possible interaction partners (Leach et al. 2013). This is possible because they contain an interaction domain that adopts variable conformations depending on the respective binding partner, thus ensuring mutually exclusive interactions. In this way ENL/AF9 contacts either one of the AFF1-4 family proteins, DOT1L, or a polycomb protein (see below). AFF proteins then serve as scaffold linking the elongation factor ELL2 and P-TEFb to ENL/AF9. It is not yet clear if and how alternative binding modes of ENL/AF9 are actively regulated. The various interactions may happen subsequently or in different complexes co-occupying a locus. MLL fusions, however, have acquired the capability to mediate all these interactions at the same time, as recent results from our lab indicate that MLL fusions can di- or mulitmerize (see below).

Still, the most counter-intuitive finding is the association of polycomb repressor proteins with the EAP activator complex. The chromobox protein CBX8 as well as the corepressor BCoR have been shown to interact directly with ENL and AF9 (Garcia-Cuellar et al. 2001; Srinivasan et al. 2003), and these proteins co-purify with endogenous EAP. One potential explanation is suggested by a report that claims that AF9 evokes a "moonlighting" function in CBX8 turning it into an activator through recruitment of the TIP60 histone acetyltransferase (Tan et al. 2011) [Protein moonlighting is a phenomenon where one protein has two, often unrelated, functions (Jeffery 2003)]. However, our own studies (Maethner et al. 2013) rather indicate that ENL (in conjunction with EAP) is able to inactivate the intrinsic repressor activity of CBX8 (and PRC1) as prerequisite for efficient activation of target loci. Particularly, the *HOX* loci as important MLL targets are normally subject to polycomb-mediated repression (see Chaps. 5 and 6 this book), and therefore this obstacle will have to be overcome before efficient transcription can commence.

In summary MLL fusion partners tie a highly active elongator complex to a vestigial MLL. In this way transcriptional elongation is efficiently promoted and endogenous repressors are removed. Whenever MLL will home in on a locus that is susceptible to this kind of stimulation (e.g., the *HOX* loci) this will elicit target activation. As an interesting side aspect, a similar mechanism has been identified for the control of the HIV major long-terminal repeat promoter. The responsible viral transactivator Tat also uses an EAP-like complex to stimulate viral transcription by elongation control (He et al. 2010; Sobhian et al. 2010), and it is expected that more transcription factors will recruit and use the help of EAP/SEC/DotCom. In this light it is important to note that also MLL-AF6 transformed cells are critically dependent on DOT1L, although a direct connection between AF6 and EAP has not yet been detected (Deshpande et al. 2013). Remarkably, and despite the multi-pronged "attack," MLL fusion proteins are not able to activate targets on their own. They are still critically dependent on the presence of normal wt-MLL (Thiel et al. 2010). Deletion of MLL in MLL fusion transformed cells leads to a loss of MLL target expression accompanied by cell cycle arrest and apoptosis underlining the general importance of H3K4 methylation for all transcription. Correspondingly, leukemia cells with MLL translocations never harbor biallelic MLL alterations and leave the second wt-MLL allele untouched.

Clearly, the various enzymes within EAP may be attractive targets for a therapeutic intervention. First preclinical experiments with specific inhibitors of the DOT1L methyltransferase have yielded promising results in animal models of MLL fusion-induced leukemia (Daigle et al. 2011). Likewise indirect inhibition of P-TEFb by disabling a bromodomain factor (BRD4) that seems to be involved in recruitment of this cyclin/CDK dimer is also effective in mixed lineage leukemia (Dawson et al. 2011). In our hands small molecule inhibitors of CDK9 show promising efficacy in animal studies (RKS unpublished). However, it is mandatory to keep in mind that all activities that are "highjacked" by MLL fusion proteins are essential cellular functions that are important for normal physiology. Therefore, severe side effects cannot be excluded as it has been demonstrated for inhibition of DOT1L (Kim et al. 2012; Nguyen et al. 2011a). The question of a "therapeutic window" will be the decisive factor if these substances will find their way into the clinic.

7.3.2 The Odd Outsiders: CBP and EEN

Different, yet related pathways seem to underlie the transforming activity of two other MLL fusion proteins. In rare cases of therapy-induced leukemia, MLL is fused to the histone acetyltransferase CBP (Satake et al. 1997). Although HAT activity has not been shown biochemically, the domains conferring acetyltransferase activity were essential for the overall transforming activity of MLL-CBP (Lavau et al. 2000). Therefore, it seems conceivable that increased acetylation at target loci leads to aberrant activation. Unfortunately it is not known if histone acetylation makes H3K79 methylation and elongation control by P-TEFb obsolete. In this case MLL-CBP cells should be relatively resistant towards the respective inhibitors.

Lastly, a single case has been reported where MLL was joined to a member of the endophilin family of proteins normally involved in endocytosis (So et al. 1997). This MLL-EEN fusion binds the protein arginine methyltransferase PRMT1 and it was suggested that histone methylation by aberrantly recruited PRMT1 may be at the basis of MLL-EEN induced transformation (Cheung et al. 2007).

7.3.3 The Silent Majority: Cytoplasmic Proteins and Dimerization as Oncogenic Event

By far most the most diverse group of MLL partners was predominantly identified in adult cases of MLL rearranged leukemia. In contrast to the "frequent" partners most of these genes have been found only once or at best a few times involved in 11q23 translocations. This particular pattern of occurrence may be a hint that these MLL rearrangements need more cooperating mutations and hence more time before they elicit acute leukemia. This is supported by the fact that many of the resulting fusions do not read out in the classical retroviral transduction assays (Fuchs et al. 2001). From a biochemical perspective almost all rare fusion partners are normally cytoplasmic proteins. Strikingly, many of these carry di- or multimeriation domains. Indeed, it could be demonstrated that fusions of MLL with an artificial dimerization domain create weakly transforming proteins (Martin et al. 2003; So et al. 2003). Di- or mulitmerization appears to be a feature of many oncogenic fusion proteins, and self-association domains have been recognized to be essential for transformation and also for the retinoic acid receptor fusions in acute promyleocytic leukemia (Sternsdorf et al. 2006; Zhou et al. 2006). From a functional point of view, dimers could act either as "enhancers" amplifying the (weak) effect of a single protein or they could allow the simultaneous occurrence of events (for example the concurrent recruitment of two different enzymatic activities) that normally have to take place consecutively. In this way normal checkpoints may be bypassed.

Indeed, there is a mechanistic connection between the function of the "frequent" MLL partners and the "dimerizers." The N terminus of MLL interacts with the PAF complex, and consequently a MLL di- or multimer would bring more PAF into the vicinity of a target locus. The PAF complex stimulates transcriptional elongation and recruits P-TEFb triggering the same mechanism as the "frequent" fusion partners. Therefore, strongly dimerizing fusion partners would indirectly acquire a function similar to EAP members. Surprisingly, all MLL fusions, regardless of the respective partner protein, have an intrinsic di/multimerization capability mediated by the intermolecular interaction domains in the MLL N terminus (Maethner et al. 2013). This correlates well with the fact that MLL-activated chromatin shows H3K79 methylation, absence of polycomb repression, and elongation stimulation at the same time. All of these events would have to happen consecutively if MLL fusions could not dimerize because the protein partners necessary for each activity bind mutually exclusively (see above).

7.3.4 A Special Case of "Fusion": MLL Tandem Duplications

Next to chromosomal translocations the MLL gene is also subject to small internal tandem duplications that create an even larger MLL protein where an N-terminal portion including the CxxC domain is repeated twice (Basecke et al. 2006). This aberration is found predominantly in adult AML. Gene expression studies in knockin animal models revealed that cases with MLL tandem duplication resemble classical reciprocal fusions as both show the typical overexpression of *HOX* cluster proteins (Dorrance et al. 2006). However, in contrast to translocations, MLL tandem duplications are non-oncogenic on their own and require a cooperating event like constitutively active signaling molecules (e.g., FLT3 internal duplications) (Zorko et al. 2012). Thus, MLL tandem duplications resemble the "dimerizer" fusions in a certain manner. It is tempting to speculate that the duplication of the PAF interaction domain equals a MLL dimer with roughly the same consequences. However, experimental proof for this concept is still missing.

7.3.5 The Other End of MLL: Reciprocal Fusion Proteins

Chromosomal translocations do not only create MLL fusions but they may also form reciprocal products. These "mirror images" consist of transcripts where the C terminus of MLL is joined to whatever N-terminal portion is left from the respective fusion partner. Clearly, this may lead to the expression of a second chimeric protein that can support the oncogenic process. In particular for translocation t (4;11) a role for the corresponding AF4-MLL product has been suggested (Bursen et al. 2010). Artificial introduction of AF4-MLL cDNAs into the hematopoietic lineage caused low-penetrance hematological disease after a long latency. This concept has been disputed because knock down of the MLL fusion but not of the reciprocal AF4 fusion affected cell survival of authentic t(4;11) leukemia cells (Kumar et al. 2011). In addition there are a sizable number of cases where the MLL fusion is created by complicated genomic aberrations that do not create a simple reciprocal event. Yet those leukemias do not necessarily show different features.

Nevertheless, a potential function for the potential AF4-MLL fusion through modulation of RUNX1 activity was described (Wilkinson et al. 2013). Precedence for the role of a reciprocal fusion product has been shown for RARα fusions where these reciprocal constructs augment transforming capacity (Rego and Pandolfi 2002). For MLL fusions the final jury is still out whether this "dual" action is of real importance for disease etiology.

7.4 Downstream of MLL Fusion Proteins: HOX Genes as Master Controllers of Hematopoiesis

7.4.1 HOX Gene Expression as Hallmark of MLL-Induced Leukemia

As all MLL fusions conserve the major features that have been implicated in target gene recognition for wt-MLL, it seems logical that both proteins should control the same downstream genes. This assumption has been proven for the clustered *HOX* homeobox genes. Array experiments proved that leukemia cases with MLL rearrangement conspicuously overexpressed members of this gene family (Armstrong et al. 2002). This is true for the majority of mixed lineage leukemia, although, it was recently noticed that a small subgroup of MLL-AF4 rearranged cases lacks this characteristic *HOX* expression signature (Starkova et al. 2010). It is not yet clear if this is a secondary event or if this reflects a different disease etiology. Unexpectedly on the genomic level, experiments with an inducible MLL fusion protein demonstrated that wt-MLL and MLL fusion co-localize only on a small subset of loci and not all genes that are recognized by MLL fusions also respond with an increased transcriptional output (Wang et al. 2011). This is likely a consequence of the fact that MLL fusions stimulate transcriptional elongation rather than initiation. Elongation may not be the limiting step for every transcript thus reducing the number of genes that actually react to MLL fusions. To date, however, no experimental studies examining this hypothesis have been published.

Irrespective of the exact number and nature of genes that are aberrantly expressed in MLL rearranged leukemic cells, it is clear that the relative overexpression of *HOX* genes and their protein interaction partners from the MEIS and PBX families must play a dominant role in the cellular transformation (Zeisig et al. 2004). It has been shown several times that increased levels of certain *HOX* genes in combination with overexpressed *MEIS1* create a very powerful transforming stimulus for hematopoietic cells [for reviews about this topic see Alharbi et al. (2012), Eklund (2011)].

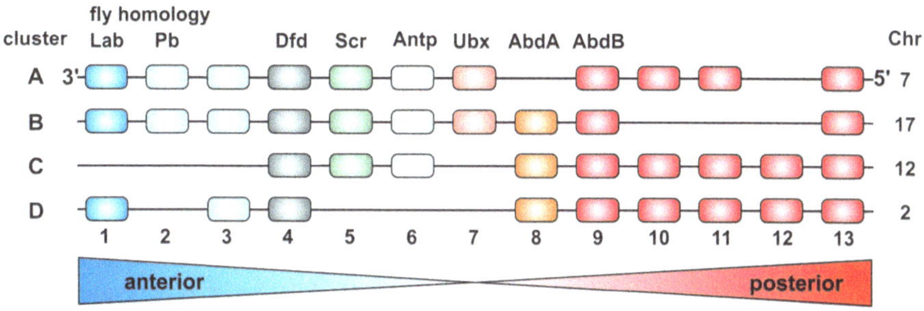

Fig. 7.3 The structure of mammalian HOX gene clusters. HOX clusters A–D are schematically depicted. Paralog groups are *shaded* according to their homology to the closest fly homolog as indicated on *top* (*Lab* labial, *Pb* proboscipedia, *Dfd* deformed, *Scr* sex comby reduced, *Antp* antennapedia, *Ubx* ultrabithorax, *AbdA* abdominal A, *AbdB* abdominal B). "Anterior" and "posterior" indicate the respective expression domain during embryogenesis

7.4.2 HOX Genes in Hematopoiesis

Under normal conditions HOX homeobox proteins are best known as master regulators of body segment identity. Mammals encode 39 *HOX* genes that are arranged in four clusters (A–D). Each of them contains between 9 and 11 *HOX* genes belonging to paralog groups 1–13 (see Fig. 7.3).

From a molecular perspective one of the most peculiar features of *HOX* genes is their stringent and conserved line-up within the genome where genes of different paralog groups follow each other head to tail in exactly the same order as the respective protein expression domains in the developing embryo. In this way paralog groups 1–6 (corresponding to the *anntennapedia* cluster in fly) are expressed in the anterior embryo whereas HOX7 to HOX13 gene products (*bithorax* in Drosophila) are located posteriorly. This phenomenon has been termed colinearity and may be based on changing DNA topology (Andrey et al. 2013), but the exact regulatory mechanism ensuring the faithful execution of this program has not yet been completely elucidated. It is clear that *HOX* expression is exquisitely sensitive to the balance of polycomb (epigenetically repressive) and trithorax (epigenetic activators like MLL) activities. In addition long noncoding RNAs (Rinn et al. 2007) and miRNAs (Popovic et al. 2009) have been implicated in *HOX* control but details are still obscure. After the essential segmented body plan has been established, *HOX*-driven regulatory circuits are "recycled" to control the development of organ systems from stem cells. In the hematopoietic lineage *HOX* genes mainly from the A and B clusters are predominantly expressed in stem- and early progenitor cells (Pineault et al. 2002). Their expression gets gradually extinguished starting with the "anterior" genes and proceeding "posteriorly" during cellular differentiation. The absence of *HOX* products is a prerequisite for proper maturation as compulsory *HOX* expression blocks differentiation and leads to the accumulation of highly proliferative precursor cells (Kroon et al. 1998; Perkins et al. 1990; Thorsteinsdottir et al. 2001).

The most well-known example for *HOX* control of hematopoietic phenotypes is HOXB4 (Antonchuk et al. 2002; Sauvageau et al. 1995). Increasing levels of HOXB4 leads to an expansion of the HSC compartment, and several groups actively pursue this pathway with the aim to generate HSCs in vitro for therapeutic purposes (Huang et al. 2010; Watts et al. 2012). Another consequence of *HOX* involvement during hematopoiesis is the high proportion of leukemia samples that show a relative overexpression of these genes. Besides MLL rearranged cases of mixed lineage leukemia, this feature is prevalent in acute myeloid leukemia with normal karyotype. Clinical studies singled out *HOXA9* as a prognostic factor negatively associated with survival (Golub et al. 1999). The *HOX*-A cluster is also targeted in acute T-cell leukemia by chromosomal translocations. These bring part of the locus under control of the strong T-cell receptor enhancer (Soulier et al. 2005; Speleman et al. 2005). Deregulated *HOX* expression has been also detected in hematological malignancies with NPM1 mutations (Vassiliou et al. 2011) and in cells overexpressing CDX, a *HOX*-upstream factor (Frohling et al. 2007). In addition *HOXA9* seems to be also involved during blast crisis in chronic myeloid leukemia (Ito et al. 2010) and in a subgroup of multiple myeloma (Chapman et al. 2011). Finally, as a consequence of chromosomal translocations occurring in AML, HOX protooncoproteins can be activated by structural rearrangements as fusions with members of the nucleopore complex (Nakamura 2005).

Despite the universal participation of *HOX* genes in the generation of leukemia, surprisingly little is known about the oncogenic pathways triggered by the respective proteins. Like classical transcription factors they are bipartite proteins containing a highly conserved homeobox DNA-binding motif at the C terminus and a very divergent N terminus that is unrelated between paralog groups. This part encodes the transcriptional effector domains and mediates recruitment of protein cofactors mainly from the TALE (three amino acid extension loop) families like MEIS and PBX. These proteins, themselves endowed with a homeobox-like DNA binding domain, increase the affinity of HOX proteins for DNA and seem to regulate the effect of HOX proteins on target transcription. In contrast, the HOX homeodomain selects for the proper, paralog-specific target genes. This has been shown in swap experiments (Breitinger et al. 2012) and is contrary to earlier in vitro studies that could detect only a very relaxed DNA binding specificity of the relatively invariant homeobox DNA binding domain (Chang et al. 1996). For efficient leukemogenesis both cofactors and HOX proteins need to be present simultaneously. This has been demonstrated for MEIS1 and PBX3 as both proteins seem to be rate limiting for efficient transformation in MLL rearranged leukemia (Li et al. 2013; Wong et al. 2007). Consequently, MEIS and PBX family members are almost always expressed together with HOX proteins in leukemia transformed by that pathway.

A systematic study revealed that the majority of *HOXA* genes with the exception of *HOXA2* and *HOXA5* has at least some degree of transforming activity for primary hematopoietic cells (Bach et al. 2010). The most potent oncogenic HOX proteins are HOXA9 and HOXA10, but also overexpression of HOXA1 and

HOXA4 in conjunction with MEIS1 could elicit acute leukemia in mice. An important emerging factor triggered by all transforming HOX proteins tested so far is *MYB*, the gene coding for the cellular counterpart of the "myeloblastosis" retroviral oncogene. The *MYB* encoded transcription factor has been shown to be essential but not sufficient for HOX/MEIS-mediated leukemia (Hess et al. 2006). This role was corroborated independently for MLL-AF9-induced disease in an unbiased shRNA screen (Zuber et al. 2011). Genomic amplification of the *MYB* locus has been detected in T-ALL (Lahortiga et al. 2007). All available data hint to an important role of MYB in a variety of hematological malignancies but, with the exception of *MYC* that seems to be under control of MYB, there is no clear understanding which pathways are responsible for the transforming activity of MYB in hematopoietic cells. The nature of the contributing factors that cooperate with MYB within the HOX controlled network is equally enigmatic. Studies employing inducible HOX derivatives to identify HOXA9 targets by global ChIP-seq and array analysis have been published (Breitinger et al. 2013; Huang et al. 2012b). Next to MYB these uncovered several genes with a known role in leukemia like *LMO2*, *FLT3*, *SOX4*, *ERG*, and *VAV*, but it is not yet established if and how these cooperate and if these are sufficient to mediate the transforming activity of HOXA9.

7.5 Uncharted Territory: Conclusions and Outlook

Research about MLL fusion proteins has come a long way in the last 21 years since the first identification of the underlying molecular lesion. Major pathways have been elucidated, and first preclinical trials are under way exploiting the potential vulnerabilities of MLL fusion-induced transformation. Still there are wide gaps in our knowledge that need to be filled for a complete understanding that will be prerequisite to make a coordinated effort for better treatments. An (incomplete) list of the unknowns is added below.

Firstly, why does MLL-induced leukemia appear so early in life? It is not clear if infant mixed lineage leukemia is the same disease as in adults. It was speculated that a primitive hematopoietic cell within the fetal liver or even a mesenchymal stem cell may be the actual target for the chromosomal translocation in pediatric cases (Bueno et al. 2011). Closely related with this problem is the issue, why secondary cooperating mutations seem to be happening so frequently within such a short time frame in very young children. Maybe, perturbation of DNA repair by MLL fusions is a possible lead for this problem. A few single studies suggest that MLL fusions inactivate P53, inhibit double strand repair, and make cells more susceptible to further genetic insult (Eguchi et al. 2006; Maki et al. 1999; Wiederschain et al. 2003). However, conclusive evidence is still missing.

The second problem touches on the mechanism that leads to gene activation downstream of MLL fusion proteins. We do not really know the exact purpose of H3K79 methylation and how this modification contributes to enhanced transcription. Likewise it is enigmatic how MLL fusions inactivate the repressor activity of the polycomb complexes to neutralize their negative influence on *HOX* expression. The regulatory landscape of the *HOX* loci is almost completely unidentified. Therefore, at present it is simply unclear where to look to solve these problems. With the advent of genome-wide techniques, this obstacle should be soon obsolete and surely a detailed picture of how MLL fusions affect the normal epigenetic control of these loci will emerge in the near future. This will enable us to tackle the next problem: Can the runaway elongation in MLL leukemia be reined in again without affecting global mechanisms and thus leading to general toxicity?

Third, although *HOX* genes undoubtedly are major players in the oncogenic pathways downstream of MLL fusions, it is by far not clear if they act on their own. A detailed census of MLL fusion targets outside of the *HOX* loci is sorely missing. We'll have to understand if these other genes are only "bystanders" or if they modulate the disease course and outcome and therefore may constitute potential therapeutic targets on their own.

Finally, the transforming program elicited by *HOX* genes only begins to be unraveled. Because *HOX* expression is the leukemogenic driver of a sizable portion of acute leukemia cases, also beyond MLL, it is of utmost importance to untangle this mechanism. This may reveal therapeutically accessible hubs as potential attack points for future therapies.

References

Adelman K, Lis JT (2012) Promoter-proximal pausing of RNA polymerase II: emerging roles in metazoans. Nat Rev Genet 13:720–731. doi:10.1038/nrg3293

Alharbi RA, Pettengell R, Pandha HS, Morgan R (2012) The role of HOX genes in normal hematopoiesis and acute leukemia. Leukemia 27(5):1000–1008. doi:10.1038/leu.2012.356

Allen MD, Grummitt CG, Hilcenko C, Min SY, Tonkin LM, Johnson CM, Freund SM, Bycroft M, Warren AJ (2006) Solution structure of the nonmethyl-CpG-binding CXXC domain of the leukaemia-associated MLL histone methyltransferase. EMBO J 25:4503–4512. doi:10.1038/sj.emboj.7601340

Anderson M, Fair K, Amero S, Nelson S, Harte PJ, Diaz MO (2002) A new family of cyclophilins with an RNA recognition motif that interact with members of the trx/MLL protein family in Drosophila and human cells. Dev Genes Evol 212:107–113. doi:10.1007/s00427-002-0213-8

Andrey G, Montavon T, Mascrez B, Gonzalez F, Noordermeer D, Leleu M, Trono D, Spitz F, Duboule D (2013) A switch between topological domains underlies HoxD genes collinearity in mouse limbs. Science 340:1234167. doi:10.1126/science.1234167

Antonchuk J, Sauvageau G, Humphries RK (2002) HOXB4-induced expansion of adult hematopoietic stem cells ex vivo. Cell 109:39–45

Ardehali MB, Mei A, Zobeck KL, Caron M, Lis JT, Kusch T (2011) Drosophila Set1 is the major histone H3 lysine 4 trimethyltransferase with role in transcription. EMBO J 30:2817–2828. doi:10.1038/emboj.2011.194

Armstrong SA, Staunton JE, Silverman LB, Pieters R, den Boer ML, Minden MD, Sallan SE, Lander ES, Golub TR, Korsmeyer SJ (2002) MLL translocations specify a distinct gene expression profile that distinguishes a unique leukemia. Nat Genet 30:41–47. doi:10.1038/ng765

Ayton PM, Chen EH, Cleary ML (2004) Binding to nonmethylated CpG DNA is essential for target recognition, transactivation, and myeloid transformation by an MLL oncoprotein. Mol Cell Biol 24:10470–10478. doi:10.1128/MCB.24.23.10470-10478.2004

Bach C, Mueller D, Buhl S, Garcia-Cuellar MP, Slany RK (2009) Alterations of the CxxC domain preclude oncogenic activation of mixed-lineage leukemia 2. Oncogene 28:815–823. doi:10.1038/onc.2008.443

Bach C, Buhl S, Mueller D, Garcia-Cuellar MP, Maethner E, Slany RK (2010) Leukemogenic transformation by HOXA cluster genes. Blood 115:2910–2918. doi:10.1182/blood-2009-04-216606

Basecke J, Whelan JT, Griesinger F, Bertrand FE (2006) The MLL partial tandem duplication in acute myeloid leukaemia. Br J Haematol 135:438–449. doi:10.1111/j.1365-2141.2006.06301.x

Berger R, Bernheim A, Sigaux F, Daniel MT, Valensi F, Flandrin G (1982) Acute monocytic leukemia chromosome studies. Leuk Res 6:17–26

Birke M, Schreiner S, Garcia-Cuellar MP, Mahr K, Titgemeyer F, Slany RK (2002) The MT domain of the proto-oncoprotein MLL binds to CpG-containing DNA and discriminates against methylation. Nucleic Acids Res 30:958–965

Bitoun E, Oliver PL, Davies KE (2007) The mixed-lineage leukemia fusion partner AF4 stimulates RNA polymerase II transcriptional elongation and mediates coordinated chromatin remodeling. Hum Mol Genet 16:92–106. doi:10.1093/hmg/ddl444

Blobel GA, Kadauke S, Wang E, Lau AW, Zuber J, Chou MM, Vakoc CR (2009) A reconfigured pattern of MLL occupancy within mitotic chromatin promotes rapid transcriptional reactivation following mitotic exit. Mol Cell 36:970–983. doi:10.1016/j.molcel.2009.12.001

Breitinger C, Maethner E, Garcia-Cuellar MP, Slany RK (2012) The homeodomain region controls the phenotype of HOX-induced murine leukemia. Blood 120:4018–4027. doi:10.1182/blood-2011-10-384685

Breitinger C, Maethner E, Garcia-Cuellar MP, Schambony A, Fischer KD, Schilling K, Slany RK (2013) HOX genes regulate Rac1 activity in hematopoietic cells through control of Vav2 expression. Leukemia 27:236–238. doi:10.1038/leu.2012.166

Bueno C, Montes R, Catalina P, Rodriguez R, Menendez P (2011) Insights into the cellular origin and etiology of the infant pro-B acute lymphoblastic leukemia with MLL-AF4 rearrangement. Leukemia 25:400–410. doi:10.1038/leu.2010.284

Bursen A, Schwabe K, Ruster B, Henschler R, Ruthardt M, Dingermann T, Marschalek R (2010) The AF4.MLL fusion protein is capable of inducing ALL in mice without requirement of MLL. AF4. Blood 115:3570–3579. doi:10.1182/blood-2009-06-229542

Chang CP, Brocchieri L, Shen WF, Largman C, Cleary ML (1996) Pbx modulation of Hox homeodomain amino-terminal arms establishes different DNA-binding specificities across the Hox locus. Mol Cell Biol 16:1734–1745

Chang PY, Hom RA, Musselman CA, Zhu L, Kuo A, Gozani O, Kutateladze TG, Cleary ML (2010) Binding of the MLL PHD3 finger to histone H3K4me3 is required for MLL-dependent gene transcription. J Mol Biol 400:137–144. doi:10.1016/j.jmb.2010.05.005

Chapman MA, Lawrence MS, Keats JJ, Cibulskis K, Sougnez C, Schinzel AC, Harview CL, Brunet JP, Ahmann GJ, Adli M, Anderson KC, Ardlie KG, Auclair D, Baker A, Bergsagel PL, Bernstein BE, Drier Y, Fonseca R, Gabriel SB, Hofmeister CC, Jagannath S, Jakubowiak AJ, Krishnan A, Levy J, Liefeld T, Lonial S, Mahan S, Mfuko B, Monti S, Perkins LM, Onofrio R, Pugh TJ, Rajkumar SV, Ramos AH, Siegel DS, Sivachenko A, Stewart AK, Trudel S, Vij R, Voet D, Winckler W, Zimmerman T, Carpten J, Trent J, Hahn WC, Garraway LA, Meyerson M, Lander ES, Getz G, Golub TR (2011) Initial genome sequencing and analysis of multiple myeloma. Nature 471:467–472. doi:10.1038/nature09837

Chen YX, Yan J, Keeshan K, Tubbs AT, Wang H, Silva A, Brown EJ, Hess JL, Pear WS, Hua X (2006) The tumor suppressor menin regulates hematopoiesis and myeloid transformation by influencing Hox gene expression. Proc Natl Acad Sci U S A 103:1018–1023. doi:10.1073/pnas.0510347103

Chen J, Santillan DA, Koonce M, Wei W, Luo R, Thirman MJ, Zeleznik-Le NJ, Diaz MO (2008) Loss of MLL PHD finger 3 is necessary for MLL-ENL-induced hematopoietic stem cell immortalization. Cancer Res 68:6199–6207. doi:10.1158/0008-5472.CAN-07-6514

Chen DY, Liu H, Takeda S, Tu HC, Sasagawa S, Van Tine BA, Lu D, Cheng EH, Hsieh JJ (2010) Taspase1 functions as a non-oncogene addiction protease that coordinates cancer cell proliferation and apoptosis. Cancer Res 70:5358–5367. doi:10.1158/0008-5472.CAN-10-0027

Chen DY, Lee Y, Van Tine BA, Searleman AC, Westergard TD, Liu H, Tu HC, Takeda S, Dong Y, Piwnica-Worms DR, Oh KJ, Korsmeyer SJ, Hermone A, Gussio R, Shoemaker RH, Cheng EH, Hsieh JJ (2012) A pharmacologic inhibitor of the protease Taspase1 effectively inhibits breast and brain tumor growth. Cancer Res 72:736–746. doi:10.1158/0008-5472.CAN-11-2584

Cheung N, Chan LC, Thompson A, Cleary ML, So CW (2007) Protein arginine-methyltransferase-dependent oncogenesis. Nat Cell Biol 9:1208–1215. doi:10.1038/ncb1642

Childs CC, Hirsch-Ginsberg C, Culbert SJ, Ahearn M, Reuben J, Trujillo JM, Cork A, Walters RR, Freireich EJ, Stass SA (1988) Lineage heterogeneity in acute leukemia with the t(4;11) abnormality: implications for acute mixed lineage leukemia. Hematol Pathol 2:145–157

Cierpicki T, Risner LE, Grembecka J, Lukasik SM, Popovic R, Omonkowska M, Shultis DD, Zeleznik-Le NJ, Bushweller JH (2010) Structure of the MLL CXXC domain-DNA complex and its functional role in MLL-AF9 leukemia. Nat Struct Mol Biol 17:62–68. doi:10.1038/nsmb.1714

Cowell IG, Sondka Z, Smith K, Lee KC, Manville CM, Sidorczuk-Lesthuruge M, Rance HA, Padget K, Jackson GH, Adachi N, Austin CA (2012) Model for MLL translocations in therapy-related leukemia involving topoisomerase IIbeta-mediated DNA strand breaks and gene proximity. Proc Natl Acad Sci U S A 109:8989–8994. doi:10.1073/pnas.1204406109

Daigle SR, Olhava EJ, Therkelsen CA, Majer CR, Sneeringer CJ, Song J, Johnston LD, Scott MP, Smith JJ, Xiao Y, Jin L, Kuntz KW, Chesworth R, Moyer MP, Bernt KM, Tseng JC, Kung AL, Armstrong SA, Copeland RA, Richon VM, Pollock RM (2011) Selective killing of mixed lineage leukemia cells by a potent small-molecule DOT1L inhibitor. Cancer Cell 20:53–65. doi:10.1016/j.ccr.2011.06.009

Dawson MA, Prinjha RK, Dittmann A, Giotopoulos G, Bantscheff M, Chan WI, Robson SC, Chung CW, Hopf C, Savitski MM, Huthmacher C, Gudgin E, Lugo D, Beinke S, Chapman TD, Roberts EJ, Soden PE, Auger KR, Mirguet O, Doehner K, Delwel R, Burnett AK, Jeffrey P, Drewes G, Lee K, Huntly BJ, Kouzarides T (2011) Inhibition of BET recruitment to chromatin as an effective treatment for MLL-fusion leukaemia. Nature 478:529–533. doi:10.1038/nature10509

Deshpande AJ, Chen L, Fazio M, Sinha AU, Bernt KM, Banka D, Dias S, Chang J, Olhava EJ, Daigle SR, Richon VM, Pollock RM, Armstrong SA (2013) Leukemic transformation by the MLL-AF6 fusion oncogene requires the H3K79 methyltransferase Dot1l. Blood 121(13): 2533–2541. doi:10.1182/blood-2012-11-465120

DiMartino JF, Miller T, Ayton PM, Landewe T, Hess JL, Cleary ML, Shilatifard A (2000) A carboxy-terminal domain of ELL is required and sufficient for immortalization of myeloid progenitors by MLL-ELL. Blood 96:3887–3893

Djabali M, Selleri L, Parry P, Bower M, Young BD, Evans GA (1992) A trithorax-like gene is interrupted by chromosome 11q23 translocations in acute leukaemias. Nat Genet 2:113–118. doi:10.1038/ng1092-113

Dorrance AM, Liu S, Yuan W, Becknell B, Arnoczky KJ, Guimond M, Strout MP, Feng L, Nakamura T, Yu L, Rush LJ, Weinstein M, Leone G, Wu L, Ferketich A, Whitman SP, Marcucci G, Caligiuri MA (2006) Mll partial tandem duplication induces aberrant Hox expression in vivo via specific epigenetic alterations. J Clin Invest 116:2707–2716. doi:10.1172/JCI25546

Eguchi M, Eguchi-Ishimae M, Knight D, Kearney L, Slany R, Greaves M (2006) MLL chimeric protein activation renders cells vulnerable to chromosomal damage: an explanation for the very short latency of infant leukemia. Genes Chromosomes Cancer 45:754–760. doi:10.1002/gcc.20338

Eklund E (2011) The role of Hox proteins in leukemogenesis: insights into key regulatory events in hematopoiesis. Crit Rev Oncog 16:65–76

Ernst P, Wang J, Huang M, Goodman RH, Korsmeyer SJ (2001) MLL and CREB bind cooperatively to the nuclear coactivator CREB-binding protein. Mol Cell Biol 21:2249–2258. doi:10.1128/MCB.21.7.2249-2258.2001

Ernst P, Fisher JK, Avery W, Wade S, Foy D, Korsmeyer SJ (2004) Definitive hematopoiesis requires the mixed-lineage leukemia gene. Dev Cell 6:437–443

Estable MC, Naghavi MH, Kato H, Xiao H, Qin J, Vahlne A, Roeder RG (2002) MCEF, the newest member of the AF4 family of transcription factors involved in leukemia, is a positive transcription elongation factor-b-associated protein. J Biomed Sci 9:234–245

Frohling S, Scholl C, Bansal D, Huntly BJ (2007) HOX gene regulation in acute myeloid leukemia: CDX marks the spot? Cell Cycle 6:2241–2245

Fuchs U, Rehkamp G, Haas OA, Slany R, Konig M, Bojesen S, Bohle RM, Damm-Welk C, Ludwig WD, Harbott J, Borkhardt A (2001) The human formin-binding protein 17 (FBP17) interacts with sorting nexin, SNX2, and is an MLL-fusion partner in acute myelogeneous leukemia. Proc Natl Acad Sci U S A 98:8756–8761. doi:10.1073/pnas.121433898

Garcia-Cuellar MP, Zilles O, Schreiner SA, Birke M, Winkler TH, Slany RK (2001) The ENL moiety of the childhood leukemia-associated MLL-ENL oncoprotein recruits human Polycomb 3. Oncogene 20:411–419. doi:10.1038/sj.onc.1204108

Golub TR, Slonim DK, Tamayo P, Huard C, Gaasenbeek M, Mesirov JP, Coller H, Loh ML, Downing JR, Caligiuri MA, Bloomfield CD, Lander ES (1999) Molecular classification of cancer: class discovery and class prediction by gene expression monitoring. Science 286: 531–537

Goto NK, Zor T, Martinez-Yamout M, Dyson HJ, Wright PE (2002) Cooperativity in transcription factor binding to the coactivator CREB-binding protein (CBP). The mixed lineage leukemia protein (MLL) activation domain binds to an allosteric site on the KIX domain. J Biol Chem 277:43168–43174. doi:10.1074/jbc.M207660200

Grembecka J, He S, Shi A, Purohit T, Muntean AG, Sorenson RJ, Showalter HD, Murai MJ, Belcher AM, Hartley T, Hess JL, Cierpicki T (2012) Menin-MLL inhibitors reverse oncogenic activity of MLL fusion proteins in leukemia. Nat Chem Biol 8:277–284. doi:10.1038/nchembio.773

Gu Y, Nakamura T, Alder H, Prasad R, Canaani O, Cimino G, Croce CM, Canaani E (1992) The t (4;11) chromosome translocation of human acute leukemias fuses the ALL-1 gene, related to Drosophila trithorax, to the AF-4 gene. Cell 71:701–708

He N, Liu M, Hsu J, Xue Y, Chou S, Burlingame A, Krogan NJ, Alber T, Zhou Q (2010) HIV-1 Tat and host AFF4 recruit two transcription elongation factors into a bifunctional complex for coordinated activation of HIV-1 transcription. Mol Cell 38:428–438. doi:10.1016/j.molcel.2010.04.013

Heisterkamp N, Stephenson JR, Groffen J, Hansen PF, de Klein A, Bartram CR, Grosveld G (1983) Localization of the c-abl oncogene adjacent to a translocation break point in chronic myelocytic leukaemia. Nature 306:239–242

Herz HM, Mohan M, Garruss AS, Liang K, Takahashi YH, Mickey K, Voets O, Verrijzer CP, Shilatifard A (2012) Enhancer-associated H3K4 monomethylation by Trithorax-related, the Drosophila homolog of mammalian Mll3/Mll4. Genes Dev 26:2604–2620. doi:10.1101/gad.201327.112

Hess JL, Bittner CB, Zeisig DT, Bach C, Fuchs U, Borkhardt A, Frampton J, Slany RK (2006) c-Myb is an essential downstream target for homeobox-mediated transformation of hematopoietic cells. Blood 108:297–304. doi:10.1182/blood-2005-12-5014

Ho LL, Sinha A, Verzi M, Bernt KM, Armstrong S, Shivdasani RA (2013) DOT1L-mediated H3K79 methylation in chromatin is dispensable for Wnt pathway-specific and other intestinal epithelial functions. Mol Cell Biol 33:1735–1745. doi:10.1128/MCB.01463-12

Hsieh JJ, Cheng EH, Korsmeyer SJ (2003a) Taspase1: a threonine aspartase required for cleavage of MLL and proper HOX gene expression. Cell 115:293–303

Hsieh JJ, Ernst P, Erdjument-Bromage H, Tempst P, Korsmeyer SJ (2003b) Proteolytic cleavage of MLL generates a complex of N- and C-terminal fragments that confers protein stability and subnuclear localization. Mol Cell Biol 23:186–194

Huang CH, Chen PM, Lu TC, Kung WM, Chiou TJ, Yang MH, Kao JY, Wu KJ (2010) Purified recombinant TAT-homeobox B4 expands CD34(+) umbilical cord blood and peripheral blood progenitor cells ex vivo. Tissue Eng Part C Methods 16:487–496. doi:10.1089/ten.TEC.2009.0163

Huang J, Gurung B, Wan B, Matkar S, Veniaminova NA, Wan K, Merchant JL, Hua X, Lei M (2012a) The same pocket in menin binds both MLL and JUND but has opposite effects on transcription. Nature 482:542–546. doi:10.1038/nature10806

Huang Y, Sitwala K, Bronstein J, Sanders D, Dandekar M, Collins C, Robertson G, MacDonald J, Cezard T, Bilenky M, Thiessen N, Zhao Y, Zeng T, Hirst M, Hero A, Jones S, Hess JL (2012b) Identification and characterization of Hoxa9 binding sites in hematopoietic cells. Blood 119: 388–398. doi:10.1182/blood-2011-03-341081

Ito T, Kwon HY, Zimdahl B, Congdon KL, Blum J, Lento WE, Zhao C, Lagoo A, Gerrard G, Foroni L, Goldman J, Goh H, Kim SH, Kim DW, Chuah C, Oehler VG, Radich JP, Jordan CT, Reya T (2010) Regulation of myeloid leukaemia by the cell-fate determinant Musashi. Nature 466:765–768. doi:10.1038/nature09171

Jeffery CJ (2003) Moonlighting proteins: old proteins learning new tricks. Trends Genet 19: 415–417. doi:10.1016/S0168-9525(03)00167-7

Jones B, Su H, Bhat A, Lei H, Bajko J, Hevi S, Baltus GA, Kadam S, Zhai H, Valdez R, Gonzalo S, Zhang Y, Li E, Chen T (2008) The histone H3K79 methyltransferase Dot1L is essential for mammalian development and heterochromatin structure. PLoS Genet 4:e1000190. doi:10.1371/journal.pgen.1000190

Kaneko Y, Maseki N, Takasaki N, Sakurai M, Hayashi Y, Nakazawa S, Mori T, Takeda T, Shikano T et al (1986) Clinical and hematologic characteristics in acute leukemia with 11q23 translocations. Blood 67:484–491

Kendall GM, Little MP, Wakeford R, Bunch KJ, Miles JC, Vincent TJ, Meara JR, Murphy MF (2013) A record-based case–control study of natural background radiation and the incidence of childhood leukaemia and other cancers in Great Britain during 1980–2006. Leukemia 27:3–9. doi:10.1038/leu.2012.151

Kerimoglu C, Agis-Balboa RC, Kranz A, Stilling R, Bahari-Javan S, Benito-Garagorri E, Halder R, Burkhardt S, Stewart AF, Fischer A (2013) Histone-methyltransferase MLL2 (KMT2B) is required for memory formation in mice. J Neurosci 33:3452–3464. doi:10.1523/JNEUROSCI.3356-12.2013

Kim W, Kim R, Park G, Park JW, Kim JE (2012) Deficiency of H3K79 histone methyltransferase Dot1-like protein (DOT1L) inhibits cell proliferation. J Biol Chem 287:5588–5599. doi:10.1074/jbc.M111.328138

Krauter J, Wagner K, Schafer I, Marschalek R, Meyer C, Heil G, Schaich M, Ehninger G, Niederwieser D, Krahl R, Buchner T, Sauerland C, Schlegelberger B, Dohner K, Dohner H, Schlenk RF, Ganser A (2009) Prognostic factors in adult patients up to 60 years old with acute myeloid leukemia and translocations of chromosome band 11q23: individual patient data-based meta-analysis of the German Acute Myeloid Leukemia Intergroup. J Clin Oncol 27: 3000–3006. doi:10.1200/JCO.2008.16.7981

Kroon E, Krosl J, Thorsteinsdottir U, Baban S, Buchberg AM, Sauvageau G (1998) Hoxa9 transforms primary bone marrow cells through specific collaboration with Meis1a but not Pbx1b. EMBO J 17:3714–3725. doi:10.1093/emboj/17.13.3714

Kumar AR, Yao Q, Li Q, Sam TA, Kersey JH (2011) t(4;11) leukemias display addiction to MLL-AF4 but not to AF4-MLL. Leuk Res 35:305–309. doi:10.1016/j.leukres.2010.08.011

Ladopoulos V, Hofemeister H, Hoogenkamp M, Riggs AD, Stewart AF, Bonifer C (2013) The histone methyltransferase KMT2B is required for RNA polymerase II association and protection from DNA methylation at the MagohB CpG island promoter. Mol Cell Biol 33: 1383–1393. doi:10.1128/MCB.01721-12

Lahortiga I, De Keersmaecker K, Van Vlierberghe P, Graux C, Cauwelier B, Lambert F, Mentens N, Beverloo HB, Pieters R, Speleman F, Odero MD, Bauters M, Froyen G, Marynen P, Vandenberghe P, Wlodarska I, Meijerink JP, Cools J (2007) Duplication of the MYB oncogene in T cell acute lymphoblastic leukemia. Nat Genet 39:593–595. doi:10.1038/ng2025

Lakhani VT, You YN, Wells SA (2007) The multiple endocrine neoplasia syndromes. Annu Rev Med 58:253–265. doi:10.1146/annurev.med.58.100305.115303

Lavau C, Szilvassy SJ, Slany R, Cleary ML (1997) Immortalization and leukemic transformation of a myelomonocytic precursor by retrovirally transduced HRX-ENL. EMBO J 16:4226–4237

Lavau C, Du C, Thirman M, Zeleznik-Le N (2000) Chromatin-related properties of CBP fused to MLL generate a myelodysplastic-like syndrome that evolves into myeloid leukemia. EMBO J 19:4655–4664. doi:10.1093/emboj/19.17.4655

Leach BI, Kuntimaddi A, Schmidt CR, Cierpicki T, Johnson SA, Bushweller JH (2013) Leukemia fusion target AF9 is an intrinsically disordered transcriptional regulator that recruits multiple partners via coupled folding and binding. Structure 21:176–183. doi:10.1016/j.str.2012.11.011

Li Z, Zhang Z, Li Y, Arnovitz S, Chen P, Huang H, Jiang X, Hong GM, Kunjamma RB, Ren H, He C, Wang CZ, Elkahloun AG, Valk PJ, Dohner K, Neilly MB, Bullinger L, Delwel R, Lowenberg B, Liu PP, Morgan R, Rowley JD, Yuan CS, Chen J (2013) PBX3 is an important cofactor of HOXA9 in leukemogenesis. Blood 121:1422–1431. doi:10.1182/blood-2012-07-442004

Lin C, Smith ER, Takahashi H, Lai KC, Martin-Brown S, Florens L, Washburn MP, Conaway JW, Conaway RC, Shilatifard A (2010) AFF4, a component of the ELL/P-TEFb elongation complex and a shared subunit of MLL chimeras, can link transcription elongation to leukemia. Mol Cell 37:429–437. doi:10.1016/j.molcel.2010.01.026

Liu H, Takeda S, Kumar R, Westergard TD, Brown EJ, Pandita TK, Cheng EH, Hsieh JJ (2010) Phosphorylation of MLL by ATR is required for execution of mammalian S-phase checkpoint. Nature 467:343–346. doi:10.1038/nature09350

Maethner E, Garcia-Cuellar MP, Breitinger C, Takacova S, Divoky V, Hess JL, Slany RK (2013) MLL-ENL inhibits polycomb repressive complex 1 to achieve efficient transformation of hematopoietic cells. Cell Rep 3:1553–1566

Maki K, Mitani K, Yamagata T, Kurokawa M, Kanda Y, Yazaki Y, Hirai H (1999) Transcriptional inhibition of p53 by the MLL/MEN chimeric protein found in myeloid leukemia. Blood 93: 3216–3224

Martin ME, Milne TA, Bloyer S, Galoian K, Shen W, Gibbs D, Brock HW, Slany R, Hess JL (2003) Dimerization of MLL fusion proteins immortalizes hematopoietic cells. Cancer Cell 4: 197–207

Meyer C, Kowarz E, Hofmann J, Renneville A, Zuna J, Trka J, Ben Abdelali R, Macintyre E, De Braekeleer E, De Braekeleer M, Delabesse E, de Oliveira MP, Cave H, Clappier E, van Dongen JJ, Balgobind BV, van den Heuvel-Eibrink MM, Beverloo HB, Panzer-Grumayer R, Teigler-Schlegel A, Harbott J, Kjeldsen E, Schnittger S, Koehl U, Gruhn B, Heidenreich O, Chan LC, Yip SF, Krzywinski M, Eckert C, Moricke A, Schrappe M, Alonso CN, Schafer BW, Krauter J, Lee DA, Zur Stadt U, Te Kronnie G, Sutton R, Izraeli S, Trakhtenbrot L, Lo Nigro L, Tsaur G, Fechina L, Szczepanski T, Strehl S, Ilencikova D, Molkentin M, Burmeister T, Dingermann T, Klingebiel T, Marschalek R (2009) New insights to the MLL recombinome of acute leukemias. Leukemia 23:1490–1499. doi:10.1038/leu.2009.33

Miller T, Krogan NJ, Dover J, Erdjument-Bromage H, Tempst P, Johnston M, Greenblatt JF, Shilatifard A (2001) COMPASS: a complex of proteins associated with a trithorax-related SET domain protein. Proc Natl Acad Sci U S A 98:12902–12907. doi:10.1073/pnas.231473398

Milne TA, Briggs SD, Brock HW, Martin ME, Gibbs D, Allis CD, Hess JL (2002) MLL targets SET domain methyltransferase activity to Hox gene promoters. Mol Cell 10:1107–1117

Milne TA, Hughes CM, Lloyd R, Yang Z, Rozenblatt-Rosen O, Dou Y, Schnepp RW, Krankel C, Livolsi VA, Gibbs D, Hua X, Roeder RG, Meyerson M, Hess JL (2005a) Menin and MLL cooperatively regulate expression of cyclin-dependent kinase inhibitors. Proc Natl Acad Sci U S A 102:749–754. doi:10.1073/pnas.0408836102

Milne TA, Martin ME, Brock HW, Slany RK, Hess JL (2005b) Leukemogenic MLL fusion proteins bind across a broad region of the Hox a9 locus, promoting transcription and multiple histone modifications. Cancer Res 65:11367–11374. doi:10.1158/0008-5472.CAN-05-1041

Milne TA, Kim J, Wang GG, Stadler SC, Basrur V, Whitcomb SJ, Wang Z, Ruthenburg AJ, Elenitoba-Johnson KS, Roeder RG, Allis CD (2010) Multiple interactions recruit MLL1 and MLL1 fusion proteins to the HOXA9 locus in leukemogenesis. Mol Cell 38:853–863. doi:10.1016/j.molcel.2010.05.011

Mo R, Rao SM, Zhu YJ (2006) Identification of the MLL2 complex as a coactivator for estrogen receptor alpha. J Biol Chem 281:15714–15720. doi:10.1074/jbc.M513245200

Mohan M, Herz HM, Takahashi YH, Lin C, Lai KC, Zhang Y, Washburn MP, Florens L, Shilatifard A (2010) Linking H3K79 trimethylation to Wnt signaling through a novel Dot1-containing complex (DotCom). Genes Dev 24:574–589. doi:10.1101/gad.1898410

Mohan M, Herz HM, Smith ER, Zhang Y, Jackson J, Washburn MP, Florens L, Eissenberg JC, Shilatifard A (2011) The COMPASS family of H3K4 methylases in Drosophila. Mol Cell Biol 31:4310–4318. doi:10.1128/MCB.06092-11

Monroe SC, Jo SY, Sanders DS, Basrur V, Elenitoba-Johnson KS, Slany RK, Hess JL (2011) MLL-AF9 and MLL-ENL alter the dynamic association of transcriptional regulators with genes critical for leukemia. Exp Hematol 39(77–86):e1–e5. doi:10.1016/j.exphem.2010.09.003

Mueller D, Bach C, Zeisig D, Garcia-Cuellar MP, Monroe S, Sreekumar A, Zhou R, Nesvizhskii A, Chinnaiyan A, Hess JL, Slany RK (2007) A role for the MLL fusion partner ENL in transcriptional elongation and chromatin modification. Blood 110:4445–4454. doi:10.1182/blood-2007-05-090514

Mueller D, Garcia-Cuellar MP, Bach C, Buhl S, Maethner E, Slany RK (2009) Misguided transcriptional elongation causes mixed lineage leukemia. PLoS Biol 7:e1000249. doi:10.1371/journal.pbio.1000249

Muntean AG, Giannola D, Udager AM, Hess JL (2008) The PHD fingers of MLL block MLL fusion protein-mediated transformation. Blood 112:4690–4693. doi:10.1182/blood-2008-01-134056

Muntean AG, Tan J, Sitwala K, Huang Y, Bronstein J, Connelly JA, Basrur V, Elenitoba-Johnson KS, Hess JL (2010) The PAF complex synergizes with MLL fusion proteins at HOX loci to promote leukemogenesis. Cancer Cell 17:609–621. doi:10.1016/j.ccr.2010.04.012

Nakamura T (2005) NUP98 fusion in human leukemia: dysregulation of the nuclear pore and homeodomain proteins. Int J Hematol 82:21–27. doi:10.1532/IJH97.04160

Nguyen AT, He J, Taranova O, Zhang Y (2011a) Essential role of DOT1L in maintaining normal adult hematopoiesis. Cell Res 21:1370–1373. doi:10.1038/cr.2011.115

Nguyen AT, Xiao B, Neppl RL, Kallin EM, Li J, Chen T, Wang DZ, Xiao X, Zhang Y (2011b) DOT1L regulates dystrophin expression and is critical for cardiac function. Genes Dev 25:263–274. doi:10.1101/gad.2018511

Novo FJ, Vizmanos JL (2006) Chromosome translocations in cancer: computational evidence for the random generation of double-strand breaks. Trends Genet 22:193–196. doi:10.1016/j.tig.2006.02.001

Okada Y, Feng Q, Lin Y, Jiang Q, Li Y, Coffield VM, Su L, Xu G, Zhang Y (2005) hDOT1L links histone methylation to leukemogenesis. Cell 121:167–178. doi:10.1016/j.cell.2005.02.020

Pekowska A, Benoukraf T, Zacarias-Cabeza J, Belhocine M, Koch F, Holota H, Imbert J, Andrau JC, Ferrier P, Spicuglia S (2011) H3K4 tri-methylation provides an epigenetic signature of active enhancers. EMBO J 30:4198–4210. doi:10.1038/emboj.2011.295

Perkins A, Kongsuwan K, Visvader J, Adams JM, Cory S (1990) Homeobox gene expression plus autocrine growth factor production elicits myeloid leukemia. Proc Natl Acad Sci U S A 87: 8398–8402

Petruk S, Sedkov Y, Johnston DM, Hodgson JW, Black KL, Kovermann SK, Beck S, Canaani E, Brock HW, Mazo A (2012) TrxG and PcG proteins but not methylated histones remain associated with DNA through replication. Cell 150:922–933. doi:10.1016/j.cell.2012.06.046

Pineault N, Helgason CD, Lawrence HJ, Humphries RK (2002) Differential expression of Hox, Meis1, and Pbx1 genes in primitive cells throughout murine hematopoietic ontogeny. Exp Hematol 30:49–57

Popovic R, Riesbeck LE, Velu CS, Chaubey A, Zhang J, Achille NJ, Erfurth FE, Eaton K, Lu J, Grimes HL, Chen J, Rowley JD, Zeleznik-Le NJ (2009) Regulation of mir-196b by MLL and its overexpression by MLL fusions contributes to immortalization. Blood 113:3314–3322. doi:10.1182/blood-2008-04-154310

Pui CH, Carroll WL, Meshinchi S, Arceci RJ (2011) Biology, risk stratification, and therapy of pediatric acute leukemias: an update. J Clin Oncol 29:551–565. doi:10.1200/JCO.2010.30.7405

Reeves R (2001) Molecular biology of HMGA proteins: hubs of nuclear function. Gene 277:63–81

Rego EM, Pandolfi PP (2002) Reciprocal products of chromosomal translocations in human cancer pathogenesis: key players or innocent bystanders? Trends Mol Med 8:396–405

Rinn JL, Kertesz M, Wang JK, Squazzo SL, Xu X, Brugmann SA, Goodnough LH, Helms JA, Farnham PJ, Segal E, Chang HY (2007) Functional demarcation of active and silent chromatin domains in human HOX loci by noncoding RNAs. Cell 129:1311–1323. doi:10.1016/j.cell.2007.05.022

Rowley JD (1973) Letter: a new consistent chromosomal abnormality in chronic myelogenous leukaemia identified by quinacrine fluorescence and Giemsa staining. Nature 243:290–293

Rowley JD, Olney HJ (2002) International workshop on the relationship of prior therapy to balanced chromosome aberrations in therapy-related myelodysplastic syndromes and acute leukemia: overview report. Genes Chromosomes Cancer 33:331–345

Satake N, Ishida Y, Otoh Y, Hinohara S, Kobayashi H, Sakashita A, Maseki N, Kaneko Y (1997) Novel MLL-CBP fusion transcript in therapy-related chronic myelomonocytic leukemia with a t(11;16)(q23;p13) chromosome translocation. Genes Chromosomes Cancer 20:60–63

Sauvageau G, Thorsteinsdottir U, Eaves CJ, Lawrence HJ, Largman C, Lansdorp PM, Humphries RK (1995) Overexpression of HOXB4 in hematopoietic cells causes the selective expansion of more primitive populations in vitro and in vivo. Genes Dev 9:1753–1765

Schuettengruber B, Martinez AM, Iovino N, Cavalli G (2011) Trithorax group proteins: switching genes on and keeping them active. Nat Rev Mol Cell Biol 12:799–814. doi:10.1038/nrm3230

Schuetz A, Allali-Hassani A, Martin F, Loppnau P, Vedadi M, Bochkarev A, Plotnikov AN, Arrowsmith CH, Min J (2006) Structural basis for molecular recognition and presentation of histone H3 by WDR5. EMBO J 25:4245–4252. doi:10.1038/sj.emboj.7601316

Sebastian S, Sreenivas P, Sambasivan R, Cheedipudi S, Kandalla P, Pavlath GK, Dhawan J (2009) MLL5, a trithorax homolog, indirectly regulates H3K4 methylation, represses cyclin A2 expression, and promotes myogenic differentiation. Proc Natl Acad Sci U S A 106:4719–4724. doi:10.1073/pnas.0807136106

Shilatifard A (2012) The COMPASS family of histone H3K4 methylases: mechanisms of regulation in development and disease pathogenesis. Annu Rev Biochem 81:65–95. doi:10.1146/annurev-biochem-051710-134100

Shilatifard A, Lane WS, Jackson KW, Conaway RC, Conaway JW (1996) An RNA polymerase II elongation factor encoded by the human ELL gene. Science 271:1873–1876

Shtivelman E, Lifshitz B, Gale RP, Canaani E (1985) Fused transcript of abl and bcr genes in chronic myelogenous leukaemia. Nature 315:550–554

Sims RJ 3rd, Millhouse S, Chen CF, Lewis BA, Erdjument-Bromage H, Tempst P, Manley JL, Reinberg D (2007) Recognition of trimethylated histone H3 lysine 4 facilitates the recruitment

of transcription postinitiation factors and pre-mRNA splicing. Mol Cell 28:665–676. doi:10.1016/j.molcel.2007.11.010
Slany RK, Lavau C, Cleary ML (1998) The oncogenic capacity of HRX-ENL requires the transcriptional transactivation activity of ENL and the DNA binding motifs of HRX. Mol Cell Biol 18:122–129
So CW, Caldas C, Liu MM, Chen SJ, Huang QH, Gu LJ, Sham MH, Wiedemann LM, Chan LC (1997) EEN encodes for a member of a new family of proteins containing an Src homology 3 domain and is the third gene located on chromosome 19p13 that fuses to MLL in human leukemia. Proc Natl Acad Sci U S A 94:2563–2568
So CW, Lin M, Ayton PM, Chen EH, Cleary ML (2003) Dimerization contributes to oncogenic activation of MLL chimeras in acute leukemias. Cancer Cell 4:99–110
Sobhian B, Laguette N, Yatim A, Nakamura M, Levy Y, Kiernan R, Benkirane M (2010) HIV-1 Tat assembles a multifunctional transcription elongation complex and stably associates with the 7SK snRNP. Mol Cell 38:439–451. doi:10.1016/j.molcel.2010.04.012
Soulier J, Clappier E, Cayuela JM, Regnault A, Garcia-Peydro M, Dombret H, Baruchel A, Toribio ML, Sigaux F (2005) HOXA genes are included in genetic and biologic networks defining human acute T-cell leukemia (T-ALL). Blood 106:274–286. doi:10.1182/blood-2004-10-3900
Speleman F, Cauwelier B, Dastugue N, Cools J, Verhasselt B, Poppe B, Van Roy N, Vandesompele J, Graux C, Uyttebroeck A, Boogaerts M, De Moerloose B, Benoit Y, Selleslag D, Billiet J, Robert A, Huguet F, Vandenberghe P, De Paepe A, Marynen P, Hagemeijer A (2005) A new recurrent inversion, inv(7)(p15q34), leads to transcriptional activation of HOXA10 and HOXA11 in a subset of T-cell acute lymphoblastic leukemias. Leukemia 19:358–366. doi:10.1038/sj.leu.2403657
Srinivasan RS, de Erkenez AC, Hemenway CS (2003) The mixed lineage leukemia fusion partner AF9 binds specific isoforms of the BCL-6 corepressor. Oncogene 22:3395–3406. doi:10.1038/sj.onc.1206361
Stark B, Umiel T, Mammon Z, Galili N, Dzaledetti M, Cohen IJ, Steinberg M, Vogel R, Zaizov R (1986) Leukemia of early infancy. Early B-cell lineage associated with t(4:11). Cancer 58:1265–1271
Starkova J, Zamostna B, Mejstrikova E, Krejci R, Drabkin HA, Trka J (2010) HOX gene expression in phenotypic and genotypic subgroups and low HOXA gene expression as an adverse prognostic factor in pediatric ALL. Pediatr Blood Cancer 55:1072–1082. doi:10.1002/pbc.22749
Stauber RH, Bier C, Knauer SK (2012) Targeting Taspase1 for cancer therapy–letter. Cancer Res 72:2912. doi:10.1158/0008-5472.CAN-12-0150, author reply 2913
Sternsdorf T, Phan VT, Maunakea ML, Ocampo CB, Sohal J, Silletto A, Galimi F, Le Beau MM, Evans RM, Kogan SC (2006) Forced retinoic acid receptor alpha homodimers prime mice for APL-like leukemia. Cancer Cell 9:81–94. doi:10.1016/j.ccr.2005.12.030
Takeda S, Chen DY, Westergard TD, Fisher JK, Rubens JA, Sasagawa S, Kan JT, Korsmeyer SJ, Cheng EH, Hsieh JJ (2006) Proteolysis of MLL family proteins is essential for taspase1-orchestrated cell cycle progression. Genes Dev 20:2397–2409. doi:10.1101/gad.1449406
Tan J, Jones M, Koseki H, Nakayama M, Muntean AG, Maillard I, Hess JL (2011) CBX8, a polycomb group protein, is essential for MLL-AF9-induced leukemogenesis. Cancer Cell 20:563–575. doi:10.1016/j.ccr.2011.09.008
Thiel AT, Blessington P, Zou T, Feather D, Wu X, Yan J, Zhang H, Liu Z, Ernst P, Koretzky GA, Hua X (2010) MLL-AF9-induced leukemogenesis requires coexpression of the wild-type Mll allele. Cancer Cell 17:148–159. doi:10.1016/j.ccr.2009.12.034
Thorsteinsdottir U, Kroon E, Jerome L, Blasi F, Sauvageau G (2001) Defining roles for HOX and MEIS1 genes in induction of acute myeloid leukemia. Mol Cell Biol 21:224–234. doi:10.1128/MCB.21.1.224-234.2001
Tkachuk DC, Kohler S, Cleary ML (1992) Involvement of a homolog of Drosophila trithorax by 11q23 chromosomal translocations in acute leukemias. Cell 71:691–700

Vassiliou GS, Cooper JL, Rad R, Li J, Rice S, Uren A, Rad L, Ellis P, Andrews R, Banerjee R, Grove C, Wang W, Liu P, Wright P, Arends M, Bradley A (2011) Mutant nucleophosmin and cooperating pathways drive leukemia initiation and progression in mice. Nat Genet 43: 470–475. doi:10.1038/ng.796

Vastenhouw NL, Schier AF (2012) Bivalent histone modifications in early embryogenesis. Curr Opin Cell Biol 24:374–386. doi:10.1016/j.ceb.2012.03.009

Vermaelen K, Barbieri D, Michaux JL, Tricot G, Casteels-Van Daele M, Noens L, Van Hove W, Drochmans A, Louwagie A, Van den Berghe H (1983) Anomalies of the long arm of chromosome 11 in human myelo- and lymphoproliferative disorders. I. Acute nonlymphocytic leukemia. Cancer Genet Cytogenet 10:105–116

Voigt P, LeRoy G, Drury WJ 3rd, Zee BM, Son J, Beck DB, Young NL, Garcia BA, Reinberg D (2012) Asymmetrically modified nucleosomes. Cell 151:181–193. doi:10.1016/j.cell.2012.09.002

Wang P, Lin C, Smith ER, Guo H, Sanderson BW, Wu M, Gogol M, Alexander T, Seidel C, Wiedemann LM, Ge K, Krumlauf R, Shilatifard A (2009) Global analysis of H3K4 methylation defines MLL family member targets and points to a role for MLL1-mediated H3K4 methylation in the regulation of transcriptional initiation by RNA polymerase II. Mol Cell Biol 29:6074–6085. doi:10.1128/MCB.00924-09

Wang QF, Wu G, Mi S, He F, Wu J, Dong J, Luo RT, Mattison R, Kaberlein JJ, Prabhakar S, Ji H, Thirman MJ (2011) MLL fusion proteins preferentially regulate a subset of wild-type MLL target genes in the leukemic genome. Blood 117:6895–6905. doi:10.1182/blood-2010-12-324699

Wang J, Muntean AG, Hess JL (2012) ECSASB2 mediates MLL degradation during hematopoietic differentiation. Blood 119:1151–1161. doi:10.1182/blood-2011-06-362079

Watts KL, Nelson V, Wood BL, Trobridge GD, Beard BC, Humphries RK, Kiem HP (2012) Hematopoietic stem cell expansion facilitates multilineage engraftment in a nonhuman primate cord blood transplantation model. Exp Hematol 40:187–196. doi:10.1016/j.exphem.2011.11.009

Wiederschain D, Kawai H, Gu J, Shilatifard A, Yuan ZM (2003) Molecular basis of p53 functional inactivation by the leukemic protein MLL-ELL. Mol Cell Biol 23:4230–4246

Wilkinson AC, Ballabio E, Geng H, North P, Tapia M, Kerry J, Biswas D, Roeder RG, Allis CD, Melnick A, de Bruijn MF, Milne TA (2013) RUNX1 is a Key target in t(4;11) leukemias that contributes to gene activation through an AF4-MLL complex interaction. Cell Rep 3:116–127. doi:10.1016/j.celrep.2012.12.016

Wong P, Iwasaki M, Somervaille TC, So CW, Cleary ML (2007) Meis1 is an essential and rate-limiting regulator of MLL leukemia stem cell potential. Genes Dev 21:2762–2774. doi:10.1101/gad.1602107

Wu M, Wang PF, Lee JS, Martin-Brown S, Florens L, Washburn M, Shilatifard A (2008) Molecular regulation of H3K4 trimethylation by Wdr82, a component of human Set1/COMPASS. Mol Cell Biol 28:7337–7344. doi:10.1128/MCB.00976-08

Wysocka J, Swigut T, Milne TA, Dou Y, Zhang X, Burlingame AL, Roeder RG, Brivanlou AH, Allis CD (2005) WDR5 associates with histone H3 methylated at K4 and is essential for H3 K4 methylation and vertebrate development. Cell 121:859–872. doi:10.1016/j.cell.2005.03.036

Wysocka J, Swigut T, Xiao H, Milne TA, Kwon SY, Landry J, Kauer M, Tackett AJ, Chait BT, Badenhorst P, Wu C, Allis CD (2006) A PHD finger of NURF couples histone H3 lysine 4 trimethylation with chromatin remodelling. Nature 442:86–90. doi:10.1038/nature04815

Xia ZB, Anderson M, Diaz MO, Zeleznik-Le NJ (2003) MLL repression domain interacts with histone deacetylases, the polycomb group proteins HPC2 and BMI-1, and the corepressor C-terminal-binding protein. Proc Natl Acad Sci U S A 100:8342–8347. doi:10.1073/pnas.1436338100

Yokoyama A, Cleary ML (2008) Menin critically links MLL proteins with LEDGF on cancer-associated target genes. Cancer Cell 14:36–46. doi:10.1016/j.ccr.2008.05.003

Yokoyama A, Kitabayashi I, Ayton PM, Cleary ML, Ohki M (2002) Leukemia proto-oncoprotein MLL is proteolytically processed into 2 fragments with opposite transcriptional properties. Blood 100:3710–3718. doi:10.1182/blood-2002-04-1015

Yokoyama A, Somervaille TC, Smith KS, Rozenblatt-Rosen O, Meyerson M, Cleary ML (2005) The menin tumor suppressor protein is an essential oncogenic cofactor for MLL-associated leukemogenesis. Cell 123:207–218. doi:10.1016/j.cell.2005.09.025

Yokoyama A, Lin M, Naresh A, Kitabayashi I, Cleary ML (2010) A higher-order complex containing AF4 and ENL family proteins with P-TEFb facilitates oncogenic and physiologic MLL-dependent transcription. Cancer Cell 17:198–212. doi:10.1016/j.ccr.2009.12.040

Yu BD, Hess JL, Horning SE, Brown GA, Korsmeyer SJ (1995) Altered Hox expression and segmental identity in Mll-mutant mice. Nature 378:505–508. doi:10.1038/378505a0

Zeisig BB, Milne T, Garcia-Cuellar MP, Schreiner S, Martin ME, Fuchs U, Borkhardt A, Chanda SK, Walker J, Soden R, Hess JL, Slany RK (2004) Hoxa9 and Meis1 are key targets for MLL-ENL-mediated cellular immortalization. Mol Cell Biol 24:617–628

Zeisig DT, Bittner CB, Zeisig BB, Garcia-Cuellar MP, Hess JL, Slany RK (2005) The eleven-nineteen-leukemia protein ENL connects nuclear MLL fusion partners with chromatin. Oncogene 24:5525–5532. doi:10.1038/sj.onc.1208699

Zeleznik-Le NJ, Harden AM, Rowley JD (1994) 11q23 translocations split the "AT-hook" cruciform DNA-binding region and the transcriptional repression domain from the activation domain of the mixed-lineage leukemia (MLL) gene. Proc Natl Acad Sci U S A 91:10610–10614

Zhou J, Peres L, Honore N, Nasr R, Zhu J, de The H (2006) Dimerization-induced corepressor binding and relaxed DNA-binding specificity are critical for PML/RARA-induced immortalization. Proc Natl Acad Sci U S A 103:9238–9243. doi:10.1073/pnas.0603324103

Zhou Q, Li T, Price DH (2012) RNA polymerase II elongation control. Annu Rev Biochem 81:119–143. doi:10.1146/annurev-biochem-052610-095910

Ziemin-van der Poel S, McCabe NR, Gill HJ, Espinosa R 3rd, Patel Y, Harden A, Rubinelli P, Smith SD, LeBeau MM, Rowley JD et al (1991) Identification of a gene, MLL, that spans the breakpoint in 11q23 translocations associated with human leukemias. Proc Natl Acad Sci U S A 88:10735–10739

Zorko NA, Bernot KM, Whitman SP, Siebenaler RF, Ahmed EH, Marcucci GG, Yanes DA, McConnell KK, Mao C, Kalu C, Zhang X, Jarjoura D, Dorrance AM, Heerema NA, Lee BH, Huang G, Marcucci G, Caligiuri MA (2012) Mll partial tandem duplication and Flt3 internal tandem duplication in a double knock-in mouse recapitulates features of counterpart human acute myeloid leukemias. Blood 120:1130–1136. doi:10.1182/blood-2012-03-415067

Zuber J, Rappaport AR, Luo W, Wang E, Chen C, Vaseva AV, Shi J, Weissmueller S, Fellmann C, Taylor MJ, Weissenboeck M, Graeber TG, Kogan SC, Vakoc CR, Lowe SW (2011) An integrated approach to dissecting oncogene addiction implicates a Myb-coordinated self-renewal program as essential for leukemia maintenance. Genes Dev 25:1628–1640. doi:10.1101/gad.17269211

Chapter 8
Role of DNA Methyltransferases and DNA Methylation in Cell Fate Decisions During Blood Cell Development and Leukemia

Grant A. Challen and Jennifer J. Trowbridge

Abstract Increasing evidence indicates that DNA methylation and the proteins responsible for catalyzing DNA methylation (the DNA methyltransferases or DNMTs) play critical roles in embryonic specification of the blood cell lineage and during differentiation of adult hematopoietic stem cells (HSCs). Furthermore, the identification of somatic mutations in DNMTs in a high frequency of human blood cancers suggests that altered DNA methylation is a critical component of leukemogenesis. This review will highlight our current understanding of the function of DNA methylation and the major DNMTs in hematopoiesis, describe the extent of characterization of mutant DNMTs in human blood cancer and other diseases, and discuss strategies to target altered DNA methylation or the activity of mutant DNMTs for more precise and effective leukemia therapy. Future studies aimed at understanding how DNA methylation regulates gene expression in concert with other epigenetic modifications, how DNMTs are directed to their cell type-specific target loci, and the exact biological functions of mutant DNMTs will be pivotal in the development of novel, targeted therapies for blood cancers.

Keywords DNA methylation • DNMT • Epigenetics • Hematopoiesis • Leukemia

G.A. Challen
Division of Oncology, Department of Internal Medicine, Washington University in St. Louis, St. Louis, MO 63117, USA

J.J. Trowbridge (✉)
The Jackson Laboratory, Bar Harbor, ME 04609, USA
e-mail: Jennifer.Trowbridge@jax.org

C. Bonifer and P.N. Cockerill (eds.), *Transcriptional and Epigenetic Mechanisms Regulating Normal and Aberrant Blood Cell Development*, Epigenetics and Human Health, DOI 10.1007/978-3-642-45198-0_8, © Springer-Verlag Berlin Heidelberg 2014

8.1 Introduction: DNA Methylation and the Roles of DNMTs

Hematopoietic stem cells (HSCs) reside in the bone marrow and cycle through stages of self-renewal, quiescence, proliferation, and differentiation to generate all the cell types of the hematopoietic system. A number of transcription factors and cytokines have been identified that regulate HSC behavior, but precisely how these factors orchestrate blood homeostasis is the subject of ongoing investigation. These processes are now recognized to include a significant epigenetic component. DNA methylation of cytosine-phosphate-guanine (CpG) dinucleotides is one of the major epigenetic modifications of the vertebrate genome. DNA methylation controls crucial cell fate decisions during ontogeny and stem cell development and has other roles in maintaining genomic integrity such as silencing of retrotransposons, X chromosome inactivation, genomic imprinting, and the regulation of tissue-specific and/or context-specific gene expression (Reik et al. 2001; Dean et al. 2003). DNA methylation can act to enhance or repress transcription of various genes in a cell-specific manner, with the influencing nature of the 5-methylcytosine residues on gene expression contextually dependent on their locations in relation to genes. Regulation of gene expression by DNA methylation is traditionally thought to occur primarily through methylation of CpG-dense regions called CpG islands (CGIs). Approximately half of all human gene promoters contain CGIs and hypermethylation of these regions classically results in gene repression (Meehan et al. 1992; Ballestar and Wolffe 2001). However, with the introduction of next-generation sequencing technologies, it is becoming apparent this is a very narrow view as to the role of DNA methylation in control of the genome. DNA methylation of other genomic features such as gene body exons or 3′ UTRs can be associated with enhanced gene expression (Yu et al. 2013).

DNA methylation is catalyzed by a family of enzymes called DNA methyltransferases (DNMTs), and mouse mutant models have demonstrated that these genes are essential for normal embryonic development with death occurring from the 10-somite stage to 4 weeks postnatally depending on which DNMT gene is inactivated and how completely (Okano et al. 1999; Attwood et al. 2002). Thus far, three mammalian genes have been shown to encode for catalytic DNA methyltransferases—DNMT1, DNMT3A, and DNMT3B. DNMT1 preferentially targets hemi-methylated DNA and is thought of as a maintenance methyltransferase that reestablishes DNA methylation marks on daughter strands during DNA replication (Okano et al. 1998). DNMT3A and DNMT3B act as de novo methyltransferases that methylate DNA in response to specific cues at distinct loci. Embryonic stem (ES) cells that lack DNMT1, DNMT3A, and DNMT3B are viable and maintain replication potential but progressively lose differentiation potential with repeated passage (Chen et al. 2003; Tsumura et al. 2006). This suggests that one of the functions of these enzymes is to progressively establish DNA methylation patterns during differentiation toward a defined lineage to restrict differentiation potential. All three functional DNMTs are highly expressed in

HSCs, and DNMT1 and DNMT3A have recently been shown to be critical for normal HSC function. Conditional knockout of DNMT1 in mouse HSCs leads to nearly immediate and complete loss of HSC activity in vivo (Trowbridge et al. 2009), and HSCs from mice with reduced DNMT1 activity through a hypomorphic allele are unable to differentiate into lymphoid progeny (Broske et al. 2009). Conditional inactivation of DNMT3A in HSCs progressively impedes differentiation into peripheral blood lineages over serial transplantation, while simultaneously expanding HSC numbers in vivo, leading to a massive accumulation of phenotypically defined HSCs in the bone marrow (Challen et al. 2012). The contrast in HSC phenotypes between the DNMT1 and DNMT3A conditional knockout models suggests that DNA methylation mediated by these enzymes has a number of critical and distinct roles in maintenance, self-renewal, and differentiation of HSCs.

8.2 DNA Methylation and Stem Cell Fate Decisions in Hematopoiesis

Differentiation of HSCs is a highly hierarchical process, producing successively more lineage-restricted progenitors that undergo extensive proliferation and ultimately give rise to the mature cells of the blood and bone marrow. Hematopoietic development and homeostasis rely on the balance between faithful stem cell self-renewal and the ordered, sequential execution of programs essential for lineage commitment. The role of DNA methylation in HSC function was until recently unexplored, but this modification has been implicated as an epigenetic mechanism that stabilizes stem cell fate decisions, and under normal circumstances, commitment is thought to be unidirectional with repressive epigenetic marks stabilizing loss of plasticity. While the DNMTs are highly expressed in HSCs and mouse mutant models have shown them to be essential for normal HSC function, there is little understanding of the mechanisms by which DNA methylation regulates gene expression during hematopoiesis. It is not known how the DNA methyltransferase enzymes are directed to their cell type-specific target loci or how DNA methylation marks influence the regulation of other epigenetic modifications such as histone methylation and acetylation.

These issues were addressed by a recent study comparing DNA methylation patterns between human hematopoietic stem and progenitor cells (HSPCs; Lineage—CD34+ CD38−), neutrophils, and B cells. In contrast to popular belief, this study noted that the promoter CGIs of lineage-specific factors (e.g., CD19 = lymphoid; CEBP = myeloid) were actually predominantly unmethylated across all samples. The level of gene expression of such genes correlated with the length of extension of the unmethylated region from the CGI promoters in the 3′ direction, with the DNA methylation of the regions flanking the CGI (termed "CGI shores") being the stronger indicator of the level of gene expression (Hodges, Mol Cell,

2011). Genes characteristic of lymphoid- or myeloid-lineage differentiation (in terms of transcript expression) showed opposing patterns of DNA methylation in mature B cells versus neutrophils. However, these loci showed an intermediate pattern in HSCs, both at the level of individual CpGs and overall methylation. These data support the notion that HSCs are epigenetically "poised" or "primed" for lineage selection depending on the type of stress or stimuli encountered. On the whole, this study and others have noted that as hematopoiesis progresses and more lineage-restricted and committed cell types are produced, global levels of DNA methylation increase compared to the primitive HSC compartment (Bock et al. 2012). This is consistent with a more multipotent HSC as the pinnacle of the hematopoietic hierarchy, being able to access multiple differentiation programs, with DNA methylation being introduced later and epigenetically repressing alternative lineage programs as hematopoiesis proceeds and differentiation commitment decisions are made.

8.3 The De Novo DNA Methyltransferases: Roles in Hematopoiesis and Human Disease

The de novo DNA methyltransferase family includes DNMT3A and DNMT3B. These enzymes participate in developmental functions such as establishment of DNA methylation patterns during pre-implantation/embryogenesis (Borgel et al. 2010), site-specific methylation of distinct loci in response to environmental cues in adult stem cells, as well as in oncogenic transformation by hypermethylation and epigenetic silencing of tumor suppressor genes (Fernandez et al. 2012). While DNMT3B is highly expressed in long-term HSCs, the precise role of this methyltransferase in hematopoiesis has proved questionable. Some studies have suggested that DNMT3B is dispensable for adult HSC maintenance (Tadokoro et al. 2007), but may be required during HSC development in ontogeny. In humans, genetic mutations of DNMT3B have not been associated with hematopoietic malignancies. However, hypomorphic germline mutations in DNMT3B are responsible for two-thirds of immunodeficiency, centromere instability, facial anomalies (ICF) syndrome cases, a rare autosomal recessive disease characterized by immune defects, instability of pericentromeric satellite 2-containing heterochromatin, facial abnormalities, and mental retardation (Xu et al. 1999). ICF syndrome is characterized by variable reductions in serum immunoglobulin levels, causing most patients to succumb to infectious diseases before adulthood. ICF patients lacking DNMT3B mutations have a distinct set of DNA methylation defects versus DNMT3B-mutant ICF patients, pointing to the existence of distinct disease subtypes (Jiang et al. 2005). The DNMT3B mutations found in ICF generally target the catalytic domain and may not always result in complete loss of enzymatic activity, with this variability in residual enzymatic activity likely accounting for the clinical variability of the disease (Ehrlich et al. 2008).

Comparison of lymphoblastoid cell lines derived from ICF patients with normal individuals showed significant changes in expression of genes critical for immune function that are likely highly relevant to the ICF phenotype. Approximately half of the genes upregulated in ICF cells showed loss of DNA methylation compared to control cells (Jin et al. 2008). Of the genes which were downregulated in ICF cells, many could be predicted to directly lead to aspects of the ICF phenotype such as downregulation of immunoglobulin heavy-chain genes as a contributor to the immune system defects (Jin et al. 2008). The mechanisms by which DNMT3B mutations lead to gene repression are unclear, but perhaps are modulated by global alterations in transcription factor binding because of altered heterochromatin structure, or changes in the histone code resulting from abnormal DNA methylation patterns. While DNMT3B appears to be dispensable for adult HSC homeostasis in the mouse, it may have distinct roles in later stages of immune cell selection and development.

Epigenetic dysfunction plays a central role in the pathology of many human cancers. Aberrant DNA methylation patterns are widely reported in a variety of human cancers, but the pathological consequences of these marks are undefined (Robertson 2005). In contrast to DNMT3B, genetic mutations in DNMT3A have been identified in a wide spectrum of blood cancers, potentially providing a mechanism for altered DNA methylation patterns in hematopoietic malignancies. DNMT3A mutations have been predominantly characterized in myeloid malignancies, although recent studies have also identified DNMT3A mutations in lymphoid diseases such as T-cell lymphoma (Couronne et al. 2012) and T-cell acute lymphoblastic leukemia (Grossmann et al. 2013) at a similar frequency (10–25 % of patients). Myeloid malignancies are clonal diseases of hematopoietic stem and progenitor cells that result from genetic and epigenetic lesions that perturb key processes such as self-renewal, differentiation, and proliferation. The evolution of next-generation sequencing technologies has allowed unparalleled analytic depth of the cancer genomes of individual patients and identified most (if not all) of the important driver mutations in myeloid malignancies (Cancer Genome Atlas Research 2013). Through such studies, several groups have independently reported DNMT3A mutations with predicted translational consequences (non-synonymous substitutions, insertions/deletions, splice-site nucleotide changes) in ~20 % of acute myeloid leukemia (AML) patients (Ley et al. 2010; Yan et al. 2011), ~10 % of myelodysplastic syndrome (MDS) patients (Ewalt et al. 2011; Walter et al. 2011), and ~10 % of patients with myeloproliferative neoplasms (Stegelmann et al. 2011). In AML, DNMT3A mutations predominantly occur in intermediate-risk cytogenetic patients and the median overall survival of patients in this group with DNMT3A mutations is significantly shorter than among those without (Ley et al. 2010). Moreover, MDS patients with DNMT3A mutations are more likely to develop a more aggressive secondary AML (Thol et al. 2011b; Walter et al. 2011). In hematopoietic cancers, DNMT3A mutations are significantly correlated with FLT3, NPM1, and IDH1 mutations and associated with poor event-free and overall survival (Ley et al. 2010; Welch et al. 2012; Cancer Genome Atlas Research 2013; Grossmann et al. 2013).

All of the described DNMT3A mutations in AML and MDS are heterozygous, but it is not yet clear whether these mutations result in loss-of-function, gain-of-function, or have a dominant-negative effect. The vast majority of genetic mutations are single nucleotide variants (SNVs) with enrichment for the mutations within the methyltransferase domain. Recurrent mutations within highly conserved regions within the catalytic domain of DNMT3A may suggest a potential gain-of-function mechanism conferring a mutant protein with novel function. The divergent frameshift and non-sense mutations that occur upstream of the methyltransferase domain are largely predicted to result in truncated proteins that eliminate or shorten the methyltransferase domain and more likely to function as loss-of-function mutations (Ley et al. 2010), a classic pattern seen for many tumor suppressor genes such as TP53 and BRCA1. The most commonly mutated residue is R882 in the DNA methyltransferase domain. In the largest survey of DNMT3A sequencing in AML, of the 62 patients identified to have any DNMT3A mutation, 37 (60 %) had SNVs at R882 (Ley et al. 2010). Of all R882 variants, the most common is the R882H substitution. Biochemical analysis of the R882H variant protein has revealed reduced methyltransferase activity and diminished ability to bind DNA (Yan et al. 2011). However, in this particular study, the mutant protein was not tested in the presence of the wild-type protein which may alter the phenotype. This is of particular relevance since all DNMT3A mutations in myeloid malignancy patients are heterozygous and sequencing studies have demonstrated that the wild-type allele remains expressed.

The exact biological functions of these DNMT3A mutations remain to be elucidated, and work is currently progressing on structural and biochemical analysis of the mutant proteins as well as modeling the disease in transgenic mice. Mice that are homozygous germline null for DNMT3A are born runted, fail to thrive, and die of aganglionic megacolon at about 1 month of age (Okano et al. 1999). However, using a conditional knockout mouse model whereby DNMT3A was specifically ablated in adult HSCs, one group showed that loss of DNMT3A progressively impeded HSC differentiation over multiple rounds of serial transplantation (Challen et al. 2012). While the functional output of DNMT3A-null HSCs was apparently normal after primary transplantation, successive passaging of the mutant HSCs in vivo led to a steady decline in peripheral blood cell output. In contrast, there was a massive accumulation of phenotypically defined DNMT3A-null HSCs in the bone marrow of transplant recipients. The functional phenotype of these mutant HSCs implied a disconnect between the delicate balance of self-renewal and differentiation, with the absence of DNMT3A resulting in a bias towards self-renewal cell fate decisions at the expense of differentiation. Transcriptome analysis revealed upregulation of genes traditionally associated with HSC self-renewal in DNMT3A-null HSCs (e.g., Runx1), likely restraining them to the stem cell state upon receiving a signal that normally cues for HSC differentiation (Challen et al. 2012). Such genes are normally epigenetically silenced by DNA methylation during hematopoietic lineage commitment, leading to hypermethylation and transcriptional repression in mature cells of the peripheral blood. DNMT3A-null HSCs were unable to epigenetically silence these genes by DNA methylation in a forced model of stem cell differentiation, leading to inefficient differentiation and manifesting hypomethylation and incomplete repression of HSC-specific genes in the mature progeny of DNMT3A-null HSCs (Challen et al. 2012). Thus, DNMT3A plays a specific role in permitting HSC differentiation during normal hematopoiesis as in its absence, phenotypically normal but functionally impotent stem cells accumulate and differentiation capacity is progressively lost.

Conditional inactivation of DNMT3A in adult mouse HSCs leads to a similar phenotype as witnessed in DNMT3A-mutant myeloid malignancy patients, namely an accumulation of relatively undifferentiated cells (HSCs in mice, blasts in patients) in the bone marrow. However, loss of DNMT3A in mouse HSCs does not appear to be sufficient for full transformation and other genetic hits are likely required. This may provide some clues towards the sequence of genetic events required for pathogenesis in humans. Acquisition of a DNMT3A mutation in a stem/progenitor cell is likely an early event in myeloid disease transformation (Welch et al. 2012). The resultant mutation may confer a clonal advantage to cells bearing this hit, or inhibit their terminal differentiation, resulting in an accumulation of an expanded "pre-leukemic" cell population in the bone marrow. In this scenario, full-blown transformation would only be achieved with the addition of co-operating mutations (e.g., FLT3, IDH1), attained stochastically via normal cellular processes leading to DNA mutations. In support of this notion, myeloid malignancies are typically diseases of the elderly patient population, providing the time required for a DNMT3A mutation clone to acquire other co-operating genetic mutations and obtain clonal dominance. Furthermore, DNMT3A mutations have not been identified in pediatric leukemia patients of either myeloid (Thol et al. 2011a) or lymphoid lineage (Paganin et al. 2011), suggesting that childhood diseases do not provide enough incubation time for a DNMT3A mutant clone to acquire enough additional mutations to expand.

The most obvious mechanism of pathogenesis for DNMT3A mutations would be altered DNA methylation patterns leading to changes in gene expression or to genomic instability. Compared to control stem cells, DNMT3A-null HSCs only show a minor reduction in total genomic 5-methylcytosine levels (Challen et al. 2012). Analysis of genomic DNA methylation patterns by reduced representation bisulfite sequencing (RRBS) showed that DNMT3A-null mouse HSCs manifest both increased and decreased methylation at distinct loci, including paradoxically a significant CpG island hypermethylation (Challen et al. 2012). This study was only able to show a direct link between changes in DNA methylation and alterations in gene expression in DNMT3A-null HSCs for select genes. There was no global correlation between loss of DNMT3A-mediated methylation and transcriptional output and the functional changes in DNMT3A-deficient HSCs cannot be explained by changes in DNA methylation alone. Similarly in AML patients, the mean 5-methylcytosine content of each genome carrying a DNMT3A mutation was indistinguishable from DNMT3A wild-type patients. MeDIP-chip analysis also showed nearly identical methylated regions between the two groups of patients, with only a small number of loci being exceptions (Ley et al. 2010). Moreover, there was no correlation between these subtle changes in DNA methylation in DNMT3A-mutation AML patients and expression of nearby genes. Gene expression analysis has also demonstrated that DNMT3A-mutation patients cannot be clustered in an unbiased fashion based on expression signatures within AML samples with a normal cytogenetic profile. Moreover, whole-genome sequencing has demonstrated that the total number of SNVs in AML genomes is not significantly influenced by DNMT3A mutation status, indicating the mutant protein does not affect chromosomal stability or genomic integrity (Ley et al. 2010). Cumulatively, mouse and human patient studies strongly suggest that DNMT3A is critical for normal HSC function and that DNMT3A mutations are likely relevant to the pathogenesis of myeloid malignancies, but the pathogenic mechanisms remain to be elucidated since the mutations do not appear to dramatically alter gene expression, DNA methylation patterns, or chromosomal stability. This may suggest that the mechanisms of malignant transformation in these patients occur independently of changes in DNA methylation or that loss of DNMT3A regulates other molecules which are the functional effectors of the phenotype. One such effect may be the ability of the mutant DNMT3A proteins to bind to interacting partners in functional complexes. One such binding protein is DNMT3L, a DNA methyltransferase-like processivity factor that forms heterotetramers with DNMT3A to modulate the catalytic mechanism of DNA methyltransferase activity (Jia et al. 2007). DNMT3L also has the potential to regulate targeting of DNA methylation because it interacts with the inactive non-methylated form of lysine 4 within histone H3, but not with the activated methylated form (Cedar and Bergman 2009). However, whether this higher order DNMT3A–DNMT3L catalytic complex is responsible for de novo DNA methylation in hematopoiesis is controversial since DNMT3L is not expressed in mouse HSCs or AML patient samples (Challen, personal observation).

Exactly how DNMT3A mutations exert an epigenetic effect (if any) remains to be defined, although they probably do so in a different way from TET2 or IDH1/2 mutations since they often co-occur with either of them in AML (whereas TET2 and IDH mutations are mutually exclusive events). This is highly relevant to the consideration of mechanisms in AML because TET2 and IDH1/2 both also feed into the regulation of DNA methylation (reviewed elsewhere in this book): TET2 catalyzes the oxidation of 5-methylcytosine to 5-hydroxymethylcytosine which appears to be an intermediate product in DNA demethylation pathways (Ko et al. 2010), whereas IDH1/2 mutations produce an oncometabolite that inhibits TET2 activity (Dang et al. 2009; Gross et al. 2010). Clearly this is a research area of high clinical relevance as hypomethylating cytosine analog drugs have clinical activity in MDS. The development of targeted epigenetic therapies for DNMT3A-mutant patients could provide a novel mechanism of therapeutic intervention for a wide variety of cancers.

8.4 DNMT1: Roles in Hematopoiesis and Human Disease

DNMT1 was the first mammalian DNA methyltransferase enzyme to be cloned and biochemically characterized. It is crucial for maintenance of DNA methylation, regulation of gene expression, and chromatin stability. DNMT1 shows a high preference for hemi-methylated DNA over unmethylated substrates and achieves maintenance methylation through cell division by catalyzing specific methylation of hemi-methylated CpG dinucleotides produced during DNA replication. Highlighting the importance of this methyltransferase, mice homozygous germline null for DNMT1 die in early embryogenesis prior to the eight-somite stage with extensive demethylation of the genome (Lei et al. 1996). To circumvent this lethality, generation of a conditional knockout allele of DNMT1 has allowed study of the cellular response to demethylation in cell lines and primary adult tissues. Initial work deriving DNMT1 conditional knockout mouse embryonic fibroblasts (MEFs) demonstrated that these cells undergo a uniform p53-dependent cell death shortly after DNMT1 deletion (Jackson-Grusby et al. 2001), highlighting the critical role of DNMT1 in somatic cells.

Recently, the functional and molecular phenotype of DNMT1 conditional knockout HSCs and hematopoietic cells has been described. Following loss of DNMT1, widespread defects in HSC self-renewal, differentiation, and bone marrow niche retention were observed (Trowbridge et al. 2009). Furthermore, loss of DNMT1 in myeloid progenitor cells resulted in inappropriate cell cycling and premature expression of genes directing maturation. In an independent study, mice carrying a hypomorphic allele resulting in reduced expression of DNMT1 were found to have defects in HSC self-renewal and inappropriate activation of myelo-erythroid genes, resulting in defects in differentiation towards the lymphoid lineage (Broske et al. 2009). These studies suggest that the gene expression program in HSCs may be more sensitive to loss of DNMT1 than in lineage-committed cells and that preservation of DNA methylation in HSCs is critical for maintaining self-renewal and proper lineage differentiation. Investigation into the molecular mechanisms underlying the DNMT1 conditional knockout or hypomorph HSC phenotype is incomplete. Differential gene expression analysis has revealed that many genes upregulated upon loss of DNMT1 in HSCs do not contain promoter CGIs, suggesting that regulation of transcription by DNA methylation outside the context of CGI-containing promoters may strongly contribute to the DNMT1-mutant phenotype. This idea is supported by a recent report that promoter CGIs of lineage-specific factors are predominantly unmethylated in HSCs (Hodges et al. 2011). The degree of DNA methylation changes induced genome wide as a consequence of DNMT1 loss or haploinsufficiency, the precise localization of these DNA methylation changes, and the correlation to gene expression has not yet been examined.

DNMT1, like the de novo methyltransferases, does not act as an enzyme in isolation. Domains in the N-terminal region of DNMT1 mediate interactions with the transcriptional repressor DMAP1, the DNA replication machinery PCNA, and

UHRF1 and target DNMT1 to the replication foci during S phase of the cell cycle (Rountree et al. 2000). DNMT1 also physically interacts with many other components of the epigenetic machinery, including HDAC1, HDAC2, DNMT3A, SUV39H1, SET7/9, G9A, and EZH2 (Fuks et al. 2000; Robertson et al. 2000). While DNMT1 is classically known as a maintenance DNA methyltransferase, additional roles for DNMT1 in regulating gene expression are not as clearly understood. Interestingly, DNMT1 engineered with a mutant catalytic domain retains the ability to deplete active histone marks and recruit the H3K4 demethylase KDM1A (LSD1)(Clements et al. 2012). Together, these studies demonstrate that DNMT1 can act as a transcriptional repressor independently of its catalytic domain, and this may involve a role in scaffolding other epigenetic regulatory complexes.

DNA methylation profiling of AMLs indicates that dysregulation of promoter cytosine methylation is a universal feature of the disease. However, abnormal DNA methylation patterning adopts distinct and specific distributions dependent at least in part on genetic background and context (Akalin et al. 2012). Consistent with these complex changes, the mechanism of action of DNMT1 alterations in driving or inhibiting tumorigenesis is complex and context dependent, and the basis remains to be fully elucidated. Although somatic mutations in DNMT1 in blood cancers appear to be extremely rare (Dolnik et al. 2012), depletion of DNMT1 using a hypomorphic allele was found to increase susceptibility to tumorigenesis, including T-cell lymphoma, likely as a result of increased chromosomal instability (Gaudet et al. 2003). In contrast, overexpression of DNMT1 has been demonstrated to result in hypermethylation and transformation of human cells (Wu et al. 1993). In a defined mouse model of AML, depletion of DNMT1 by haploinsufficiency delayed progression of leukemogenesis and impaired leukemia stem cell self-renewal (Trowbridge et al. 2012). In this model, DNMT1 haploinsufficiency resulted in derepression of tumor suppressor genes, reduced DNA methylation, and reduced bivalent chromatin marks, suggesting that DNA methylation mediates silencing of bivalent domains to enforce transcriptional repression. Other human diseases driven by mutation or loss of DNMT1 highlight the complex pathogenesis of aberrant methylation. Mutations within DNMT1 are the only recurrent mutation found in autosomal dominant cerebellar ataxia, deafness and narcolepsy (ADCA-DN), and hereditary sensory and autonomic neuropathy (HSAN1)(Winkelmann et al. 2012). These mutations cause premature degradation of mutant DNMT1, reduced DNA methyltransferase activity, and impaired heterochromatin binding during the G2 cell cycle phase. Deciphering the context-dependent roles of DNA methylation mediated by DNMT1, and specific interactions of DNMT1 with other epigenetic regulators, will provide needed insight into the complexity of DNMT1 in human leukemogenesis and other diseases.

8.5 DNMT Mutations and Altered DNA Methylation in Cancer: Targets for Epigenetic Therapies?

Recent results from The Cancer Genome Atlas (TCGA) Research Network suggest that mutations in DNA methylation-related genes (including DNMT3A, TET2, IDH1/2) occur in up to 44 % of human AML (Cancer Genome Atlas Research 2013). Many of these mutations are mutually exclusive of transcription factor fusions and, in some cases, are sufficient to induce leukemogenesis. Given the high frequency of these mutations, new therapies targeting mutant proteins or the functions of mutant proteins should improve outcomes for patients with MDS and AML. For this purpose, it will be critical to ascribe gains, losses, or changes in protein function resulting from these commonly recurring mutations, as well as characterize specific DNA methylation changes that occur as a result of these mutations.

Even in the absence of mutations in DNA methylation-related genes, consistent changes in DNA methylation patterns have been observed in leukemia compared to normal cellular counterparts. For example, leukemia cells demonstrate regional DNA hypermethylation of tumor suppressor genes (Akalin et al. 2012). Thus, small molecule drugs that induce DNA hypomethylation have been attractive targets to induce re-expression of these tumor suppressors in leukemia cells. Supportive of this strategy, two DNA hypomethylating drugs, 5-azadeoxyctidine [decitabine (DAC)] and 5-azacitidine (AZA), have demonstrated a favorable therapeutic index and efficacy in some MDS patients (Wijermans et al. 2005; Kantarjian et al. 2006). However, this treatment is not effective in all patients, and the precise mechanism of action is not clear. It was anticipated that these agents would be incorporated within DNA and suppress maintenance methylation during DNA replication on the basis that they are not substrates for DNMTs. However, after incorporation into DNA they can also become covalently linked to DNMTs, leading to DNA damage and DNMT degradation (Ghoshal et al. 2005). In addition to inducing DNA hypomethylation, DAC and AZA were also found to modulate the immune system, depending on the dose administered. Understanding how these drugs work is a pivotal step in furthering epigenetic therapy. Clinical clues suggest that, in MDS, low doses of these drugs provide antitumor effects over time, including sustained genome-wide changes in promoter DNA methylation and gene expression, rather than acutely exerting cytotoxic effects (Tsai et al. 2012). It is also appealing to consider combination therapy, where use of inhibitors such as DAC and AZA could sensitize leukemia cells to other drugs and allow use of less toxic doses for these other agents (Juergens et al. 2011). A large AML cohort study from TCGA has revealed that the strongest DNA methylation signatures in leukemias occur in CpG-sparse regions of the genome, supporting the concept that methylation in gene bodies and intergenic regions is important for the regulation of gene expression (Cancer Genome Atlas Research 2013). However, the significance of this finding with respect to therapeutic targeting remains unclear.

Discovery and development of therapeutic targets to specifically alter the activity of individual epigenetic regulators, rather than generic DNA hypomethylating agents such as DAC and AZA, will offer greater specificity in therapeutic response and greater therapeutic index. An improved understanding of how the DNA methylation machinery interacts with other proteins, how this machinery is targeted to genomic DNA, and the identification of compounds that interfere with catalytic activity all will inform development of additional therapeutic targets in MDS and AML. In a recent example highlighting a novel approach, two micro-RNAs (MiRs) were reported to effectively target DNMT1 activity in ovarian cancer cells (Xiang et al. 2013). Development of novel strategies will be important both in the context of tumors carrying mutations in DNA methylation-related genes and to those that carry epigenetic alterations independent of specific mutations.

8.6 Concluding Remarks

The question is not so much whether epigenetic regulation plays an important role in guiding hematopoietic development and differentiation, but how it occurs—what are the modifications, machinery, and genes that are involved at any given stage of hematopoietic development. Moreover, the question naturally turns to whether these phenomena can be manipulated for therapeutic intervention. Understanding the role of the DNMTs during normal hematopoietic differentiation is crucial for understanding the significance of abnormal DNA methylation in hematopoietic malignancies and the development of targeted therapeutics for DNMT3A mutation-driven cancers.

References

Akalin A, Garrett-Bakelman FE, Kormaksson M et al (2012) Base-pair resolution DNA methylation sequencing reveals profoundly divergent epigenetic landscapes in acute myeloid leukemia. PLoS Genet 8:e1002781. doi:10.1371/journal.pgen.1002781

Attwood JT, Yung RL, Richardson BC (2002) DNA methylation and the regulation of gene transcription. Cell Mol Life Sci 59:241–257

Ballestar E, Wolffe AP (2001) Methyl-CpG-binding proteins. Targeting specific gene repression. Eur J Biochem 268:1–6

Bock C, Beerman I, Lien WH et al (2012) DNA methylation dynamics during in vivo differentiation of blood and skin stem cells. Mol Cell 47:633–647. doi:10.1016/j.molcel.2012.06.019

Borgel J, Guibert S, Li Y et al (2010) Targets and dynamics of promoter DNA methylation during early mouse development. Nat Genet 42:1093–1100. doi:10.1038/ng.708

Broske AM, Vockentanz L, Kharazi S et al (2009) DNA methylation protects hematopoietic stem cell multipotency from myeloerythroid restriction. Nat Genet 41:1207–1215. doi:10.1038/ng.463

Cancer Genome Atlas Research N (2013) Genomic and epigenomic landscapes of adult de novo acute myeloid leukemia. N Engl J Med 368:2059–2074. doi:10.1056/NEJMoa1301689

Cedar H, Bergman Y (2009) Linking DNA methylation and histone modification: patterns and paradigms. Nat Rev Genet 10:295–304
Challen GA, Sun D, Jeong M et al (2012) Dnmt3a is essential for hematopoietic stem cell differentiation. Nat Genet 44:23–31. doi:10.1038/ng.1009
Chen T, Ueda Y, Dodge JE, Wang Z, Li E (2003) Establishment and maintenance of genomic methylation patterns in mouse embryonic stem cells by Dnmt3a and Dnmt3b. Mol Cell Biol 23:5594–5605
Clements EG, Mohammad HP, Leadem BR, Easwaran H, Cai Y, Van Neste L, Baylin SB (2012) DNMT1 modulates gene expression without its catalytic activity partially through its interactions with histone-modifying enzymes. Nucleic Acids Res 40:4334–4346. doi:10.1093/nar/gks031
Couronne L, Bastard C, Bernard OA (2012) TET2 and DNMT3A mutations in human T-cell lymphoma. N Engl J Med 366:95–96. doi:10.1056/NEJMc1111708
Dang L, White DW, Gross S et al (2009) Cancer-associated IDH1 mutations produce 2-hydroxyglutarate. Nature 462:739–744. doi:10.1038/nature08617
Dean W, Santos F, Reik W (2003) Epigenetic reprogramming in early mammalian development and following somatic nuclear transfer. Semin Cell Dev Biol 14:93–100
Dolnik A, Engelmann JC, Scharfenberger-Schmeer M et al (2012) Commonly altered genomic regions in acute myeloid leukemia are enriched for somatic mutations involved in chromatin remodeling and splicing. Blood 120:e83–e92. doi:10.1182/blood-2011-12-401471
Ehrlich M, Sanchez C, Shao C et al (2008) ICF, an immunodeficiency syndrome: DNA methyltransferase 3B involvement, chromosome anomalies, and gene dysregulation. Autoimmunity 41:253–271. doi:10.1080/08916930802024202
Ewalt M, Galili NG, Mumtaz M et al (2011) DNMT3a mutations in high-risk myelodysplastic syndrome parallel those found in acute myeloid leukemia. Blood Cancer J 1:e9. doi:10.1038/bcj.2011.7
Fernandez AF, Huidobro C, Fraga MF (2012) De novo DNA methyltransferases: oncogenes, tumor suppressors, or both? Trends Genet 28:474–479. doi:10.1016/j.tig.2012.05.006
Fuks F, Burgers WA, Brehm A, Hughes-Davies L, Kouzarides T (2000) DNA methyltransferase Dnmt1 associates with histone deacetylase activity. Nat Genet 24:88–91. doi:10.1038/71750
Gaudet F, Hodgson JG, Eden A et al (2003) Induction of tumors in mice by genomic hypomethylation. Science 300:489–492. doi:10.1126/science.1083558
Ghoshal K, Datta J, Majumder S, Bai S, Kutay H, Motiwala T, Jacob ST (2005) 5-Aza-deoxycytidine induces selective degradation of DNA methyltransferase 1 by a proteasomal pathway that requires the KEN box, bromo-adjacent homology domain, and nuclear localization signal. Mol Cell Biol 25:4727–4741. doi:10.1128/MCB.25.11.4727-4741.2005
Gross S, Cairns RA, Minden MD et al (2010) Cancer-associated metabolite 2-hydroxyglutarate accumulates in acute myelogenous leukemia with isocitrate dehydrogenase 1 and 2 mutations. J Exp Med 207:339–344. doi:10.1084/jem.20092506
Grossmann V, Haferlach C, Weissmann S et al (2013) The molecular profile of adult T-cell acute lymphoblastic leukemia: mutations in RUNX1 and DNMT3A are associated with poor prognosis in T-ALL. Genes Chromosomes Cancer 52:410–422. doi:10.1002/gcc.22039
Hodges E, Molaro A, Dos Santos CO et al (2011) Directional DNA methylation changes and complex intermediate states accompany lineage specificity in the adult hematopoietic compartment. Mol Cell 44:17–28. doi:10.1016/j.molcel.2011.08.026
Jackson-Grusby L, Beard C, Possemato R et al (2001) Loss of genomic methylation causes p53-dependent apoptosis and epigenetic deregulation. Nat Genet 27:31–39. doi:10.1038/83730
Jia D, Jurkowska RZ, Zhang X, Jeltsch A, Cheng X (2007) Structure of Dnmt3a bound to Dnmt3L suggests a model for de novo DNA methylation. Nature 449:248–251. doi:10.1038/nature06146
Jiang YL, Rigolet M, Bourc'his D et al (2005) DNMT3B mutations and DNA methylation defect define two types of ICF syndrome. Hum Mutat 25:56–63. doi:10.1002/humu.20113

Jin B, Tao Q, Peng J et al (2008) DNA methyltransferase 3B (DNMT3B) mutations in ICF syndrome lead to altered epigenetic modifications and aberrant expression of genes regulating development, neurogenesis and immune function. Hum Mol Genet 17:690–709. doi:10.1093/hmg/ddm341

Juergens RA, Wrangle J, Vendetti FP et al (2011) Combination epigenetic therapy has efficacy in patients with refractory advanced non-small cell lung cancer. Cancer Discov 1:598–607. doi:10.1158/2159-8290.CD-11-0214

Kantarjian H, Issa JP, Rosenfeld CS et al (2006) Decitabine improves patient outcomes in myelodysplastic syndromes: results of a phase III randomized study. Cancer 106:1794–1803. doi:10.1002/cncr.21792

Ko M, Huang Y, Jankowska AM et al (2010) Impaired hydroxylation of 5-methylcytosine in myeloid cancers with mutant TET2. Nature 468:839–843. doi:10.1038/nature09586

Lei H, Oh SP, Okano M, Juttermann R, Goss KA, Jaenisch R, Li E (1996) De novo DNA cytosine methyltransferase activities in mouse embryonic stem cells. Development 122:3195–3205

Ley TJ, Ding L, Walter MJ et al (2010) DNMT3A mutations in acute myeloid leukemia. N Engl J Med 363:2424–2433. doi:10.1056/NEJMoa1005143

Meehan R, Lewis J, Cross S, Nan X, Jeppesen P, Bird A (1992) Transcriptional repression by methylation of CpG. J Cell Sci Suppl 16:9–14

Okano M, Xie S, Li E (1998) Cloning and characterization of a family of novel mammalian DNA (cytosine-5) methyltransferases. Nat Genet 19:219–220. doi:10.1038/890

Okano M, Bell DW, Haber DA, Li E (1999) DNA methyltransferases Dnmt3a and Dnmt3b are essential for de novo methylation and mammalian development. Cell 99:247–257. doi:10.1016/S0092-8674(00)81656-6 [pii]

Paganin M, Pigazzi M, Bresolin S et al (2011) DNA methyltransferase 3a hot-spot locus is not mutated in pediatric patients affected by acute myeloid or T-cell acute lymphoblastic leukemia: an Italian study. Haematologica 96:1886–1887. doi:10.3324/haematol.2011.049825

Reik W, Dean W, Walter J (2001) Epigenetic reprogramming in mammalian development. Science 293:1089–1093

Robertson KD (2005) DNA methylation and human disease. Nat Rev Genet 6:597–610. doi:10.1038/nrg1655

Robertson KD, Ait-Si-Ali S, Yokochi T, Wade PA, Jones PL, Wolffe AP (2000) DNMT1 forms a complex with Rb, E2F1 and HDAC1 and represses transcription from E2F-responsive promoters. Nat Genet 25:338–342. doi:10.1038/77124

Rountree MR, Bachman KE, Baylin SB (2000) DNMT1 binds HDAC2 and a new co-repressor, DMAP1, to form a complex at replication foci. Nat Genet 25:269–277. doi:10.1038/77023

Stegelmann F, Bullinger L, Schlenk RF et al (2011) DNMT3A mutations in myeloproliferative neoplasms. Leukemia 25:1217–1219. doi:10.1038/leu.2011.77

Tadokoro Y, Ema H, Okano M, Li E, Nakauchi H (2007) De novo DNA methyltransferase is essential for self-renewal, but not for differentiation, in hematopoietic stem cells. J Exp Med 204:715–722. doi:10.1084/jem.20060750, jem.20060750 [pii]

Thol F, Heuser M, Damm F, Klusmann JH, Reinhardt K, Reinhardt D (2011a) DNMT3A mutations are rare in childhood acute myeloid leukemia. Haematologica 96:1238–1240. doi:10.3324/haematol.2011.046839

Thol F, Winschel C, Ludeking A et al (2011b) Rare occurrence of DNMT3A mutations in myelodysplastic syndromes. Haematologica 96:1870–1873. doi:10.3324/haematol.2011.045559

Trowbridge JJ, Snow JW, Kim J, Orkin SH (2009) DNA methyltransferase 1 is essential for and uniquely regulates hematopoietic stem and progenitor cells. Cell Stem Cell 5:442–449. doi:10.1016/j.stem.2009.08.016

Trowbridge JJ, Sinha AU, Zhu N, Li M, Armstrong SA, Orkin SH (2012) Haploinsufficiency of Dnmt1 impairs leukemia stem cell function through derepression of bivalent chromatin domains. Genes Dev 26:344–349. doi:10.1101/gad.184341.111

Tsai HC, Li H, Van Neste L et al (2012) Transient low doses of DNA-demethylating agents exert durable antitumor effects on hematological and epithelial tumor cells. Cancer Cell 21:430–446. doi:10.1016/j.ccr.2011.12.029

Tsumura A, Hayakawa T, Kumaki Y et al (2006) Maintenance of self-renewal ability of mouse embryonic stem cells in the absence of DNA methyltransferases Dnmt1, Dnmt3a and Dnmt3b. Genes Cells 11:805–814. doi:10.1111/j.1365-2443.2006.00984.x

Walter MJ, Ding L, Shen D et al (2011) Recurrent DNMT3A mutations in patients with myelodysplastic syndromes. Leukemia 25:1153–1158. doi:10.1038/leu.2011.44

Welch JS, Ley TJ, Link DC et al (2012) The origin and evolution of mutations in acute myeloid leukemia. Cell 150:264–278. doi:10.1016/j.cell.2012.06.023

Wijermans PW, Lubbert M, Verhoef G, Klimek V, Bosly A (2005) An epigenetic approach to the treatment of advanced MDS; the experience with the DNA demethylating agent 5-aza-2-′-deoxycytidine (decitabine) in 177 patients. Ann Hematol 84(Suppl 1):9–17. doi:10.1007/s00277-005-0012-1

Winkelmann J, Lin L, Schormair B et al (2012) Mutations in DNMT1 cause autosomal dominant cerebellar ataxia, deafness and narcolepsy. Hum Mol Genet 21:2205–2210. doi:10.1093/hmg/dds035

Wu J, Issa JP, Herman J, Bassett DE Jr, Nelkin BD, Baylin SB (1993) Expression of an exogenous eukaryotic DNA methyltransferase gene induces transformation of NIH 3T3 cells. Proc Natl Acad Sci U S A 90:8891–8895

Xiang Y, Ma N, Wang D, et al (2013) MiR-152 and miR-185 co-contribute to ovarian cancer cells cisplatin sensitivity by targeting DNMT1 directly: a novel epigenetic therapy independent of decitabine. Oncogene. doi:10.1038/onc.2012.575

Xu GL, Bestor TH, Bourc'his D et al (1999) Chromosome instability and immunodeficiency syndrome caused by mutations in a DNA methyltransferase gene. Nature 402:187–191. doi:10.1038/46052

Yan XJ, Xu J, Gu ZH et al (2011) Exome sequencing identifies somatic mutations of DNA methyltransferase gene DNMT3A in acute monocytic leukemia. Nat Genet 43:309–315. doi:10.1038/ng.788

Yu DH, Ware C, Waterland RA et al (2013) Developmentally programmed 3′ CpG island methylation confers tissue- and cell-type-specific transcriptional activation. Mol Cell Biol 33:1845–1858. doi:10.1128/MCB.01124-12

Part III
Epigenetic and Transcriptional Mechanisms Regulating Cell Fate Decisions and Blood Cell Lineage Development

Chapter 9
Transcriptional and Epigenetic Regulation in the Development of Myeloid Cells: Normal and Diseased Myelopoiesis

Peter Laslo and Tomas Stopka

Abstract Myeloid cells constitute the innate arm of the vertebrate immune system and arise from haematopoietic stem cells being committed to their cell fate through a series of lineage restrictions regulated by a gene regulatory network. This gene network consists of transcription factors as well as components of the epigenetic machinery that, in cooperation with one another, will programme progenitors to adopt and differentiate along a certain lineage programme. By virtue of their obligatory function, dysregulation in the activity of these regulatory factors can contribute to the pathogenesis of myeloid leukaemias. To understand the molecular aetiology of myeloid dysplasias it is imperative to first study and model the network that regulates normal development. Equipped with this crucial understanding we can then begin to decipher what, how and why things have gone wrong in the pathology of myeloid leukaemias.

Keywords Transcription factor • Epigenetics • Myeloid • Gene Network • HSC • LSC • Differentiation • Leukaemia • AML • MDS • CML • APL • PU.1 • C/EBPA • RUNX1 • GATA-2 • RARα • MLL • TET2 • IDH1/2 • DNMT3A • EZH2

P. Laslo (✉)
Section of Experimental Haematology, Leeds Institute of Cancer and Pathology, St. James's University Hospital, University of Leeds, Leeds, UK
e-mail: p.laslo@leeds.ac.uk

T. Stopka
Institute of Pathophysiology and First Department of Medicine-Haematology, First Faculty of Medicine, Charles University in Prague, Prague, Czech Republic
e-mail: tstopka@lf1.cuni.cz

C. Bonifer and P.N. Cockerill (eds.), *Transcriptional and Epigenetic Mechanisms Regulating Normal and Aberrant Blood Cell Development*, Epigenetics and Human Health, DOI 10.1007/978-3-642-45198-0_9, © Springer-Verlag Berlin Heidelberg 2014

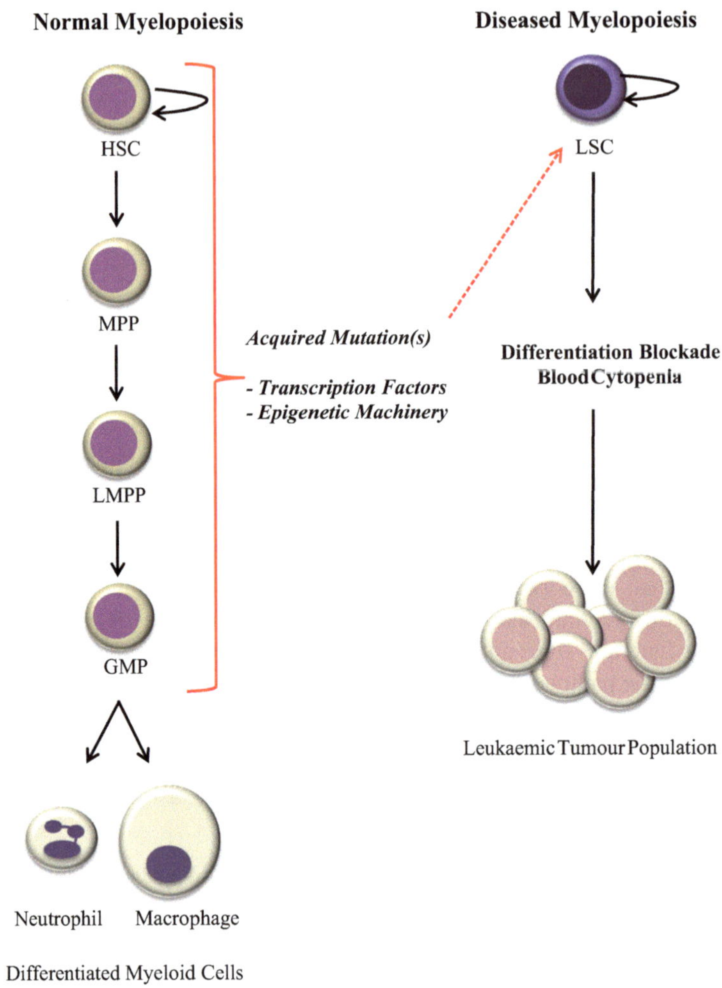

Fig. 9.1 Dysregulation of normal myelopoiesis resulting in the generation of the leukaemic stem cell and the onset of myeloid dysplasia characterised by a block of cellular differentiation and peripheral blood cytopenia. Please refer to main text for details. *HSC* Haematopoietic stem cell, *MPP* Multipotential progenitor, *LMPP* Lymphoid-primed multipotenital progenitor, *GMP* Granulocyte-macrophage progenitor, *LSC* Leukaemic stem cell

9.1 Introduction

9.1.1 Myelopoiesis

Haematopoietic stem cells (HSCs) reside in the bone marrow and are empowered with self-renewal capacity and the ability to reconstitute all lineages of the blood system. Generation of myeloid cells from the HSC involves a series of sequential cell fate decisions that transit the HSC through a hierarchically organised cascade of progenitors (HSC > MPP > LMPP > GMP) culminating in a final bimodal choice of the GMP progenitor to differentiate into either a macrophage or neutrophil (Fig. 9.1).

Transcription factors have long been recognised as major regulators of myeloid development and primarily function to specify and re-enforce each cell fate decision, thus ensuring the successful developmental transition from one progenitor to the next. Transcription factors do not act alone but work in combination as exemplified by the functions of PU.1 and C/EBPa in regulating the expression of many myeloid-specific genes [as reviewed in Friedman (2007)]. Moreover, transcription factors often cross-antagonise one another's activity to suppress an alternate cell fate choice. This lineage restriction commits the progenitor along the chosen cell fate, as exemplified by PU.1 and GATA1 in the choice between the myeloid and erythroid lineages [as reviewed in Burda et al. (2010)]. As such, transcription factors regulate a cell fate decision by controlling the expression of lineage-specific target genes as well as suppressing the unwanted genetic programmes of the alternate fates.

Mounting evidence demonstrates the importance of the epigenetic machinery in programming the chromatin structure (histone modification or DNA methylation) of lineage-specific genes, thus setting the stage for transcription factors to bind and function. Coinciding with the role of transcription factors, the activation and silencing of lineage-specific genes during myelopoiesis require dynamic alteration of their chromatin structure in a developmentally regulated manner [as reviewed in Bonifer et al. (2008)].

Ineffective myelopoiesis is a hallmark of myeloid leukaemias characterised by the accumulation of immature myeloid cells in the bone marrow and blood (Tenen 2003). This block in myeloid development arises by acquired mutations in the activity of either transcription factor(s) and/or components of the epigenetic machinery resulting in a failure of the progenitor cell to properly execute the genetic programme necessary to transit onto the next cell fate. As such, lineage-specific genes fail to be correctly activated or silenced thus upsetting the balance of cell death, proliferation, and differentiation resulting in the shift towards cancer. Consequently, the stem or progenitor cell is transformed into a leukaemic stem cell (Fig. 9.1).

Essentially functioning as normal HSCs, these leukaemic stem cells (LSCs) have the potential to both self-renew and differentiate, albeit in a dysregulated manner, thereby propagating themselves and giving rise to 'differentiated' progeny that represent the bulk tumour population. While tumour burden is responsible for clinical symptoms the persistent propagation of LSCs is accountable for disease maintenance. To cure patients the LSCs must be completely eliminated.

9.1.2 Myeloid Leukaemias

Myeloid disorders can be distinguished into three main types: Acute Myeloid Leukaemia (AML), Myeloproliferative Neoplasms (MPN) and Myelodysplastic Syndromes (MDS). AML disorders are characterised by an arrest in early differentiation resulting in a blast cell tumour population while MPN is initially marked by enhanced cellular proliferation and a blockade that allows production of multiple differentiation intermediates. In turn, MDS is distinguished by inefficient haematopoiesis. Notably, both MPN and MDS progress to AML. Dysplasias from each type include AML: AML itself with its unique subtype Acute Promyelocytic Leukaemia (APL), MPN: Chronic Myeloid Leukaemia (CML) and MDS: aggressive MDS-secondary AML (MDS/AML).

AML is a heterogeneous group of diseases all of which are defined by rapid cycling and invading blasts and is sub-categorized based on which myeloid cell(s) is afflicted: spectrum of earlier myeloid G/M progenitors (M0-2, defined by Cooperative FAB Group), promyelocytes (M3, APL), monocytes (M4-5) or red (M6) and platelet (M7) precursors. Cytogenetic examination of AML patients is currently the preferred method to detect chromosomal aberrations and translocations. Combinations of these abnormalities, designated as ‘complex aberrations’, usually mark aggressive AML forms often with dysplastic features (MDS/AML). Furthermore, their presence is associated with poor patient survival and serves as a prognostic factor.

APL represents one of several subtypes of AML and is defined by presence of the *PML-RARα* fusion and blast cells blocked at the promyelocytic cell stage, a relatively late stage of granulocytic development. APL responds exceptionally to therapy, currently ATRA and arsenic, which specifically target the fusion oncogene and induce terminal granulocytic differentiation of the leukaemic cells.

MDS is primarily defined by a lack of normal mature blood cells (cytopenia) characterised by accumulation of myeloid precursors in bone marrow. Cytopenias can be either uni-lineage (i.e. refractory anaemia with defects in erythroid lineage) or multi-lineage (i.e. refractory cytopenia with multi-lineage dysplasia and defects in erythroid, megakaryocyte and neutrophil lineages). MDS can transform into an AML-like disease accompanied by an increase in the blast cell population.

CML is a clonal stem cell malignancy associated with the t(9:22) chromosome translocation which fuses the *BCR* and *ABL* genes. The oncogene product, BCR-ABL, displays constitutive tyrosine kinase and transforming activity. CML exhibits biphasic pathology. The initial ‘chronic phase’ (CP) is characterised by a leukaemic proliferation of myeloid progenitors resulting in the expansion of terminally differentiated myeloid cells within the periphery. Without clinical intervention, such as continuous administration of tyrosine kinase inhibitors, the disease course changes into ‘blast crisis’ (BC) distinguished by an arrest in myeloid differentiation and accumulation of immature blasts cells in bone marrow and periphery.

9.2 How Components of the Myeloid Gene Network Function: Normal and Diseased Myelopoiesis

9.2.1 PU.1

Originally identified as a target gene of proviral integration of the spleen focus forming virus in erythroleukaemias (Moreau-Gachelin et al. 1988), PU.1 is a member of the Ets family of transcription factors. PU.1 contains various functional domains namely an N-terminal transactivation domain and an Ets domain recognising the core purine-rich GGAA DNA motif (Kodandapani et al. 1996).

Expression of PU.1 is restricted to the haematopoietic system where it exhibits a highly dynamic pattern. Aided by GFP reporter mice (Back et al. 2005; Nutt et al. 2005), activity of the PU.1 locus has been precisely mapped with low expression within long-term HSCs and sequentially increasing along the myeloid lineage (ST-HSCs > LMMP > GMP > macrophages and neutrophils). Targeted disruption of *PU.1* within mice results in fetal (Scott et al. 1994) or perinatal (McKercher et al. 1996) lethality and severe impairment of haematopoiesis. *PU.1*$^{-/-}$ mice have a reduced pool of HSC and progenitors (Dakic et al. 2005; Iwasaki et al. 2005) with slightly decreased MEP population and a loss of mature macrophages, neutrophils, B cells, T cells and mast cells (Scott et al. 1994; Walsh et al. 2002; Kim et al. 2004). Conditional deletion of *PU.1* in adult mice results in the failure of CMPs and GMPs to be generated (Dakic et al. 2005; Iwasaki et al. 2005), establishing the necessity of this transcription factor for specification of the myeloid lineages.

Myeloid specification and commitment are sensitive to, and regulated by, PU.1 in a dose-dependent manner. Using *PU.1*$^{-/-}$ progenitors, sub-threshold activity of PU.1 can transcriptionally prime and specify early myeloid progenitors where upon reaching a critical threshold of activity, increased PU.1 levels promote macrophage differentiation (Laslo et al. 2006). Genome-wide expression studies demonstrate that PU.1 can modulate the expression of at least 1,000 genes of the myeloid network (Laslo et al. 2006; Burda et al. 2009) including known cell surface receptor proteins (CD11b, CD16 and CD64) as well as all three myeloid cytokines (M-CSF, G-CSF and GM-CSF) and their respective receptors. Recent ChIP-Seq studies have provided a greater insight into how PU.1 regulates the myeloid network (Ghisletti et al. 2010; Heinz et al. 2010). With over 45,000 PU.1-binding sites identified, the majority were associated with intra- and extra-genic regions. In combination with other transcription factors (C/EBPa, AP-1 and IRF-4) PU.1 establishes and defines the macrophage-specific cistrome and initiates nucleosome remodelling at these sites as well as deposition of H3K4me1 histone mark culminating in the establishment of promoter-distal cis-regulatory elements (Heinz et al. 2010).

PU.1 expression is dependent upon its proximal promoter (Chen et al. 1995), although it alone cannot drive transcription in a correct lineage-specific manner unless several distal enhancers are present (Li et al. 2001). A key upstream regulator enhancer (URE) is located within a DNase1 hypersensitivity site located −14 kb or −17 kb in mice and humans, respectively (Rosenbauer et al. 2004; Okuno et al. 2005; Leddin et al. 2011). Using murine models, deletion of the URE results in an 80 % reduction in PU.1 expression within bone marrow cells and the onset of AML (Rosenbauer et al. 2004). Importantly, restoring PU.1 within these mutant cells rescues the leukaemic block. As such, an AML leukaemic state is established and maintained by the inability of low PU.1 concentrations to further promote differentiation. Restoring PU.1 to higher concentrations reinstates the underlying gene network enabling these cells to continue along their developmental pathway.

PU.1 mutations were originally reported in 7 % of M4/5-AML patients with defects resulting in abrogation of its function (Mueller et al. 2002). Subsequent studies failed to demonstrate such a similar association (Lamandin et al. 2002). However, it is noted that the earlier work utilised AML patients from a Japanese only cohort and therefore the discrepancies observed could be due to population-specific differences.

As demonstrated by animal models, dysregulation in the precise expression levels of PU.1 during myeloid development leads to the onset of a leukaemic state. Notably, several oncogenic fusion products, including RUNX1-ETO and PML-RARα, target PU.1 expression and function. RUNX1-ETO physically interacts with the DNA-binding domain of the PU.1 protein resulting in the displacement of the c-jun coactivator and subsequent downregulation of PU.1 transactivation activity (Vangala et al. 2003). Notably, overexpression of PU.1 within the Kasumi cell line, established from a t(8:21) AML patient, could rescue myeloid differentiation. The *PU.1* gene is a direct target of PML-RARα (Martens et al. 2010) whereby conditional expression of this oncogene within U937 myeloid progenitors suppresses PU.1 expression resulting in a block in myeloid differentiation (Mueller et al. 2006). Treatment of primary APL leukaemic blasts with the RARα ligand ATRA can restore PU.1 expression and rescue granulocyte differentiation (Mueller et al. 2006). Importantly, these observations demonstrate that suppression of PU.1 activity is a primary mechanism in AML pathology.

PU.1 expression is downregulated in a subset of MDS patients due to DNA hypermethylation of the URE (Curik et al. 2012). Treatment of primary MDS samples with the demethylating agent 5-azacitidine can partially restore myeloid differentiation by specifically demethylating the URE leading to the upregulation of PU.1 expression. Recent studies have focussed on MDS patients (cytogenetically normal refractory cytopenia with multi-lineage dysplasia) and identified >4,500 gene promoters that have increased H3K27me3 histone methylation in comparison to age-sex matched healthy controls (Cheng et al. 2013). Intriguingly, the PU.1 motif was significantly enriched within these regions. Treatment of primary MDS bone marrow cells with H3K27me3 inhibitors increased the expression of PU.1 target genes and promoted cell differentiation. Collectively, these observations propose a role for of epigenetic modifications of the PU.1 pathway in MDS pathology involving DNA methylation at the *PU.1* locus as well as chromatin changes of its target genes.

9.2.2 C/EBPa

C/EBPa (CCAAT/Enhancer-binding protein a) is the founding member of the C/EBP family of transcription factors containing two N-terminal transactivation domains and a C-terminal basic leucine zipper DNA-binding region. C/EBPa is expressed at low levels within HSCs and increases as these cells develop into CMPs and GMPs while repressed within MEPs and early-lymphoid lineages (Akashi et al. 2000). Of the mature blood cells, expression of C/EBPa is restricted to granulocytes, monocytes, basophils and eosinophils.

Full-length C/EBPa protein is 42 kDa in molecular weight (p42) while translation from a downstream in-frame start codon leads to co-expression of a 30 kDa (p30) isoform. Notably, while p30 retains the ability to bind to DNA it lacks the N-terminal transactivation domain and thus functions as a dominant negative to the p42 C/EBPa isoform (Pabst et al. 2001b).

C/EBPa binds to DNA as a homodimer, or heterodimer with other C/EBP proteins, and recognises the motif sequence TTgcgcAA within mouse myeloid progenitors (Heinz et al. 2010). Initial studies suggested an important role for C/EBPa in myelopoiesis as it regulated the receptors for all three myeloid cytokines. This observation was corroborated in $CEBPA^{-/-}$ knockout mice which has a selective block in granulocyte differentiation (Zhang et al. 1997). Conditional deletion of *CEBPA* in adult mice further articulated the myeloid block to occur between the developmental transition of CMPs to GMPs (Zhang et al. 2004). In addition to its role in myeloid development, C/EBPa regulates the self-renewal of adult HSCs (Zhang et al. 2004; Ye et al. 2013).

Collectively, these observations demonstrate a crucial regulatory role of C/EBPa as a myeloid lineage determinant. Indeed, C/EBPa, in collaboration with other transcription factors, delineates and primes potential cis-regulatory elements required for macrophage identity (Heinz et al. 2010). Moreover, ectopic expression of C/EBPa within B-lymphocytes initiates trans-differentiation of these cells into myeloid cells, namely that of macrophages (Xie et al. 2004).

Sporadic *CEBPA* mutations in AML with normal cytogenetics were first reported in 2001 (Pabst et al. 2001b). Several large population studies now report the frequency of these mutations to be between 5 and 14 % of AML cases and are prevalent in M1, M2 and M4 FAB subtypes (Gombart et al. 2002; Snaddon et al. 2003). Two general categories of *CEBPA* mutations were identified: (1) N-terminal mutations that inhibit translation of p42 while allowing expression of p30 protein and (2) C-terminal in-frame mutations within the basic leucine zipper DNA-binding domain that disrupt dimerization and DNA binding of both isoforms. Mice engineered to express either type of these mutated *CEBPA* alleles develop a transplantable AML-like disease (Kirstetter et al. 2008; Bereshchenko et al. 2009). Interestingly, the majority of *CEBPA* mutations in AML are bi-allelic with a combination of both N- and C-terminal mutations, each located on different alleles. Is there a leukaemic advantage in having bi-allelic mutations? Recent animal modelling demonstrates that in comparison to the respective individual mutation, having both N- and C-terminal *CEBPA* mutations develops a more aggressive leukaemia (Bereshchenko et al. 2009).

AML patients with the t(8:21) translocation do not harbour *CEBPA* mutations, yet expression of both C/EBPa mRNA and protein is significantly downregulated within blast cells (Pabst et al. 2001a). Similarly, the Kasumi cell line has negligible expression of C/EBPa. Restoring C/EBPa expression within Kasumi cells was sufficient to override the developmental arrest and induce granulocytic differentiation (Pabst et al. 2001a). The fusion protein arising from the t(8:21) translocation, RUNX1-ETO, can physically interact with C/EBPa and suppress its transactivation function (Pabst et al. 2001a). Notably, of the genes regulated by C/EBPa is *CEBPA* itself, thus establishing an auto-regulatory loop. As such, RUNX1-ETO disrupts this loop leading to transcriptional suppression of *CEBPA*. Overall, these observations underlie the importance of C/EBPa dysregulation as the primary mechanism for the leukaemic myeloid block as seen in t(8:21) AML.

Disruption of myeloid differentiation is a characteristic feature of CML disease progression and is correlated with a loss in C/EBPa expression (Perrotti et al. 2002). Specifically, C/EBPa protein is readily detected in CP-CML samples yet absent in samples isolated from BC-CML patients. Restoring C/EBPa within BC-CML blast cells rescued their differentiation demonstrating that loss of C/EBPa is an essential molecular event for the disease progression of CML (Ferrari-Amorotti et al. 2006). However, unlike previous mechanisms, the loss of C/EBPa function is not associated with transcription repression or acquired mutations but rather a block in protein translation (Perrotti et al. 2002). Kinase activity originating from BCR-ABL regulates the activity of the RNA-binding protein heterogenous nuclear ribonucleoprotein E2 (hnRNPE2) which subsequently targets C/EBPa transcripts and prevents protein translation. Notably, expression of hnRNPE2 is low in CP-CML but becomes readily detectable upon BC-CML.

9.2.3 *RUNX1*

In 1973, Janet Rowley made the seminal observation of a reciprocal translocation occurring between chromosomes 8 and 21 in AML patients (Rowley 1973). Almost two decades later, one of the genes involved in this translocation was identified as *RUNX1* on chromosome 21 (Miyoshi et al. 1991). Subsequent studies defined the fusion partner as the co-repressor *ETO* (Miyoshi et al. 1993).

RUNX1 (AML1) is a transcription factor characterised by a C-terminal transactivation domain as well as a N-terminal Runt domain that recognises the consensus AACCACA sequence (Meyers et al. 1993) and enables heterodimerization with the core-binding factor beta (CBFβ) protein. Notably, CBFβ itself does not bind DNA yet upon dimerization with RUNX1 stabilises the conformation of the Runt domain and increases the binding affinity of RUNX1 to DNA (Tahirov et al. 2001). Expression of RUNX1 is detected at the earliest stages of definitive haematopoiesis, namely in the mesoderm-derived hemogenic endothelium, and plays an essential role in specifying the HSC lineage (North et al. 1999; Chen et al. 2009; Lancrin et al. 2009). In the absence of *RUNX1*, mice die in utero at E12.5 and HSCs fail to be generated (Okuda et al. 1996; Chen et al. 2009). Circumventing the embryonic lethality, conditional deletion of *RUNX1* in adult mice demonstrated that once the HSC compartment is established, RUNX1 is no longer required for its maintenance (Ichikawa et al. 2004; Growney et al. 2005). Moreover, the conditional deletion of *RUNX1* induces a MDS-like myeloproliferative phenotype within the spleen and thymus (Growney et al. 2005; Putz et al. 2006), and although this itself is not an overt AML disease, the HSC progenitors are predisposed to full-blown leukaemia following secondary mutations (Motoda et al. 2007; Jacob et al. 2010).

Various mutations of *RUNX1* were first described in AML-M0 patients (Osato et al. 1999) and subsequently in MDS (Imai et al. 2000). Majority of these mutations are located in the Runt domain resulting in abnormal DNA binding and altered transactivation of target genes such as *CSFR1* (Osato et al. 1999).

While >15 somatically acquired chromosome translocations involving *RUNX1* have been molecularly defined in AML the most common fusion partner is chromosome 8 being associated with 30 % of AML-M2 patients (Peterson et al. 2007). The t(8;21) translocation results in the production of the RUNX1-ETO fusion protein consisting of the N-terminal portion of RUNX1 (including the Runt domain but not the transactivation region) joined to most of the ETO transcriptional repressor. The 'classic model' of RUNX1-ETO-induced leukaemia proposes that RUNX1-ETO binds to RUNX1 target genes and represses transcription by ETO-mediated recruitment of co-repressors (Hart and Foroni 2002). Indeed, direct gene targets that are predominantly repressed by RUNX1-ETO include CSFR1 (Follows et al. 2003) and p14ARF (Linggi et al. 2002). However, RUNX1-ETO can also mediate gene expression activation (Wang et al. 2011; Shia et al. 2012; Sun et al. 2013).

Blast cells of t(8;21) AML frequently express B-cell-specific genes including the transcription factor Pax5 and its direct target CD19 (Walter et al. 2010). This mixed lineage phenotype observation is rarely found in other AML (Tiacci et al. 2004), suggesting a specific role of Pax5 in the pathology of t(8;21) AML. As *PAX5* is uniquely expressed within B cells and functions to repress the myeloid programme (Heavey et al. 2003) its aberrant expression within t(8;21) blasts cells is detrimental to cellular differentiation. Yet how is *PAX5* dysregulated in AML? Surprisingly, *PAX5* expression is not directly dependent upon the actions of RUNX1-ETO but rather that of constitutive MAP kinase signalling. Specifically, the *PAX5* locus is associated with the polycomb-repressive complex in myeloid progenitors. In t(8;21) AML the polycomb repression is relieved by the actions of MAP kinase signalling derived from additional mutation(s) that constitutively activate kinase signalling pathway(s). As such, mutations in growth factor receptors not only provide a growth advantage to the leukaemic cell but can contribute to the dysregulation of the gene network (Ray et al. 2013).

9.2.4 GATA-2

GATA-2 (GATA-binding protein 2) is a member of the GATA family of zinc-finger transcription factors and contains the canonical GATA DNA-binding domain: two evolutionary conserved zinc fingers located at the -NH_3 and -COOH ends, respectively (Evans and Felsenfeld 1989). Expression of GATA-2 within the haematopoietic system is restricted to the HSCs and early-uncommitted progenitors (Tsai et al. 1994) while it is decreased in early-committed precursors of erythroid, megakaryocyte and mast cells. Genetic ablation of *GATA-2* in mice results in severe anaemia and embryonic lethality and has extensive defects in definitive haematopoiesis and in the proliferation and maintenance of HSCs (Tsai and Orkin 1997; Rodrigues et al. 2005). Within myeloid-committed progenitors, GATA-2 facilitates self-renewal of the GMP population (Rodrigues et al. 2008).

Initial observations associated the inappropriate overexpression of GATA-2 with several myeloid disorders, namely in MDS (Fadilah et al. 2002), in myeloid neoplasias having 3q rearrangements (Lahortiga et al. 2004) and AML where it is an indicator of poor prognosis (Luesink et al. 2012; Vicente et al. 2012). In addition to being overexpressed, somatic *GATA-2* mutations have been identified in CML patients correlating with progression to BC (Zhang et al. 2008). Of 85 BC-CML patients analysed, 8 were identified as carrying a *GATA-2* mutation within the COOH zinc finger (nucleotide substitution; L395V). Notably no such mutation was identified within CP-CML patients. Interestingly, patients harbouring this mutation displayed a shorter duration for the disease to transit from CP to BC, suggesting an accelerated disease transformation. The L359V mutation creates a dominant-active GATA-2 protein with higher affinity to bind DNA and augmented GATA-2 transactivation activity. Strikingly, the L395V mutation was not identified within AML or MDS patients (Zhang et al. 2009) underlining the specificity of this mutation to CML pathology. Increased functional activity of GATA-2 within human haematopoietic progenitors can modulate the cell cycle (Tipping et al. 2009). As such, the increase in GATA-2 activity can promote disease severity by endowing leukaemic stem cells with quiescence and providing a possible mechanism by which they can escape from the cytotoxicity of chemotherapy.

While MDS and AML are typically sporadic there are reported cases, although rare, of families developing MDS/AML in a heritable fashion. The recent work of Hahn et al. sequenced 50 candidate genes within affected individuals from five MDS/AML pedigrees (Hahn et al. 2011). This analysis revealed a missense *GATA-2* mutation (T354M) in three families and a 3-bp deletion (T355del) in another. Notably both mutations are adjacent to one another and reside within the COOH zinc-finger domain. These mutations were not identified within sporadic AML patients. Molecular characterisation of these mutants demonstrated a loss-of-function phenotype with (1) decreased affinity for DNA, (2) reduced GATA-2 transactivation activity and (3) failure to optimally activate genes when expressed in HL-60 cells. Although these mutations are heritable, not every carrier developed MDS/AML. As such, the acquisition of *GATA-2* mutation(s) does not itself induce

MDS/AML but rather predispose the myeloid progenitors and requires subsequent 'second hits' to induce the pathology.

9.2.5 *RARα*

Retinoic acid (RA) exerts a wide range of functions during vertebrate development and acts as a ligand for three closely related members of the nuclear receptor superfamily; RARα, RARβ and RARγ (Niederreither and Dolle 2008). The RARs function as transcriptional regulators and bind to specific retinoic acid response cis-elements (RAREs). RARs bind to RAREs as heterodimers with members of the RXR nuclear receptor family (α, β or γ) and this RAR/RXR complex is required for receptor function (Kastner et al. 1997). Canonical RAREs consist of direct repeats of the consensus AGGTCA sequence separated by a spacing of 1 (D1), 2 (D2) or 5 (D5) nucleotides (Balmer and Blomhoff 2005).

Mutant $RAR\alpha^{-/-}$ mice are viable with no overt haematopoietic defect (Kastner et al. 2001). Although the loss of RARα function does not significantly impact myeloid development, it is a common target for chromosomal rearrangements in all cases of APL. Of these chromosome translocations, *RARα* is partnered with one of five genes; *PML*, *PLZF*, *NuMa*, *Nucleophosmin* or *STAT5b*. The fusion proteins consist of a variable N terminus (partner protein) with a fixed C-terminal portion that contains both DNA binding and ligand binding domains of RARα. The most frequent translocation (>97 %) detected in APL patients is the t(15:17) fusion which involves the promyelocytic leukaemia (PML) gene.

The *PML-RARα* translocation is a key initiating factor in APL pathology (Grignani et al. 1993; He et al. 1997). Using either established APL cell lines or inducible expression of PML-RARα in U937 cells, global analysis identified approximately 3,000 PML-RARα-binding sites (Martens et al. 2010; Wang et al. 2010). Several different classes of genes were targeted including genes encoding transcription factors (PU.1, Gfi1 and RUNX1) and various chromatin-modifying enzymes (JMJD3, HDAC9 and DNMT3A). Interestingly, the PML-RARα protein bound not only to the expected D1, D2 and D5 RAREs but also to novel motifs that contained altered spacing between the direct repeats (between 0 and 13) as well as their respective orientation (direct, invert or everted repeats). The ability of PML-RARα to bind to a wider selection of degenerate RARE motifs emphasises that this gain of DNA-binding capacity is an essential feature of its leukaemogenesis.

PML-RARα can alter the epigenetic landscape on target genes by the recruitment of DNA and histone methyltransferases (Di Croce et al. 2002; Villa et al. 2007). Yet these findings were demonstrated on a limited set of genes, mainly that of the *RARβ* promoter. However, on a global scale these epigenetic modifications were not associated with PML-RARα occupancy (Martens et al. 2010) suggesting loci selectivity in PML-RARα recruitment of these factors. Interestingly, H3 acetylation was substantially depleted at PML-RARα-binding sites

(Martens et al. 2010) corroborating earlier findings that PML-RARα can recruit HDACs (Grignani et al. 1998). This negative correlation between H3 acetylation and PML-RARα occupancy highlights the significance of histone deacetylases in establishing and or maintaining a repressive chromatin structure, making HDACs an attractive target as a therapeutic treatment.

Using a bioinformatics approach, the majority of PML-RARα-bound regions contained one-half RARE site (RAREh) together with a PU.1 consensus site (RAREh-PU.1) suggesting a composite element bound by both factors (Wang et al. 2010). Molecular analysis demonstrated that PML-RARα could physically interact with PU.1 leading to the repression of PU.1-mediated transactivation. Interestingly, recruitment of PML-RARα to the RAREh-PU.1 motifs requires PU.1 to be pre-bound to DNA. As such, PU.1 functions as a molecular beacon for PML-RARα occupancy resulting in the formation of a protein–protein complex that suppresses the PU.1-regulated gene network.

9.2.6 Mixed Lineage Leukaemia

MLL, the mammalian orthologue of the Drosophila Trithorax gene, is a histone H3K4 methyltransferase functioning via its C-terminal SET domain. MLL has a complex protein structure and physically interacts with many haematopoietic transcription factors including RUNX1, at the URE of *PU.1*, and thus has a key regulatory role in early myeloid differentiation (Huang et al. 2011). *MLL* is involved in several chromosome rearrangements involving the balanced translocation of N-terminal MLL with C-terminal portion of more than 60 individual partners. The result of these translocations leads to a disruption of the H3K4 methyltransferase activity of MLL and the onset of acute leukaemias and also MDS (see Chap. 7). In addition to being involved in chromosome translocations, MLL is mutated (internal tandem duplication) in approximately 10 % of AML patients with normal cytogenetics (Caligiuri et al. 1998) resulting in defective transactivation properties and is associated with poor prognosis (Martin et al. 2003). Notably, all MLL mutations are strongly associated with aggressive forms of acute leukaemias (ALL, AML or mixed phenotype acute leukaemia). Moreover, in adult leukaemias, MLL mutations can appear as a secondary mutation acquired during drug treatment, also termed therapy-related mutations. For example, some Hodgkin's lymphoma patients who are treated with topoisomerase II inhibitors will acquire MLL mutations and the development of a very aggressive AML with markedly unfavourable prognosis (Mosad et al. 2012).

9.2.7 *TET2 and IDH1/2*

TET2 (Ten-eleven-translocation-2) belongs to a family of three dioxygenase enzymes that catalyse the hydroxylation of 5-methylcytosine to 5-hydroxymethylcytosine on DNA. This conversion ultimately leads to the erasure of DNA methylation marks and is crucial for epigenetic maintenance of the genome. TET2 is ubiquitously expressed in all blood lineages and its conditional deletion within the haematopoietic compartment increases the self-renewal capacity of HSCs and the onset of a myeloid leukaemia (Moran-Crusio et al. 2011).

Notably, the catalytic activity of TET2 is dependent upon α-ketoglutarate. The IDH1 and IDH2 enzymes, which are components of the Krebs cycle, function to convert isocitrate to α-ketoglutarate in the cytoplasm and mitochondria, respectively. As such, any defect in IDH1/2 leads to a loss in α-ketoglutarate production, thus indirectly hindering TET2 function.

TET2 mutations are frequently observed in myeloid malignancies including MDS (24 %) (Kosmider et al. 2009), MPN (14 %) (Schaub et al. 2010) and AML (12 %) (Abdel-Wahab et al. 2009). In some MDS subtypes, such as CMML, the mutation rate of *TET2* exceeds 50 % (Kosmider et al. 2009). Over 200 different *TET2* mutations have been identified and are heterogeneous in nature (missense, nonsense and frame shifts). Overall, each mutation decreases TET2 activity by either decreasing TET2 expression, reducing catalytic activity or acting as a dominant negative (Mohr et al. 2011). Interestingly, most *TET2* mutations are heterozygous with expression of the wild-type allele retained.

Interestingly, and somewhat perplexing, the global DNA methylation status of *TET2* mutations is not always in concordance to the predicted phenotype as in some myeloid tumours the loss of function was associated with widespread hypomethylation (Ko et al. 2010). However, others have reported that a decrease in TET2 activity, mediated by IDH1/2 mutations, results in a hypermethylated DNA phenotype (Figueroa et al. 2010). As myeloid malignancies often harbour additional mutations, particularly those of the epigenetic machinery such as *DMNT3A*, the cumulative effect of these mutations combined could be obscuring any linear correlations.

Given the indirect dependency of TET2 activity upon a functioning IDH1/2, mutations in these enzymes were identified in myeloid leukaemias (Figueroa et al. 2010). Unlike TET2, most of the mutations in *IDH1* are almost exclusively found at a single amino acid and results in a gain of function (Dang et al. 2009). Specifically, *IDH1* mutants fail to catalyse the conversion of isocitrate to α-ketoglutarate but rather generate 2-hydroxyglutarate, which is a competitive inhibitor of α-ketoglutarate-dependent enzymes including TET2. In AML, mutations in *IDH1/2* and *TET2* are mutually exclusive and while they both function in the same biochemical pathway, the pathology of the mutations is not equivalent noting that *IDH1/2* mutations are preferentially found in acute rather than chronic leukaemias while those of *TET2* are evenly distributed.

9.2.8 *DNMT3A*

Covalent methylation of DNA is catalysed by a family of DNA Methyltransferases (DNMTs) and is a two-step process initially established as hemi-methylated DNA by *de novo* DNA methyltransferases DNMT3A and DNMT3B. Once marked, the methylation pattern is faithfully copied during cell division within replication foci by DMNT1. Somatic mutations in *DMNT3A* were first identified by exome sequencing of AML-M5 patients with normal karyotype (Yan et al. 2011). Mutations in *DMNT3B* were not detected suggesting that *DMNT3A* is the preferential target in AML. Large cohort studies have since detected recurrent *DNMT3A* mutations in cytogenetic normal AML (20–23 %) and were only identified in defined intermediate-risk AML of M4-M5 subtype (Ley et al. 2010; Ribeiro et al. 2012). *DNMT3A* mutations correlated with unfavourable prognosis and decreased patient survival. For further information the authors refer the readers to Chap. 8.

9.2.9 *EZH2*

Polycomb group (PcG) proteins function to efficiently repress the expression of unwanted genetic programmes including those of the *Hox* genes. Functioning by two distinct multiprotein complexes, PRC1 and PRC2, respectively, the PcG proteins maintain the transcriptional silence of target genes throughout cell division (Wang et al. 2004).

Enhancer of Zeste 2 (EZH2) is a component of the PRC2 complex and functions as a histone methyltransferase that catalyses the methylation of H3K27. Inactivating mutations in EZH2 were found in 6–12 % of MDS and some MPN patients (Ernst et al. 2010; Nikoloski et al. 2010). Notably these loss-of-function mutations consisted of missense, nonsense and premature stop codons spread throughout the gene or were somatic mutations arising from the loss of chromosome 7 or 7q. For further information the authors refer the readers to Chap. 6.

9.3 Perspectives

9.3.1 *How Many Mutations Does a Leukaemia Need?*

Whole-genome sequencing of individual AML patients has identified hundreds of mutations, yet must this many genetic alterations be accumulated for a leukaemic cell to develop? To address this issue, recent studies have sequenced the genomes from 24 AML patients. To define which mutations were relevant to AML pathology, exome sequencing of in vitro expanded multi-potential progenitor cells from healthy volunteers was used as control. Surprisingly, the total number of mutations identified in the healthy control was similar to that identified in AML patients demonstrating that normal haematopoietic progenitors acquire random benign mutations. More striking was the finding that just one or two mutations are needed to initiate a pre-leukaemic disease requiring at least four additional mutations for overt AML (Welch et al. 2012).

Genetic mutations associated with myeloid malignancies can be categorized into three types. Driver mutations are necessary for initiation of disease and primarily function to confer a growth advantage and establish a pre-leukaemic clone. Driver mutations themselves can be confirmed in mouse models as exemplified by the *PML-RARA* fusion that could recapitulate the disease in transgenic mice (He et al. 1997).

Cooperating mutations are pre-existing or often acquired after the initiating event and function to promote and/or accelerate the disease. Interestingly, detection of cooperating mutations can have clinical prognostic value and help indicate worsening of disease, as exemplified by acquisition of the *FLT3-ITD* mutation which is indicative of MDS transformation to AML (Dicker et al. 2010).

Finally, passenger mutations do not contribute or alter the fitness of the tumour cell and are often pre-existing or acquired during establishment of a leukaemic clone due to genomic instability.

9.3.2 Leukaemic Stem Cells

It is now evident that myeloid leukaemias arise from a rare population of leukaemia-initiating stem cells (LSCs). LSC activity of primary human AML cells resides within a $CD34^{+}CD38^{-}$ population (Bonnet and Dick 1997). As HSCs are the only long-lived population and also share $CD34^{+}CD38^{-}$ phenotype it was proposed that LSCs arise from transformed HSCs. However, even committed myeloid progenitors (GMPs) that have no stem cell properties can be transformed into LSCs by acquisition of driver mutations including MOZ-TIF2 (Huntly et al. 2004) and MLL-AF9 (Krivtsov et al. 2006). Notably, MLL-AF9-transformed myeloid LSCs do not undergo any wide-scale genetic reprogramming but rather acquire a stem cell-specific gene signature (Krivtsov et al. 2006). Interestingly, other driver mutations, such as the N-terminal *CEBPA* mutation (Kirstetter et al. 2008), also activate genes that confer self-renewal properties, some of which are also regulated by MLL-AF9. These observations suggest that although distinct, a primary directive of driver mutations is to activate a stem cell-specific gene signature. Yet the molecular mechanism of how different driver mutations are capable of activating the same set of genes is yet to be determined.

In light of all these findings, which compartment do LSCs actually reside; are they HSC-like or committed GMP progenitors? By immune-phenotyping human AML patients, recent studies demonstrate that LSCs simultaneously reside in both populations (Goardon et al. 2011). Specifically, $CD34^{+}$ AML cells consist of two independent molecularly distinct populations; both having LSC properties. Based on their in vitro differentiation potential, one population was defined as the lymphoid-primed multi-potential progenitor (LMPP) capable of giving rise to both myeloid and lymphoid cells but not erythrocyte and megakaryocytes. The other population displayed a more restrictive capacity generating only myeloid cells and classified as a GMP population. Strikingly, these two populations, much like their normal counterparts, are hierarchically ordered with the LMPP-LSCs giving rise to GMP-LSCs but not the converse. Notably, a similar hierarchy was noted in an animal model of the N-terminal *CEBPA* AML whereby a HSC-like LSC generated GMP-LSCs (Bereshchenko et al. 2009).

By accurately identifying these AML LSCs population we can now begin to grasp a better knowledge of how haematopoietic progenitors are transformed and identify novel targeted therapies. Furthermore, they also provide novel prognostic measures to determine disease resistance relapse after therapy treatment and confidently establish whether the LSCs have truly been eradicated (Craddock et al. 2013).

9.3.3 Oncogene Dependence and Therapy of Myeloid Leukaemias

Despite acquiring cooperating and passenger mutations, the survival of cancer cells can be impaired by the inactivation of the initial driver mutation. This phenomenon, called 'oncogene dependence', provides a rationale for molecular targeted therapy (Weinstein 2002). The role of oncogene dependence in the pathology of myeloid leukaemias is epitomised by CML and APL forming the rationale for current clinical therapies.

BCR-ABL is the driving mutation of CML and its constitutive tyrosine kinase activity is central for its disease pathology. Targeting the tyrosine kinase activity of BCR-ABL for therapeutic application was first suggested in 1992 (Anafi et al. 1992) and these findings spawned the generation of the ST1571 compound 4 years later (Druker et al. 1996). ST1571 (Imatinib) interacts with BCR-ABL and functions as a competitive inhibitor of the ATP-binding site and blocks the catalytic kinase activity. Promising preclinical observations led to clinical trials in the use of Imatinib in CML and given its unprecedented success it is now the first-line treatment of CML.

PML-RARα is the driving mutation for APL, and similar to BCR-ABL, its activity is central for the disease. Currently there are two therapeutic approaches in treating APL, both specifically targeting PML-RARα. First, in the absence of the all-trans-RA (ATRA) ligand, the RAR/RXR heterodimers bind to RAREs and form a repressive complex (Nagy et al. 1997). Upon ligand binding, induced conformational changes in RAR/RXR displace the co-repressor proteins enabling recruitment of co-activator complexes and activation of gene transcription. PML-RARα has an increased binding efficiency to the co-repressors and fails to displace them under physiological ATRA concentrations. However, increased doses of ATRA (100–1,000-fold) can induce disassociation of the co-repressors from PML-RARα and induce terminal neutrophil differentiation. The use of this 'differentiation therapy' approach has led to the clinical use of ATRA for APL (Huang et al. 1988). The second treatment uses arsenic trioxide (ATO) which targets PML proteins, including PML-RARα, and induces their degradation (Zhu et al. 1997). As with ATRA, the ability of ATO to induce neutrophil differentiation of APL cells has led to its clinical use in the treatment of this disorder (Mathews et al. 2010).

These observations underscore the clinical relevance of oncogene dependence in treating myeloid leukaemias. Whether this holds promise for all myelodysplasias or only some is unclear; however, preliminary evidence from targeting IDH2 (Wang et al. 2013), MLL fusions (Grembecka et al. 2012) as well as RUNX1-ETO (Ptasinska et al. 2012; Sun et al. 2013) provides strong support for this therapeutic approach.

Acknowledgements We apologise to the many authors whose work we could not cite owing to space constraints. This work was supported by grants from Leukaemia and Lymphoma Research (PL), Royal Society (PL), Grant Agency of the Czech Republic # P305/12/1033 (TS) Institutional (TS): Charles University: UNCE 204021, First Medical Faculty: PRVOUK-P24/LF1/3 & SVV-2013-264507, BIOCEV—Biotechnology and Biomedicine Centre of the Academy of Sciences and Charles University (CZ.1.05/1.1.00/02.0109), from the European Regional Development Fund.

References

Abdel-Wahab O, Mullally A, Hedvat C et al (2009) Genetic characterization of TET1, TET2, and TET3 alterations in myeloid malignancies. Blood 114:144–147. doi:10.1182/blood-2009-03-210039

Akashi K, Traver D, Miyamoto T, Weissman IL (2000) A clonogenic common myeloid progenitor that gives rise to all myeloid lineages. Nature 404:193–197

Anafi M, Gazit A, Gilon C, Ben-Neriah Y, Levitzki A (1992) Selective interactions of transforming and normal abl proteins with ATP, tyrosine-copolymer substrates, and tyrphostins. J Biol Chem 267:4518–4523

Back J, Allman D, Chan S, Kastner P (2005) Visualizing PU.1 activity during hematopoiesis. Exp Hematol 33:395–402

Balmer JE, Blomhoff R (2005) A robust characterization of retinoic acid response elements based on a comparison of sites in three species. J Steroid Biochem Mol Biol 96:347–354. doi:10.1016/j.jsbmb.2005.05.005

Bereshchenko O, Mancini E, Moore S et al (2009) Hematopoietic stem cell expansion precedes the generation of committed myeloid leukemia-initiating cells in C/EBPalpha mutant AML. Cancer Cell 16:390–400. doi:10.1016/j.ccr.2009.09.036

Bonifer C, Hoogenkamp M, Krysinska H, Tagoh H (2008) How transcription factors program chromatin–lessons from studies of the regulation of myeloid-specific genes. Semin Immunol 20:257–263. doi:10.1016/j.smim.2008.05.001, S1044-5323(08)00045-6 [pii]

Bonnet D, Dick JE (1997) Human acute myeloid leukemia is organized as a hierarchy that originates from a primitive hematopoietic cell. Nat Med 3:730–737

Burda P, Curik N, Kokavec J et al (2009) PU.1 activation relieves GATA-1-mediated repression of Cebpa and Cbfb during leukemia differentiation. Mol Cancer Res 7:1693–1703. doi:10.1158/1541-7786.MCR-09-0031

Burda P, Laslo P, Stopka T (2010) The role of PU.1 and GATA-1 transcription factors during normal and leukemogenic hematopoiesis. Leukemia 24:1249–1257. doi:10.1038/leu.2010.104

Caligiuri MA, Strout MP, Lawrence D et al (1998) Rearrangement of ALL1 (MLL) in acute myeloid leukemia with normal cytogenetics. Cancer Res 58:55–59

Chen HM, Zhang P, Voso MT et al (1995) Neutrophils and monocytes express high levels of PU.1 (Spi-1) but not Spi-B. Blood 85:2918–2928

Chen MJ, Yokomizo T, Zeigler BM, Dzierzak E, Speck NA (2009) Runx1 is required for the endothelial to haematopoietic cell transition but not thereafter. Nature 457:887–891. doi:10.1038/nature07619

Cheng JX, Anastasi J, Watanabe K et al (2013) Genome-wide profiling reveals epigenetic inactivation of the PU.1 Pathway by histone H3 lysine 27 tri-methylation in cytogenetically normal myelodysplastic syndrome. Leukemia 27(6):1291–1300. doi:10.1038/leu.2013.45

Craddock C, Quek L, Goardon N et al (2013) Azacitidine fails to eradicate leukemic stem/progenitor cell populations in patients with acute myeloid leukemia and myelodysplasia. Leukemia 27:1028–1036. doi:10.1038/leu.2012.312

Curik N, Burda P, Vargova K et al (2012) 5-Azacitidine in aggressive myelodysplastic syndromes regulates chromatin structure at PU.1 gene and cell differentiation capacity. Leukemia 26:1804–1811. doi:10.1038/leu.2012.47

Dakic A, Metcalf D, Di Rago L, Mifsud S, Wu L, Nutt SL (2005) PU.1 regulates the commitment of adult hematopoietic progenitors and restricts granulopoiesis. J Exp Med 201:1487–1502. doi:10.1084/jem.20050075

Dang L, White DW, Gross S et al (2009) Cancer-associated IDH1 mutations produce 2-hydroxyglutarate. Nature 462:739–744. doi:10.1038/nature08617

Di Croce L, Raker VA, Corsaro M et al (2002) Methyltransferase recruitment and DNA hypermethylation of target promoters by an oncogenic transcription factor. Science 295:1079–1082. doi:10.1126/science.1065173

Dicker F, Haferlach C, Sundermann J et al (2010) Mutation analysis for RUNX1, MLL-PTD, FLT3-ITD, NPM1 and NRAS in 269 patients with MDS or secondary AML. Leukemia 24:1528–1532. doi:10.1038/leu.2010.124

Druker BJ, Tamura S, Buchdunger E et al (1996) Effects of a selective inhibitor of the Abl tyrosine kinase on the growth of Bcr-Abl positive cells. Nat Med 2:561–566

Ernst T, Chase AJ, Score J et al (2010) Inactivating mutations of the histone methyltransferase gene EZH2 in myeloid disorders. Nat Genet 42:722–726. doi:10.1038/ng.621

Evans T, Felsenfeld G (1989) The erythroid-specific transcription factor Eryf1: a new finger protein. Cell 58:877–885

Fadilah SA, Cheong SK, Roslan H, Rozie-Hanisa M, Yen GK (2002) GATA-1 and GATA-2 gene expression is related to the severity of dysplasia in myelodysplastic syndrome. Leukemia 16:1563–1565. doi:10.1038/sj.leu.2402517

Ferrari-Amorotti G, Keeshan K, Zattoni M et al (2006) Leukemogenesis induced by wild-type and STI571-resistant BCR/ABL is potently suppressed by C/EBPalpha. Blood 108:1353–1362

Figueroa ME, Abdel-Wahab O, Lu C et al (2010) Leukemic IDH1 and IDH2 mutations result in a hypermethylation phenotype, disrupt TET2 function, and impair hematopoietic differentiation. Cancer Cell 18:553–567. doi:10.1016/j.ccr.2010.11.015

Follows GA, Tagoh H, Lefevre P, Hodge D, Morgan GJ, Bonifer C (2003) Epigenetic consequences of AML1-ETO action at the human c-FMS locus. EMBO J 22:2798–2809. doi:10.1093/emboj/cdg250

Friedman AD (2007) Transcriptional control of granulocyte and monocyte development. Oncogene 26:6816–6828. doi:10.1038/sj.onc.1210764, 1210764 [pii]

Ghisletti S, Barozzi I, Mietton F et al (2010) Identification and characterization of enhancers controlling the inflammatory gene expression program in macrophages. Immunity 32:317–328. doi:10.1016/j.immuni.2010.02.008

Goardon N, Marchi E, Atzberger A et al (2011) Coexistence of LMPP-like and GMP-like leukemia stem cells in acute myeloid leukemia. Cancer Cell 19:138–152. doi:10.1016/j.ccr.2010.12.012

Gombart AF, Hofmann WK, Kawano S et al (2002) Mutations in the gene encoding the transcription factor CCAAT/enhancer binding protein alpha in myelodysplastic syndromes and acute myeloid leukemias. Blood 99:1332–1340

Grembecka J, He S, Shi A et al (2012) Menin-MLL inhibitors reverse oncogenic activity of MLL fusion proteins in leukemia. Nat Chem Biol 8:277–284. doi:10.1038/nchembio.773

Grignani F, Ferrucci PF, Testa U et al (1993) The acute promyelocytic leukemia-specific PML-RAR alpha fusion protein inhibits differentiation and promotes survival of myeloid precursor cells. Cell 74:423–431

Grignani F, De Matteis S, Nervi C et al (1998) Fusion proteins of the retinoic acid receptor-alpha recruit histone deacetylase in promyelocytic leukaemia. Nature 391:815–818. doi:10.1038/35901

Growney JD, Shigematsu H, Li Z et al (2005) Loss of Runx1 perturbs adult hematopoiesis and is associated with a myeloproliferative phenotype. Blood 106:494–504. doi:10.1182/blood-2004-08-3280

Hahn CN, Chong CE, Carmichael CL et al (2011) Heritable GATA2 mutations associated with familial myelodysplastic syndrome and acute myeloid leukemia. Nat Genet 43:1012–1017. doi:10.1038/ng.913

Hart SM, Foroni L (2002) Core binding factor genes and human leukemia. Haematologica 87:1307–1323

He LZ, Tribioli C, Rivi R et al (1997) Acute leukemia with promyelocytic features in PML/RARalpha transgenic mice. Proc Natl Acad Sci USA 94:5302–5307

Heavey B, Charalambous C, Cobaleda C, Busslinger M (2003) Myeloid lineage switch of Pax5 mutant but not wild-type B cell progenitors by C/EBPalpha and GATA factors. EMBO J 22:3887–3897. doi:10.1093/emboj/cdg380

Heinz S, Benner C, Spann N et al (2010) Simple combinations of lineage-determining transcription factors prime cis-regulatory elements required for macrophage and B cell identities. Mol Cell 38:576–589. doi:10.1016/j.molcel.2010.05.004

Huang ME, Ye YC, Chen SR et al (1988) Use of all-trans retinoic acid in the treatment of acute promyelocytic leukemia. Blood 72:567–572

Huang G, Zhao X, Wang L et al (2011) The ability of MLL to bind RUNX1 and methylate H3K4 at PU.1 regulatory regions is impaired by MDS/AML-associated RUNX1/AML1 mutations. Blood 118:6544–6552. doi:10.1182/blood-2010-11-317909

Huntly BJ, Shigematsu H, Deguchi K et al (2004) MOZ-TIF2, but not BCR-ABL, confers properties of leukemic stem cells to committed murine hematopoietic progenitors. Cancer Cell 6:587–596. doi:10.1016/j.ccr.2004.10.015

Ichikawa M, Asai T, Saito T et al (2004) AML-1 is required for megakaryocytic maturation and lymphocytic differentiation, but not for maintenance of hematopoietic stem cells in adult hematopoiesis. Nat Med 10:299–304. doi:10.1038/nm997

Imai Y, Kurokawa M, Izutsu K et al (2000) Mutations of the AML1 gene in myelodysplastic syndrome and their functional implications in leukemogenesis. Blood 96:3154–3160

Iwasaki H, Somoza C, Shigematsu H et al (2005) Distinctive and indispensable roles of PU.1 in maintenance of hematopoietic stem cells and their differentiation. Blood 106:1590–1600

Jacob B, Osato M, Yamashita N et al (2010) Stem cell exhaustion due to Runx1 deficiency is prevented by Evi5 activation in leukemogenesis. Blood 115:1610–1620. doi:10.1182/blood-2009-07-232249

Kastner P, Mark M, Ghyselinck N, Krezel W, Dupe V, Grondona JM, Chambon P (1997) Genetic evidence that the retinoid signal is transduced by heterodimeric RXR/RAR functional units during mouse development. Development 124:313–326
Kastner P, Lawrence HJ, Waltzinger C, Ghyselinck NB, Chambon P, Chan S (2001) Positive and negative regulation of granulopoiesis by endogenous RARalpha. Blood 97:1314–1320
Kim HG, de Guzman CG, Swindle CS, Cotta CV, Gartland L, Scott EW, Klug CA (2004) The ETS family transcription factor PU.1 is necessary for the maintenance of fetal liver hematopoietic stem cells. Blood 104:3894–3900. doi:10.1182/blood-2002-08-2425
Kirstetter P, Schuster MB, Bereshchenko O et al (2008) Modeling of C/EBPalpha mutant acute myeloid leukemia reveals a common expression signature of committed myeloid leukemia-initiating cells. Cancer Cell 13:299–310. doi:10.1016/j.ccr.2008.02.008
Ko M, Huang Y, Jankowska AM et al (2010) Impaired hydroxylation of 5-methylcytosine in myeloid cancers with mutant TET2. Nature 468:839–843. doi:10.1038/nature09586
Kodandapani R, Pio F, Ni CZ et al (1996) A new pattern for helix-turn-helix recognition revealed by the PU.1 ETS-domain-DNA complex. Nature 380:456–460. doi:10.1038/380456a0
Kosmider O, Gelsi-Boyer V, Ciudad M et al (2009) TET2 gene mutation is a frequent and adverse event in chronic myelomonocytic leukemia. Haematologica 94:1676–1681. doi:10.3324/haematol.2009.011205
Krivtsov AV, Twomey D, Feng Z et al (2006) Transformation from committed progenitor to leukaemia stem cell initiated by MLL-AF9. Nature 442:818–822. doi:10.1038/nature04980
Lahortiga I, Vazquez I, Agirre X et al (2004) Molecular heterogeneity in AML/MDS patients with 3q21q26 rearrangements. Genes Chromosomes Cancer 40:179–189. doi:10.1002/gcc.20033
Lamandin C, Sagot C, Roumier C et al (2002) Are PU.1 mutations frequent genetic events in acute myeloid leukemia (AML)? Blood 100:4680–4681. doi:10.1182/blood-2002-08-2563
Lancrin C, Sroczynska P, Stephenson C, Allen T, Kouskoff V, Lacaud G (2009) The haemangioblast generates haematopoietic cells through a haemogenic endothelium stage. Nature 457:892–895. doi:10.1038/nature07679
Laslo P, Spooner CJ, Warmflash A et al (2006) Multilineage transcriptional priming and determination of alternate hematopoietic cell fates. Cell 126:755–766
Leddin M, Perrod C, Hoogenkamp M et al (2011) Two distinct auto-regulatory loops operate at the PU.1 locus in B cells and myeloid cells. Blood 117:2827–2838. doi:10.1182/blood-2010-08-302976
Ley TJ, Ding L, Walter MJ et al (2010) DNMT3A mutations in acute myeloid leukemia. N Engl J Med 363:2424–2433. doi:10.1056/NEJMoa1005143
Li Y, Okuno Y, Zhang P et al (2001) Regulation of the PU.1 gene by distal elements. Blood 98: 2958–2965
Linggi B, Muller-Tidow C, van de Locht L et al (2002) The t(8;21) fusion protein, AML1 ETO, specifically represses the transcription of the p14(ARF) tumor suppressor in acute myeloid leukemia. Nat Med 8:743–750. doi:10.1038/nm726
Luesink M, Hollink IH, van der Velden VH et al (2012) High GATA2 expression is a poor prognostic marker in pediatric acute myeloid leukemia. Blood 120:2064–2075. doi:10.1182/blood-2011-12-397083
Martens JH, Brinkman AB, Simmer F et al (2010) PML-RARalpha/RXR alters the epigenetic landscape in acute promyelocytic leukemia. Cancer Cell 17:173–185. doi:10.1016/j.ccr.2009.12.042
Martin ME, Milne TA, Bloyer S et al (2003) Dimerization of MLL fusion proteins immortalizes hematopoietic cells. Cancer Cell 4:197–207
Mathews V, George B, Chendamarai E et al (2010) Single-agent arsenic trioxide in the treatment of newly diagnosed acute promyelocytic leukemia: long-term follow-up data. J Clin Oncol 28:3866–3871. doi:10.1200/JCO.2010.28.5031
McKercher SR, Torbett BE, Anderson KL et al (1996) Targeted disruption of the PU.1 gene results in multiple hematopoietic abnormalities. EMBO J 15:5647–5658

Meyers S, Downing JR, Hiebert SW (1993) Identification of AML-1 and the (8;21) translocation protein (AML-1/ETO) as sequence-specific DNA-binding proteins: the runt homology domain is required for DNA binding and protein-protein interactions. Mol Cell Biol 13:6336–6345
Miyoshi H, Shimizu K, Kozu T, Maseki N, Kaneko Y, Ohki M (1991) t(8;21) breakpoints on chromosome 21 in acute myeloid leukemia are clustered within a limited region of a single gene, AML1. Proc Natl Acad Sci USA 88:10431–10434
Miyoshi H, Kozu T, Shimizu K et al (1993) The t(8;21) translocation in acute myeloid leukemia results in production of an AML1-MTG8 fusion transcript. EMBO J 12:2715–2721
Mohr F, Dohner K, Buske C, Rawat VP (2011) TET genes: new players in DNA demethylation and important determinants for stemness. Exp Hematol 39:272–281. doi:10.1016/j.exphem.2010.12.004
Moran-Crusio K, Reavie L, Shih A et al (2011) Tet2 loss leads to increased hematopoietic stem cell self-renewal and myeloid transformation. Cancer Cell 20:11–24. doi:10.1016/j.ccr.2011.06.001
Moreau-Gachelin F, Tavitian A, Tambourin P (1988) Spi-1 is a putative oncogene in virally induced murine erythroleukaemias. Nature 331:277–280. doi:10.1038/331277a0
Mosad E, Abdou M, Zaky AH (2012) Rearrangement of the myeloid/lymphoid leukemia gene in therapy-related myelodysplastic syndrome in patients previously treated with agents targeting DNA topoisomerase II. Oncology 83:128–134. doi:10.1159/000338769
Motoda L, Osato M, Yamashita N et al (2007) Runx1 protects hematopoietic stem/progenitor cells from oncogenic insult. Stem Cells 25:2976–2986. doi:10.1634/stemcells.2007-0061
Mueller BU, Pabst T, Osato M et al (2002) Heterozygous PU.1 mutations are associated with acute myeloid leukemia. Blood 100:998–1007
Mueller BU, Pabst T, Fos J et al (2006) ATRA resolves the differentiation block in t(15;17) acute myeloid leukemia by restoring PU.1 expression. Blood 107:3330–3338
Nagy L, Kao HY, Chakravarti D et al (1997) Nuclear receptor repression mediated by a complex containing SMRT, mSin3A, and histone deacetylase. Cell 89:373–380
Niederreither K, Dolle P (2008) Retinoic acid in development: towards an integrated view. Nat Rev Genet 9:541–553. doi:10.1038/nrg2340
Nikoloski G, Langemeijer SM, Kuiper RP et al (2010) Somatic mutations of the histone methyltransferase gene EZH2 in myelodysplastic syndromes. Nat Genet 42:665–667. doi:10.1038/ng.620
North T, Gu TL, Stacy T et al (1999) Cbfa2 is required for the formation of intra-aortic hematopoietic clusters. Development 126:2563–2575
Nutt SL, Metcalf D, D'Amico A, Polli M, Wu L (2005) Dynamic regulation of PU.1 expression in multipotent hematopoietic progenitors. J Exp Med 201:221–231
Okuda T, van Deursen J, Hiebert SW, Grosveld G, Downing JR (1996) AML1, the target of multiple chromosomal translocations in human leukemia, is essential for normal fetal liver hematopoiesis. Cell 84:321–330
Okuno Y, Huang G, Rosenbauer F et al (2005) Potential autoregulation of transcription factor PU.1 by an upstream regulatory element. Mol Cell Biol 25:2832–2845
Osato M, Asou N, Abdalla E et al (1999) Biallelic and heterozygous point mutations in the runt domain of the AML1/PEBP2alphaB gene associated with myeloblastic leukemias. Blood 93:1817–1824
Pabst T, Mueller BU, Harakawa N et al (2001a) AML1-ETO downregulates the granulocytic differentiation factor C/EBPalpha in t(8;21) myeloid leukemia. Nat Med 7:444–451. doi:10.1038/86515
Pabst T, Mueller BU, Zhang P et al (2001b) Dominant-negative mutations of CEBPA, encoding CCAAT/enhancer binding protein-alpha (C/EBPalpha), in acute myeloid leukemia. Nat Genet 27:263–270. doi:10.1038/85820
Perrotti D, Cesi V, Trotta R et al (2002) BCR-ABL suppresses C/EBPalpha expression through inhibitory action of hnRNP E2. Nat Genet 30:48–58

Peterson LF, Boyapati A, Ahn EY et al (2007) Acute myeloid leukemia with the 8q22;21q22 translocation: secondary mutational events and alternative t(8;21) transcripts. Blood 110:799–805. doi:10.1182/blood-2006-11-019265

Ptasinska A, Assi SA, Mannari D et al (2012) Depletion of RUNX1/ETO in t(8;21) AML cells leads to genome-wide changes in chromatin structure and transcription factor binding. Leukemia 26:1829–1841. doi:10.1038/leu.2012.49

Putz G, Rosner A, Nuesslein I, Schmitz N, Buchholz F (2006) AML1 deletion in adult mice causes splenomegaly and lymphomas. Oncogene 25:929–939. doi:10.1038/sj.onc.1209136

Ray D, Kwon SY, Tagoh H, Heidenreich O, Ptasinska A, Bonifer C (2013) Lineage inappropriate PAX5 expression in t(8;21) acute myeloid leukemia requires signalling mediated abrogation of polycomb repression. Blood 122(5):759–69. doi:10.1182/blood-2013-02-482497

Ribeiro AF, Pratcorona M, Erpelinck-Verschueren C et al (2012) Mutant DNMT3A: a marker of poor prognosis in acute myeloid leukemia. Blood 119:5824–5831. doi:10.1182/blood-2011-07-367961

Rodrigues NP, Janzen V, Forkert R et al (2005) Haploinsufficiency of GATA-2 perturbs adult hematopoietic stem-cell homeostasis. Blood 106:477–484. doi:10.1182/blood-2004-08-2989

Rodrigues NP, Boyd AS, Fugazza C et al (2008) GATA-2 regulates granulocyte-macrophage progenitor cell function. Blood 112:4862–4873. doi:10.1182/blood-2008-01-136564

Rosenbauer F, Wagner K, Kutok JL et al (2004) Acute myeloid leukemia induced by graded reduction of a lineage-specific transcription factor, PU.1. Nat Genet 36:624–630

Rowley JD (1973) Identificaton of a translocation with quinacrine fluorescence in a patient with acute leukemia. Ann Genet 16:109–112

Schaub FX, Looser R, Li S, Hao-Shen H, Lehmann T, Tichelli A, Skoda RC (2010) Clonal analysis of TET2 and JAK2 mutations suggests that TET2 can be a late event in the progression of myeloproliferative neoplasms. Blood 115:2003–2007. doi:10.1182/blood-2009-09-245381

Scott EW, Simon MC, Anastasi J, Singh H (1994) Requirement of transcription factor PU.1 in the development of multiple hematopoietic lineages. Science 265:1573–1577

Shia WJ, Okumura AJ, Yan M et al (2012) PRMT1 interacts with AML1-ETO to promote its transcriptional activation and progenitor cell proliferative potential. Blood 119:4953–4962. doi:10.1182/blood-2011-04-347476

Snaddon J, Smith ML, Neat M et al (2003) Mutations of CEBPA in acute myeloid leukemia FAB types M1 and M2. Genes Chromosomes Cancer 37:72–78. doi:10.1002/gcc.10185

Sun XJ, Wang Z, Wang L et al (2013) A stable transcription factor complex nucleated by oligomeric AML1-ETO controls leukaemogenesis. Nature 500:93–97. doi:10.1038/nature12287

Tahirov TH, Inoue-Bungo T, Morii H et al (2001) Structural analyses of DNA recognition by the AML1/Runx-1 Runt domain and its allosteric control by CBFbeta. Cell 104:755–767

Tenen DG (2003) Disruption of differentiation in human cancer: AML shows the way. Nat Rev Cancer 3:89–101. doi:10.1038/nrc989, nrc989 [pii]

Tiacci E, Pileri S, Orleth A et al (2004) PAX5 expression in acute leukemias: higher B-lineage specificity than CD79a and selective association with t(8;21)-acute myelogenous leukemia. Cancer Res 64:7399–7404. doi:10.1158/0008-5472.CAN-04-1865

Tipping AJ, Pina C, Castor A et al (2009) High GATA-2 expression inhibits human hematopoietic stem and progenitor cell function by effects on cell cycle. Blood 113:2661–2672. doi:10.1182/blood-2008-06-161117

Tsai FY, Orkin SH (1997) Transcription factor GATA-2 is required for proliferation/survival of early hematopoietic cells and mast cell formation, but not for erythroid and myeloid terminal differentiation. Blood 89:3636–3643

Tsai FY, Keller G, Kuo FC et al (1994) An early haematopoietic defect in mice lacking the transcription factor GATA-2. Nature 371:221–226. doi:10.1038/371221a0

Vangala RK, Heiss-Neumann MS, Rangatia JS et al (2003) The myeloid master regulator transcription factor PU.1 is inactivated by AML1-ETO in t(8;21) myeloid leukemia. Blood 101:270–277

Vicente C, Vazquez I, Conchillo A et al (2012) Overexpression of GATA2 predicts an adverse prognosis for patients with acute myeloid leukemia and it is associated with distinct molecular abnormalities. Leukemia 26:550–554. doi:10.1038/leu.2011.235

Villa R, Pasini D, Gutierrez A et al (2007) Role of the polycomb repressive complex 2 in acute promyelocytic leukemia. Cancer Cell 11:513–525. doi:10.1016/j.ccr.2007.04.009

Walsh JC, DeKoter RP, Lee HJ et al (2002) Cooperative and antagonistic interplay between PU.1 and GATA-2 in the specification of myeloid cell fates. Immunity 17:665–676

Walter K, Cockerill PN, Barlow R et al (2010) Aberrant expression of CD19 in AML with t(8;21) involves a poised chromatin structure and PAX5. Oncogene 29(20):2927–2937

Wang L, Brown JL, Cao R, Zhang Y, Kassis JA, Jones RS (2004) Hierarchical recruitment of polycomb group silencing complexes. Mol Cell 14:637–646. doi:10.1016/j.molcel.2004.05.009

Wang K, Wang P, Shi J et al (2010) PML/RARalpha targets promoter regions containing PU.1 consensus and RARE half sites in acute promyelocytic leukemia. Cancer Cell 17:186–197. doi:10.1016/j.ccr.2009.12.045

Wang L, Gural A, Sun XJ et al (2011) The leukemogenicity of AML1-ETO is dependent on site-specific lysine acetylation. Science 333:765–769. doi:10.1126/science.1201662

Wang F, Travins J, Delabarre B et al (2013) Targeted inhibition of mutant IDH2 in leukemia cells induces cellular differentiation. Science 340:622–626. doi:10.1126/science.1234769

Weinstein IB (2002) Cancer. Addiction to oncogenes–the Achilles heal of cancer. Science 297:63–64. doi:10.1126/science.1073096

Welch JS, Ley TJ, Link DC et al (2012) The origin and evolution of mutations in acute myeloid leukemia. Cell 150:264–278. doi:10.1016/j.cell.2012.06.023

Xie H, Ye M, Feng R, Graf T (2004) Stepwise reprogramming of B cells into macrophages. Cell 117:663–676

Yan XJ, Xu J, Gu ZH et al (2011) Exome sequencing identifies somatic mutations of DNA methyltransferase gene DNMT3A in acute monocytic leukemia. Nat Genet 43:309–315. doi:10.1038/ng.788

Ye M, Zhang H, Amabile G et al (2013) C/EBPa controls acquisition and maintenance of adult haematopoietic stem cell quiescence. Nat Cell Biol 15:385–394. doi:10.1038/ncb2698

Zhang DE, Zhang P, Wang ND, Hetherington CJ, Darlington GJ, Tenen DG (1997) Absence of granulocyte colony-stimulating factor signaling and neutrophil development in CCAAT enhancer binding protein alpha- deficient mice. Proc Natl Acad Sci USA 94:569–574

Zhang P, Iwasaki-Arai J, Iwasaki H et al (2004) Enhancement of hematopoietic stem cell repopulating capacity and self-renewal in the absence of the transcription factor C/EBP alpha. Immunity 21:853–863

Zhang SJ, Ma LY, Huang QH et al (2008) Gain-of-function mutation of GATA-2 in acute myeloid transformation of chronic myeloid leukemia. Proc Natl Acad Sci USA 105:2076–2081. doi:10.1073/pnas.0711824105

Zhang SJ, Shi JY, Li JY (2009) GATA-2L359V mutation is exclusively associated with CML progression but not other hematological malignancies and GATA-2 P250A is a novel single nucleotide polymorphism. Leuk Res 33:1141–1143. doi:10.1016/j.leukres.2009.02.025

Zhu J, Koken MH, Quignon F et al (1997) Arsenic-induced PML targeting onto nuclear bodies: implications for the treatment of acute promyelocytic leukemia. Proc Natl Acad Sci USA 94:3978–3983

Chapter 10
The Molecular Basis of Normal Erythroid/Megakaryocyte Development and Mechanisms of Epigenetic/Transcriptional Deregulation Leading to Erythroleukemia and Thalassaemia

Douglas Vernimmen

Abstract The recent introduction of high-throughput sequencing technology has now provided a very broad picture of the chromatin landscape and transcription factor binding throughout the genome in many cell types, including the different haematopoietic lineages. The epigenetic machinery affecting DNA methylation and histone modification has been well characterised, and large scale screening for inhibitors has already led to the production of specific drugs used for treating patients. Moreover, it appears that many epigenetic regulators are mutated in various diseases, particularly in the haematopoietic compartment. I discuss here the accessible models used to study erythropoiesis and megakaryopoiesis and their associated transcription and epigenetic programme. Finally, I will describe the current approaches used for epigenetic therapy in myeloid malignancies and haemoglobinopathies such as thalassaemia.

Keywords Differentiation • Transcription • Epigenetics • Chromatin • Globin • Thalassemia • Erythropoiesis • Megakaryopoiesis

10.1 Introduction

The erythroid lineage has been a major centre of interest because of the very strong expression of the globin genes, which are producing the major proteins found in red blood cells. The adult haemoglobin (HbA) is made by the formation of a tetramer containing two α chains and two β chains. The level of expression of these proteins needs to be equimolar to ensure the formation of this tetramer, without which an imbalance would create insoluble homotetramers, the key common feature of thalassaemia.

D. Vernimmen (✉)
The Roslin Institute, Developmental Biology Division, University of Edinburgh, Easter Bush, Midlothian EH25 9RG, UK
e-mail: douglas.vernimmen@roslin.ed.ac.uk

C. Bonifer and P.N. Cockerill (eds.), *Transcriptional and Epigenetic Mechanisms Regulating Normal and Aberrant Blood Cell Development*, Epigenetics and Human Health, DOI 10.1007/978-3-642-45198-0_10, © Springer-Verlag Berlin Heidelberg 2014

Over the last 3 decades, the globin genes have been key model systems used to study gene regulation and have so far elucidated many major principles of how mammalian genes are regulated during development and differentiation. A large community of scientists have used several models such as using different sources, species and cell types representing different stages of erythropoiesis but also material from patients bearing different types of mutations involved in the downregulation of globin expression, leading to thalassaemias.

Being derived from a common precursor, the megakaryocyte and erythrocyte lineages share a common number of transcription factors (TFs) critical for their development, including GATA1, FOG1, SCL, NF-E2 and GFI1b. Other factors like EKLF (Erythroid Kruppel-like Factor/KLF1), GABPa and FLI1 (megakaryocytic) are tissue specific. These TFs are also differentially expressed during differentiation, suggesting a different role for these TFs (Anguita et al. 2004). For example, GATA2 is expressed early, in common myeloid progenitors (CMP), whereas GATA1 is expressed later, in megakaryocytic and erythroid progenitors (MEP). It is of interest that GATA2 regulates the expression of GATA1, which in turn, represses GATA2 via a negative feedback loop (Grass et al. 2003). The later specification into the erythroid or the megakaryocytic lineage therefore relies on the other lineage-specific TFs like EKLF and FLI1. For these reasons, abnormalities in erythroid development present in myelodysplasias (MDS) and acute myeloid leukaemias (AML) are often found in parallel with abnormalities in the megakaryocytic lineage.

10.2 Accessible Systems for Studying the Key Stages of Erythroid and Megakaryocytic Differentiation

The process of haematopoiesis is largely conserved throughout evolution and in mammals has been most extensively studied in mice and in humans (Cantor and Orkin 2002; Orkin 2000). The epigenetic and transcriptional mechanisms underlying haematopoietic development are best defined for the more mature populations within the lymphoid and myeloid/erythroid compartments. This bias is mainly due to the high number of cells required for some experiments such as chromatin immunoprecipitation (ChIP). For this main reason, progenitors of the megakaryocyte and erythroid lineage can't be analysed in the same depth as in fully differentiated cells. Alternative sources of different progenitors are now under development by immortalising cells by viral infection (Houston et al. 2007; Ney and D'Andrea 2000; Oakley et al. 2012).

As a source of pluripotent cells, human (h) and mouse (m) embryonic stem (ES) cells are often used, since these can be abundantly produced. Umbilical cord blood mononuclear CD34+ cells (haematopoietic progenitor cells—HPCs) are commonly used to produce fully differentiated haematopoietic cells (Fig. 10.1). These can be isolated by cell sorting in the laboratory and they are also commercially available.

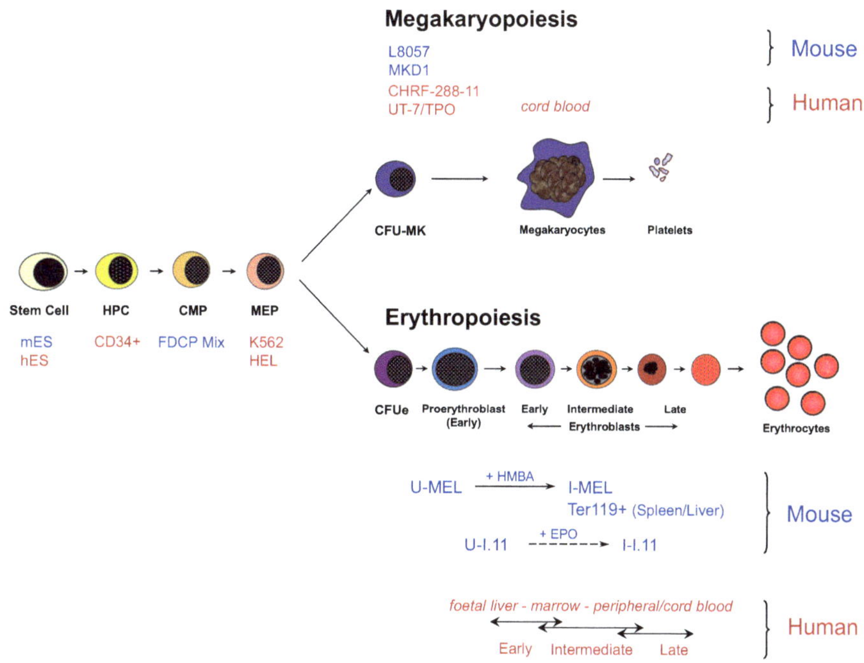

Fig. 10.1 Erythropoiesis and megakaryopoiesis and the different cell types available. Please note that differentiation is a continuous process and that each stage is not a steady state. Cell lines have been associated with one of these stages according to some (but not all) of their features. Mouse cells are shown in *blue* and human in *red*

CD34+ HPC also contain multipotential progenitors (including the common myeloid progenitor, CMP, or colony-forming unit-granulocyte, erythrocyte, monocyte, megakaryocyte, CFU-GEMM) that can differentiate into a wide variety of mature blood cells, including erythroid. In the mouse, the factor-dependent cell Patterson (FDCP)-mix cells (Spooncer et al. 1986) have been considered to be equivalent of CMP suitable for differentiation studies (Fig. 10.1). In humans, progenitors can also be obtained from bone marrow samples, which can be collected from individuals undergoing total hip replacement for osteoarthritis (De Gobbi et al. 2011) or after mobilisation for bone marrow transplant. However, because human bone marrow samples are difficult to obtain routinely, alternative species such as pigs could offer a greater advantage for studies of mammalian haematopoiesis. Indeed, pigs seems to be quite similar to humans in many aspects (Aigner et al. 2010; Prather 2013) and the number of cells that can be obtained from full bone marrow is therefore unlimited.

The earliest progenitors that are entirely restricted to the red cell lineage produce large erythroid colonies in vitro, consisting of several "bursts" of smaller colonies known as burst-forming unit-erythroid (so-called BFU-E). Late erythroid progenitors (identified in clonal assays and called colony-forming units, CFU-E, Fig. 10.1) correspond to the earliest recognisable erythroid precursor in the bone marrow (the proerythroblast). As these erythroid precursors progress through maturation (intermediate erythroblasts), the nucleus becomes progressively condensed, (intermediate and late erythroblasts) and is eventually expelled, producing the mature red cell (Fig. 10.1). For both the human and mouse models, large numbers of definitive primary erythroid cells can be obtained from various sources, including foetal liver, bone marrow, cord blood and circulating blood. In the mouse, an extra source can be obtained from adult spleen after phenylhydrazine treatment (Spivak et al. 1973). In addition to these primary cells, immortalised mouse erythroid cells such as mouse erythroleukemia (MEL) cells and I-11 cells also offer a strong advantage (Dolznig et al. 2001; Ney and D'Andrea 2000) (Fig. 10.1). MEL cells are blocked at the proerythroblast stage of differentiation and can be induced by a variety of chemical agents to undergo the terminal stages of differentiation, thus providing a useful cellular model for this stage of erythropoiesis in mouse (Marks et al. 1987). I-11 erythroblasts cells were isolated from p53-mutated mice and can be differentiated in culture after exposure to different hormones such as erythropoietin (EPO) (Dolznig et al. 2001). Unfortunately, there are still no simple, equivalent cellular models of human erythropoiesis, although K562 cells are often used to represent an "erythroid" cell line (e.g. ENCODE consortium). This cell line is however an unsatisfactory model since K562 cells also have megakaryocytic precursor properties and are pseudo-triploid (originally derived from a patient with chronic myeloid leukaemia) that produces very small amounts of embryonic globins (Lozzio and Lozzio 1975; Rutherford et al. 1981).

A valuable additional resource is interspecific hybrids of MEL cells containing a human chromosome (16 for α-globin locus and 11 for β-globin locus). On induction, these hybrids mimic the terminal stages of erythropoiesis, expressing not just the endogenous mouse globin genes but also the human globin genes (Deisseroth and Hendrick 1978; Zeitlin and Weatherall 1983). Therefore hybrids derived from normal individuals and from patients with previously characterised, natural mutations of the two globin clusters can be analysed (Craddock et al. 1995; Forrester et al. 1990). These mutant chromosomes may include a deletion of one or more of the remote upstream elements whilst keeping the globin genes intact or a chromosome in which the globin genes are deleted but the upstream elements remain intact (Vernimmen et al. 2007).

Finally, megakaryocytes, the precursors of platelets, constitute less than about 0.05 % of all nucleated cells in the bone marrow and are therefore difficult to isolate in large numbers. Nevertheless, human cell lines with megakaryocyte potential have been established from the blood of patients with leukaemia. Many of these cell lines have primitive multiphenotypic properties of erythroid, myeloid and megakaryocytic cells, while some show more restricted megakaryocytic-specific markers (Saito 1997). In fact, megakaryocytes/erythroid progenitors (MEPs) express both thrombopoietin (TpoR/c-Mpl) and erythropoietin (EpoR) receptors (Pronk et al. 2007), reflecting their bipotential stage. The human cell lines CHRF-288-11 (Fugman et al. 1990), and UT-7/TPO (Komatsu et al. 1996; Komatsu et al. 1993) were established from patients with acute megakaryocytic leukaemia. Of interest, K562 and HEL cell lines are enabled to differentiate along either an erythroid or a megakaryocytic lineage using a variety of inducing agents (Pencovich et al. 2011); but without producing cells that closely resemble their normal counterparts. In the mouse, the megakaryoblastic cell line L8057 (Ishida et al. 1993) was derived from an irradiated C3H/He mouse and MKD1 was derived from ES cells and is considered as a primary megakaryocyte progenitor (Chagraoui and Porcher 2012) (Fig. 10.1). A wide variety of haematopoietic cell types have now been immortalised using these procedures, including erythroid, megakaryocytic, monocytic, myelocytic and multipotential cell types. As for the erythroid lineage (Pope et al. 2000), a well-characterised two-phase liquid culture system has been developed for the *in vitro* differentiation of human CD34+ cells into megakaryocytes (Cortin et al. 2005; Tijssen et al. 2011). Methods for culturing megakaryocytes from (m) and (h) ES cells and to induce transdifferentiation of other lineages have been also published (Fujimoto et al. 2003; Gaur et al. 2006; Ono et al. 2012).

10.3 Transcription and Epigenetic Regulation of Erythroid and Megakaryocytic Differentiation

Transcription factors involved in erythropoiesis and megakaryopoiesis are very similar as mentioned earlier. In addition to these lineage-affiliated transcription factors, many widely expressed TFs such as Sp/X-KLFs (e.g. Sp1, Sp3, BKLF and ZBP-89) have been previously implicated in erythropoiesis and detected on the globin loci (Higgs et al. 2008). The availability of a variety of primary cells and cell lines has enabled many laboratories to analyse mainly globin gene expression at many stages of differentiation and, therefore, provides a way to investigate the order of events leading to gene activation. With these various resources, we can examine the binding of TFs, epigenetic regulators and their associated histone modifications to the various cis-acting elements, not only at the time the globin genes are transcribed but also at previous steps in the differentiation pathway. In this way, we have analysed the sequential order of events required to prime the α-globin domain for transcription (Anguita et al. 2004; De Gobbi et al. 2007; Vernimmen et al. 2007).

Overall, the α and β loci share the common characteristic of being sequentially activated by priming first at the remote regulatory sequences, with sites of TF binding and chromatin modifications subsequently moving closer towards the promoter (usually downstream), in a manner suggesting that there is a polarity in the spread of these activation marks. This phenomenon of polarity is interesting because it is generally accepted that enhancers function independently of their orientation with respect to the promoter. However, regulatory elements such as the β-globin LCR consist of a cluster of several enhancers and other DNaseI Hypersensitive Sites (DHSs), and this combination may account for the polarity phenomenon. Of interest, we recently found that the removal of an upstream element of the human α-globin locus most strongly affects TF binding at the downstream DHS, with a maximal reduction at the target promoter, suggesting again that the activating signal is unidirectional and propagates from the upstream enhancers towards the downstream promoter (Vernimmen et al. 2011).

Recently, the teams of Bertie Göttgens and Willem Ouwehand at the University of Cambridge published a very elegant study with human primary megakaryocytes (Tijssen et al. 2011). Using cord blood CD34+ cells, up to 10^8 primary megakaryocytes have been produced by induction using TPO and IL-1β. In this study, a genome-wide analysis has been made possible and the binding of several key TFs (GATA1, GATA2, RUNX1, FLI1 and SCL), together with a few histone modifications (H3K4me3, H3K27me3 and H3ac) was analysed. This study showed that binding of all five factors is observed in 30 % of megakaryocyte-specific genes, suggesting that alternative TF combinations mediating megakaryocyte-specific expression remain to be discovered. Additionally, the PDZK1IP1 protein (also known as MAP17), which encodes a membrane-associated protein, was also found to be bound to these elements. It is interesting to note that the Ouwehand laboratory also observed that subtle variation in the transcript level of a member of the jumonji family of demethylases, named *JMJD1C*, modifies platelet count and volume (Gieger et al. 2011).

10.4 Epigenetic Control of Globin Genes Expression

The literature on the β-globin locus is very abundant, but not always consistent, and is dependent on the model used. I will therefore summarise what we know about the epigenetic control of the α-globin locus during erythroid differentiation and the role of remote enhancers in this process.

The human α-globin genes are associated with a CpG island, whereas in rodents (e.g. mouse and rat) this CpG island has been eroded during evolution (Lynch et al. 2012). In hES cells and non-erythroid cells, we found that this CpG island is associated with polycomb repressive complex 2 (PRC2) recruitment and its associated chromatin signature (H3K27me3), whereas these are not found in the rodent α-globin genes (Garrick et al. 2008). In hES cells, the α-globin regions are spanned by both repressive (H3K27me3) and active (H3K4me3) chromatin marks, referred to as bivalent domains. However, these two domains don't entirely overlap and the ratio observed changes progressively during differentiation. This trend led us to investigate if the progressive increase of H3K4me3 could be due to the expression of the gene at irrelevant stages (De Gobbi et al. 2011) rather than a signature of priming, suggested by previous studies (Bernstein et al. 2006). Therefore we have measured mRNA expression at a single cell level and found that although full length α-globin mRNA was detected in a small number of cells, this proportion increased during differentiation. We therefore suggested that bivalent domains might be the consequence of a subpopulation phenomenon rather than true priming in all cells (De Gobbi et al. 2011).

Interestingly, when the main regulatory sequence of the human α-globin cluster (MCS-R2) is removed, the establishment of H3K4me3 still reaches close to normal levels, whereas only 1 % of normal expression is observed in terminally differentiated erythroblasts (Vernimmen et al. 2011). Therefore, without the enhancer, the status of the α-globin locus in erythroblasts is strikingly similar to that seen in hES cells (Garrick et al. 2008; Vernimmen et al. 2011). Since the work in Adrian Bird's group showing that a CXXC domain protein, cfp1 (also called CGBP), was involved in the deposition of H3K4me3 at the CpG island of target genes regardless of their transcription status (Thomson et al. 2010), we investigated its potential role. CGBP is part of a Trithorax Group (TrxG) complex (human Set1), and therefore would generate H3K4me3 at any CpG island. However, by ChIP analysis, we didn't detect CGBP at the α-globin CpG island in the absence of MCS-R2, suggesting an alternative mechanism for H3K4me3 deposition. We also showed that CGBP was excluded from CpG islands bound by PcG by comparing human non-expressing (i.e. non-erythroid) *versus* expressing cells (erythroid). We therefore investigated if the lack of CGBP in the ΔMCS-R2 mutant could be also correlated to persistent PcG binding. This is precisely what we found: in the absence of MCS-R2, PRC2 remained at the α-globin CpG island throughout the whole process of erythroid differentiation, whereas this complex was removed in the presence of the remote enhancer. Of importance, the removal of PRC2 and H3K27me3 was associated with the recruitment of the H3K27me3 demethylase JMJD3 (Vernimmen et al. 2011). We therefore showed that the enhancer is required for PRC2 removal and this involves the recruitment of JMJD3. Hence, our research presents new views about how enhancers work: they are involved in the recruitment of key activating transcription factors (TFs) and the pre-initiation complex (PIC), but also they facilitate the removal of repressor complexes such as PRC2. Further studies of other loci will be required to determine if PRC2 removal from target genes is a common function of enhancers.

10.5 Epigenetic Deregulation in Myeloid Malignancies and Haemoglobinopathies: Solutions and Therapies

10.5.1 Myeloid Malignancies

The recent finding of mutations in epigenetic regulators has accounted for a new and common class of mutant disease alleles that contribute to the pathogenesis of myeloid malignancies in addition to the classical class I genes that affect proliferation and class II that affect differentiation (reviewed in (Shih et al. 2012). These include mutations in tet methylcytosine dioxygenase 2 (*TET2*), isocitrate dehydrogenase -1 (*IDH1*) and -2 (*IDH2*), additional sex combs-like 1 (*ASXL1*), enhancer of zeste homologue 2 (*EZH2*), mixed-lineage leukaemia (*MLL*) and DNA methyltransferase 3A (*DNMT3A*), which have recently been shown to have biological, clinical and potential therapeutic relevance to myeloid malignancies (Shih et al. 2012). There are also very intriguing correlations between cancer-associated DNA hypermethylation and genes marked with "bivalent" histone modifications in multipotent cells (Deneberg et al. 2011; Iliou et al. 2011; Rakyan et al. 2010; Rodriguez et al. 2008).

Today, at least four different DNA modifications and 16 classes of histone modifications have been described (Dawson and Kouzarides 2012). The methylation of the 5-carbon on cytosine residues (5mC) in CpG dinucleotides was the first described covalent modification of DNA. The cytosine is methylated by a family of DNA methyltransferases (DMTs). One of these, DNMT3A, is mutated in acute myeloid leukaemia (AML) (Ley et al. 2010), myeloproliferative diseases (MPD) and myelodysplasic syndromes (MDS). Hypomethylating agents such as azacitidine (5-azacytidine) and decitabine (5-aza-deoxycytidine, DAC) have shown promising results for treatment of MDS (Fenaux et al. 2009). The nature of the strong response seen in MDS patients is, however, not well understood, but it seems that low doses of such agents hold the key to therapeutic benefit (Tsai et al. 2012). It is also emerging that the combinatorial use of DNMT and HDAC inhibitors may offer a superior therapeutic outcomes (Gore 2011). These agents are also used for the treatment of β thalassaemia (see below).

However, 5mC is not the only DNA modification. Indeed, 5mC oxidative intermediates such as 5-hydroxymethylcytosine (5hmC), 5-formylcytosine (5fC) and 5-carboxylcytosine (5caC) are other metabolites found at CG sequences (Dawson and Kouzarides 2012). Although the biological significance of these intermediates is not known, their associated respective enzymes represent other potential targets for treatment in AML. For example, the ten-eleven translocation (TET 1-3) family of proteins are the mammalian DNA hydroxylases responsible for catalytically converting 5mC to 5hmC (Wu and Zhang 2011). TET1 and TET2 expression is mainly confined to ES cells, whereas TET3 expression is upregulated in differentiated cells. Studies have linked TET1 to epigenetic repression complexes such as SIN3A and PRC2 (Williams et al. 2011; Wu et al. 2011), and mouse Tet2 has been implicated in haematopoiesis by reducing multilineage repopulation capacity of haematopoietic stem cells (HSC) and limiting myeloid differentiation potential (Ko et al. 2011; Ko and Rao 2011). Studies have suggested that 5hmC inhibits the methyl-CpG-binding protein MeCP2 from binding DNA and therefore DNA methylation. Of interest, a few studies showed that 5hmC was associated with both gene bodies and enhancers in hES cells (Stroud et al. 2011) and mES cells (Song et al. 2013). As discussed in chapter 9 of this volume, *TET1* gene is subjected to fusion with *MLL* gene in translocations found in AML (Lorsbach et al. 2003) and recurrent mutations in TET2 were also found numerous haematological malignancies (Cimmino et al. 2011; Delhommeau et al. 2009; Langemeijer et al. 2009).

The 16 different classes of histone modifications involve many different amino acids in each histone protein and encompass a total of about 150 known specific modifications which vary greatly in their properties (Tan et al. 2011). Histone acetylation is associated with open chromatin conformation and therefore also with transcription activation. Histone lysine acetyltransferases (KAT) were the first enzymes shown to modify histones and are often involved in chromosomal translocations in different malignancies including blood. For example, a translocation involving MOZ (KAT6A) and nuclear receptor TIF2 has been found in AML (Huntly et al. 2004).

BAP1 (BRCA1-associated protein1) was originally identified as an ubiquitin hydrolase that binds the RING finger domain of BRCA1 (Jensen et al. 1998). BAP1 contains numerous functional domains, including a ubiquitin C-terminal hydrolase (UCH) domain, a host cell factor-1 (HCF-1)-binding domain and binding domains for BRCA1 and BARD1. BAP1 is involved in numerous biological processes including chromatin dynamics, DNA damage response and regulation of the cell cycle and cell growth. Although deletions and/or point mutations in the BAP1 gene are present in melanoma, mesothelioma, lung adenocarcinoma, meningioma and renal cell carcinoma (Goldstein 2011), nonsense mutations in the hydrolase domain has been found in one patient with MDS (Dey et al. 2012). Dey et al. also found that loss of BAP1 in adult mouse hematopoietic lineages leads to a myeloproliferative/ myelodysplasic disorder with features of human chronic myelomonocytic leukaemia (CMML). BAP1 also interacts with ASXL1 and ASXL2 (Scheuermann et al. 2010). Interestingly, ASXL1 has been also found mutated in CMML (Dey et al. 2012). The orthologues of BAP1 and ASXL1 in Drosophila (Calypso and Asx) are components of the PcG repressor complex (PRC1), suggesting that BAP1 could be a histone deubiquitinase and therefore a PRC1 antagonist (Scheuermann et al. 2010).

Kinase JAK2 specifically phosphorylates H3Y41, disrupts the binding of the chromatin repressor HP1α and activates the expression of haematopoietic oncogenes such as Lmo2 (Dawson et al. 2009). A number of JAK2 inhibitors have been discovered and are currently being developed as therapeuthics for myeloproliferative neoplasms (Atallah and Verstovsek 2009; Chen and Prchal 2010; Chen et al. 2010; Verstovsek 2009). The modulation of the activity of demethylases by small molecules in high-throughput screens is now also possible (Kruidenier et al. 2012; McCabe et al. 2012). However, the difficulty in obtaining specificity for compounds directed against different related enzymes has not been solved (e.g. GSK-J3 and GSK-J4 inhibit both JMJD3 and UTX; Kruidenier et al. 2012).

10.5.2 Haemoglobinopathies

In humans, two loci direct the synthesis of haemoglobin: the α locus, which contains the embryonic ζ gene, followed by two adult α genes (α2 and α1), and the β locus, which contains the embryonic ε gene, two foetal γ genes and a single adult β gene. The β locus also contains an additional gene δ, located just upstream the β gene and is expressed at low levels for a short period of time just after birth (Fig. 10.2a). The gene clusters are both arranged in the order in which they are expressed during development. Whereas the embryonic genes (ζ and ε) are expressed during primitive erythropoiesis (yolk sack), the foetal (α and γ) and

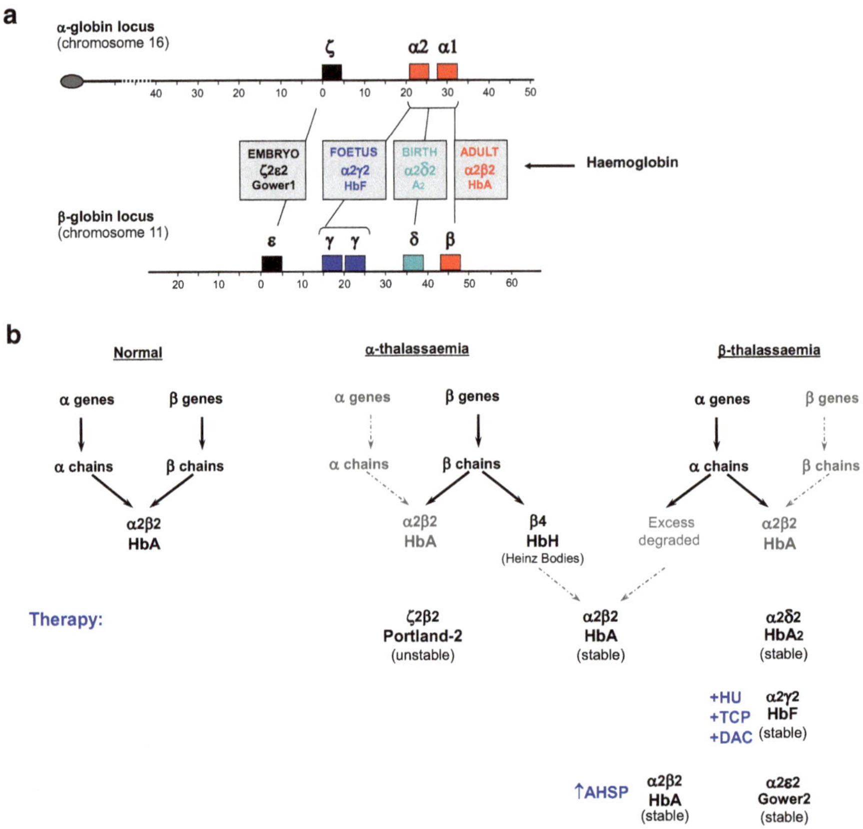

Fig. 10.2 Structure of the human globin loci and the molecular basis of thalassaemias. (**a**) structure of the α- and β-globin locus. Scale is represented in kilobases and the position zero is attributed to the first gene for each locus (ζ and ε respectively). Note that haemoglobin A2 is only expressed during a few weeks after birth. (**b**) Molecular basis of α and β thalassaemia. Thalassaemia are due to a decrease of production of the globin genes (more often deletions removing the genes). The therapeutic approaches (*blue*) consist in reactivating the expression of the silenced (embryonic or foetal) genes to restore a stable tetramer of haemoglobin

adult (α and β) genes are expressed during definitive erythropoiesis (foetal liver and then bone marrow). Note that erythroid cells in the yolk sac may have a different origin from those in the bone marrow (Palis 2008; Sankaran et al. 2010). Therefore, the two switches of expression on the β locus should occur in two separate cell types: $\varepsilon \rightarrow \gamma$ in the yolk sac and foetal liver (cells originated from haemangioblasts) and $\gamma \rightarrow \beta$ in the foetal liver and bone marrow (cells originated from HSC).

Sickle cell anaemia (SCA) and thalassaemia together comprise the most commonly inherited diseases in humans. Thalassaemia are characterised by inherited mutations leading to a reduction (e.g. deletion of remote regulatory sequences) or absence (e.g. deletion of one gene) of the synthesis of α- (α-thalassaemia) or β-globin (β-thalassaemia) chains from one allele. Mutations of TFs regulating globin expression have also been observed. These mutations in *cis* and *trans* have been reviewed by (Fucharoen and Viprakasit 2009; Higgs et al. 2012; Higgs et al. 2005). Sickle-cell anaemia is another inherited disorder whereby a mutation occurs in the coding sequence of the β-globin gene and this changes the solubility of the protein (Haemoglobin HbS). Under hypoxic conditions, deoxy HbS molecules polymerise inside the cells, forming rigid, sickled cells (Kutlar 2007).

Observations made in a group of patients with a disorder called hereditary persistence of foetal haemoglobin (HPFH), where the levels of foetal haemoglobin (HbF) are high, raised the idea that reactivating the silenced globin genes would be an elegant strategy for the treatment of β thalassaemia and SCA (Forget 1998; Thein and Menzel 2009). In SCA, HbF not only dilutes HbS, thereby decreasing the molecular contact and polymerisation for HbS, but it also inhibits the latter event through formation of $\alpha_2\beta^S\gamma$ hybrids (Noguchi et al. 1988). This has launched intensive research into globin transcription regulation and testing a variety of compounds to increase the production of HbF. The reactivation of the adult δ genes never generated a centre of interest, probably because its normal level of expression is very low. However it would be an interesting challenge to achieve its overexpression. Any attempts to accomplish the reactivation of embryonic ζ and ε genes may be even more problematic since these are originally expressed in cells from a different origin (Palis 2008; Sankaran et al. 2010). For example, these embryonic genes might require additional tissue-specific TFs. However they seem to also share some TFs in common and use the same set of enhancers (Palstra et al. 2003).

Many promising inducers of HbF have been tested, but with limited success due to weak effects, lack of specificity, cytotoxicity and mutagenicity (Bianchi et al. 2009; Musallam et al. 2013). Importantly, up until now these compounds have been tested on a few target genes only, and therefore without taking into account the genome-wide effect in the cells tested. Nowadays, genome-wide testing of effects is made possible with the use of high-throughput sequencing technology. It is also important to mention that these drugs aim to specifically increase levels of an alternative β-like globin chain to complement the affected β-globin gene. For this reason it is important that these inducers don't also increase any of the α-like chains, which would maintain an imbalance in α- and β-globin chains expression. For example, both loci (ζ and γ genes) are targeted by the repressive H3K27me3 mark (Sankaran et al. 2011; Vernimmen et al. 2011). Therefore, the new compound directed against the H3K27 methyltransferase EZH2 (McCabe et al. 2012) would be an inappropriate approach. Among the inducers of HbF, some have an effect on epigenetic regulators, others have various effect, not only on transcription but also probably on mRNA processing and stability, etc. Since the early observation that DNA methylation was involved in the silencing of the human γ-globins, inhibitors such as azacitidine and decitabine (see above) have been tested for the treatment of β-thalassaemia (Musallam et al. 2013). The only current therapy for SCA and β-thalassaemia is treatment with hydroxyurea (HU). HU induces foetal haemoglobin (HbF) synthesis in some patients (~50 %), but the mechanisms of action of HU are largely unknown.

In adult cells, the silenced foetal γ-globin gene is repressed by different protein complexes. Among these, DNMT3A/PRMT5, DRED (direct repeat erythroid definitive, comprising TR2, TR4 and DNMT1), CoREST and IKAROS have been found to be associated with BCL11A (Bottardi et al. 2009; Sankaran et al. 2008; Xu et al. 2013). DRED and CoREST are also composed of LSD1, a H3K4me2 demethylase found at many enhancers and associated with H3K4me1 (Whyte et al. 2012). DNMT3A (see above) is involved in methylation of the γ-globin gene promoter and is recruited by the arginine methyltransferase PRMT5 (Rank et al. 2010). Therefore, inhibitors against these enzymes would be good drug candidates for restoration of HbF expression in patients. In fact, a recent study using a specific inhibitor for LSD1 called tranylcypromine (TCP) (Binda et al. 2010) gave promising results in increasing HbF to therapeutic levels, in a dose-dependent manner (Shi et al. 2013) (Fig. 10.2b). Moreover, the toxic accumulation of free α chains observed in β thalassaemias could be neutralised by overexpression of the alpha-haemoglobin-stabilising protein (AHSP), which is an erythroid-specific protein that acts as a molecular chaperone for the free α chains of haemoglobin. The mechanisms of epigenetic regulation of this chaperone could also open new horizons for epigenetic therapy in thalassaemia (Khandros et al. 2012).

10.6 Conclusions

Since the original observation made by Alfred Mirsky about half a century ago (Allfrey et al. 1964; Allfrey and Mirsky 1964; Black et al. 2012), the last decade has led to the identification of numerous enzymes required for transcription regulation, progression through the cell cycle and differentiation (Black et al. 2012). Most of these enzymes seem to be mutated in many different diseases including haematological disorders (Dawson and Kouzarides 2012). At the moment, none of these alterations have been shown to be involved in any form of thalassaemia, but the use of epigenetic signatures allowed us to identify a new molecular mechanism for thalassaemia (De Gobbi et al. 2006). Furthermore, the manipulation of the enzymatic activity of these epigenetic regulators has launched the development of new therapeutic approaches.

Previously, the ENCODE project has been a tremendous input in the field with the annotation of epigenetic marks and the recruitment of some associated enzymes (Dunham et al. 2012). However, most of these studies have been done in cell lines that might not entirely reflect primary cells (see comment above regarding K562). More recently, the BLUEPRINT consortium has been initiated to develop such studies by using human primary cells. The BLUEPRINT consortium aims to further the understanding of how genes are activated or repressed using distinct types of primary haematopoietic cells from healthy individuals and from their malignant leukaemic counterparts. This should build a full picture of the dynamics of epigenetic changes during haematopoiesis and mechanistic insights should emerge. However, to interpret these studies, other technical issues will have to be considered such as the fact that epigenetic regulators are also differentially expressed through the cell cycle (Black et al. 2012). The combination of epigenetic modifications found at a given locus might not coexist, but would reflect a combination of cells at a different time of the cell cycle (reviewed in Black et al. 2012; Dawson and Kouzarides 2012).

To date, scientists have been able to address which specific epigenetic modification complexes are linked to the chromatin state associated with activation and repression. However, we still don't know how epigenetics controls and determines cell fate ("The epigenetic programme"). The current model is that in multipotent stem cells, all developmentally regulated genes are primed by multipotent TFs, so-called pioneer TFs, bookmarking the genome to maintain this priming for gene expression after cell division (Caravaca et al. 2013; Kadauke and Blobel 2013; Kadauke et al. 2012; Rada-Iglesias 2013; Zaret and Carroll 2011). However, it still needs to be determined whether these TFs recruit the epigenetic regulators or whether inherited (and therefore pre-established) epigenetic marks (or histone variants) recruit these pioneer TFs. A decade ago, single-cell RNA PCR studies found promiscuous expression of myeloid, but not lymphoid, lineage-restricted genes and markers in HSCs and CMPs suggesting that myeloid priming precedes lymphoid priming (Hu et al. 1997; Miyamoto et al. 2002). This would suggest that other factors influence the timing towards a given lineage commitment. What is the role of CpG islands in this process? Originally it was believed that CpG islands on housekeeping genes had the intrinsic ability to keep their promoters accessible in any cell type, allowing basal transcription to occur. However, we now know that CpG islands are also found in the promoters of developmentally regulated genes which are not always active. For this group of CpG island promoters, Polycomb group Proteins (PcG) act to block the promoter accessibility in inappropriate lineages or in at other stages during the differentiation programme. CpG islands have been found to be directly involved not only in the recruitment of repressive PcG complexes (Lynch et al. 2012; Mendenhall et al. 2010) but also in the recruitment of MLL complexes through CXXC domain proteins such as Cfp1 (Clouaire et al. 2012; Thomson et al. 2010). Nevertheless, it is not at all clear why developmentally regulated genes should have CpG islands in the first instance? The mouse α- and β- globin genes and the human β-globin gene don't have any CpG islands in their promoters and these are perfectly well regulated during erythropoiesis.

Although the epigenetic machinery might be impaired in some genetic diseases and therefore a "druggable" target, SCA and thalassaemia involve only a dysfunctional globin gene, which doesn't affect the epigenetic balance of the cell. Therefore, how can ubiquitously expressed epigenetic regulators serve as selective targets? The answer may lie in the fact that some epigenetic components control a small number of genes instead of having global effects on gene expression (are all peaks observed in genome-wide studies functional?). However, even if successful in tissue culture, inhibitors could act on a different population of genes in other cell types once in human body. Although it seems that low doses of these agents would be the key, the blood still has the unique advantage of being a liquid tissue and it is therefore worth considering an ex vivo approach, as in regenerative medicine: treat the cells in vitro with the right epigenetic drug and then reintroduce these cells back into the organism. This would surely avoid toxicity in the whole patient.

Acknowledgements I would like to thank Mattia Frontini and Ailbhe Brazel for critically reviewing the manuscript. The Vernimmen lab benefits funding from by the British Society for Haematology (BSH), the Lady Tata Memorial Trust and the Roslin Foundation. Douglas Vernimmen is supported by a Chancellor's Fellowship at The University of Edinburgh. The Roslin Institute receives Institute Strategic Grant funding from the BBSRC.

References

Aigner B, Renner S, Kessler B, Klymiuk N, Kurome M, Wunsch A, Wolf E (2010) Transgenic pigs as models for translational biomedical research. J Mol Med (Berl) 88:653–664

Allfrey VG, Mirsky AE (1964) Structural modifications of histones and their possible role in the regulation of RNA synthesis. Science 144:559

Allfrey VG, Faulkner R, Mirsky AE (1964) Acetylation and methylation of histones and their possible role in the regulation of Rna synthesis. Proc Natl Acad Sci USA 51:786–794

Anguita E, Hughes J, Heyworth C, Blobel GA, Wood WG, Higgs DR (2004) Globin gene activation during haemopoiesis is driven by protein complexes nucleated by GATA-1 and GATA-2. EMBO J 23:2841–2852

Atallah E, Verstovsek S (2009) Prospect of JAK2 inhibitor therapy in myeloproliferative neoplasms. Expert Rev Anticancer Ther 9:663–670

Bernstein BE, Mikkelsen TS, Xie X, Kamal M, Huebert DJ, Cuff J, Fry B, Meissner A, Wernig M, Plath K, Jaenisch R, Wagschal A, Feil R, Schreiber SL, Lander ES (2006) A bivalent chromatin structure marks key developmental genes in embryonic stem cells. Cell 125:315–326

Bianchi N, Zuccato C, Lampronti I, Borgatti M, Gambari R (2009) Fetal hemoglobin inducers from the natural world: a novel approach for identification of drugs for the treatment of {beta}-thalassemia and sickle-cell anemia. Evid Based Complement Alternat Med 6:141–151

Binda C, Valente S, Romanenghi M, Pilotto S, Cirilli R, Karytinos A, Ciossani G, Botrugno OA, Forneris F, Tardugno M, Edmondson DE, Minucci S, Mattevi A, Mai A (2010) Biochemical, structural, and biological evaluation of tranylcypromine derivatives as inhibitors of histone demethylases LSD1 and LSD2. J Am Chem Soc 132:6827–6833

Black JC, Van Rechem C, Whetstine JR (2012) Histone lysine methylation dynamics: establishment, regulation, and biological impact. Mol Cell 48:491–507

Bottardi S, Ross J, Bourgoin V, Fotouhi-Ardakani N, el Affar B, Trudel M, Milot E (2009) Ikaros and GATA-1 combinatorial effect is required for silencing of human gamma-globin genes. Mol Cell Biol 29:1526–1537

Cantor AB, Orkin SH (2002) Transcriptional regulation of erythropoiesis: an affair involving multiple partners. Oncogene 21:3368–3376

Caravaca JM, Donahue G, Becker JS, He X, Vinson C, Zaret KS (2013) Bookmarking by specific and nonspecific binding of FoxA1 pioneer factor to mitotic chromosomes. Genes Dev 27:251–260

Chagraoui H, Porcher C (2012) Establishment of an ES cell-derived murine megakaryocytic cell line, MKD1, with features of primary megakaryocyte progenitors. PLoS One 7:e32981

Chen AT, Prchal JT (2010) JAK2 kinase inhibitors and myeloproliferative disorders. Curr Opin Hematol 17:110–116

Chen Q, Lu Z, Jin Y, Wu Y, Pan J (2010) Triptolide inhibits Jak2 transcription and induces apoptosis in human myeloproliferative disorder cells bearing Jak2V617F through caspase-3-mediated cleavage of Mcl-1. Cancer Lett 291:246–255

Cimmino L, Abdel-Wahab O, Levine RL, Aifantis I (2011) TET family proteins and their role in stem cell differentiation and transformation. Cell Stem Cell 9:193–204

Clouaire T, Webb S, Skene P, Illingworth R, Kerr A, Andrews R, Lee JH, Skalnik D, Bird A (2012) Cfp1 integrates both CpG content and gene activity for accurate H3K4me3 deposition in embryonic stem cells. Genes Dev 26:1714–1728

Cortin V, Garnier A, Pineault N, Lemieux R, Boyer L, Proulx C (2005) Efficient in vitro megakaryocyte maturation using cytokine cocktails optimized by statistical experimental design. Exp Hematol 33:1182–1191

Craddock CF, Vyas P, Sharpe JA, Ayyub H, Wood WG, Higgs DR (1995) Contrasting effects of alpha and beta globin regulatory elements on chromatin structure may be related to their different chromosomal environments. EMBO J 14:1718–1726

Dawson MA, Kouzarides T (2012) Cancer epigenetics: from mechanism to therapy. Cell 150:12–27

Dawson MA, Bannister AJ, Gottgens B, Foster SD, Bartke T, Green AR, Kouzarides T (2009) JAK2 phosphorylates histone H3Y41 and excludes HP1alpha from chromatin. Nature 461:819–822

De Gobbi M, Viprakasit V, Hughes JR, Fisher C, Buckle VJ, Ayyub H, Gibbons RJ, Vernimmen D, Yoshinaga Y, de Jong P, Cheng JF, Rubin EM, Wood WG, Bowden D, Higgs DR (2006) A regulatory SNP causes a human genetic disease by creating a new transcriptional promoter. Science 312:1215–1217

De Gobbi M, Anguita E, Hughes J, Sloane-Stanley JA, Sharpe JA, Koch CM, Dunham I, Gibbons RJ, Wood WG, Higgs DR (2007) Tissue-specific histone modification and transcription factor binding in alpha globin gene expression. Blood 110:4503–4510

De Gobbi M, Garrick D, Lynch M, Vernimmen D, Hughes JR, Goardon N, Luc S, Lower KM, Sloane-Stanley JA, Pina C, Soneji S, Renella R, Enver T, Taylor S, Jacobsen SE, Vyas P, Gibbons RJ, Higgs DR (2011) Generation of bivalent chromatin domains during cell fate decisions. Epigenetics Chromatin 4:9

Deisseroth A, Hendrick D (1978) Human alpha-globin gene expression following chromosomal dependent gene transfer into mouse erythroleukemia cells. Cell 15:55–63

Delhommeau F, Dupont S, Della Valle V, James C, Trannoy S, Masse A, Kosmider O, Le Couedic JP, Robert F, Alberdi A, Lecluse Y, Plo I, Dreyfus FJ, Marzac C, Casadevall N, Lacombe C, Romana SP, Dessen P, Soulier J, Viguie F, Fontenay M, Vainchenker W, Bernard OA (2009) Mutation in TET2 in myeloid cancers. N Engl J Med 360:2289–2301

Deneberg S, Guardiola P, Lennartsson A, Qu Y, Gaidzik V, Blanchet O, Karimi M, Bengtzen S, Nahi H, Uggla B, Tidefelt U, Hoglund M, Paul C, Ekwall K, Dohner K, Lehmann S (2011) Prognostic DNA methylation patterns in cytogenetically normal acute myeloid leukemia are predefined by stem cell chromatin marks. Blood 118:5573–5582

Dey A, Seshasayee D, Noubade R, French DM, Liu J, Chaurushiya MS, Kirkpatrick DS, Pham VC, Lill JR, Bakalarski CE, Wu J, Phu L, Katavolos P, LaFave LM, Abdel-Wahab O, Modrusan Z, Seshagiri S, Dong K, Lin Z, Balazs M, Suriben R, Newton K, Hymowitz S, Garcia-Manero G, Martin F, Levine RL, Dixit VM (2012) Loss of the tumor suppressor BAP1 causes myeloid transformation. Science 337:1541–1546

Dolznig H, Boulme F, Stangl K, Deiner EM, Mikulits W, Beug H, Mullner EW (2001) Establishment of normal, terminally differentiating mouse erythroid progenitors: molecular characterization by cDNA arrays. FASEB J 15:1442–1444

Dunham I, Kundaje A, Aldred SF, Collins PJ, Davis CA, Doyle F, Epstein CB, Frietze S, Harrow J, Kaul R, Khatun J, Lajoie BR, Landt SG, Lee BK, Pauli F, Rosenbloom KR, Sabo P, Safi A, Sanyal A, Shoresh N, Simon JM, Song L, Trinklein ND, Altshuler RC, Birney E, Brown JB, Cheng C, Djebali S, Dong X, Ernst J, Furey TS, Gerstein M, Giardine B, Greven M, Hardison RC, Harris RS, Herrero J, Hoffman MM, Iyer S, Kelllis M, Kheradpour P, Lassmann T, Li Q,

Lin X, Marinov GK, Merkel A, Mortazavi A, Parker SC, Reddy TE, Rozowsky J, Schlesinger F, Thurman RE, Wang J, Ward LD, Whitfield TW, Wilder SP, Wu W, Xi HS, Yip KY, Zhuang J, Bernstein BE, Green ED, Gunter C, Snyder M, Pazin MJ, Lowdon RF, Dillon LA, Adams LB, Kelly CJ, Zhang J, Wexler JR, Good PJ, Feingold EA, Crawford GE, Dekker J, Elinitski L, Farnham PJ, Giddings MC, Gingeras TR, Guigo R, Hubbard TJ, Kellis M, Kent WJ, Lieb JD, Margulies EH, Myers RM, Starnatoyannopoulos JA, Tennebaum SA, Weng Z, White KP, Wold B, Yu Y, Wrobel J, Risk BA, Gunawardena HP, Kuiper HC, Maier CW, Xie L, Chen X, Mikkelsen TS et al (2012) An integrated encyclopedia of DNA elements in the human genome. Nature 489:57–74

Fenaux P, Mufti GJ, Hellstrom-Lindberg E, Santini V, Finelli C, Giagounidis A, Schoch R, Gattermann N, Sanz G, List A, Gore SD, Seymour JF, Bennett JM, Byrd J, Backstrom J, Zimmerman L, McKenzie D, Beach C, Silverman LR (2009) Efficacy of azacitidine compared with that of conventional care regimens in the treatment of higher-risk myelodysplastic syndromes: a randomised, open-label, phase III study. Lancet Oncol 10:223–232

Forget BG (1998) Molecular basis of hereditary persistence of fetal hemoglobin. Ann N Y Acad Sci 850:38–44

Forrester WC, Epner E, Driscoll MC, Enver T, Brice M, Papayannopoulou T, Groudine M (1990) A deletion of the human beta-globin locus activation region causes a major alteration in chromatin structure and replication across the entire beta-globin locus. Genes Dev 4:1637–1649

Fucharoen S, Viprakasit V (2009) Hb H disease: clinical course and disease modifiers. Hematology Am Soc Hematol Educ Program 26–34

Fugman DA, Witte DP, Jones CL, Aronow BJ, Lieberman MA (1990) In vitro establishment and characterization of a human megakaryoblastic cell line. Blood 75:1252–1261

Fujimoto TT, Kohata S, Suzuki H, Miyazaki H, Fujimura K (2003) Production of functional platelets by differentiated embryonic stem (ES) cells in vitro. Blood 102:4044–4051

Garrick D, De Gobbi M, Samara V, Rugless M, Holland M, Ayyub H, Lower K, Sloane-Stanley J, Gray N, Koch C, Dunham I, Higgs DR (2008) The role of the polycomb complex in silencing alpha-globin gene expression in nonerythroid cells. Blood 112:3889–3899

Gaur M, Kamata T, Wang S, Moran B, Shattil SJ, Leavitt AD (2006) Megakaryocytes derived from human embryonic stem cells: a genetically tractable system to study megakaryocytopoiesis and integrin function. J Thromb Haemost 4:436–442

Gieger C, Radhakrishnan A, Cvejic A, Tang W, Porcu E, Pistis G, Serbanovic-Canic J, Elling U, Goodall AH, Labrune Y, Lopez LM, Magi R, Meacham S, Okada Y, Pirastu N, Sorice R, Teumer A, Voss K, Zhang W, Ramirez-Solis R, Bis JC, Ellinghaus D, Gogele M, Hottenga JJ, Langenberg C, Kovacs P, O'Reilly PF, Shin SY, Esko T, Hartiala J, Kanoni S, Murgia F, Parsa A, Stephens J, van der Harst P, Ellen van der Schoot C, Allayee H, Attwood A, Balkau B, Bastardot F, Basu S, Baumeister SE, Biino G, Bomba L, Bonnefond A, Cambien F, Chambers JC, Cucca F, D'Adamo P, Davies G, de Boer RA, de Geus EJ, Doring A, Elliott P, Erdmann J, Evans DM, Falchi M, Feng W, Folsom AR, Frazer IH, Gibson QD, Glazer NL, Hammond C, Hartikainen AL, Heckbert SR, Hengstenberg C, Hersch M, Illig T, Loos RJ, Jolley J, Khaw KT, Kuhnel B, Kyrtsonis MC, Lagou V, Lloyd-Jones H, Lumley T, Mangino M, Maschio A, Mateo Leach I, McKnight B, Memari Y, Mitchell BD, Montgomery GW, Nakamura Y, Nauck M, Navis G, Nothlings U, Nolte IM, Porteous DJ, Pouta A, Pramstaller PP, Pullat J, Ring SM, Rotter JI, Ruggiero D, Ruokonen A, Sala C, Samani NJ, Sambrook J, Schlessinger D et al (2011) New gene functions in megakaryopoiesis and platelet formation. Nature 480:201–208

Goldstein AM (2011) Germline BAP1 mutations and tumor susceptibility. Nat Genet 43:925–926

Gore SD (2011) New ways to use DNA methyltransferase inhibitors for the treatment of myelodysplastic syndrome. Hematology Am Soc Hematol Educ Program 2011:550–555

Grass JA, Boyer ME, Pal S, Wu J, Weiss MJ, Bresnick EH (2003) GATA-1-dependent transcriptional repression of GATA-2 via disruption of positive autoregulation and domain-wide chromatin remodeling. Proc Natl Acad Sci U S A 100:8811–8816

Higgs DR, Garrick D, Anguita E, De Gobbi M, Hughes J, Muers M, Vernimmen D, Lower K, Law M, Argentaro A, Deville MA, Gibbons R (2005) Understanding alpha-globin gene regulation: aiming to improve the management of thalassemia. Ann N Y Acad Sci 1054:92–102

Higgs DR, Vernimmen D, Wood B (2008) Long-range regulation of alpha-globin gene expression. Adv Genet 61:143–173

Higgs DR, Engel JD, Stamatoyannopoulos G (2012) Thalassaemia. Lancet 379:373–383

Houston IB, Huang KJ, Jennings SR, DeKoter RP (2007) PU.1 immortalizes hematopoietic progenitors in a GM-CSF-dependent manner. Exp Hematol 35:374–384

Hu M, Krause D, Greaves M, Sharkis S, Dexter M, Heyworth C, Enver T (1997) Multilineage gene expression precedes commitment in the hemopoietic system. Genes Dev 11:774–785

Huntly BJ, Shigematsu H, Deguchi K, Lee BH, Mizuno S, Duclos N, Rowan R, Amaral S, Curley D, Williams IR, Akashi K, Gilliland DG (2004) MOZ-TIF2, but not BCR-ABL, confers properties of leukemic stem cells to committed murine hematopoietic progenitors. Cancer Cell 6:587–596

Iliou MS, Lujambio A, Portela A, Brustle O, Koch P, Andersson-Vincent PH, Sundstrom E, Hovatta O, Esteller M (2011) Bivalent histone modifications in stem cells poise miRNA loci for CpG island hypermethylation in human cancer. Epigenetics 6:1344–1353

Ishida Y, Levin J, Baker G, Stenberg PE, Yamada Y, Sasaki H, Inoue T (1993) Biological and biochemical characteristics of murine megakaryoblastic cell line L8057. Exp Hematol 21:289–298

Jensen DE, Proctor M, Marquis ST, Gardner HP, Ha SI, Chodosh LA, Ishov AM, Tommerup N, Vissing H, Sekido Y, Minna J, Borodovsky A, Schultz DC, Wilkinson KD, Maul GG, Barlev N, Berger SL, Prendergast GC, Rauscher FJ 3rd (1998) BAP1: a novel ubiquitin hydrolase which binds to the BRCA1 RING finger and enhances BRCA1-mediated cell growth suppression. Oncogene 16:1097–1112

Kadauke S, Blobel GA (2013) Mitotic bookmarking by transcription factors. Epigenetics Chromatin 6:6

Kadauke S, Udugama MI, Pawlicki JM, Achtman JC, Jain DP, Cheng Y, Hardison RC, Blobel GA (2012) Tissue-specific mitotic bookmarking by hematopoietic transcription factor GATA1. Cell 150:725–737

Khandros E, Thom CS, D'Souza J, Weiss MJ (2012) Integrated protein quality-control pathways regulate free alpha-globin in murine beta-thalassemia. Blood 119:5265–5275

Ko M, Rao A (2011) TET2: epigenetic safeguard for HSC. Blood 118:4501–4503

Ko M, Bandukwala HS, An J, Lamperti ED, Thompson EC, Hastie R, Tsangaratou A, Rajewsky K, Koralov SB, Rao A (2011) Ten-Eleven-Translocation 2 (TET2) negatively regulates homeostasis and differentiation of hematopoietic stem cells in mice. Proc Natl Acad Sci USA 108:14566–14571

Komatsu N, Yamamoto M, Fujita H, Miwa A, Hatake K, Endo T, Okano H, Katsube T, Fukumaki Y, Sassa S et al (1993) Establishment and characterization of an erythropoietin-dependent subline, UT-7/Epo, derived from human leukemia cell line, UT-7. Blood 82:456–464

Komatsu N, Kunitama M, Yamada M, Hagiwara T, Kato T, Miyazaki H, Eguchi M, Yamamoto M, Miura Y (1996) Establishment and characterization of the thrombopoietin-dependent megakaryocytic cell line, UT-7/TPO. Blood 87:4552–4560

Kruidenier L, Chung CW, Cheng Z, Liddle J, Che K, Joberty G, Bantscheff M, Bountra C, Bridges A, Diallo H, Eberhard D, Hutchinson S, Jones E, Katso R, Leveridge M, Mander PK, Mosley J, Ramirez-Molina C, Rowland P, Schofield CJ, Sheppard RJ, Smith JE, Swales C, Tanner R, Thomas P, Tumber A, Drewes G, Oppermann U, Patel DJ, Lee K, Wilson DM (2012) A selective jumonji H3K27 demethylase inhibitor modulates the proinflammatory macrophage response. Nature 488:404–408

Kutlar A (2007) Sickle cell disease: a multigenic perspective of a single gene disorder. Hemoglobin 31:209–224

Langemeijer SM, Kuiper RP, Berends M, Knops R, Aslanyan MG, Massop M, Stevens-Linders E, van Hoogen P, van Kessel AG, Raymakers RA, Kamping EJ, Verhoef GE, Verburgh E, Hagemeijer A, Vandenberghe P, de Witte T, van der Reijden BA, Jansen JH (2009) Acquired mutations in TET2 are common in myelodysplastic syndromes. Nat Genet 41:838–842

Ley TJ, Ding L, Walter MJ, McLellan MD, Lamprecht T, Larson DE, Kandoth C, Payton JE, Baty J, Welch J, Harris CC, Lichti CF, Townsend RR, Fulton RS, Dooling DJ, Koboldt DC, Schmidt H, Zhang Q, Osborne JR, Lin L, O'Laughlin M, McMichael JF, Delehaunty KD, McGrath SD, Fulton LA, Magrini VJ, Vickery TL, Hundal J, Cook LL, Conyers JJ, Swift GW, Reed JP, Alldredge PA, Wylie T, Walker J, Kalicki J, Watson MA, Heath S, Shannon WD, Varghese N, Nagarajan R, Westervelt P, Tomasson MH, Link DC, Graubert TA, DiPersio JF, Mardis ER, Wilson RK (2010) DNMT3A mutations in acute myeloid leukemia. N Engl J Med 363:2424–2433

Lorsbach RB, Moore J, Mathew S, Raimondi SC, Mukatira ST, Downing JR (2003) TET1, a member of a novel protein family, is fused to MLL in acute myeloid leukemia containing the t (10;11)(q22;q23). Leukemia 17:637–641

Lozzio CB, Lozzio BB (1975) Human chronic myelogenous leukemia cell-line with positive Philadelphia chromosome. Blood 45:321–334

Lynch MD, Smith AJ, De Gobbi M, Flenley M, Hughes JR, Vernimmen D, Ayyub H, Sharpe JA, Sloane-Stanley JA, Sutherland L, Meek S, Burdon T, Gibbons RJ, Garrick D, Higgs DR (2012) An interspecies analysis reveals a key role for unmethylated CpG dinucleotides in vertebrate Polycomb complex recruitment. EMBO J 31(2):317–329

Marks PA, Sheffery M, Rifkind RA (1987) Induction of transformed cells to terminal differentiation and the modulation of gene expression. Cancer Res 47:659–666

McCabe MT, Ott HM, Ganji G, Korenchuk S, Thompson C, Van Aller GS, Liu Y, Graves AP, Della Pietra A 3rd, Diaz E, LaFrance LV, Mellinger M, Duquenne C, Tian X, Kruger RG, McHugh CF, Brandt M, Miller WH, Dhanak D, Verma SK, Tummino PJ, Creasy CL (2012) EZH2 inhibition as a therapeutic strategy for lymphoma with EZH2-activating mutations. Nature 492:108–112

Mendenhall EM, Koche RP, Truong T, Zhou VW, Issac B, Chi AS, Ku M, Bernstein BE (2010) GC-rich sequence elements recruit PRC2 in mammalian ES cells. PLoS Genet 6:e1001244

Miyamoto T, Iwasaki H, Reizis B, Ye M, Graf T, Weissman IL, Akashi K (2002) Myeloid or lymphoid promiscuity as a critical step in hematopoietic lineage commitment. Dev Cell 3:137–147

Musallam KM, Taher AT, Cappellini MD, Sankaran VG (2013) Clinical experience with fetal hemoglobin induction therapy in patients with beta-thalassemia. Blood 121:2199–2212, quiz 2372

Ney PA, D'Andrea AD (2000) Friend erythroleukemia revisited. Blood 96:3675–3680

Noguchi CT, Rodgers GP, Serjeant G, Schechter AN (1988) Levels of fetal hemoglobin necessary for treatment of sickle cell disease. N Engl J Med 318:96–99

Oakley K, Han Y, Vishwakarma BA, Chu S, Bhatia R, Gudmundsson KO, Keller J, Chen X, Vasko V, Jenkins NA, Copeland NG, Du Y (2012) Setbp1 promotes the self-renewal of murine myeloid progenitors via activation of Hoxa9 and Hoxa10. Blood 119:6099–6108

Ono Y, Wang Y, Suzuki H, Okamoto S, Ikeda Y, Murata M, Poncz M, Matsubara Y (2012) Induction of functional platelets from mouse and human fibroblasts by p45NF-E2/Maf. Blood 120:3812–3821

Orkin SH (2000) Diversification of haematopoietic stem cells to specific lineages. Nat Rev Genet 1:57–64

Palis J (2008) Ontogeny of erythropoiesis. Curr Opin Hematol 15:155–161

Palstra RJ, Tolhuis B, Splinter E, Nijmeijer R, Grosveld F, de Laat W (2003) The beta-globin nuclear compartment in development and erythroid differentiation. Nat Genet 35:190–194

Pencovich N, Jaschek R, Tanay A, Groner Y (2011) Dynamic combinatorial interactions of RUNX1 and cooperating partners regulates megakaryocytic differentiation in cell line models. Blood 117:e1–e14

Pope SH, Fibach E, Sun J, Chin K, Rodgers GP (2000) Two-phase liquid culture system models normal human adult erythropoiesis at the molecular level. Eur J Haematol 64:292–303

Prather RS (2013) Pig genomics for biomedicine. Nat Biotechnol 31:122–124

Pronk CJ, Rossi DJ, Mansson R, Attema JL, Norddahl GL, Chan CK, Sigvardsson M, Weissman IL, Bryder D (2007) Elucidation of the phenotypic, functional, and molecular topography of a myeloerythroid progenitor cell hierarchy. Cell Stem Cell 1:428–442

Rada-Iglesias A (2013) Pioneering barren land: mitotic bookmarking by transcription factors. Dev Cell 24:342–344

Rakyan VK, Down TA, Maslau S, Andrew T, Yang TP, Beyan H, Whittaker P, McCann OT, Finer S, Valdes AM, Leslie RD, Deloukas P, Spector TD (2010) Human aging-associated DNA hypermethylation occurs preferentially at bivalent chromatin domains. Genome Res 20:434–439

Rank G, Cerruti L, Simpson RJ, Moritz RL, Jane SM, Zhao Q (2010) Identification of a PRMT5-dependent repressor complex linked to silencing of human fetal globin gene expression. Blood 116:1585–1592

Rodriguez J, Munoz M, Vives L, Frangou CG, Groudine M, Peinado MA (2008) Bivalent domains enforce transcriptional memory of DNA methylated genes in cancer cells. Proc Natl Acad Sci U S A 105:19809–19814

Rutherford T, Clegg JB, Higgs DR, Jones RW, Thompson J, Weatherall DJ (1981) Embryonic erythroid differentiation in the human leukemic cell line K562. Proc Natl Acad Sci USA 78:348–352

Saito H (1997) Megakaryocytic cell lines. Baillieres Clin Haematol 10:47–63

Sankaran VG, Menne TF, Xu J, Akie TE, Lettre G, Van Handel B, Mikkola HK, Hirschhorn JN, Cantor AB, Orkin SH (2008) Human fetal hemoglobin expression is regulated by the developmental stage-specific repressor BCL11A. Science 322:1839–1842

Sankaran VG, Xu J, Orkin SH (2010) Advances in the understanding of haemoglobin switching. Br J Haematol 149:181–194

Sankaran VG, Xu J, Byron R, Greisman HA, Fisher C, Weatherall DJ, Sabath DE, Groudine M, Orkin SH, Premawardhena A, Bender MA (2011) A functional element necessary for fetal hemoglobin silencing. N Engl J Med 365:807–814

Scheuermann JC, de Ayala Alonso AG, Oktaba K, Ly-Hartig N, McGinty RK, Fraterman S, Wilm M, Muir TW, Muller J (2010) Histone H2A deubiquitinase activity of the Polycomb repressive complex PR-DUB. Nature 465:243–247

Shi L, Cui S, Engel JD, Tanabe O (2013) Lysine-specific demethylase 1 is a therapeutic target for fetal hemoglobin induction. Nat Med 19:291–294

Shih AH, Abdel-Wahab O, Patel JP, Levine RL (2012) The role of mutations in epigenetic regulators in myeloid malignancies. Nat Rev Cancer 12:599–612

Song CX, Szulwach KE, Dai Q, Fu Y, Mao SQ, Lin L, Street C, Li Y, Poidevin M, Wu H, Gao J, Liu P, Li L, Xu GL, Jin P, He C (2013) Genome-wide profiling of 5-formylcytosine reveals its roles in epigenetic priming. Cell 153:678–691

Spivak JL, Toretti D, Dickerman HW (1973) Effect of phenylhydrazine-induced hemolytic anemia on nuclear RNA polymerase activity of the mouse spleen. Blood 42:257–266

Spooncer E, Heyworth CM, Dunn A, Dexter TM (1986) Self-renewal and differentiation of interleukin-3-dependent multipotent stem cells are modulated by stromal cells and serum factors. Differentiation 31:111–118

Stroud H, Feng S, Morey Kinney S, Pradhan S, Jacobsen SE (2011) 5-Hydroxymethylcytosine is associated with enhancers and gene bodies in human embryonic stem cells. Genome Biol 12: R54

Tan M, Luo H, Lee S, Jin F, Yang JS, Montellier E, Buchou T, Cheng Z, Rousseaux S, Rajagopal N, Lu Z, Ye Z, Zhu Q, Wysocka J, Ye Y, Khochbin S, Ren B, Zhao Y (2011) Identification of 67 histone marks and histone lysine crotonylation as a new type of histone modification. Cell 146:1016–1028

Thein SL, Menzel S (2009) Discovering the genetics underlying foetal haemoglobin production in adults. Br J Haematol 145:455–467

Thomson JP, Skene PJ, Selfridge J, Clouaire T, Guy J, Webb S, Kerr AR, Deaton A, Andrews R, James KD, Turner DJ, Illingworth R, Bird A (2010) CpG islands influence chromatin structure via the CpG-binding protein Cfp1. Nature 464:1082–1086

Tijssen MR, Cvejic A, Joshi A, Hannah RL, Ferreira R, Forrai A, Bellissimo DC, Oram SH, Smethurst PA, Wilson NK, Wang X, Ottersbach K, Stemple DL, Green AR, Ouwehand WH, Gottgens B (2011) Genome-wide analysis of simultaneous GATA1/2, RUNX1, FLI1, and SCL binding in megakaryocytes identifies hematopoietic regulators. Dev Cell 20:597–609

Tsai HC, Li H, Van Neste L, Cai Y, Robert C, Rassool FV, Shin JJ, Harbom KM, Beaty R, Pappou E, Harris J, Yen RW, Ahuja N, Brock MV, Stearns V, Feller-Kopman D, Yarmus LB, Lin YC, Welm AL, Issa JP, Minn I, Matsui W, Jang YY, Sharkis SJ, Baylin SB, Zahnow CA (2012) Transient low doses of DNA-demethylating agents exert durable antitumor effects on hematological and epithelial tumor cells. Cancer Cell 21:430–446

Vernimmen D, De Gobbi M, Sloane-Stanley JA, Wood WG, Higgs DR (2007) Long-range chromosomal interactions regulate the timing of the transition between poised and active gene expression. EMBO J 26:2041–2051

Vernimmen D, Lynch MD, De Gobbi M, Garrick D, Sharpe JA, Sloane-Stanley JA, Smith AJ, Higgs DR (2011) Polycomb eviction as a new distant enhancer function. Genes Dev 25:1583–1588

Verstovsek S (2009) Therapeutic potential of JAK2 inhibitors. Hematology Am Soc Hematol Educ Program 636–42

Whyte WA, Bilodeau S, Orlando DA, Hoke HA, Frampton GM, Foster CT, Cowley SM, Young RA (2012) Enhancer decommissioning by LSD1 during embryonic stem cell differentiation. Nature 482:221–225

Williams K, Christensen J, Pedersen MT, Johansen JV, Cloos PA, Rappsilber J, Helin K (2011) TET1 and hydroxymethylcytosine in transcription and DNA methylation fidelity. Nature 473:343–348

Wu H, Zhang Y (2011) Mechanisms and functions of Tet protein-mediated 5-methylcytosine oxidation. Genes Dev 25:2436–2452

Wu H, D'Alessio AC, Ito S, Xia K, Wang Z, Cui K, Zhao K, Sun YE, Zhang Y (2011) Dual functions of Tet1 in transcriptional regulation in mouse embryonic stem cells. Nature 473:389–393

Xu J, Bauer DE, Kerenyi MA, Vo TD, Hou S, Hsu YJ, Yao H, Trowbridge JJ, Mandel G, Orkin SH (2013) Corepressor-dependent silencing of fetal hemoglobin expression by BCL11A. Proc Natl Acad Sci U S A 110:6518–6523

Zaret KS, Carroll JS (2011) Pioneer transcription factors: establishing competence for gene expression. Genes Dev 25:2227–2241

Zeitlin HC, Weatherall DJ (1983) Selective expression within the human alpha globin gene complex following chromosome-dependent transfer into diploid mouse erythroleukaemia cells. Mol Biol Med 1:489–500

Chapter 11
The Molecular Basis of T Cell Development and How Epigenetic/Transcriptional Deregulation Leads to T-ALL

Will Bailis and Warren S. Pear

Abstract A wide network of transcription factors orchestrates T lineage commitment. These factors are essential regulators of differentiation, survival, and cell growth in thymocytes. In order to maintain developmental fidelity, the expression of these transcriptional factors is tightly associated with a series of developmental checkpoints that ensure the termination of aberrant differentiation. Failure to maintain this tight control can result in neoplastic transformation. This chapter will review these key transcriptional regulators and discuss their contributions to both normal and malignant thymocyte development.

Keywords Thymopoiesis • Thymocyte • T cell • Notch • TCF1 • Gata3 • Bcl11b • T-ALL

T cell development occurs in an oncogenic brew of profligate proliferation, extensive growth factor signaling, and genetic instability. The mammalian immune system is faced with the daunting challenge of generating a T cell receptor (TCR) repertoire large enough to cover every potential antigen a host will encounter in their lifetime from a limited number of progenitors. To accomplish this herculean task, the few progenitors that enter the thymus must expand by several orders of magnitude in order to generate sufficient genetic material to survive the ravages of antigen receptor recombination and TCR selection. Developing thymocytes must endure two in-frame rearrangements of the TCR β-chain (TCRβ) and an additional in-frame rearrangement of the TCR α-chain (TCRα), before being subjected to positive and negative selection, which eliminates all TCRs incapable of interacting with host-MHC and all self-reactive T cells.

W. Bailis • W.S. Pear (✉)
Department of Pathology and Laboratory Medicine, Abramson Family Cancer Research Institute, and Institute for Immunology, The Perelman School of Medicine at the University of Pennsylvania, 654 BRB II/III, 421 Curie Blvd, Philadelphia, PA 19104, USA
e-mail: wpear@mail.med.upenn.edu

C. Bonifer and P.N. Cockerill (eds.), *Transcriptional and Epigenetic Mechanisms Regulating Normal and Aberrant Blood Cell Development*, Epigenetics and Human Health, DOI 10.1007/978-3-642-45198-0_11, © Springer-Verlag Berlin Heidelberg 2014

This process of rampant proliferation interspersed between periods of DNA recombination and somatic mutation requires tight regulation in order to avoid oncogenic transformation. To this end, T cell development involves a series of checkpoints that move thymocytes from one survival program to another, such that development is intimately linked to survival signal addiction. At the earliest stages of development, thymocytes rely on a network of stem cell-like genes and growth factor signals through Kit and the IL-7R, which allow them to expand to sufficient cell numbers. Once these cells migrate to the thymic cortex and receive Notch signals, a transcriptional switch occurs that terminates the stem/progenitor program, initiates the T cell program, and establishes Notch-dependent survival. Finally, successful TCRβ rearrangement results in the formation of the pre-TCR (a complex of TCRβ and the invariant pre-TCRα), and the resulting signals extinguish Notch expression and start a survival program driven by pre-TCR and TCR signaling. Failure to properly terminate these survival programs results in the accumulation of pro-growth signals, which may ultimately tip the balance from normal to malignant development. Accordingly, acute T cell lymphoblastic leukemia (T-ALL) frequently arises from inappropriate mutations that disrupt the ordered progression of these programs, making the transcriptional control of the early stages of T cell development a critical process. This chapter will review the key transcriptional regulators that govern the balance between normal and malignant thymocyte development.

11.1 T Cell Development Overview

Although many hematopoietic progenitor cells have the capacity to become T cells, only a few migrate into the thymus (Bhandoola et al. 2007; Ng et al. 2009). Upon entry, these thymic immigrants respond to their new environment by proliferating extensively and initiating the T cell developmental program. The engagement of the T cell lineage program involves the inactivation of stem cell and progenitor genes, the activation of lineage-specific factors, and the restriction of alternate fates (Rothenberg et al. 2012). Intimately, intertwined with the transcriptional reprogramming of thymocytes, the ongoing process of TCR rearrangement provides an additional level of regulatory complexity to T cell development and can begin even before progenitors enter the thymus (Allman et al. 2003; Igarashi et al. 2002).

In this manner, T cell development can be divided into multiple stages by the initiation of the T cell transcriptional program and then several developmental checkpoints governed by the ordered rearrangement of the TCR and the selection of a thymocyte's antigen receptor against self-peptides (Fig. 11.1). These developmental stages are defined by a set of surface markers that correlate with T cell maturation. The earliest populations of thymocytes are defined by their lack of expression of the co-receptors CD4 and CD8 and thus termed double negative (DN) cells. DN cells can be further subfractionated based upon their expression of CD25, the IL-2 receptor α-chain, and CD44, a cell adhesion receptor. The first cells to enter the thymus, DN1, are defined as $CD25^{-}CD44^{+}$ and progress to $CD25^{+}CD44^{+}$ DN2 cells, $CD25^{+}CD44^{-}$ DN3 cells, and then $CD25^{-}CD44^{-}$ DN4

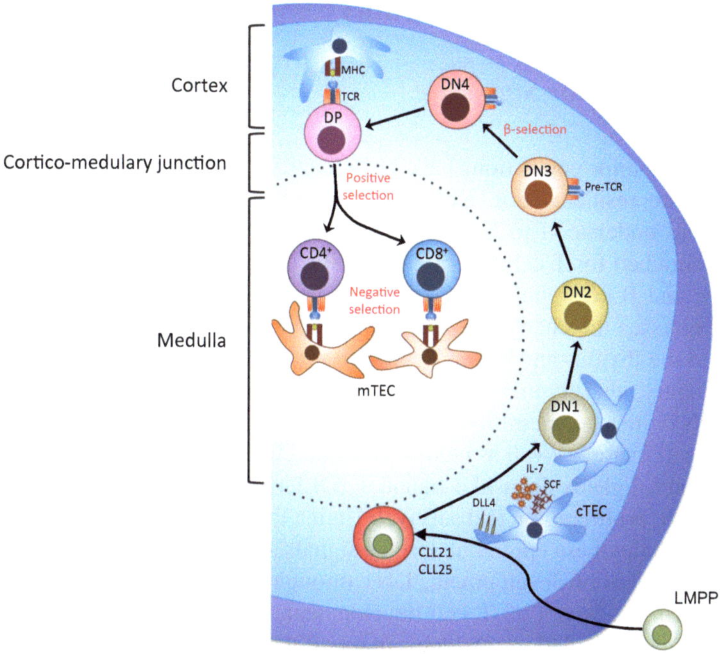

Fig. 11.1 Stages of thymocyte development. Multipotent precursors are recruited to the thymus in a CLL21- and CLL25-dependent manner and are initially maintained by IL-7 and SCF survival signals, supplied by cortical thymic epithelial cells (cTECs). These early precursors, DN1 cells, then initiate the T lineage program upon Notch ligand engagement and progress to the DN2 stage. At the DN3 stage, thymocytes begin to express the pre-TCR complex, composed of pre-TCRα and TCRβ, and are licensed to undergo β-selection and proceed to the DN4 stage. Once the TCRα locus is functionally rearranged and a mature TCR complex is expressed on the cell surface, DP thymocytes are subjected to positive selection by interacting with MHC bearing cTECs at the cortico-medulary junction. DP thymocytes then progress to either the $CD4^+$ SP or $CD8^+$ SP stage, depending on the class of MHC their TCR recognizes. Thymocyte development is completed after self-reactive SP thymocytes are eliminated by negative selection through their recognitions of self-peptides presented by medullarly thymic epithelial cells

cells. From DN1 through DN3, thymocytes rearrange the TCRβ, in a recombination activating gene (RAG) 1 and 2 dependent manner, and are held at a DN3 checkpoint that ensures functional TCRβ expression, termed β-selection. Subsequent to the DN stages of T cell development, thymocytes become double positive (DP) cells, characterized by the surface expression of CD4 and CD8. At the DP stage, thymocytes test a fully rearranged TCR against self-peptides presented on the major histocompatibility complex (MHC), such that all thymocytes are capable of recognizing self-MHC (positive selection), but do not activate upon engagement of a self-peptide:MHC complex (negative selection). T cell development in the thymus ends when thymocytes successfully undergo selection and become single positive (SP) for either CD4 or CD8, as determined by the class of MHC that a thymocyte recognizes (MHC-I for $CD8^+$ SP cells and MHC-II for $CD4^+$ SP cells).

11.2 Building the T Cell Identity

11.2.1 Progenitor Signature

Multiple progenitor populations have been identified that can develop into T cells. However, the resident bone marrow lymphoid primed multipotent progenitors (LMPP) are understood to be the primary source of T cell precursors. Though they are enriched for T cell potential, LMPP's can also give rise to myeloid cells, NK cells, and B cells, but not erythrocytes or megakaryoctyes. It is not until an LMPP expresses the chemokine receptors CCR7 and CCR9 that they are able to migrate to the thymus and begin T cell differentiation.

Upon entry, the thymic epithelium provides these recent thymic immigrants (DN1 cells) with a potent milieu of both soluble factors, such as interleukin 7 (IL-7) and stem cell factor, as well as membrane bound ligands, the most critical of which are the Notch ligands delta-like ligand 1 (DLL1) and DLL4, that support T cell development, proliferation, and survival. Until a thymocyte successfully rearranges the TCR β-chain, these epithelial derived signals sustain the T cell program during the earliest stages of development, DN1 through DN3. Though the thymic environment strongly favors αβ T cell development, other lineages can emerge from these pro-T cell populations and are supported within the thymus. The potential to differentiate into these alternate lineages is strongest at the DN1 stage and is almost completely lost by the DN3 stage.

This loss of multipotentiality is mirrored by a progressive termination of a stem cell-like transcriptional identity and the emergence of a T cell-specific transcriptional program, composed of Notch1, TCF1, GATA3, and Bcl11b. Global mapping of histones and RNAseq analysis of developing thymocytes have revealed that this progressive loss of stem/progenitor-specific gene expression results from a variety of histone transformations at multiple time points during development. These data suggest that multiple biochemically and temporally distinct mechanisms govern the transition from a stem-like program to a T cell program, rather than a single switch. Although much of the stem cell gene expression program is turned off, it is important to note that some genes sustain their expression throughout later stages of T cell development, such as Myb and Ikaros. While these sustained genes may play important regulatory roles, it has been suggested that they are unlikely to be rate-limiting factors for commitment.

11.3 A Haunting Legacy: The Stem Cell Program and T-ALL

Among the most dynamically regulated of all genes in the T lineage, Tal1, Lmo2, Lyl1, and Hhex are highly expressed early in T cell development and dramatically downregulated as thymocyte development proceeds (Fig. 11.2) (Rothenberg et al. 2012). Several of these factors are known to interact, such as Tal1 and Lmo2, and many of the stem legacy genes act as potent regulators of cell survival and proliferation in both normal and oncogenic development. While the exact role that these factors play in early T cell development remains unclear, data suggest that the termination of their expression is essential for normal developmental progression and to prevent transformation. Corroborating this, artificially sustaining expression of these factors results in a block in T cell development and the onset of leukemia.

The basic helix-loop-helix transcription (bHLH) factor Tal1 is a member of the E-box-binding protein family and functions as a heterodimer with either of the E-box proteins E2A or HEB (Hsu et al. 1994; Ono et al. 1998). Tal1 antagonizes E2A factors by converting E47 and HEB dimers from activators to repressors (Fig. 11.3a) (Herblot et al. 2000; O'Neil et al. 2004). The extent to which Tal1/E2A heterodimers occupy similar sites both early in thyomcyte development and in the context of T-ALL, where Tal1 expression is sustained, remains unclear; however, Tal1 antagonism of E2A is likely an important mechanism for Tal1-mediated T-ALL. In addition to the inhibition of E2A factors, Tal1 has the ability to synergize with members of the Lmo and GATA families, where they form discrete multi-protein complexes in concert with Lbd1, and bind cooperatively to E box and GATA motifs (Fig. 11.3) (Meier et al. 2006; Soler et al. 2010). These Tal1/GATA complexes function in diverse hemopoietic and endothelial lineages, with GATA3 the dominant GATA factor in T lineage cells (Coma et al. 2013; Ono et al. 1998; Wadman et al. 1997; Wilson et al. 2010). Assembly of these complexes can involve either Lmo1 or Lmo2 (Ono et al. 1998), but requires TAL, as E2A homodimers cannot interact with Lmo/GATA complexes (Wadman et al. 1997). While Tal1 and Gata3 expression do not normally coincide in thymocyte development, enduring Tal1 expression permits novel complex formation and likely promotes T-ALL progression (Fig. 11.3b).

Tal1 is normally expressed in hematopoietic progenitors and continues to be expressed up until the DN2 stage of T cell development (Rothenberg et al. 2008). Ectopic expression of Tal1 leads to a developmental block at the DN1 stage, suggesting that the subsequent Tal1 downregulation is essential for developmental progression (Cheng et al. 2007). Persistent Tal1 expression features in up to 25 % of T-ALL cases, suggesting that sustained activation can lead to transformation (Carroll et al. 1990). A small proportion of T-ALL with Tal1 gene activation arise from translocation of *Tal1* to the TCR loci, while the bulk of mutations arise from a deletion that results in the removal of the *Tal1* 5′ regulatory sequence

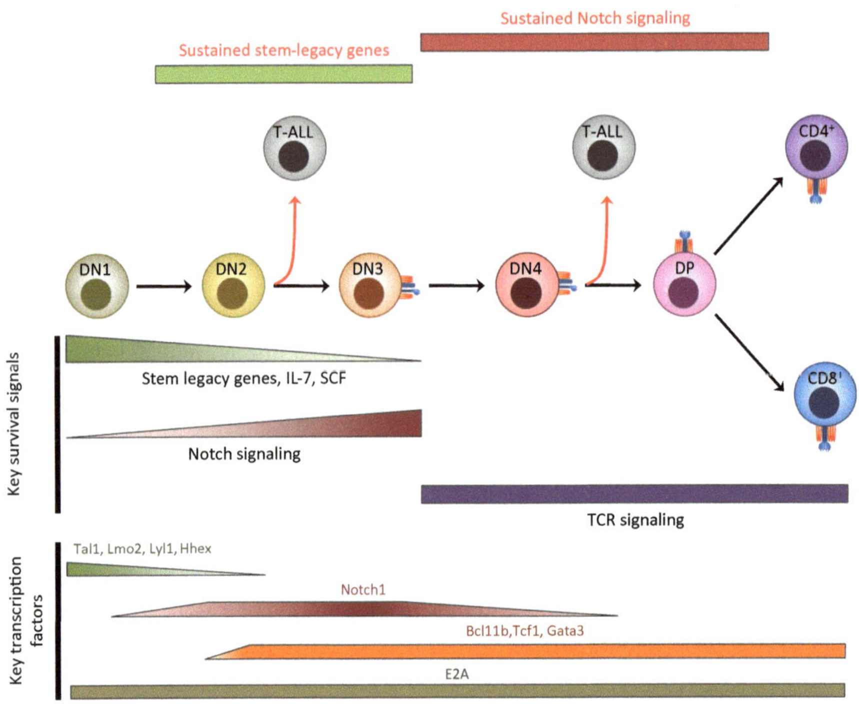

Fig. 11.2 Survival signals in normal and malignant thymocytes. Early thymocyte development is supported by a combination of growth factor signaling and the expression of stem cell-associated survival program. Exposure to Notch ligands then reprograms T lineage precursors away from a stem-associated program and toward a Notch-dependent state. Finally, Notch signaling promotes the expression of the pre-TCR complex, initiating β-selection, which terminates Notch-dependence and activates a TCR-dependent survival program. Failure to properly disengage a preceding survival program results in the accumulation of pro-survival signals that contributes to oncogenesis

and the juxtaposition with the upstream gene *Sil*, resulting in gene dysregulation (Brown et al. 1990; Carroll et al. 1990; Janssen et al. 1993).

Despite the high rate of Tal1 activation in human T-ALL, data from mouse models suggest that Tal1 activation alone is not sufficient for transformation. Tal1-driven leukemia depends on both the timing of its expression and the presence of additional oncogenic cofactors. Transgenic expression of Tal1 using a *Cd2* promoter, which is expressed at the DN4 stage following β-selection, failed to induce T-ALL in mice (Curtis et al. 1997; Larson et al. 1996; Robb et al. 1995). In contrast, transgenic Tal1 expression driven by the *Lck* promoter, expressed at the DN2 stage, was sufficient to permit thymocyte transformation in 30 % of mice, at a long latency (Condorelli et al. 1996). These findings suggest that there is a Tal1 oncogenic window during T cell development that coincides with the stage of development where Tal1 expression is normally downregulated. This may indicate that other factors and pathways are active during this developmental period that are capable of

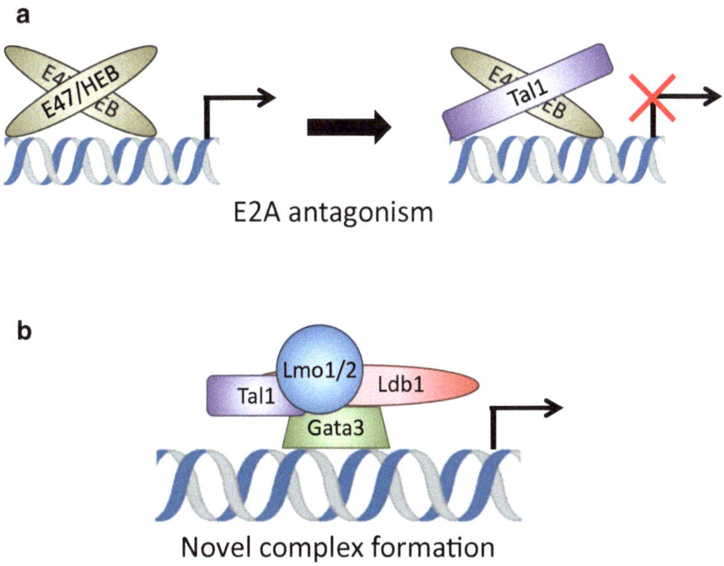

Fig. 11.3 The mechanism of Tal1 action during thymocyte development is unknown. This figure shows suggested mechanisms for Tal1 in T-ALL. Tal1 is highly expressed early in T cell development but is rapidly and robustly downregulated by the DN2 stage. Sustained Tal1 expression promotes T-ALL and is understood to transform thymocytes by two mechanisms: (**a**) Tal1 can heterodimerize with the E2A factors E47 and HEB, where it converts E47 and HEB dimers from activators to repressors; (**b**) continued Tal1 expression allows for the formation of novel activating transcriptional complexes involving Lbd1, Gata3, and either Lmo1 or Lmo2. Collectively, these aberrant transcriptional changes dysregulate T lineage commitment and heighten the oncogenic potential of thymocytes

synergizing with Tal1 to mediate transformation but are later downregulated after T lineage commitment has occurred.

Consistent with this, multiple reports demonstrate that TAL1 expression can lead to rapid onset T-ALL in the context of additional genetic hits, such as activation LMO2 gene expression. Similar to TAL1, the transcriptional regulator LMO2 is a member of the early thymocyte stem-legacy signature and is markedly downregulated at the DN2 stage. Sustained expression of LMO2 leads to a DN block, indicating that, like TAL1, downregulation of LMO2 is a rate-limiting step in thymocyte development (Larson et al. 1996). Moreover, maintaining elevated expression of LMO2, or the related protein LMO1, results in T-ALL development, with a long latency (Chervinsky et al. 1999; Larson et al. 1996). Although LMO1, LMO2, and TAL1 each individually drive low penetrance and long latency T-ALL when expressed alone, combined expression of TAL1 with one of the LMO proteins induces rapid and aggressive T-ALL in all mice (Chervinsky et al. 1999; Larson et al. 1996; Ono et al. 1997). TAL1 and LMO proteins have both been shown to displace factors required for T cell developmental progression, such as other E-proteins and LMO4, promoting a state of heightened proliferation and impaired differentiation (Grutz et al. 1998; Herblot et al. 2000; Park and Sun 1998). It is important to note that these findings are not unique to TAL1 and the LMO proteins. Hhex- and Tal1-related gene Lyl1 are both capable of promoting T-ALL and the failure to downregulate these factors disrupts normal T lineage differentiation (George et al. 2003; Mack et al. 2002; Zhong et al. 2007). Like Tal1, Lyl1 is also an E-protein that interacts with LMO2, where the factors cooperate to promote murine T-ALL (McCormack et al. 2013). Taken together, these studies suggest that termination of the stem/progenitor program in early thymocytes is not only rate-limiting for T cell development but also prevents the emergence of populations with elevated oncogenic potential (Fig. 11.2).

11.3.1 Notch Signaling

Notch receptors are a family of evolutionarily conserved type-I transmembrane glycoproteins that play a critical role in large array of developmental processes, including multiple stages of T cell development (Bailis et al. 2013; Fang et al. 2007; Han et al. 2002; Maillard et al. 2004; Radtke et al. 1999). Mammals possess four different Notch receptors (Notch1–4) that share homology with *Drosophila* Notch and five Notch ligands. The mammalian ligands Delta-like 1, 3, and 4 (Dll1, 3, 4) are homologous to the *Drosophila* Delta, while Jagged1 and 2 are structurally similar to the *Drosophila* Serrate. The extracellular portion of Notch is responsible for ligand interaction and is composed of epidermal growth factor (EGF-like) repeats, LIN12 repeats (LNRs), and a membrane-proximal heterodimerization domain (HD). In the absence of ligand, the LNR and the HD domains constitute a negative regulatory region (NRR) that ensures Notch receptors remain membrane-bound and transcriptionally inactive. In response to ligand the Notch intracellular domain (ICN) is released by proteolytic cleavage by gamma secretase and directly mediates signal transduction. ICN consists of a RAM domain, 7 ankyrin-like repeats, nuclear localization signal sequences, and a C-terminal PEST domain that regulates protein stability.

While there are structural differences amongst the four Notch receptors, all are activated in a ligand-dependent manner and converge on the same core transcriptional complex. Upon Notch receptor–ligand interaction between two neighboring cells, a series of proteolytic cleavage sites become exposed that lead to cleavage of an extracellular site by ADAM family metalloproteases and subsequent cleavage of a transmembrane site by a gamma-secretase complex. This proteolytic processing frees ICN from the membrane, which then translocates to the nucleus and complexes with the transcription factor RBPJ and MAML. MAML then recruits co-activators such as p300 and PCAF to form a large multiprotein transcriptional activation complex. In this manner, Notch signaling results from the conversion of a transmembrane receptor into a transcriptional activator, allowing for rapid changes in target gene expression.

11.4 Notch Ligands and T Cell Development

Of the five mammalian Notch ligands, only four are expressed in the thymus: Jag1, Jag2, Dll1, and Dll4 (Felli et al. 1999; Harman et al. 2003; Schmitt et al. 2004; Schmitt and Zuniga-Pflucker 2002). Although all four have the capacity to engage the Notch pathway, different cell populations have varying capacities to "see" a given ligand, due to posttranslational modification of Notch receptors or anatomical restriction of ligand expression (Kopan and Ilagan 2009). Stromal cell systems have been used to test the capacity of different Notch ligands to support T cell development in vitro. Work using human cord blood progenitors on the S17 murine bone marrow stromal cell line found that while Jag1 could not support T cell development or restrict B cell potential, Dll1 expression promoted T cell differentiation at the expense of B cell development from a multipotent precursor (Jaleco et al. 2001). Corroborating human studies, expression of either Dll1 or Dll4 on OP9 stromal cells is sufficient to support T cell development from mouse hematopoietic progenitors, whereas Jag1 could only promote Notch signaling in thymocytes during a narrow window between the DN1 and DN3 stages, which directed cells toward the NK and $\gamma\delta$ T cell lineages (Besseyrias et al. 2007; Hozumi et al. 2004; Lehar et al. 2005; Schmitt and Zuniga-Pflucker 2002).

Though many of these in vitro findings have been validated in vivo, Notch ligand knockout models revealed distinct ligand requirements. Consistent with in vitro data, mice deficient in either Jag1 or Jag2 display normal $\alpha\beta$ T cell development. However, Jag2 knockouts exhibit a deficiency in $\gamma\delta$ T cell numbers, suggesting that Delta and Jagged ligands may have distinct capacities to support $\alpha\beta$ versus $\gamma\delta$ T cell development, respectively (Jiang et al. 1998; Mancini et al. 2005). In contrast to stromal cell studies, loss of Dll1 in vivo did not disrupt thymocyte development, but instead was found to be essential for marginal zone B cells (Hozumi et al. 2004). Subsequent work revealed that Dll4 is the relevant Notch ligand in the thymus. Conditional deletion of Dll4 in thymic epithelial cells results in a complete block in thymocyte development and the emergence of immature thymic B cells (Hozumi et al. 2008a; Koch et al. 2008). These findings are supported by data showing Dll1 is primarily expressed in thymic blood vessels and not in the thymic epithelium, while Dll4 is highly expressed by thymic epithelial cells (Hozumi et al. 2008a; Koch et al. 2008). Thus, Dll4 is the relevant Notch ligand for in vivo T cell development. However, both Dll1 and Dll4 have the capacity to engage Notch activation and the T cell program in vitro.

11.5 Notch Signaling and T Cell Development

Although Notch1, 2, and 3 are all expressed in developing thymocytes, only Notch1 is both necessary and sufficient for T cell lineage commitment. Overexpression of the ICN1 fragment of Notch1 in bone marrow progenitors causes ectopic T cell development and impairs B cell development, indicating a role for Notch in instructing the T rather than B fate decision (Pui et al. 1999). Consistent with gain-of-function data, reciprocal loss-of-function experiments revealed that the inducible deletion of Notch1 or RBPJ or expressing a GFP-tagged, dominant-negative form of MAML (DNMAML) resulted in a severe block in T cell development at the DN stage and the accumulation of thymic B cells (Han et al. 2002; Maillard et al. 2004; Radtke et al. 1999). Conditional deletion models further mapped the stage-specific requirements for Notch signaling. Deletion of a floxed RBPJ gene at the DN2 stage, using Lck-cre, resulted in a developmental block at the DN3 stage, whereas use of a CD4-cre, which allows for deletion between the DN3 and DN4 stages, did not alter T cell development (Tanigaki et al. 2004). Moreover, the development of the DNMAML transgenic mouse, in which the DNMAML transgene is inserted into the *Rosa26* locus and preceded by a floxed STOP cassette, provided the ability to track Notch inhibition at the single cell level via the GFP tag on DNMAML. Studies using these mice revealed that the developmental block occurred in DN3a cells at the β-selection checkpoint. Importantly, when DN3 cells from Lck-cre × DNMAML mice are intrathymically injected into recipient mice, DNMAML expressing thymocytes failed to give rise to appreciable numbers of DP or SP cells, clearly illustrating an absolute requirement for Notch signaling in the DN compartment (Maillard et al. 2006). Together, these data favor a model in which Notch signaling is essential up until a functional TCRβ is rearranged, but that persistent Notch signals are not essential thereafter.

11.6 Notch and Pre-TCR

The timing of the transition from Notch-dependent to Notch-independent development suggests cross talk between pre-TCR and Notch signaling, and research has revealed significant interaction between these two signaling pathways (Fig. 11.4). ChIP and reporter assays have shown that Notch signaling governs pre-TCR signaling by regulating *Ptcra* gene expression, which encodes a surrogate TCRα (pTa) that complexes with a functionally rearranged β-chain to facilitate pre-TCR signaling (Bellavia et al. 2007; Reizis and Leder 2002). In addition to directly activating *Ptcra* expression, the Notch pathway also indirectly promotes *Ptcra* gene expression. RBPJ shares a DNA-binding motif with the transcription repressor Ikaros, and data suggest there is the potential for these two proteins to compete for binding sites (Hsu et al. 1994; Ono et al. 1998). Signaling through Notch3 promotes the expression of the RNA-binding protein HuD, which results in enhanced generation of Ikaros isoforms that lack the capacity to bind DNA, through alternative splicing. In turn, these isoforms disrupt the activity of full-length Ikaros and mediate enhanced RBPJ binding at pTa and increased Notch-dependent expression (Bellavia et al. 2007). Finally, Notch signaling is also implicated in promoting rearrangement of the *Tcrb* loci, suggesting that Notch might regulate pre-TCR

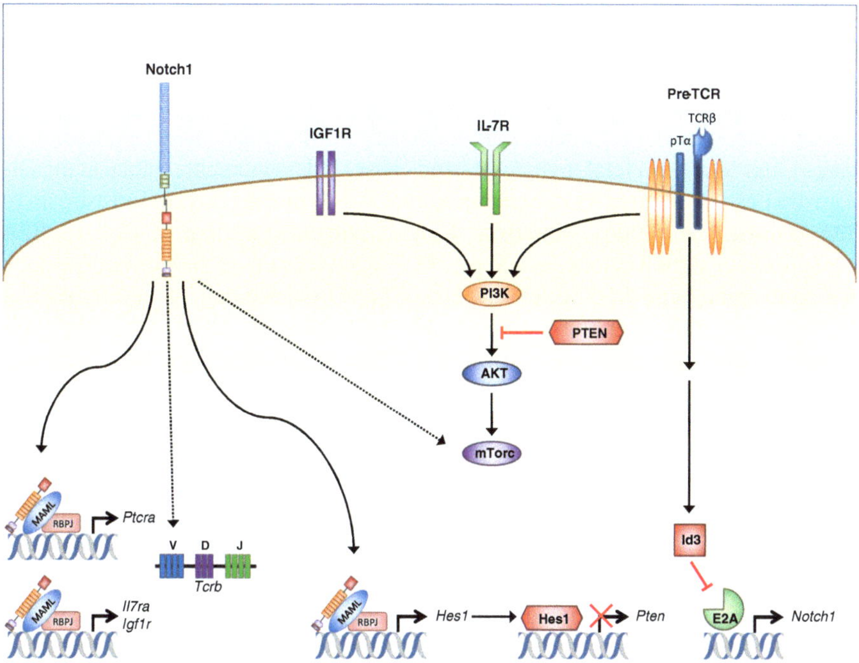

Fig. 11.4 Cross talk between Notch and pre-TCR signaling. Although each pathway utilizes a unique mechanism of signal transduction, both converge upon a common set of downstream targets and regulate gene expression of the other pathway. Notch signaling is essential for the assembly of the pre-TCR, as it is an obligate activator of *Ptcra* transcription and has been implicated in promoting *Tcrb* gene rearrangements. Moreover, Notch and pre-TCR signaling each feed into the PI3K/AKT pathway. The pre-TCR complex directly activates PI3K/AKT signaling, while the Notch pathway indirectly regulates PI3K/AKT by repressing *Pten* expression via Hes1 and promoting the expression of *Il7ra* and *Igf1r*, receptors that directly activate PI3K/AKT. In response to pre-TCR signaling, *Id3* expression increases and provides negative feedback on the Notch pathway, by inhibiting E2A-mediated activation of *Notch1* transcription

signaling at the level of both pre-TCRα and TCRβ expression (Ciofani et al. 2006; Wolfer et al. 2002).

While Notch signaling plays a key role in regulating components of the pre-TCR machinery, the resulting pre-TCR signals are essential for extinguishing *Notch1* expression and mediating the progression to Notch-independent development. From the DN1 stage up through the DN3a stage of thymocyte development, *Notch1* mRNA expression increases and is regulated both by Notch1 itself and the E-proteins E2A and HEB (Ikawa et al. 2006; Yashiro-Ohtani et al. 2009). After β-selection, *Notch1* expression is sharply decreased in DN3b cells and remains low for the remainder of thymic development (Taghon et al. 2006; Yashiro-Ohtani et al. 2009). Accordingly, *Notch1* expression is rapidly and robustly downregulated upon PMA and ionmycin stimulation of Rag2-deficient DN3 cells, which mimics pre-TCR signaling, demonstrating that pre-TCR signals are capable of directly regulating *Notch1*.

This interplay between pre-TCR signaling and *Notch1* expression is regulated by a complex signaling network mediated by the cross talk of these two pathways with E-proteins. While Notch signaling is the primary driver of the T lineage program, E-proteins play an essential role in orchestrating the kinetics of Notch activity during thymocyte development. *E2a* deficiency results in decreased expression of *Notch1* and *Notch3* in fetal thymocytes and causes a developmental block at the DN stage that can be rescued by overexpression of ICN1, placing Notch downstream of E2A (Ikawa et al. 2006). Moreover, E2A activity mirrors that of Notch1. E2A remains active until β-selection, and pre-TCR signals promote the expression of the E-protein antagonist *Id3*, which disrupts E2A DNA binding in later stages of development (Bain et al. 2001; Engel et al. 2001; Xi et al. 2006). Altogether, this creates a signal transduction circuit that consists of a feedforward loop involving E2A-driven Notch1 expression, followed by Notch1 autoregulation, that is then abolished by a negative feedback loop originating with pre-TCR signaling and terminating with Id3 antagonism of E2A and downregulation of Notch1 expression (Yashiro-Ohtani et al. 2009).

In addition to the direct cross talk between Notch and pre-TCR signaling, there is significant interplay between the two pathways at the level of survival and proliferation signals in DN3 cells. Both pathways display significant overlap with the AKT pathway, and AKT signals can bypass the requirement for both Notch and pre-TCR signaling at the DN3 stage. Use of myristoylated AKT (myr-AKT), a membrane tethered and constitutively active form of AKT, can partially bypass the requirement for Notch signaling during β-selection (Ciofani and Zuniga-Pflucker 2005). Furthermore, deletion of the PI3K/AKT signaling antagonist, phosphatase and tensin homologue (PTEN), permits the development of thymocytes from DN3 to DP cells, even in the absence of IL-7 or pre-TCR signals (Hagenbeek 2004). While AKT is downstream of pre-TCR signaling, Notch signaling does not directly activate the AKT pathway. Notch has been suggested to regulate AKT through multiple overlapping mechanisms. The canonical Notch target Hes1 binds the *Pten* locus and negatively regulates *Pten* in both normal and oncogenic T cell development (Palomero et al. 2007; Wong et al. 2012). In a separate arm of Notch-AKT cross talk, components of the mammalian target of rapamycin (mTOR) complex have been implicated as Notch pathway targets in the context of T-ALL (Chan et al. 2007; Lee et al. 2012). Finally, Notch has also been found to regulate the expression of receptors on thymocytes that engage the AKT pathway, such as the insulin-like growth factor 1 receptor and the IL-7Ra (Gonzalez-Garcia et al. 2009; Medyouf et al. 2011). Collectively, these data support a model in which Notch and pre-TCR signaling both provide critical inputs into the AKT pathway at the DN3 stage, leading to a transition from a Notch-mediated to a TCR-driven survival program.

Although there is overlap between Notch and pre-TCR signaling, the two pathways are not redundant in thymocyte development. Multiple groups have found that pre-TCR and TCR signaling are incapable of compensating for a loss of Notch. In an in vitro T cell differentiation system, retroviral expression of TCRβ or TCRαβ could not support T cell development in the absence of Notch signals.

Moreover, the same study found that anti-CD3e treatment or expression TCRβ in conjunction with constitutively active Lck, Fyn, Ras, or PKCα all failed to bypass the requirement for Notch, demonstrating that the inability to overcome a lack of Notch signaling doesn't stem from a quantitative deficiency in pre-TCR signaling (Ciofani et al. 2004). Corroborating these in vitro findings, the DN3 block observed in Lck-cre x DNMAML mice cannot be rescued by the enforced expression of *Tcrb* or *Tcra/b* transgenes (Maillard et al. 2006). Conversely, Notch signaling by itself is incapable of overcoming the loss of pre-TCR signaling. RAG2-deficient mice are incapable of generating the double-stranded breaks necessary for functional TCRβ rearrangement and thus display a block at the β-selection checkpoint; overexpression of ICN1 in these mice fails to restore development into the DP stage. Similarly, overexpression of ICN1 in SLP76-deficient mice, an essential downstream component of pre-TCR signaling, fails to drive the development of DP cells (Allman et al. 2001).

While it remains unclear how these two pathways provide qualitatively distinct inputs into T cell differentiation, recent work has begun to clarify this issue. Detailed analysis of thymocyte expansion in the presence of cell cycle inhibitors has evinced an essential role for proliferation in the DN to DP progression. In the absence of Notch signaling, inducing proliferation by overexpression of either cell cycle machinery components or the Notch target Myc was capable of partially restoring in vitro development up to the DP stage. In contrast, inducing proliferation by overexpressing cell cycle machinery failed to rescue development in the absence of a pre-TCR, as seen in Rag2-deficient thymocytes (Kreslavsky et al. 2012). Thus, the proliferative burst observed post-β-selection is an essential downstream function of Notch during T cell development, whereas the pre-TCR pathway provides distinct, cell cycle-independent signals. Altogether, these findings indicate that Notch and pre-TCR signaling each provide essential and nonredundant inputs into thymocye development.

11.7 Notch and T-ALL

The robust pro-survival and proliferative signals provided by the Notch pathway during normal thymopoiesis also make it a potent oncogene. Indeed, Notch was first discovered through the analysis of T-ALLs with a (7;9) translocation that resulted in the fusion of *NOTCH1* on chromosome 9 to *TCRB* enhancer/promoter elements and the aberrant expression of a constitutively active, truncated nuclear Notch1 (Ellisen et al. 1991). When intracellular Notch is expressed in murine bone marrow progenitors, a rapid onset T-ALL emerges, demonstrating that Notch activation is sufficient for transformation (Pear et al. 1996).

While originally identified in tumors containing a *TCRB* translocation, this mutation was later found to be rare. Instead, the predominant *NOTCH1* gain of function mutations, which are found in over 60 % of human T-ALLs (Weng et al. 2004), occur via point and truncation mutations to the *NOTCH1* gene itself.

The majority of *NOTCH1* mutations arise from point mutations to the NRR, which either destabilize the extracellular domain and permit ligand-independent cleavage of Notch1 and release of the ICN signaling peptide or destroy the PEST domain, which prolongs the half-life of nuclear NOTCH1 (Weng et al. 2004). Both classes of mutations allow cells to both increase the level of Notch signal received and sustain Notch signaling in environments that no longer support it. Although this increased "portability" of Notch signaling may the play a part in allowing transformed cells to escape the thymus, the dose of Notch signaling is a critical determinant of whether Notch acts as an oncogene or an effector of T cell development. Weak activating mutations to the NRR or PEST domain are sufficient to drive ectopic T cell development in transduced bone marrow cells and suppress B and myeloid differentiation, but fail to give rise to leukemia in the absence of additional genetic lesions. However, when multiple activating mutations to Notch are present or when ICN1 is expressed, the resulting signals promote both ectopic T cell development and leukemia (Chiang et al. 2008). Thus, normal T cell development must walk a tenuous line between the ability of Notch1 to initiate and drive the T cell program and the capacity of Notch to act as a potent oncogene.

This oncogenic potential of Notch signaling likely stems from its positioning as a key regulator of multiple pro-survival and pro-proliferative pathways. ChIP-on-chip profiling after Notch inhibition in gamma-secretase inhibitor (GSI) treated T-ALL cell lines revealed that more than 40 % of Notch-responsive loci were effectors of cell metabolism and protein biosynthesis (Palomero et al. 2006). Among these genes, c-Myc has been identified as a critical Notch target in T-ALL (Palomero et al. 2006; Sharma et al. 2006; Weng et al. 2006). Expression of c-Myc rescues T-ALL cell lines from Notch withdrawal, and c-Myc inhibitors block the ability of ICN1 to rescue T-ALL cell lines from GSI treatment, demonstrating that c-Myc is both necessary and sufficient for Notch1's leukemogenic functions (Palomero et al. 2006; Sharma et al. 2006; Weng et al. 2006).

In addition to c-Myc, Notch signaling has also been suggested to regulate the NF-κB and AKT pathways. Ectopic expression of ICN1 in bone marrow progenitors promotes the nuclear localization of NF-κB, the expression of NF-κB components, and the expression of NF-κB target genes. Although inhibition of NF-κB impairs the growth of Notch1-dependent human T-ALL cell lines, NF-κB is not sufficient to drive T-ALL (Vilimas et al. 2007). Given that NF-κB signaling is necessary but not sufficient for T-ALL, NF-κB may act in collaboration with Notch1 to promote c-Myc expression, as studies have identified NF-κB-binding sites in the *Myc* promoter (Park and Wei 2003). In parallel to its affects on NF-κB, Notch signaling has also been implicated in regulating the AKT pathway in cancer. The Notch-Hes1-PTEN axis seen during thymocyte development is an important mediator of Notch-driven T-ALL. Loss of PTEN is a frequent occurrence in GSI-resistant T-ALL's and is sufficient to rescue the survival and proliferation defects observed in GSI treated, Notch-dependent T-ALL cell lines, indicating AKT is an important downstream effector of Notch in T-ALL (Palomero et al. 2007). Accordingly, AKT signaling has been found to be a critical downstream mediator of Notch1-driven glycolysis (Landor et al. 2011). Moreover, multiple reports have linked Notch signaling to the AKT pathway member mTOR. Loss of Rictor, a component of mTORC2, significantly increases median survival and reduces leukemic cell organ infiltration in a murine model of Notch1-induced T-ALL (Lee et al. 2012). Corroborating this finding, GSI treatment or expression of DNMAML in T-ALL cell lines results in the inhibition of multiple components of mTOR signaling that could be rescued by expression of ICN1 (Chan et al. 2007). Altogether these data indicate that while c-Myc is the primary mediator of Notch-driven T-ALL, Notch signaling regulates multiple pathways with known oncogenic roles and likely promotes transformation by the overlapping pro-survival and pro-proliferative inputs from these signals.

11.8 Core T Cell Identity Factors

As immature T cells transit through the thymus, moving into niches of varying Notch ligand density, and eventually deactivate Notch signaling, T cell development requires a more durable and self-sustaining T lineage transcriptional network. Accordingly, Notch signaling not only plays an essential role in regulating thymocyte survival and proliferation but also is critical for promoting the expression of the core T lineage factors GATA3, TCF1, and Bcl11b. All three factors have been identified as targets of the Notch pathway and are sharply upregulated after Notch signaling is initiated. Together, they form a transcriptional circuit that regulates the expression of T cell identity genes, such as the proximal TCR signaling machinery and downstream signal transducers, and stabilizes the T cell program by cross-regulating one another.

11.9 TCF1

TCF1 (*Tcf7*) is a high mobility group box containing DNA-binding transcription factor and a member of the canonical Wnt signaling pathway, where it interacts with the transcription factors β- and γ-catenin. While TCF1 is widely expressed during embryonic development, its expression is largely restricted to hematopoietic progenitors and the T lineage after birth (Oosterwegel et al. 1993; Weber et al. 2011). After being turned off in the LMPP compartment, *Tcf1* expression is induced early during thymocyte development (Fig. 11.2), an event initiated by Notch signaling through its binding to an enhancer 31.5 kb upstream of the *Tcf7* promoter (Germar et al. 2011; Weber et al. 2011). Early work on TCF1 found that knockout mice and cells treated with antisense oligonucleotides displayed greatly reduced thymocyte numbers and a partial block after the DN4 stage, characterized by a lack of cycling cells. However, these mice remained immunocompetent because they were still capable of generating low numbers of functional mature T cells (Hattori et al. 1996; Schilham et al. 1998; Verbeek et al. 1995). Further characterization of TCF1-deficient thymocytes revealed an additional block at the DN3 stage, although these cells are capable of generating a functional pre-TCR. Moreover, TCF1$^{-/-}$ thymocytes are capable of proliferating at a rate comparable to WT cells up until the DN4 stage, after which cycling cell numbers are reduced, and display increased apoptosis. Notably, forced expression of TCRβ or anti-CD3e stimulation results in normal proliferation by TCF1-deficient cells, but the survival defect is exacerbated, indicating that TCF1 plays an essential role in coupling the survival and proliferation signals downstream of the pre-TCR (Goux et al. 2005).

More recent studies revealed an essential role for TCF1 at the earliest stages of T cell development. Detailed analysis of early thymocyte populations and the use of competitive chimeras identified a cell autonomous role for TCF1 during T cell differentiation starting at the DN1 stage (Germar et al. 2011; Goux et al. 2005; Weber et al. 2011). In the absence of TCF1, thymocytes fail to progress even to the DN2 stage under competitive settings. Gain and loss of function microarray studies further illustrated a critical role for TCF1 in regulating a large number of core T cell lineage genes, including *Gata3*, *Bcl1b*, *Cd3e*, *Cd3g*, *Lat*, *Lck*, and *Rag2* (Germar et al. 2011; Weber et al. 2011). These findings suggest that TCF1 has the capacity to launch a substantial portion of the T cell program by itself. In keeping with this, ectopic expression of TCF1 in hematopoietic progenitors allows for commitment to the T cell lineage, in vitro, even in the absence of Notch ligand or in the presence of Notch inhibitors (Weber et al. 2011). To date, Notch1 and TCF1 are the only factors identified with the capacity to induce T cell commitment alone.

Although TCF1 is capable of initiating many of the key events involved in T cell lineage commitment, it is not sufficient to activate all of the essential T cell development factors, such as the Notch targets *Hes1* or *Ptcra* (Germar et al. 2011; Weber et al. 2011). Furthermore, dual deficiency for TCF1 and its homologue LEF1, which has been suggested to compensate for the absence TCF1, results in a phenotype identical to TCF1$^{-/-}$ mice. These double knockouts display a partial block at the DN1 stage but are capable of generating DN3 cells expressing normal levels of functionally rearranged TCRβ and are capable of undergoing β-selection (Yu et al. 2012). Thus, while TCF1 provides a competitive advantage to early thymocytes and is sufficient for entry into the T cell lineage, unlike Notch1, it does not appear necessary for T cell commitment.

The mechanism of TCF1's action during T cell development remains controversial. Although it is recognized as an effector of the canonical Wnt signaling pathway and the β-catenin-binding domain of TCF1 is essential for its function during T cell differentiation, TCF1 is thought to act in a β-catenin-independent manner in thymocytes (Goux et al. 2005). In contrast to the phenotype of TCF1 knockout mice, β- and γ-catenin deficiency has no effect on thymic development (Cobas et al. 2004; Goux et al. 2005; Jeannet et al. 2008; Weber et al. 2011). Moreover, expression of the β- and γ-catenin inhibitor ICAT fails to impair in vitro T cell development (Weber et al. 2011). While the exact mechanism of TCF1 function during thymopoiesis remains unclear, TCF1 acts to mobilize and reinforce the expression of a large cohort of T cell program genes, downstream of Notch signals. Furthermore, in contrast to Notch1, TCF1 expression remains high throughout T cell development, suggesting that TCF1 exerts additional functions.

11.10 GATA3

GATA3 is a zinc finger transcription factor that plays a dynamic and highly context-dependent role in T cell development. It has been found to regulate three critical stages of T cell development: initiating the early T cell program, promoting $CD4^+$ SP thymocyte development during TCR-mediated selection, and establishing the helper T cell type 2 program during mature $CD4^+$ T cell differentiation. Consistent with the wide variety of roles GATA3 plays during T cell development, ChIPseq analysis of GATA3 binding has revealed that genome occupancy is dictated by cell development (Wei et al. 2011; Zhang et al. 2012).

Protein concentration has been suggested to play a large role in determining the stage-specific activities of GATA3. GATA3 binds to genes encoding multiple components of the TCR complex, including the *Cd3* genes, *Tcra* and *Tcrb*, and loss of GATA3 results in decreased *Cd3d* and *Cd3e* expression (Wei et al. 2011); however, ectopic expression of GATA3 fails to activate these targets in thymocytes and results in a block in T cell development (Hozumi et al. 2008b; Taghon et al. 2007). While site affinity and protein concentration can partially explain the context-specific behavior of GATA3, divergent-binding patterns are observed even in cell states with comparable levels of GATA3 expression. The remainder of these binding pattern differences is likely accounted for by GATA3's differential association with cofactors and cooperating transcription factors such as LMO and E-proteins. Indeed, motif enrichment analysis of GATA3 ChIPseq data reveals that GATA3 binding accompanies unique sets of secondary transcription factor motifs depending on the stage of T cell development (Wei et al. 2011). Thus, GATA3 activity is regulated by multiple mechanisms that in turn permit a continuum of GATA3 function.

At the earliest stages of T cell development, GATA3 is absolutely required for T lineage commitment. $GATA3^{-/-}$ mice fail to develop mature T cells and display a nearly complete block at the DN1 stage of development (Hendriks et al. 1999; Ting et al. 1996). Similarly, mice expressing hypomorphic GATA3 alleles were found to have normal numbers of LMPP's but displayed a severe cell-autonomous defect in DN1 cell generation, demonstrating that GATA3 is essential for the earliest stages of differentiation of hematopoietic progenitors into thymoyctes (Hosoya et al. 2009). The use of conditional deletion models allowed for analysis of GATA3 at later stages of thymic development and revealed a critical role for the transcription factor at β-selection. Lck-cre x $Gata3^{FL/FL}$ mice display a three to four fold reduction in thymic cellularity and a block at the DN3a stage of development. This developmental block is accompanied by an increase in apoptosis and decreased intracellular TCRβ expression; however neither TCRβ gene rearrangement nor RNA expression was found to be defective, suggesting that

GATA3 regulates TCRβ expression posttranscriptionally at the DN3 stage of development. Importantly, expression of a TCR transgene rescued neither the DN3a arrest nor the survival defects observed, indicating that GATA3 also regulates downstream events during pre-TCR signaling (Pai et al. 2003).

This dynamic nature of GATA3 requires multiple levels of regulation. *Gata3* expression can be driven by two promoters and is influenced by inputs from an intronic enhancer and an upstream silencer (Asnagli et al. 2002; Gregoire and Romeo 1999; Hwang et al. 2002). However, a YAC transgene containing these regulatory regions failed to promote GATA3 expression in the T lineage (Lakshmanan et al. 1999). Recent work identified a long range enhancer, located 280 kb downstream of *Gata3* in a gene desert, that is both necessary and sufficient for the expression of a *Gata3* BAC transgene. Though the pattern of expression driven by this element does not precisely conform to the normal developmental expression of *Gata3*, it is the first positive cis-regulatory region identified for *Gata3* in the T lineage (Hosoya-Ohmura et al. 2012).

As precise control of Gata3 protein concentration is required for normal T lineage development, *Gata3* transcription is regulated by multiple inputs. E2A factors are essential negative regulators of *Gata3* expression. Loss of E2A results in a block in thymocyte development and lineage diversion, resulting from elevated Gata3 expression; knockdown of *Gata3* mRNA in *E2A*$^{-/-}$ thymocytes is sufficient to restore progression to the DN3 stage (Xu et al. 2013). In contrast to E2A factors, Tcf1 promotes Gata3 expression via binding to the *Gata3* promoter, both early in T cell development as well as in mature peripheral T cells (Weber et al. 2011; Yu et al. 2009). Notably, loss of Tcf1 results in a failure to properly initiate *Gata3* expression, even in the presence of Notch ligand, and ectopic expression of Tcf1 drives normal Gata3 mRNA levels in the absence of Notch ligand, illustrating that Tcf1 is both necessary and sufficient for Gata3 expression in thymocytes (Weber et al. 2011). While Notch1 positively regulates transcripts generated from the *Gata3* exon-1a start site in mature CD4^{+} T cells, inhibiting Notch signaling does not affect Gata3 protein levels and it is unclear if the Notch pathway directly regulates *Gata3* expression during thymopoiesis (Amsen et al. 2007; Bailis et al. 2013; Fang et al. 2007).

In addition to transcriptional regulation, GATA3 is regulated both at the level of protein stability and translation. MAP-kinase signaling stabilizes GATA3 protein in response to TCR activation by inhibiting the ubiquitin-proteosome pathway (Yamashita et al. 2005). TCR activation also functions to promote *Gata3* translation, even when the pool of RNA is declining, through an mTOR- and PI3K-AKT-dependent pathway (Cook and Miller 2010). In this manner, GATA3 may function as an integrator of signals from the pre-TCR, IL-7R, and Notch pathways, all of which converge on PI3K-AKT signaling, and thus act as a sensor of the cumulative survival and proliferation signals a developing thymocyte receives (Rothenberg 2012).

11.11 Bcl11b

Bcl11b is a zinc finger transcription factor known to associate with both Sirt-type HDAC's and the NuRD complex to repress gene expression, though gene activating roles have also been identified (Cismasiu et al. 2005, 2009; Desplats et al. 2008; Senawong et al. 2003). Within the hematopoietic compartment, Bcl11b expression is largely restricted to the T lineage and is acutely upregulated at the DN2 stage (Fig. 11.2). Whereas Notch signaling expression initiates the T lineage program and TCF1 and GATA3 support and stabilize the transcriptional network, Bcl11b enforces commitment and ensures lineage fidelity. Loss of Bcl11b results in a severe block at the DN3 stage, increased apoptosis, and a 90 % reduction in thymic cellularity. Moreover, thymocytes from Bcl11b$^{-/-}$ mice display very low TCRβ expression, impaired β-chain rearrangements, and lack CD3e (Inoue et al. 2006; Wakabayashi et al. 2003). Despite the defects in pre-TCR assembly, expression of a rearranged TCRβ or TCRαβ transgene fails to rescue the DN3 block or restore thymic cellularity, indicating Bcl11b's functions extend beyond β-selection (Inoue et al. 2006).

In addition to its role at the DN3 stage, inducible deletion and fetal liver chimera models have revealed that Bcl11b is essential for terminating the stem/progenitor program and for maintaining T cell lineage integrity at early stages of T cell development. Early DN2 cells are highly proliferative, retain myeloid, B, and NK potential, and have a transcriptional signature reminiscent of hematopoietic precursor populations (Rothenberg et al. 2008). Induction of Bcl11b deletion in bone marrow progenitors results in a nearly absolute block in in vitro T cell development at the DN2 stage and a similar block is observed in Bcl11b$^{-/-}$ fetal liver chimera mice. Although these cells fail to differentiate into DN3 cells, they remain proliferative and their progeny resemble more DN2- and DN1-like cells for long culture durations (Ikawa et al. 2010; Li et al. 2010a). Moreover, deleted cells had a survival and proliferative advantage over non-deleted cells, suggesting loss of Bcl11b locks thymocytes into a self-renewing, stem-like state. Consistent with their self-renewal capacity, Bcl11b-deficient cells fail to downregulate many of the stem-legacy genes of early thymocytes, including *Tal1*, *Lyl1*, *Sfpi1*, *Gfi1b*, and *Bcl11a* (Li et al. 2010a).

The switch from this stem-like program to that of T lineage program appears to be regulated by an IL-7/Bcl11b axis. Culturing cells in high levels of IL-7 with Notch ligand, in the absence of feeder cells, is capable of recapitulating the stem-like phenotype and developmental block observed in Bcl11b$^{-/-}$ cells. Notably, overexpression of Bcl11b was sufficient to overcome the DN2 arrest, turn off several stem-legacy genes, and permit the expression of T lineage signature genes, such as *Ptcra* and *Cd3e*. Similarly, withdrawal of IL-7 led to DN3 progression and a robust upregulation of Bcl11b (Ikawa et al. 2010). These data support a model in which high levels of IL-7 permit the expansion of rare thymic emigrants early in development, and subsequent reduction of these signals induces Bcl11b expression, the termination of a progenitor program, and the initiation of T cell commitment.

Bcl11b not only limits progenitor gene expression and the capacity of early thymocytes to self-renew, but it is also critical for restricting the multilineage potential of these T cell precursors. In vitro culture of Bcl11b-deficient thymocytes results in a significant increase in NK, myeloid, and B cell differentiation from DN1, DN2, and the few DN3 cells present (Li et al. 2010a, b). This promiscuous differentiation is most pronounced with regard to NK potential, as Bcl11b$^{-/-}$ cells have a greater capacity to generate NK cells than T lineage cells (Li et al. 2010b). Corroborating these findings, transcriptional profiling of these cells revealed elevated expression of many NK lineage genes such as *Id2*, *Plzf*, and *Nfil3* (Li et al. 2010a, b). Despite the inability of *Bcl11b*$^{-/-}$ cells to repress progenitor and NK cell signature genes, mutant cells remained capable of initiating the T cell lineage program, with *Gata3*, *Tcf7*, *Notch1*, and *Hes1* expression levels comparable to that of a DN3 cell (Li et al. 2010a, b). These findings illustrate how the process of T lineage specification is separable from that of T lineage commitment. Whereas Notch, GATA3, and TCF1 are all required to launch the T cell program, Bcl11b represents an essential transcriptional switch that dismantles the stem-legacy network and protects the T lineage circuitry from instability.

11.12 Conclusions

The mammalian immune system must balance competing demands for antigen receptor diversity and genetic stability. Adaptive immunity requires a pool of TCR's diverse enough to recognize all potential antigens encountered during the lifetime of a host. In order to produce sufficient genetic material, T cell development engages a program of high metabolic and proliferative activity coupled with genetic rearrangements, creating an environment ripe for transformation. Despite this perfect storm of pro-oncogenic processes, mammals generate millions of T cells every day without regularly succumbing to T-ALL. This high level of developmental stability speaks to the exquisite transcriptional control over survival programs exercised by developing T cells. By linking T cell differentiation to an ordered series of survival pathways that serially extinguish the proceeding program, thymocytes are able to limit an accumulation of pro-growth signals and ensure that there is a limited window for oncogenic mutation to any one survival program.

Attesting to the stability of T cell development, initiation of T cell leukemia requires multiple oncogenic hits. Although over half of T-ALL cases exhibit activating Notch mutations, these naturally arising mutant forms of Notch are not sufficient to generate high penetrance T-ALL in animal models (Chiang et al. 2008). Moreover, while mutations to closely cooperating factors such as LMO1, LMO2, and TAL1 are all frequently associated with T-ALL, rapid onset and lethal T-ALL cannot be generated by single hits to these targets (Chervinsky et al. 1999; Condorelli et al. 1996; Curtis et al. 1997; Larson et al. 1996; Robb et al. 1995). These findings argue that there is a long latency period in which multiple oncogenic events must accumulate before T-ALL onset. More detailed analysis is needed to determine the sequence in which these genetic lesions accumulate and what additional events are required to permit genetic instability.

Acknowledgments We apologize to those whose work was not cited due to space limitations. This work was supported by T32HD007516 and 1F31CA165813 predoctoral fellowships to W.B. Research in the Pear lab is supported by grants from the National Institutes of Health (R01AI047833, PO1CA119070), the Waxman Foundation, and the Leukemia and Lymphoma Society.

References

Allman D, Karnell FG, Punt JA, Bakkour S, Xu L, Myung P, Koretzky GA, Pui JC, Aster JC, Pear WS (2001) Separation of Notch1 promoted lineage commitment and expansion/transformation in developing T cells. J Exp Med 194:99–106

Allman D, Sambandam A, Kim S, Miller JP, Pagan A, Well D, Meraz A, Bhandoola A (2003) Thymopoiesis independent of common lymphoid progenitors. Nat Immunol 4:168–174

Amsen D, Antov A, Jankovic D, Sher A, Radtke F, Souabni A, Busslinger M, McCright B, Gridley T, Flavell RA (2007) Direct regulation of Gata3 expression determines the T helper differentiation potential of Notch. Immunity 27:89–99

Asnagli H, Afkarian M, Murphy KM (2002) Cutting edge: identification of an alternative GATA-3 promoter directing tissue-specific gene expression in mouse and human. J Immunol 168:4268–4271

Bailis W, Yashiro-Ohtani Y, Fang TC, Hatton RD, Weaver CT, Artis D, Pear WS (2013) Notch simultaneously orchestrates multiple helper T cell programs independently of cytokine signals. Immunity 39:148–159

Bain G, Cravatt CB, Loomans C, Alberola-Ila J, Hedrick SM, Murre C (2001) Regulation of the helix-loop-helix proteins, E2A and Id3, by the Ras-ERK MAPK cascade. Nat Immunol 2:165–171

Bellavia D, Mecarozzi M, Campese AF, Grazioli P, Talora C, Frati L, Gulino A, Screpanti I (2007) Notch3 and the Notch3-upregulated RNA-binding protein HuD regulate Ikaros alternative splicing. EMBO J 26:1670–1680

Besseyrias V, Fiorini E, Strobl LJ, Zimber-Strobl U, Dumortier A, Koch U, Arcangeli ML, Ezine S, Macdonald HR, Radtke F (2007) Hierarchy of Notch-Delta interactions promoting T cell lineage commitment and maturation. J Exp Med 204:331–343

Bhandoola A, von Boehmer H, Petrie HT, Zuniga-Pflucker JC (2007) Commitment and developmental potential of extrathymic and intrathymic T cell precursors: plenty to choose from. Immunity 26:678–689

Brown L, Cheng JT, Chen Q, Siciliano MJ, Crist W, Buchanan G, Baer R (1990) Site-specific recombination of the tal-1 gene is a common occurrence in human T cell leukemia. EMBO J 9:3343–3351

Carroll AJ, Crist WM, Link MP, Amylon MD, Pullen DJ, Ragab AH, Buchanan GR, Wimmer RS, Vietti TJ (1990) The t(1;14)(p34;q11) is nonrandom and restricted to T-cell acute lymphoblastic leukemia: a Pediatric Oncology Group study. Blood 76:1220–1224

Chan SM, Weng AP, Tibshirani R, Aster JC, Utz PJ (2007) Notch signals positively regulate activity of the mTOR pathway in T-cell acute lymphoblastic leukemia. Blood 110:278–286

Cheng Y, Zhang Z, Slape C, Aplan PD (2007) Cre-loxP-mediated recombination between the SIL and SCL genes leads to a block in T-cell development at the CD4–CD8- to CD4+ CD8+ transition. Neoplasia 9:315–321
Chervinsky DS, Zhao XF, Lam DH, Ellsworth M, Gross KW, Aplan PD (1999) Disordered T-cell development and T-cell malignancies in SCL LMO1 double-transgenic mice: parallels with E2A-deficient mice. Mol Cell Biol 19:5025–5035
Chiang MY, Xu L, Shestova O, Histen G, L'Heureux S, Romany C, Childs ME, Gimotty PA, Aster JC, Pear WS (2008) Leukemia-associated NOTCH1 alleles are weak tumor initiators but accelerate K-ras-initiated leukemia. J Clin Invest 118:3181–3194
Ciofani M, Zuniga-Pflucker JC (2005) Notch promotes survival of pre-T cells at the beta-selection checkpoint by regulating cellular metabolism. Nat Immunol 6:881–888
Ciofani M, Schmitt TM, Ciofani A, Michie AM, Cuburu N, Aublin A, Maryanski JL, Zuniga-Pflucker JC (2004) Obligatory role for cooperative signaling by pre-TCR and Notch during thymocyte differentiation. J Immunol 172:5230–5239
Ciofani M, Knowles GC, Wiest DL, von Boehmer H, Zuniga-Pflucker JC (2006) Stage-specific and differential notch dependency at the alphabeta and gammadelta T lineage bifurcation. Immunity 25:105–116
Cismasiu VB, Adamo K, Gecewicz J, Duque J, Lin Q, Avram D (2005) BCL11B functionally associates with the NuRD complex in T lymphocytes to repress targeted promoter. Oncogene 24:6753–6764
Cismasiu VB, Duque J, Paskaleva E, Califano D, Ghanta S, Young HA, Avram D (2009) BCL11B enhances TCR/CD28-triggered NF-kappaB activation through up-regulation of Cot kinase gene expression in T-lymphocytes. Biochem J 417:457–466
Cobas M, Wilson A, Ernst B, Mancini SJ, MacDonald HR, Kemler R, Radtke F (2004) Beta-catenin is dispensable for hematopoiesis and lymphopoiesis. J Exp Med 199:221–229
Coma S, Allard-Ratick M, Akino T, van Meeteren LA, Mammoto A, Klagsbrun M (2013) GATA2 and Lmo2 control angiogenesis and lymphangiogenesis via direct transcriptional regulation of neuropilin-2. Angiogenesis 16:939–952
Condorelli GL, Facchiano F, Valtieri M, Proietti E, Vitelli L, Lulli V, Huebner K, Peschle C, Croce CM (1996) T-cell-directed TAL-1 expression induces T-cell malignancies in transgenic mice. Cancer Res 56:5113–5119
Cook KD, Miller J (2010) TCR-dependent translational control of GATA-3 enhances Th2 differentiation. J Immunol 185:3209–3216
Curtis DJ, Robb L, Strasser A, Begley CG (1997) The CD2-scl transgene alters the phenotype and frequency of T-lymphomas in N-ras transgenic or p53 deficient mice. Oncogene 15:2975–2983
Desplats PA, Lambert JR, Thomas EA (2008) Functional roles for the striatal-enriched transcription factor, Bcl11b, in the control of striatal gene expression and transcriptional dysregulation in Huntington's disease. Neurobiol Dis 31:298–308
Ellisen LW, Bird J, West DC, Soreng AL, Reynolds TC, Smith SD, Sklar J (1991) TAN-1, the human homolog of the Drosophila notch gene, is broken by chromosomal translocations in T lymphoblastic neoplasms. Cell 66:649–661
Engel I, Johns C, Bain G, Rivera RR, Murre C (2001) Early thymocyte development is regulated by modulation of E2A protein activity. J Exp Med 194:733–745
Fang TC, Yashiro-Ohtani Y, Del Bianco C, Knoblock DM, Blacklow SC, Pear WS (2007) Notch directly regulates Gata3 expression during T helper 2 cell differentiation. Immunity 27:100–110
Felli MP, Maroder M, Mitsiadis TA, Campese AF, Bellavia D, Vacca A, Mann RS, Frati L, Lendahl U, Gulino A, Screpanti I (1999) Expression pattern of notch1, 2 and 3 and Jagged1 and 2 in lymphoid and stromal thymus components: distinct ligand-receptor interactions in intrathymic T cell development. Int Immunol 11:1017–1025
George A, Morse HC 3rd, Justice MJ (2003) The homeobox gene Hex induces T-cell-derived lymphomas when overexpressed in hematopoietic precursor cells. Oncogene 22:6764–6773

Germar K, Dose M, Konstantinou T, Zhang J, Wang H, Lobry C, Arnett KL, Blacklow SC, Aifantis I, Aster JC, Gounari F (2011) T-cell factor 1 is a gatekeeper for T-cell specification in response to Notch signaling. Proc Natl Acad Sci USA 108:20060–20065

Gonzalez-Garcia S, Garcia-Peydro M, Martin-Gayo E, Ballestar E, Esteller M, Bornstein R, de la Pompa JL, Ferrando AA, Toribio ML (2009) CSL-MAML-dependent Notch1 signaling controls T lineage-specific IL-7R{alpha} gene expression in early human thymopoiesis and leukemia. J Exp Med 206:779–791

Goux D, Coudert JD, Maurice D, Scarpellino L, Jeannet G, Piccolo S, Weston K, Huelsken J, Held W (2005) Cooperating pre-T-cell receptor and TCF-1-dependent signals ensure thymocyte survival. Blood 106:1726–1733

Gregoire JM, Romeo PH (1999) T-cell expression of the human GATA-3 gene is regulated by a non-lineage-specific silencer. J Biol Chem 274:6567–6578

Grutz G, Forster A, Rabbitts TH (1998) Identification of the LMO4 gene encoding an interaction partner of the LIM-binding protein LDB1/NLI1: a candidate for displacement by LMO proteins in T cell acute leukaemia. Oncogene 17:2799–2803

Hagenbeek TJ, Naspetti M, Malergue F, Garcon F, Nunes JA, Cleutjens KB, Trapman J, Krimpenfort P, Spits H (2004) The loss of PTEN allows TCR alphabeta lineage thymocytes to bypass IL-7 and Pre-TCR-mediated signaling. J Exp Med 200:883–894

Han H, Tanigaki K, Yamamoto N, Kuroda K, Yoshimoto M, Nakahata T, Ikuta K, Honjo T (2002) Inducible gene knockout of transcription factor recombination signal binding protein-J reveals its essential role in T versus B lineage decision. Int Immunol 14:637–645

Harman BC, Jenkinson EJ, Anderson G (2003) Microenvironmental regulation of Notch signalling in T cell development. Semin Immunol 15:91–97

Hattori N, Kawamoto H, Fujimoto S, Kuno K, Katsura Y (1996) Involvement of transcription factors TCF-1 and GATA-3 in the initiation of the earliest step of T cell development in the thymus. J Exp Med 184:1137–1147

Hendriks RW, Nawijn MC, Engel JD, van Doorninck H, Grosveld F, Karis A (1999) Expression of the transcription factor GATA-3 is required for the development of the earliest T cell progenitors and correlates with stages of cellular proliferation in the thymus. Eur J Immunol 29:1912–1918

Herblot S, Steff AM, Hugo P, Aplan PD, Hoang T (2000) SCL and LMO1 alter thymocyte differentiation: inhibition of E2A-HEB function and pre-T alpha chain expression. Nat Immunol 1:138–144

Hosoya T, Kuroha T, Moriguchi T, Cummings D, Maillard I, Lim KC, Engel JD (2009) GATA-3 is required for early T lineage progenitor development. J Exp Med 206:2987–3000

Hosoya-Ohmura S, Lin YH, Herrmann M, Kuroha T, Rao A, Moriguchi T, Lim KC, Hosoya T, Engel JD (2012) An NK and T cell enhancer lies 280 kilobase pairs 3′ to the gata3 structural gene. Mol Cell Biol 31:1894–1904

Hozumi K, Negishi N, Suzuki D, Abe N, Sotomaru Y, Tamaoki N, Mailhos C, Ish-Horowicz D, Habu S, Owen MJ (2004) Delta-like 1 is necessary for the generation of marginal zone B cells but not T cells in vivo. Nat Immunol 5:638–644

Hozumi K, Mailhos C, Negishi N, Hirano K, Yahata T, Ando K, Zuklys S, Hollander GA, Shima DT, Habu S (2008a) Delta-like 4 is indispensable in thymic environment specific for T cell development. J Exp Med 205:2507–2513

Hozumi K, Negishi N, Tsuchiya I, Abe N, Hirano K, Suzuki D, Yamamoto M, Engel JD, Habu S (2008b) Notch signaling is necessary for GATA3 function in the initiation of T cell development. Eur J Immunol 38:977–985

Hsu HL, Wadman I, Tsan JT, Baer R (1994) Positive and negative transcriptional control by the TAL1 helix-loop-helix protein. Proc Natl Acad Sci USA 91:5947–5951

Hwang ES, Choi A, Ho IC (2002) Transcriptional regulation of GATA-3 by an intronic regulatory region and fetal liver zinc finger protein 1. J Immunol 169:248–253

Igarashi H, Gregory SC, Yokota T, Sakaguchi N, Kincade PW (2002) Transcription from the RAG1 locus marks the earliest lymphocyte progenitors in bone marrow. Immunity 17:117–130

Ikawa T, Kawamoto H, Goldrath AW, Murre C (2006) E proteins and Notch signaling cooperate to promote T cell lineage specification and commitment. J Exp Med 203:1329–1342
Ikawa T, Hirose S, Masuda K, Kakugawa K, Satoh R, Shibano-Satoh A, Kominami R, Katsura Y, Kawamoto H (2010) An essential developmental checkpoint for production of the T cell lineage. Science 329:93–96
Inoue J, Kanefuji T, Okazuka K, Watanabe H, Mishima Y, Kominami R (2006) Expression of TCR alpha beta partly rescues developmental arrest and apoptosis of alpha beta T cells in Bcl11b-/- mice. J Immunol 176:5871–5879
Jaleco AC, Neves H, Hooijberg E, Gameiro P, Clode N, Haury M, Henrique D, Parreira L (2001) Differential effects of Notch ligands Delta-1 and Jagged-1 in human lymphoid differentiation. J Exp Med 194:991–1002
Janssen JW, Ludwig WD, Sterry W, Bartram CR (1993) SIL-TAL1 deletion in T-cell acute lymphoblastic leukemia. Leukemia 7:1204–1210
Jeannet G, Scheller M, Scarpellino L, Duboux S, Gardiol N, Back J, Kuttler F, Malanchi I, Birchmeier W, Leutz A et al (2008) Long-term, multilineage hematopoiesis occurs in the combined absence of beta-catenin and gamma-catenin. Blood 111:142–149
Jiang R, Lan Y, Chapman HD, Shawber C, Norton CR, Serreze DV, Weinmaster G, Gridley T (1998) Defects in limb, craniofacial, and thymic development in Jagged2 mutant mice. Genes Dev 12:1046–1057
Koch U, Fiorini E, Benedito R, Besseyrias V, Schuster-Gossler K, Pierres M, Manley NR, Duarte A, Macdonald HR, Radtke F (2008) Delta-like 4 is the essential, nonredundant ligand for Notch1 during thymic T cell lineage commitment. J Exp Med 205:2515–2523
Kopan R, Ilagan MX (2009) The canonical Notch signaling pathway: unfolding the activation mechanism. Cell 137:216–233
Kreslavsky T, Gleimer M, Miyazaki M, Choi Y, Gagnon E, Murre C, Sicinski P, von Boehmer H (2012) beta-Selection-induced proliferation is required for alphabeta T cell differentiation. Immunity 37:840–853
Lakshmanan G, Lieuw KH, Lim KC, Gu Y, Grosveld F, Engel JD, Karis A (1999) Localization of distant urogenital system-, central nervous system-, and endocardium-specific transcriptional regulatory elements in the GATA-3 locus. Mol Cell Biol 19:1558–1568
Landor SK, Mutvei AP, Mamaeva V, Jin S, Busk M, Borra R, Gronroos TJ, Kronqvist P, Lendahl U, Sahlgren CM (2011) Hypo- and hyperactivated Notch signaling induce a glycolytic switch through distinct mechanisms. Proc Natl Acad Sci USA 108:18814–18819
Larson RC, Lavenir I, Larson TA, Baer R, Warren AJ, Wadman I, Nottage K, Rabbitts TH (1996) Protein dimerization between Lmo2 (Rbtn2) and Tal1 alters thymocyte development and potentiates T cell tumorigenesis in transgenic mice. EMBO J 15:1021–1027
Lee K, Nam KT, Cho SH, Gudapati P, Hwang Y, Park DS, Potter R, Chen J, Volanakis E, Boothby M (2012) Vital roles of mTOR complex 2 in Notch-driven thymocyte differentiation and leukemia. J Exp Med 209:713–728
Lehar SM, Dooley J, Farr AG, Bevan MJ (2005) Notch ligands Delta 1 and Jagged1 transmit distinct signals to T-cell precursors. Blood 105:1440–1447
Li L, Leid M, Rothenberg EV (2010a) An early T cell lineage commitment checkpoint dependent on the transcription factor Bcl11b. Science 329:89–93
Li P, Burke S, Wang J, Chen X, Ortiz M, Lee SC, Lu D, Campos L, Goulding D, Ng BL et al (2010b) Reprogramming of T cells to natural killer-like cells upon Bcl11b deletion. Science 329:85–89
Mack DL, Leibowitz DS, Cooper S, Ramsey H, Broxmeyer HE, Hromas R (2002) Down-regulation of the myeloid homeobox protein Hex is essential for normal T-cell development. Immunology 107:444–451
Maillard I, Weng AP, Carpenter AC, Rodriguez CG, Sai H, Xu L, Allman D, Aster JC, Pear WS (2004) Mastermind critically regulates Notch-mediated lymphoid cell fate decisions. Blood 104:1696–1702

Maillard I, Tu L, Sambandam A, Yashiro-Ohtani Y, Millholland J, Keeshan K, Shestova O, Xu L, Bhandoola A, Pear WS (2006) The requirement for Notch signaling at the beta-selection checkpoint in vivo is absolute and independent of the pre-T cell receptor. J Exp Med 203:2239–2245

Mancini SJ, Mantei N, Dumortier A, Suter U, MacDonald HR, Radtke F (2005) Jagged1-dependent Notch signaling is dispensable for hematopoietic stem cell self-renewal and differentiation. Blood 105:2340–2342

McCormack MP, Shields BJ, Jackson JT, Nasa C, Shi W, Slater NJ, Tremblay CS, Rabbitts TH, Curtis DJ (2013) Requirement for Lyl1 in a model of Lmo2-driven early T-cell precursor ALL. Blood 122:2093–2103

Medyouf H, Gusscott S, Wang H, Tseng JC, Wai C, Nemirovsky O, Trumpp A, Pflumio F, Carboni J, Gottardis M et al (2011) High-level IGF1R expression is required for leukemia-initiating cell activity in T-ALL and is supported by Notch signaling. J Exp Med 208:1809–1822

Meier N, Krpic S, Rodriguez P, Strouboulis J, Monti M, Krijgsveld J, Gering M, Patient R, Hostert A, Grosveld F (2006) Novel binding partners of Ldb1 are required for haematopoietic development. Development 133:4913–4923

Ng SY, Yoshida T, Zhang J, Georgopoulos K (2009) Genome-wide lineage-specific transcriptional networks underscore Ikaros-dependent lymphoid priming in hematopoietic stem cells. Immunity 30:493–507

O'Neil J, Shank J, Cusson N, Murre C, Kelliher M (2004) TAL1/SCL induces leukemia by inhibiting the transcriptional activity of E47/HEB. Cancer Cell 5:587–596

Ono Y, Fukuhara N, Yoshie O (1997) Transcriptional activity of TAL1 in T cell acute lymphoblastic leukemia (T-ALL) requires RBTN1 or -2 and induces TALLA1, a highly specific tumor marker of T-ALL. J Biol Chem 272:4576–4581

Ono Y, Fukuhara N, Yoshie O (1998) TAL1 and LIM-only proteins synergistically induce retinaldehyde dehydrogenase 2 expression in T-cell acute lymphoblastic leukemia by acting as cofactors for GATA3. Mol Cell Biol 18:6939–6950

Oosterwegel M, van de Wetering M, Timmerman J, Kruisbeek A, Destree O, Meijlink F, Clevers H (1993) Differential expression of the HMG box factors TCF-1 and LEF-1 during murine embryogenesis. Development 118:439–448

Pai SY, Truitt ML, Ting CN, Leiden JM, Glimcher LH, Ho IC (2003) Critical roles for transcription factor GATA-3 in thymocyte development. Immunity 19:863–875

Palomero T, Lim WK, Odom DT, Sulis ML, Real PJ, Margolin A, Barnes KC, O'Neil J, Neuberg D, Weng AP et al (2006) NOTCH1 directly regulates c-MYC and activates a feed-forward-loop transcriptional network promoting leukemic cell growth. Proc Natl Acad Sci USA 103:18261–18266

Palomero T, Sulis ML, Cortina M, Real PJ, Barnes K, Ciofani M, Caparros E, Buteau J, Brown K, Perkins SL et al (2007) Mutational loss of PTEN induces resistance to NOTCH1 inhibition in T-cell leukemia. Nat Med 13:1203–1210

Park ST, Sun XH (1998) The Tal1 oncoprotein inhibits E47-mediated transcription. Mechanism of inhibition. J Biol Chem 273:7030–7037

Park SW, Wei LN (2003) Regulation of c-myc gene by nitric oxide via inactivating NF-kappa B complex in P19 mouse embryonal carcinoma cells. J Biol Chem 278:29776–29782

Pear WS, Aster JC, Scott ML, Hasserjian RP, Soffer B, Sklar J, Baltimore D (1996) Exclusive development of T cell neoplasms in mice transplanted with bone marrow expressing activated Notch alleles. J Exp Med 183:2283–2291

Pui JC, Allman D, Xu L, DeRocco S, Karnell FG, Bakkour S, Lee JY, Kadesch T, Hardy RR, Aster JC, Pear WS (1999) Notch1 expression in early lymphopoiesis influences B versus T lineage determination. Immunity 11:299–308

Radtke F, Wilson A, Stark G, Bauer M, van Meerwijk J, MacDonald HR, Aguet M (1999) Deficient T cell fate specification in mice with an induced inactivation of Notch1. Immunity 10:547–558

Reizis B, Leder P (2002) Direct induction of T lymphocyte-specific gene expression by the mammalian Notch signaling pathway. Genes Dev 16:295–300
Robb L, Rasko JE, Bath ML, Strasser A, Begley CG (1995) scl, a gene frequently activated in human T cell leukaemia, does not induce lymphomas in transgenic mice. Oncogene 10:205–209
Rothenberg EV (2012) Transcriptional drivers of the T-cell lineage program. Curr Opin Immunol 24:132–138
Rothenberg EV, Moore JE, Yui MA (2008) Launching the T-cell-lineage developmental programme. Nat Rev Immunol 8:9–21
Rothenberg EV, Zhang J, Li L (2012) Multilayered specification of the T-cell lineage fate. Immunol Rev 238:150–168
Schilham MW, Wilson A, Moerer P, Benaissa-Trouw BJ, Cumano A, Clevers HC (1998) Critical involvement of Tcf-1 in expansion of thymocytes. J Immunol 161:3984–3991
Schmitt TM, Zuniga-Pflucker JC (2002) Induction of T cell development from hematopoietic progenitor cells by delta-like-1 in vitro. Immunity 17:749–756
Schmitt TM, Ciofani M, Petrie HT, Zuniga-Pflucker JC (2004) Maintenance of T cell specification and differentiation requires recurrent notch receptor-ligand interactions. J Exp Med 200:469–479
Senawong T, Peterson VJ, Avram D, Shepherd DM, Frye RA, Minucci S, Leid M (2003) Involvement of the histone deacetylase SIRT1 in chicken ovalbumin upstream promoter transcription factor (COUP-TF)-interacting protein 2-mediated transcriptional repression. J Biol Chem 278:43041–43050
Sharma VM, Calvo JA, Draheim KM, Cunningham LA, Hermance N, Beverly L, Krishnamoorthy V, Bhasin M, Capobianco AJ, Kelliher MA (2006) Notch1 contributes to mouse T-cell leukemia by directly inducing the expression of c-myc. Mol Cell Biol 26:8022–8031
Soler E, Andrieu-Soler C, de Boer E, Bryne JC, Thongjuea S, Stadhouders R, Palstra RJ, Stevens M, Kockx C, van Ijcken W et al (2010) The genome-wide dynamics of the binding of Ldb1 complexes during erythroid differentiation. Genes Dev 24:277–289
Taghon T, Yui MA, Pant R, Diamond RA, Rothenberg EV (2006) Developmental and molecular characterization of emerging beta- and gammadelta-selected pre-T cells in the adult mouse thymus. Immunity 24:53–64
Taghon T, Yui MA, Rothenberg EV (2007) Mast cell lineage diversion of T lineage precursors by the essential T cell transcription factor GATA-3. Nat Immunol 8:845–855
Tanigaki K, Tsuji M, Yamamoto N, Han H, Tsukada J, Inoue H, Kubo M, Honjo T (2004) Regulation of alphabeta/gammadelta T cell lineage commitment and peripheral T cell responses by Notch/RBP-J signaling. Immunity 20:611–622
Ting CN, Olson MC, Barton KP, Leiden JM (1996) Transcription factor GATA-3 is required for development of the T-cell lineage. Nature 384:474–478
Verbeek S, Izon D, Hofhuis F, Robanus-Maandag E, te Riele H, van de Wetering M, Oosterwegel M, Wilson A, MacDonald HR, Clevers H (1995) An HMG-box-containing T-cell factor required for thymocyte differentiation. Nature 374:70–74
Vilimas T, Mascarenhas J, Palomero T, Mandal M, Buonamici S, Meng F, Thompson B, Spaulding C, Macaroun S, Alegre ML et al (2007) Targeting the NF-kappaB signaling pathway in Notch1-induced T-cell leukemia. Nat Med 13:70–77
Wadman IA, Osada H, Grutz GG, Agulnick AD, Westphal H, Forster A, Rabbitts TH (1997) The LIM-only protein Lmo2 is a bridging molecule assembling an erythroid, DNA-binding complex which includes the TAL1, E47, GATA-1 and Ldb1/NLI proteins. EMBO J 16:3145–3157
Wakabayashi Y, Watanabe H, Inoue J, Takeda N, Sakata J, Mishima Y, Hitomi J, Yamamoto T, Utsuyama M, Niwa O et al (2003) Bcl11b is required for differentiation and survival of alphabeta T lymphocytes. Nat Immunol 4:533–539
Weber BN, Chi AW, Chavez A, Yashiro-Ohtani Y, Yang Q, Shestova O, Bhandoola A (2011) A critical role for TCF-1 in T-lineage specification and differentiation. Nature 476:63–68

Wei G, Abraham BJ, Yagi R, Jothi R, Cui K, Sharma S, Narlikar L, Northrup DL, Tang Q, Paul WE et al (2011) Genome-wide analyses of transcription factor GATA3-mediated gene regulation in distinct T cell types. Immunity 35:299–311

Weng AP, Ferrando AA, Lee W, Morris JPT, Silverman LB, Sanchez-Irizarry C, Blacklow SC, Look AT, Aster JC (2004) Activating mutations of NOTCH1 in human T cell acute lymphoblastic leukemia. Science 306:269–271

Weng AP, Millholland JM, Yashiro-Ohtani Y, Arcangeli ML, Lau A, Wai C, Del Bianco C, Rodriguez CG, Sai H, Tobias J et al (2006) c-Myc is an important direct target of Notch1 in T-cell acute lymphoblastic leukemia/lymphoma. Genes Dev 20:2096–2109

Wilson NK, Foster SD, Wang X, Knezevic K, Schutte J, Kaimakis P, Chilarska PM, Kinston S, Ouwehand WH, Dzierzak E et al (2010) Combinatorial transcriptional control in blood stem/progenitor cells: genome-wide analysis of ten major transcriptional regulators. Cell Stem Cell 7:532–544

Wolfer A, Wilson A, Nemir M, MacDonald HR, Radtke F (2002) Inactivation of Notch1 impairs VDJbeta rearrangement and allows pre-TCR-independent survival of early alpha beta Lineage Thymocytes. Immunity 16:869–879

Wong GW, Knowles GC, Mak TW, Ferrando AA, Zuniga-Pflucker JC (2012) HES1 opposes a PTEN-dependent check on survival, differentiation, and proliferation of TCRbeta-selected mouse thymocytes. Blood 120:1439–1448

Xi H, Schwartz R, Engel I, Murre C, Kersh GJ (2006) Interplay between RORgammat, Egr3, and E proteins controls proliferation in response to pre-TCR signals. Immunity 24:813–826

Xu W, Carr T, Ramirez K, McGregor S, Sigvardsson M, Kee BL (2013) E2A transcription factors limit expression of Gata3 to facilitate T lymphocyte lineage commitment. Blood 121:1534–1542

Yamashita M, Shinnakasu R, Asou H, Kimura M, Hasegawa A, Hashimoto K, Hatano N, Ogata M, Nakayama T (2005) Ras-ERK MAPK cascade regulates GATA3 stability and Th2 differentiation through ubiquitin-proteasome pathway. J Biol Chem 280:29409–29419

Yashiro-Ohtani Y, He Y, Ohtani T, Jones ME, Shestova O, Xu L, Fang TC, Chiang MY, Intlekofer AM, Blacklow SC et al (2009) Pre-TCR signaling inactivates Notch1 transcription by antagonizing E2A. Genes Dev 23:1665–1676

Yu Q, Sharma A, Oh SY, Moon HG, Hossain MZ, Salay TM, Leeds KE, Du H, Wu B, Waterman ML et al (2009) T cell factor 1 initiates the T helper type 2 fate by inducing the transcription factor GATA-3 and repressing interferon-gamma. Nat Immunol 10:992–999

Yu S, Zhou X, Steinke FC, Liu C, Chen SC, Zagorodna O, Jing X, Yokota Y, Meyerholz DK, Mullighan CG et al (2012) The TCF-1 and LEF-1 transcription factors have cooperative and opposing roles in T cell development and malignancy. Immunity 37:813–826

Zhang JA, Mortazavi A, Williams BA, Wold BJ, Rothenberg EV (2012) Dynamic transformations of genome-wide epigenetic marking and transcriptional control establish T cell identity. Cell 149:467–482

Zhong Y, Jiang L, Hiai H, Toyokuni S, Yamada Y (2007) Overexpression of a transcription factor LYL1 induces T- and B-cell lymphoma in mice. Oncogene 26:6937–6947

Chapter 12
Epigenetic Control of T-Cell Receptor Locus Rearrangements in Normal and Aberrant Conditions

Beatriz del Blanco, Úrsula Angulo, and Cristina Hernández-Munain

Abstract V(D)J recombination is the process responsible for the exclusive expression of one antigen receptor form, either T-cell receptor (TCR) or immunoglobulin (Ig), per individual T or B lymphocyte, respectively. This process is, therefore, essential for adaptive immune responses in vertebrates and it consists of the assembly of disperse gene segments present at the TCR and Ig loci to obtain a genetic structure that is able to encode a functional protein. V(D)J recombination is highly regulated during lymphocyte development and depends on the activation of accessibility control elements (ACEs) to direct the recruitment of the specific endonuclease RAG-1/2 to its site-specific target sequences that flank the gene segments. The RAG-1/2-mediated DNA cleavage at these loci is controlled by changes in the epigenome. These changes are derived from chromatin modification, transcriptional elongation, the location of the loci within the nucleus, and the three-dimensional architecture of the loci, which is controlled by functional interplay among ACEs and their bound trans-factors. This epigenetic control ultimately leads to the juxtapositioning or synapsis of two gene segments that are normally distantly located and the recombination reaction itself. This process must be completed without any errors to avoid the potential risk of generating aberrant translocations that could result in the generation of leukemia. This chapter summarizes the advances in this area at the TCR loci and the importance of regulating this potentially risky process during thymocyte development.

Keywords Allelic exclusion • Ataxia telangiectasia • Chromatin • Enhancer • Epigenetic • Histone • Immunodeficiency • Leukemia • Immunoglobulin • Nucleosome • Omenn syndrome • Promoter • T-cell development • T-cell receptor • Thymocyte • Thymus • Transcription • V(D)J recombination

B. del Blanco • Ú. Angulo • C. Hernández-Munain (✉)
Department of Cellular Biology and Immunology, Instituto de Parasitología y Biomedicina "López-Neyra" (IPBLN-CSIC), Consejo Superior de Investigaciones Científicas, Avenida del Conocimiento s/n, 18016-Armilla, Granada, Spain
e-mail: chmunain@ipb.csic.es

C. Bonifer and P.N. Cockerill (eds.), *Transcriptional and Epigenetic Mechanisms Regulating Normal and Aberrant Blood Cell Development*, Epigenetics and Human Health, DOI 10.1007/978-3-642-45198-0_12, © Springer-Verlag Berlin Heidelberg 2014

Abbreviations

ACE	Accessibility control element
AgR	Antigen receptor
AT	Ataxia telangiectasia
ATM	Ataxia telangiectasia mutated kinase
BCR	B-cell receptor
bHLH	Basic helix-loop-helix
C	Constant
CTCF	CCCTC-binding factor; constant region
CLP	Common lymphoid progenitor
D	Diversity
DN	Double negative
DNA-PKcs	DNA-dependent protein kinase
DP	Double positive
Eα	*Tcra* enhancer
Eβ	*Tcrb* enhancer
Eδ	*Tcrd* enhancer
Eγ	*Tcrg* enhancer
FISH	Fluorescence in situ hyhbridization
H3ac	Histone H3 acetylation
H3K4me2/3	Dimethylation or trimethylation of lysine 4 of histone H3
H3K9me2/3	Dimethylation or trimethylation of lysine 9 of histone H3
H3K27me2/3	Dimethylation or trimethylation of lysine 27 of histone H3
Ig	Immunoglobulin
J	Joining
LCR	Locus control region
Lig4	Cernunnos/XLF-Xrcc4/Dna ligase IV complex
N	Non-template
NHEJ	Nonhomologous end-joining
OS	Omenn syndrome
RAG	Recombination activating gene
RSS	Recombination signal sequence
SCID	Severe combined immunodeficiency disease
SP	Single positive
T-ALL	T-cell acute leukemia
TCR	T-cell receptor
TEA	T early α
TF	Transcription factor
V	Variable

12.1 Molecular Basis of V(D)J Recombination at the TCR Loci During T Lymphocyte Development

12.1.1 V(D)J Recombination: Assembly of the TCR Loci to Generate Functional Proteins

Lymphocyte differentiation constitutes one of the best examples for investigating the molecular mechanisms of epigenetic control during cell development because it occurs through a series of very well-characterized stages. The generation of T and B lymphocytes requires precise orchestration of the expression of their antigen receptors (AgRs) in the context of a highly ordered program of cellular differentiation (Cobb et al. 2006; Osipovich and Oltz 2010). Each mature lymphocyte expresses a different AgR, consisting of either a T-cell receptor (TCR) in T lymphocytes or a B-cell receptor (BCR) in B lymphocytes. TCRs are composed of either α and β chains (αβTCR) or γ and δ chains (γδTCR), whereas BCRs are composed of two immunoglobulin (Ig) heavy chains (IgH) and two Ig light chains (Igκ or Igλ). The genes that encode these chains are formed by dispersed V (variable), D (diversity), and J (joining) gene segments. Although the number of V, D, and J segments is limited, the number of different AgRs generated through the random combination of V(D)J segments is enormous, resulting in a very vast repertoire of different receptors to afford adaptive immunity. To generate a different gene structure (VDJ or VJ) that permits the expression of a functional AgR, these gene segments rearrange through a cutting and pasting process known as V(D)J recombination. The V and J segments are present in all seven AgR loci, whereas the D segments are present only in the *Tcrb*, *Tcrd*, and *Igh* loci. Hence, to produce the TCRγ, TCRα, Igκ, and Igλ chains, one of several V gene segments is joined to one of several J gene segments in only one recombination event to generate a VJ genomic structure. However, to produce the TCRβ, TCRδ, and IgH chains, two recombination events are required to ultimately join the V, D, and J gene segments in a VDJ genomic structure.

A brief description of the basic features of V(D)J recombination is provided below. V(D)J recombination is initiated through the action of a specific endonuclease that is formed by the protein products of recombination activation gene (*RAG*) 1 and 2. Together, these proteins form a complex (herein called RAG-1/2) that is specifically expressed during lymphocyte development. RAG-1/2 recognizes the conserved recombination signal sequences (RSSs) that flank the V, D, and J gene segments, inducing double-strand cleavage at the precise border between the gene segment and the RSS. Each V, D, and J gene segment is flanked by one or two RSSs, which allows for their specific recognition by RAG-1/2. These RSSs consist of conserved heptamer and nonamer elements that are separated by a less conserved spacer region of 12 or 23 base pairs. Recombination between two RSSs is limited by a rule known as 12–23, in which rearrangement is limited to a RSS containing a 12-bp spacer with another RSS containing a 23-bp spacer. Subsequently, the combined action of the DNA damage response pathway and the classic nonhomologous end-joining (NHEJ) machinery repairs the breaks (Lieber 2010; Helmink and Sleckman 2012). RAG-1/2-induced cleavage leads to the formation of two hairpin-sealed coding ends that require modification prior to ligation and two blunt RSSs ends that are directly ligated, as follows. First, the heterodimer Ku70/Ku80 binds to the coding ends and recruits the catalytic subunit of the DNA-dependent kinase (DNA-PKcs), which interacts with Artemis and confers it with a DNA endonuclease activity that is capable of opening the RAG-1/2-generated hairpins. Then, the deoxyribonucleotidyl transferase adds non-template (N) nucleotides to the newly formed junction, which is ligated by the Cernunnos/XLF-Xrcc4/DNA ligase IV (Lig4) complex. After this process, two products with different structures are generated, a signal joint and a coding joint, with the former corresponding to the ligation of the two RSSs that results in an extrachromosomal circle and the latter corresponding to the ligation of the two rearranged gene segments with junction variability due to the deletion and addition of the N nucleotides.

12.1.2 Genomic Architecture of the TCR Loci

The sequencing analysis of the genomic regions encompassing the human and mouse TCR loci revealed a strong similarity between the two species (Glusman et al. 2001). Although some differences have been found between the human and mouse sequences, each locus conserves the basic structure, as well as gene associations with other gene families such as the olfactory receptor genes that are interspersed among the 5′ Vα gene segments and the trypsinogen genes that flank the 5′ and 3′ ends of the Vβ gene segments. Similarly, the identities of the immediate neighboring genes are similarly conserved across species. For clarity, here we describe the genomic architecture of the mouse TCR loci. A comprehensive comparative study of the human and mouse TCR loci has been published elsewhere (Glusman et al. 2001).

Tcra and *Tcrd* are organized in a single genetic locus, *Tcra/Tcrd*, which spans 1.7 Mb on mouse chromosome 14 (Bosc and Lefranc 2003; Cobb et al. 2006; Genolet et al. 2012). The 1-Mb 5′ region includes a large pool of Vα gene segments and a small pool of Vδ gene segments (132V gene segments including pseudogenes) (Fig. 12.1). Some of the V gene segments can rearrange to either Jα or Dδ, contributing to both the TCRα and the TCRδ repertoires (Krangel et al. 2004). The 3′ region of the locus includes 61Jα gene segments as well as the constant region (C) Cα exons. Nested between these two regions are a small number of Dδ and Jδ gene segments as well as the Cδ exons. The nested organization of *Tcra/Tcrd* dictates that VδDδJδ and VαJα rearrangements cannot coexist on the same chromosome. In fact, Vα-to-Jα rearrangement results in the deletion of *Tcrd*.

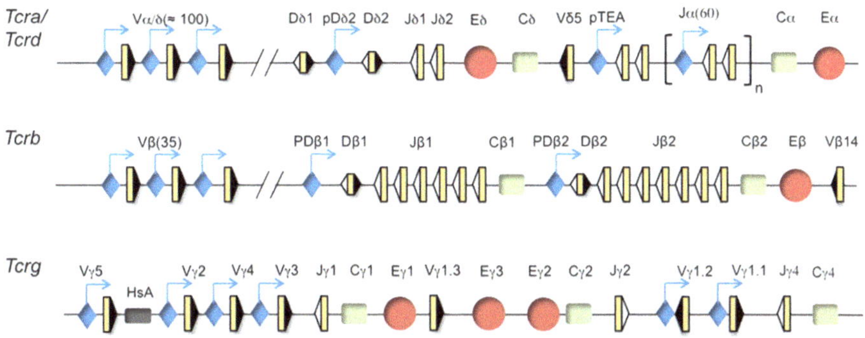

Fig. 12.1 Genomic structure of the mouse TCR loci. The V, D, and J gene segments are represented by *yellow rectangles*, and the RSSs are shown as *black* (containing a 23-bp spacer) or *white* (containing a 12-bp spacer) *triangles*. C regions are represented by *green rectangles*. Promoters are represented as *blue diamonds*. *Blue arrows* represent active sites for germline transcription. Enhancers are represented as *red circles*. *Tcra*/*Tcrd*: The 1-Mb 5′ region includes Vα and Vδ gene segments. All Vα/δ gene segments are flanked by a RSS containing a 23-bp spacer at their 3′ ends. The 3′ region of the locus includes the Jα gene segments as well as Cα. All Jα gene segments are flanked by an RSS containing a 12-bp spacer at their 5′ ends. Nested between these two regions are a small number of Dδ and Jδ gene segments as well as Cδ. The Dδ gene segments are flanked by an RSS containing a 12-bp spacer at their 5′ ends and by an RSS containing a 23-bp spacer at their 3′ ends. The Jδ gene segments are flanked by an RSS containing a 12-bp spacer at their 5′ ends. Between Cδ and the cluster of Jα gene segments, there is a single Vδ gene segment, Vδ5, which is inversely oriented with respect to the other Vα/δ gene segments and can rearrange to Dδ by inversion. *Tcrb*: The 5′ region is composed of Vβ gene segments. All Vβ gene segments are flanked by an RSS containing a 23-bp spacer at their 3′ ends. The 3′ region contains two DβJβ clusters associated with one Cβ region, Cβ1 or Cβ2. The Dβ gene segments are flanked by an RSS containing a 12-bp spacer at their 5′ ends and by an RSS containing a 23-bp spacer at their 3′ ends. The Jβ gene segments are flanked by an RSS containing a 12-bp spacer at their 5′ ends. In addition, there is a single Vβ element, Vβ14, which lies downstream of Cβ2 and can rearrange to Dβ1 or Dβ2 by inversion. *Tcrg*: This locus contains Vγ gene segments that are interspersed among three functional and one nonfunctional JγCγ clusters containing one Jγ gene segment and one Cγ region. The Vγ gene segments are flanked by an RSS containing a 23-bp spacer at their 3′ ends, whereas the Jγ gene segments are flanked by an RSS containing a 12-bp spacer at their 5′ ends

Tcrb spans 1 Mb on mouse chromosome 6 (Cobb et al. 2006). The 5′ region is composed of 35Vβ gene segments (21 are functional, and the remaining 14 are nonfunctional) (Fig. 12.1). The 3′ region contains two clusters with one Dβ and six Jβ gene segments. Each DβJβ cluster is associated with coding exons for one Cβ region, Cβ1 or Cβ2. Although the distribution of RSSs with 12-bp and 23-bp spacers should permit direct V-to-J rearrangements, these joints are very rare in vivo due to a restriction known as the “beyond 12/23 restriction” (Bassing et al. 2000).

Tcrg spans a short region of 0.2 Mb on mouse chromosome 13 (Cobb et al. 2006). This locus contains seven functional and one nonfunctional Vγ gene segments that are interspersed among three functional and one nonfunctional clusters containing one Jγ gene segment and the Cγ exons (Fig. 12.1).

12.1.3 Regulation of V(D)J Recombination at the TCR Loci During T Lymphocyte Development

Both T and B lymphocytes develop from common lymphoid progenitor (CLP) cells in the bone marrow (Fig. 12.2). CLPs lack lymphocyte surface markers, although they present incomplete D_HJ_H rearrangements within *Igh*, but they can differentiate into both T and B lymphocytes. Some precursors migrate to the thymus (thymocytes) where they receive the signaling mediated by the Notch receptors to commence their differentiation into the T lymphocyte lineage, whereas the lymphoid precursors that remain in the bone marrow differentiate into the B lymphocyte lineage (Hayday and Pennington 2007). The TCR loci are specifically activated and repressed at each of these stages, influencing cellular identity (Cobb et al. 2006). The specific expression of these loci permits the generation of two different lineages of T lymphocytes, αβ and γδ T lymphocytes.

Most immature thymocytes are known as double-negative (DN) thymocytes, as they are $CD4^-CD8^-$. DN thymocytes can be classified into four populations (DN1-4) based on the expression of CD25 and CD44: DN1 ($CD25^-CD44^+$), DN2 ($CD25^+CD44^+$), DN3 ($CD25^+CD44^-$), and DN4 ($CD25^-CD44^-$) (Pearse et al. 1989; Godfrey et al. 1993). *Tcrd* becomes transcriptionally active in late DN1 thymocytes (Dik et al. 2005; Prinz et al. 2006). Recently completed VγJγ and VδDδJδ rearrangements at *Tcrg* and *Tcrd*, respectively, and incomplete DβJβ rearrangements at *Tcrb* are detected in DN2 thymocytes, whereas extensive rearrangements at *Tcrg* and *Tcrd* and completed VβDβJβ rearrangements at *Tcrb* are detected in DN3 thymocytes (Godfrey et al. 1994; Tourigny et al. 1997; Capone et al. 1998; Livak et al. 1999; Cobb et al. 2006). Based on the expression of the activation marker CD27, DN3 thymocytes can be subdivided into two populations: DN3a ($CD27^-$) and DN3b ($CD27^+$) (Taghon et al. 2006). DN2 and DN3a thymocytes that have successfully rearranged their *Tcrg* and *Tcrd* genes express a γδ TCR and normally differentiate along the γδ T-cell pathway in a process known as γδ-selection (Fig. 12.2) (Prinz et al. 2006; Taghon et al. 2006). DN3a thymocytes that express a TCRβ chain differentiate into DN3b, DN4, and $CD4^+CD8^+$ double-positive (DP) thymocytes in a process known as β-selection (Fig. 12.2). The β-selection process is driven by signaling mediated by the pre-TCR (composed of a TCRβ and the invariant pre-Tα chain) and Notch receptors (Fehling et al. 1995; Ciofani et al. 2004). Pre-TCR signaling mediates maintenance of *Tcrb* allelic exclusion by the feedback inhibition of Vβ to DβJβ recombination as well as the activation of transcription and rearrangement of *Tcra* in DP thymocytes (Ciofani et al 2004; Jackson and Krangel 2006). In addition to these effects on *Tcra* and *Tcrb*, the signaling mediated by the pre-TCR inhibits the expression of the Notch receptors and pre-Tα (Taghon et al. 2006). In DP thymocytes, the pre-Tα present in the pre-TCRs is substituted by a TCRα chain to constitute an αβ TCR, which allows positive and negative selection processes to produce a pool of $CD4^+$ and $CD8^+$ single-positive (SP) thymocytes that migrate to the periphery as mature αβ T lymphocytes.

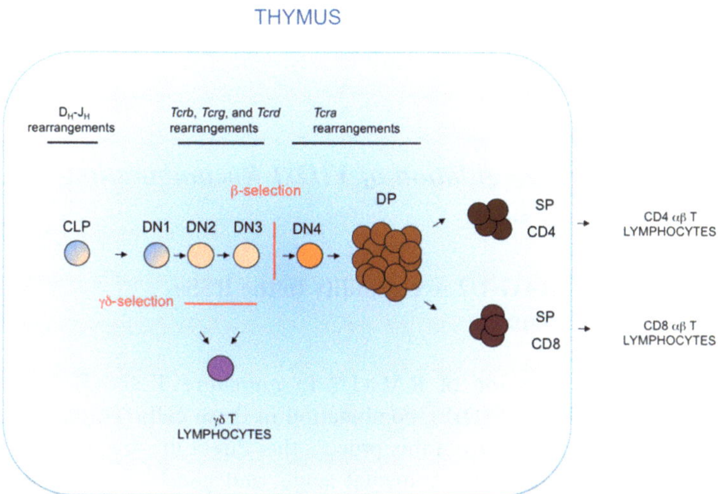

Fig. 12.2 V(D)J recombination at the TCR loci during thymocyte development. This schematic diagram of thymocyte development depicts the various developmental stages and the periods of TCR locus rearrangements. β- and γδ-selection, which are processed based on expression of a pre-TCR or a γδ TCR, are indicated in *red*. Commitment to the T-cell lineage is indicated by transition from *blue* to *salmon-pink*. αβ T lymphocyte maturation is indicated by transition to *brown*, and γδ T lymphocytes are represented in *purple*

In DN3a thymocytes, allelic exclusion at *Tcrb* restricts the assembly of a productively rearranged Vβ to DβJβ recombination to one single allele (Brady et al. 2010). The assembly and expression of a functional TCRβ in DN3a thymocytes activate intracellular pathways that signal feddback inhibition of further *Tcrb* rearrangements, cessation of RAG-1/2 expression, and differentiation to DP thymocytes. To maintain *Tcrb* allelic exclusion in DP thymocytes, Vβ-to-DβJβ rearrangements are suppressed on DβJβ rearranged alleles following RAG-1/2 re-expression in these cells. The molecular mechanisms by which it is achieved will be discussed later in the section corresponding to epigenetic regulation at *Tcrb*. In contrast, *Tcrb*, *Tcrd*, and *Tcra* are not subjected to allelic exclusion and productively assembled genes are frequently biallelically expressed (Malissen et al. 1992; Corthay et al. 2001). Although both *Tcra* alleles simultaneously attempt to create a productive VαJα rearrangement (Mauvieux et al. 2001), some data suggest that rearrangement at *Tcra* occurs in a monoallelic fashion at a given time in DP thymocytes (Chen et al. 2000; Chaumeil et al. 2013). In any case, TCRαβ monospecificity is maintained by the preferential pairing of one of the two TCRαs with the expressed TCRβ in a significant proportion of T lymphocytes that express two functional TCRαs in the cytoplasm. Expression of a functional αβ TCR on DP thymocytes terminates *Tcra* rearrangements by downregulating RAG-1/2 expression. DP thymocytes that are unable to express a functional αβTCR will die by neglect. In this way, DP thymocytes can undergo multiple rounds of Vα-to-Jα recombination until these cells express an αβ TCR that progresses through the selection checkpoints (Huang et al. 2005).

12.1.4 Epigenetic Regulation of V(D)J Recombination at the TCR Loci

12.1.4.1 Control of RAG-1/2 Accessibility to the RSSs and RSS Synapsis

The restriction of expression of RAG-1/2 to immature T and B lymphocytes explains the specificity of V(D)J recombination in these cells. Furthermore, there are three additional restrictions in this process that direct the reaction in a manner that is specific to lineage, developmental stage, and gene segment order (Cobb et al. 2006; Osipovich and Oltz 2010). The lineage specificity determines that the complete rearrangement of the TCR and Ig loci is restricted to T and B lymphocytes, respectively. The developmental stage specificity determines that during T lymphocyte development, *Tcrb*, *Tcrg*, and *Tcrd* rearrange earlier than *Tcra*, whereas during B lymphocyte development, *Igh* rearranges earlier than *Igl* and *Igk*. The order of rearrangement among the different gene segments within each locus determines that the D-to-J rearrangements proceed before the V-to-DJ rearrangements in *Tcrb* and *Igh*. These three additional levels of regulation are achieved by controlling the accessibility to RAG-1/2 in native chromatin (Stanhope-Baker et al. 1996; Ji et al. 2010b). This is the basis for the "accessibility model" for the control of V(D)J recombination that was proposed more than 25 years ago (Yancopulos and Alt 1985).

It is known that gene activation and silencing correlate with the presence of specific epigenetic marks in chromatin, including histone tail modification, nucleosome positioning, and methylation of CpG nucleotides. Not surprisingly, the epigenetic changes correlate with RAG-1/2 accessibility at the AgR loci; the accessible loci are enriched in epigenetic marks associated with gene activation, whereas the inaccessible loci are enriched in epigenetic marks associated with gene silencing (del Blanco et al. 2011; Jaeger et al. 2013). In this way, the accessibility of RAG-1/2 to the AgR loci chromatin correlates with histone H3 (H3ac) and H4 acetylation, di- and trimethylation of lysine 4 of histone H3 (H3K4me2/3), nuclease accessibility, DNA hypomethylation, and changes in nucleosomal structure, while the inaccessibility of RAG-1/2 to the AgR loci correlates with the di- and trimethylation of lysine 9 and 27 of histone H3 (H3K9me2/3 and H3K27me2/3) (Cobb et al. 2006; Kondilis-Mangum et al. 2010; Osipovich and Oltz 2010; Jaeger et al. 2013). Furthermore, it has been established that inducing the methylation of lysine 9 of histone H3 (H3K9me), which is a hallmark for silent chromatin, suppresses its Dβ-to-Jβ recombination (Osipovich et al. 2004). Moreover, blocking transcriptional elongation by introducing a terminator into the Jα gene cluster at

Tcra effectively suppresses chromatin remodeling and Vα-to-Jα recombination when Jα is positioned 3′ from the terminator (Abarrategui and Krangel 2006, 2007; del Blanco et al. 2011). All of these studies indicate that the epigenetic chromatin modifications associated with gene activation are necessary to provide RAG-1/2 with accessibility to RSSs, whereas those that are associated with gene silencing are inhibitory of RAG-1/2 recruitment. These findings correlate with in vitro studies that demonstrate that the assembly of RSSs into nucleosomes, compared with nude DNA, inhibits V(D)J recombination (Kwon et al. 1998; Golding et al. 1999), supporting the notion that nucleosomes impede the access of RAG-1/2 to chromatin. This barrier to RAG-1/2 binding imposed by nucleosomes can be surmounted by ATP-dependent chromatin remodeling complexes, such as SWI/SNF, and by histone acetylation (Kwon et al. 2000; McBlane and Boyes 2000; McMurry and Krangel 2000; Osipovich et al. 2007; Collins et al. 2013). Together, these data indicate that open chromatin facilitates the binding of RAG-1/2 to RSSs to permit its function.

These epigenetic changes are reversible and dynamic and are set through the recruitment of specific enzymes by specific transcription factors (TFs) bound to ACEs, such as enhancers and promoters (Cobb et al. 2006; Osipovich and Oltz 2010). In fact, the binding of RAG-1 to specific RSSs in TCR loci chromatin is directly mediated through the activation of transcriptional enhancers and promoters as well as by germline transcription (Ji et al. 2010a), whereas RAG-2 is generally recruited to open and active chromatin through its binding to H3K4me3 (Liu et al. 2007; Ramón-Maiques et al. 2007; Ji et al. 2010b). This mark is highly enriched near transcription initiation sites at the promoters of actively transcribed genes (Wang et al. 2008). The recruitment of RAG-1 to RSSs seems to occur in a focal manner to small regions called recombination centers, which display an enrichment of histone modifications related to the activation of gene expression, such as H3ac and H3K4me3 (Ji et al. 2010b). These recombination centers are defined as sites with a high local RAG-1/2 concentration and are localized at the 3′ end of each AgR locus to presumably facilitate RSS synapsis and V(D)J recombination. This model is in agreement with recent data demonstrating that PAXIP1, a protein associated with MLL3 and MLL4 methyltransferases and the DNA damage response, and TRIM28 (also named KAP1 or TIF1β), a protein associated with nucleosome remodeling, regulate RAG-1/2-mediated cleavage and repair during VαJα recombination in DP thymocytes through Jα H3K4me3, germline transcription, and double-strand break formation (Callen et al. 2012; Zhou et al. 2012).

Each TCR locus is equipped with at least one transcriptional enhancer near the 3′ end of the C region and numerous promoters that are associated with the V, D, and J gene segments or groups of gene segments (Fig. 12.1). These promoters drive sterile transcripts in germline loci. However, upon productive rearrangements, the V-associated promoters drive the transcription of the rearranged loci. Numerous studies using transgenic miniloci as recombination reporters as well as directed mutagenesis at the endogenous loci have demonstrated the essential role of enhancers and promoters in controlling the accessibility of RAG-1/2 to TCR locus chromatin (Cobb et al. 2006; Osipovich and Oltz 2010). The first evidence for this function of ACEs was obtained from early studies using *Tcrb*- and *Tcrd*-based transgenic substrates (Ferrier et al. 1990; Capone et al. 1993; Lauzurica and Krangel 1994a). These pioneer experiments demonstrated a direct role for transcriptional enhancers in targeting efficient recombination. Consistent with this idea, the germline deletion of specific enhancers and promoters within a particular TCR locus inhibits its rearrangement. The specific deletion of each of the enhancers present at *Tcra* (Eα), *Tcrb* (Eβ), *Tcrd* (Eδ), and *Tcrg* (Eγ) has been performed, as well as the elimination of some promoters associated with specific gene segments, such as the promoters associated with the Jα61, known as T early α (TEA) promoter, Jα49, Dβ1, Vβ13, Vγ2, and Vγ3 gene segments (Bories et al. 1996; Bouvier et al. 1996; Villey et al. 1996; Sleckman et al. 1997; Monroe et al. 1999; Whitehurst et al. 2000; Xiong et al. 2002; Ryu et al. 2004; Hawwari et al. 2005). These experiments have demonstrated that enhancers are responsible for the lineage and developmental stage specificity of V(D)J recombination through the general regulation of chromatin at multiple gene segments that are separated by large distances, whereas promoters locally mediate the accessibility of RAG-1/2 to a gene segment or group of gene segments. In addition to the requirement of opening the chromatin structure via the activation of enhancers and promoters for the induction of V(D)J recombination, in some cases, such as the activation of *Tcra* rearrangements, germline transcription is also required (Abarrategui and Krangel 2006, 2007; Ji et al. 2010a). These findings indicate that Vα-to-Jα recombination requires the elongation machinery to travel through the Jα RSSs to allow RAG-1/2 to access to the RSS chromatin, at least for Jα RSSs that are positioned far away from a promoter. However, this is not the case for RSSs that are tightly associated with a promoter, such as the Dβ1 gene segment, which depends solely on promoter activation through the recruitment of its specific TFs (Whitehurst et al. 1999; Sikes et al. 2002).

ACEs function according to their occupancy by specific TFs. The importance of TFs in controlling V(D)J recombination and transcription is derived from their specific roles in recruiting histone-modifying enzymes and the RNAPII complex as well as in their direct participation in the establishment of long-range physical interactions between ACEs. The importance of specific TFs in controlling V(D)J recombination has been addressed by several approaches, such as the mutation of DNA binding sites and the elimination or expression of the specific factor. Several TFs, including both constitutive TFs, such as the basic helix-loop-helix (bHLH) factors E2A and HEB, Ikaros, Ets-1, Sp1, GATA-3, Runx factors, TCF-1/LEF-1, CREB/ATF factors, and c-Myb, and inducible TFs, such as Notch1, STAT5, NFκB, NFAT, AP-1, and Egr-1, are involved in the regulation of the ACEs responsible for recombination of the different TCR loci (Hsiang et al. 1993, 1995; Hernández-Munain and Krangel 1994; Giese et al. 1995; Sun et al. 1995; Hernández-Munain et al. 1996, 1999; Roberts et al. 1997; Sikes et al. 1998; Bain et al. 1999; Spicuglia et al. 2000, 2002; Tripathi et al. 2000; Ghosh et al. 2001; Ye et al. 2001; Wolfer et al. 2002; Eyquem et al. 2004; Carabana et al. 2005; Agata et al. 2007; McMillan and Sikes 2008, 2009; del Blanco et al. 2009, 2012; Tani-ichi et al. 2011; Collins et al. 2013). Many of these DNA binding proteins are members of larger families that can have different or redundant functions during T-cell development. For example, Runx1 and Runx3 play critical different roles during thymocyte development (Kohu et al. 2008). Runx1 is required for the DN to DP transition and maturation of $CD4^+$ SP thymocytes, while Runx3 is important for $CD8^+$ SP thymocyte differentiation (Egawa et al. 2007). The role of each Runx protein is dependent on its expression pattern during the different stages of thymocyte differentiation. Runx1 expression is high in DN3a thymocytes and it is downregulated by pre-TCR signaling, remaining present in DP and SP thymocytes, while Runx3 expression is produced mainly in $CD8^+$ SP thymocytes (Sato et al. 2005; Taghon et al. 2006). All TCR enhancers contain essential binding sites for Runx proteins. Although both Runx1 and Runx3 can activate TCR locus transcription when they are ectopically expressed (Tani-ichi et al. 2011), the expression patterns and functions of Runx1 and Runx3 during thymocyte development suggest that Runx1 is the factor that activates all TCR enhancers.

In addition to the chromatin structure imposed by the activation of ACEs and transcriptional elongation, several studies have demonstrated that AgR loci suffer from changes in their location within the nucleus as well as changes in their three-dimensional architecture that are important to allow for and facilitate the formation of synapses between distant RSSs. These synapses are a real challenge because the participating RSSs might be separated by 1–2 Mb on the chromosome. Using 3D fluorescence in situ hybridization (3D-FISH), two dynamic phenomena have been observed during thymocyte development that are relevant to the formation of RSS synapses: (1) changes in localization of the AgR alleles within the nucleus relative to repressive compartments, such as heterochromatin and nuclear lamina, facilitate or inhibit RSS synapses, and (2) changes in the locus configuration that result in a large-scale compacted structure bring together distant gene segments to facilitate their synapsis or, in contrast, that result in a de-contracted structure separate gene segments to avoid their synapsis. The former phenomenon is involved in controlling allelic exclusion at *Tcrb*, whereas the latter one is involved in the control of Vβ-to-DβJβ rearrangements at *Tcrb* and Vα-to-Jα rearrangements at *Tcra* during T-cell development (Skok et al. 2007; Schlimgen et al. 2008; Shih and Krangel 2010; Kondilis-Mangum et al. 2011). The details of how these phenomena are thought to be involved in the regulation of recombination at these specific loci will be discussed later in the sections corresponding to epigenetic regulation at *Tcra* and *Tcrb*.

12.1.4.2 Epigenetic Regulation of Tcra/Tcrd Recombination

Tcrd and *Tcra* have distinct developmental programs of recombination and expression, which is particularly striking given the organization of their gene segments within a single genetic locus, *Tcra/Tcrd* (Fig. 12.1). *Tcrd* germline transcription and rearrangement occur in DN2/DN3a thymocytes, whereas *Tcra* germline transcription and rearrangement occur in DN4/DP thymocytes (Capone et al. 1998; Livak et al. 1999; Dik et al. 2005). Because Vα-to-Jα recombination induces the deletion of the *Tcrd* locus, this process irreversibly commits those thymocytes to the αβ T lymphocyte lineage. Hence, the regulation of *Tcrd* and *Tcra* rearrangements during thymocyte development is a critical component of αβ versus γδ T lymphocyte commitment. The two enhancers present in the locus, Eδ and Eα, control the rearrangements of *Tcrd* and *Tcra*, respectively (Sleckman et al. 1997; Monroe et al. 1999). Germline deletion of Eα results in the severe reduction of germline Jα transcription and Vα-to-Jα recombination, but it does not alter *Tcrd* recombination although it attenuates *Tcrd* transcription in γδ T lymphocytes, whereas elimination of Eδ severely impairs *Tcrd* recombination without affecting transcription and recombination at *Tcra* (Sleckman et al. 1997; Monroe et al. 1999). Eα is part of a locus control region (LCR) containing eight T-cell-specific DNaseI hypersensitive sites located between the *Tcra* locus and the ubiquitously expressed *Dad1* gene (Diaz et al. 1994). Eα activates *Tcra* germline transcription and Vα-to-Jα rearrangement in DP thymocytes, whereas other elements at the LCR act as insulator sequences that block enhancer activity to presumably manage the separate regulatory programs of *Tcra* and *Dad1* genes (Ortiz et al. 1999, 2001).

These two enhancers also orchestrate the different developmental programs at *Tcrd* and *Tcra*; Eδ is active in DN3a thymocytes and inactive in DP thymocytes, and Eα is inactive in DN3a thymocytes and active in DP thymocytes (Lauzurica and Krangel 1994b; Hernández-Munain et al. 1999). Inactivation of Eδ in DP thymocytes is mediated through the dissociation of its bound TFs due presumably to the pre-TCR-induced inhibition of Runx1 expression, which functions as a structural factor in the formation of the Eδ enhanceosome (Hernández-Munain et al. 1999; Hernández-Munain and Krangel 2002; Taghon et al. 2006). Eδ functions as a local enhancer whose influence is limited to a 10- to 20-kb region that includes the Dδ, Jδ, and Vδ5 gene segments in adult DN3a thymocytes, although it has a long-distance effect over 55 kb in which it promotes the accessibility of the Vδ4 gene segment in fetal DN3a thymocytes (Hao and Krangel 2011). Mutational analysis of Eδ revealed a critical function for c-Myb and Runx1 in Vδ-to-DδJδ recombination (Hernández-Munain et al. 1996; Lauzurica et al. 1997). However, although necessary, these two TFs are not sufficient and must collaborate with other Eδ binding factors to establish Jδ gene segment accessibility to RAG-1/2 (Lauzurica et al. 1997).

One interesting aspect in the regulation of *Tcrd* and *Tcrg* recombination is related to the sequential rearrangements of specific Vδ and Vγ gene segments during development. This regulation results in the generation of distinct waves of γδ T lymphocytes that express specific Vδ and Vγ gene segments over the life of an organism (Elliot et al. 1988; Korman et al. 1988). Rearrangements involving Vδ1 and Vδ4 are characteristic of fetal γδ T lymphocytes, whereas those involving Vδ5 are characteristic of adult γδ T lymphocytes. These specific rearrangements are evident not only in functional alleles but also in out-of-frame rearrangements and alleles that cannot encode a functional protein due to the disruption of Cδ, indicating that this regulation occurs at the level of recombination rather than being derived from cellular selection (Itohara et al. 1993). These developmentally programmed rearrangements are controlled by the activation of promoters associated with specific V gene segments, as has been demonstrated for promoters associated with the Vγ gene segments (Xiong et al. 2004). It has been proposed that the bHLH TF E2A is an important regulator in the activation of adult versus fetal Vδ and Vγ gene segments, as adult thymocytes from E2A-deficient mice exhibit increased usage of fetal Vδ/Vγ gene segments and suppressed rearrangements of those Vδ/Vγ gene segments normally used in adults (Bain et al. 1999).

Tcra recombination depends on Eα activation in DN4/DP thymocytes (Sleckman et al. 1997; Hernández-Munain et al. 1999). Eα is inactive in DN3a thymocytes and becomes activated in DN4 thymocytes upon pre-TCR signaling (Hernández-Munain et al. 1999; Ciofani et al. 2004; del Blanco et al. 2012). Eα controls Vα-to-Jα recombination by affecting the chromatin modification of the Jα cluster and the subset of proximal Vα gene segments, extending its influence to over 500 kb, which is mediated by the formation of a chromatin hub (Hawwari and Krangel 2005; Shih et al. 2012). The rearrangement of more distally Vα gene segments appears to be regulated by their associated promoters that function in an enhancer-independent fashion (Hawwari and Krangel 2005). The minimal Eα segment showing proper developmental regulation resides in a 275-bp fragment that contains four regions for TF binding, Tα1–Tα4 (Balmelle et al. 2004). Tα1–Tα2 contains essential binding sites for CREB, TCF-1/LEF-1, Runx1, and Ets-1 (Giese et al. 1995; Roberts et al. 1997). Interestingly, cooperative binding among multiple TFs is required to form a functional Tα1–Tα2 enhanceosome in vivo (Hernández-Munain et al. 1998). Other relevant binding sites within Tα1–Tα4, but outside of Tα1–Tα2, include those for NFAT, AP-1, Egr-1, Sp1, GATA-3, and the bHLH TFs E2A/HEB (Hernández-Munain et al. 1999; del Blanco et al. 2009, 2012). Recent studies on the occupancy of this enhancer during β-selection in thymocyte development have demonstrated that the combinatorial assembly of constitutive TFs in DN3a thymocytes and pre-TCR-induced TFs dictates Eα activation (del Blanco et al. 2012). In DN3a thymocytes, only constitutive TFs occupy their binding sites at the enhancer. After β-selection, during the first proliferative phase, the induced NFAT, AP-1, and Egr-1 TFs occupy their sites within the enhancer in DN4 and early proliferating DP thymocytes. This results in the recruitment of the histone acetylases CBP and p300 and in productive Eα-primary Jα promoter interactions that activate germline transcription and Vα-to-Jα recombination. During the second non-proliferative phase in small resting DP thymocytes, when extensive Vα-to-Jα recombination occurs, Eα is occupied by the constitutive TFs and remains active through the assembly of a new enhanceosome, which is characterized by strong factor binding and the further recruitment of CBP and p300 (Balmelle et al. 2004; del Blanco et al. 2012).

The genomic structure of *Tcra* permits successive VαJα rearrangements to optimize the expression of a functional TCRα chain on DP thymocytes (Fig. 12.3). This is accomplished through multiple cycles of Vα-to-Jα recombination (Huang et al. 2005). Early DP thymocytes only rearrange their most 3′ Vα gene segments to their most 5′ Jα gene segments in what it is known as primary VαJα rearrangements (Villey et al. 1996; Hawwari et al. 2005). Accordingly, DP thymocytes with limited survival exhibit a skewed TCRα repertoire that contains mostly

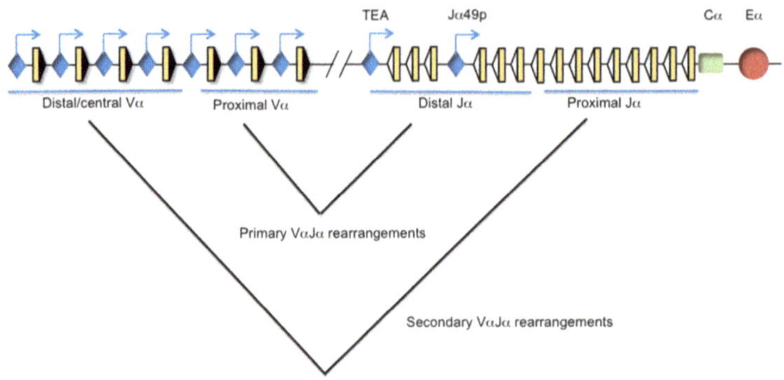

Fig. 12.3 Primary and secondary Vα-to-Jα rearrangements at *Tcra*. The Vα and Jα gene segments are represented by *yellow rectangles*, and the RSSs are shown as *black* (containing a 23-bp spacer) or *white* (containing a 12-bp spacer) *triangles*. The Cα region is represented by a *green rectangle*. The positions of the TEA and the Jα49 promoters as well as the promoters associated with Vα gene segments are represented as *blue diamonds*. *Blue arrows* represent active germline transcription. Eα is represented as a *red circle*

5′ Jα gene segments (Guo et al. 2002). The promoters associated with the most 5′ Jα gene segments (TEA and Jα49 promoters) serve to localize rearrangement of only their nearby Jα gene segments, whereas Eα has a long-range effect on locus accessibility (Villey et al. 1996; Sleckman et al. 1997; Hawwari et al. 2005). The activation of TEA and the Jα49 promoters is responsible for all primary Vα-to-Jα rearrangements at *Tcra* (Hawwari et al. 2005). Similar to Eα, the TEA promoter has also intrinsic features that allow it to developmentally regulate Vα-to-Jα recombination, as it is unable to be activated by Eδ in DN3a thymocytes when it is positioned in the same location as the Vδ5 gene segment (Huang and Sleckman 2007). To activate primary VαJα recombination, physical interactions between Eα, the TEA promoter, the Jα49 promoter, and 3′ Vα promoters are induced in early DP thymocytes (Fig. 12.4) (Seitan et al. 2011; Shih et al. 2012). These interactions mediate the formation of a chromatin hub within the 3′ region of the locus (Shih et al. 2012). The formation of this chromatin hub depends on Eα because the long-distance contacts established between the 3′ Vα and 5′ Jα promoters are lost in Eα-deleted alleles. In addition to Eα, the CCCTC-binding factor (CTCF) also contributes in part to the formation of this chromatin hub, as the interactions of Eα with the 3′ Vα and 5′ Jα promoters are reduced in CTCF-deficient DP thymocytes (Shih et al. 2012; Shih and Krangel 2013). In corroboration with these data, the conditional knockout of CTCF in DN cells also reduces Vα-to-Jα primary rearrangements, whereas the absence of Eα totally abrogates these rearrangements (Sleckman et al. 1997; Shih et al. 2012). CTCF sites generally mark ACEs in the *Tcra/Tcrd* locus, including most Vα promoters, the TEA promoter, and Eα (Shih et al. 2012). Interestingly, CTCF is constitutively bound to Eα and the TEA promoter in DN3a and DP thymocytes, whereas its binding to several sites present at Vα proximal promoters is induced by Eα in DP thymocytes (Shih et al. 2012). Because the absence of CTCF inhibits but does not abrogate the formation of the chromatin hub and Vα-to-Jα recombination, the protein complex assembled on Eα is essential for the establishment of such functional physical interactions between the enhancer and the 3′ Vα and 5′ Jα promoters (Fig. 12.4). Interestingly, CTCF deficiency or deletion of the TEA promoter and its associated CTCF-binding site caused increased interaction of Eα with the 3′ *Tcrd* region and increased transcription and recombination of the Dδ and Jδ gene segments. Based on these results, it has been proposed that the TEA promoter- and Eα-bound CTCF serves to target Eα function at the Jα gene segments (Shih et al. 2012; Shih and Krangel 2013).

These initial rearrangements can be replaced by subsequent Vα-to-Jα rearrangements, which are known as secondary VαJα rearrangements, that join progressively more 5′ Vα gene segments to progressively more 3′ Jα gene segments (Pasqual et al. 2002; Jouvin-Marche et al. 2009). Although the deletion of the Jα primary promoters induces the transcriptional activity of promoters associated with the more 3′ Jα gene segments (Fig. 12.3) (Hawwari et al. 2005; Abarrategui and Krangel 2007), this is not the mechanism that progressively targets RAG-1/2 to downstream Jα gene segments during secondary Vα-to-Jα rearrangement. The 3′ Jα gene segments are progressively opened through the activity of promoters associated with the rearranged Vα gene segments and are brought in proximity to the Jα cluster by primary VαJα rearrangements (Hawwari and Krangel 2007). Recent deep-sequencing analyses of *Tcra* transcripts in pre-immune $CD8^+$ T lymphocytes have suggested that secondary VαJα rearrangements do not occur in a coordinated sequential manner, but rather they occur randomly between all possible Vα and Jα gene segments (Genolet et al. 2012). The fact that the mentioned study analyzed a post-selected lymphocyte repertoire rather than the pre-selected repertoire present in DP thymocytes might explain this controversy. Further analyses are clearly required to resolve this issue.

To date, little is known about how secondary Vα-to-Jα rearrangements are regulated. By analyzing the three-dimensional structure of the AgR loci by 3D-FISH, it has been clearly established that changes in gene conformation bring pairs of distant RSSs into proximity (Jhunjhunwala et al. 2009). Gene contraction correlates with ongoing recombination at a particular AgR locus during lymphocyte development but is not mediated by RAG-1/2 synapsis between two recombining RSSs because it occurs in the absence of RAG-1/2. In fact, gene contraction at the AgR loci occurs at the same moment that a particular locus is transcribed and is ready to be recombined. In the case of *Tcra/Tcrd*, the locus conformation has being carefully compared between B lymphocytes and DN3a and DP thymocytes (Shih and Krangel 2010). Compared with B lymphocytes, this locus is contracted in both DN3a and DP thymocytes. However, two distinct modes of contraction can be distinguished in DN3a and DP thymocytes (Fig. 12.4). In DN3a thymocytes, the locus is fully contracted, whereas in DP thymocytes, *Tcra/Tcrd* displays a contracted conformation only in its 3′ end (including the proximal but not central or distal Vα gene segments) and not in its 5′ end. These results suggest that the extended 5′ configuration observed in DP thymocytes forces primary recombinations to use the proximal Vα gene segments. Although Eα-bound TFs and CTCF are critical to drive transcription and primary Vα-to-Jα recombination and are essential

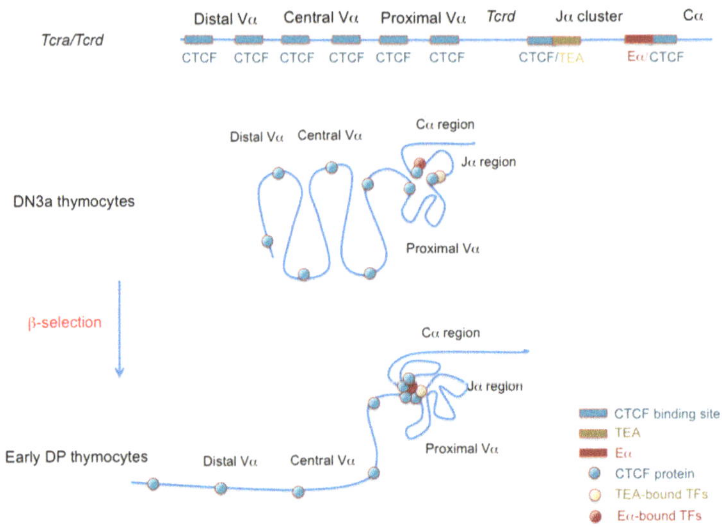

Fig. 12.4 Representation of location of CTCF-binding sites, the TEA promoter, and Eα and changes in the physical interactions among TFs and CTCF and the 3D architecture at *Tcra/Tcrd* during thymocyte development. CTCF-binding sites and protein are represented as *blue rectangles* and *circles*, respectively; the TEA promoter and the TEA promoter-bound TFs are represented as *yellow rectangles* and *circles*, respectively; Eα and Eα-bound TFs are represented as *red rectangles* and *circles*, respectively. Physical interactions at the *Tcra* 3′ region are mediated by Eα-bound TFs and CTCF/cohesin bound to Eα and the proximal Vα and distal Jα promoters in DP thymocytes. These interactions are mediated by protein–protein interactions among Eα- and promoter-bound TFs as well as by the constitutive binding of CTCF/cohesin to Eα and the TEA promoter and the Eα-dependent induced binding of CTCF/cohesin to the proximal Vα promoters. In DN3a thymocytes, Eα is inactive; CTCF/cohesin that is constitutively bound to Eα and the TEA promoter is not sufficient to form an active chromatin hub in these cells. In DP thymocytes, Eα-bound TFs activate the enhancer and induces binding of CTCF/cohesin to several binding sites at the proximal Vα promoters. Eα-bound TFs, together with Jα and Vα promoter-bound TFs and CTCF/cohesin, promote the formation of an active chromatin hub that permits the induction of germline transcription and the opening of chromatin, allowing RAG-1/2 to access to the Vα and Jα RSSs to generate synapses and subsequent VαJα recombination in DP thymocytes. In addition to these functional interactions, the genomic configuration of the locus changes during thymocyte development; it is fully contracted in DN3a thymocytes and contracted at the 3′ end of *Tcra* in DP thymocytes. This figure is based on previously published data (Shih and Krangel 2010; Shih et al. 2012)

to establish a chromatin hub between the enhancer and the Jα and Vα promoters at the 3′ end of the locus that participates in these initial rearrangements, they are not responsible for the contraction of the 3′ region (Shih and Krangel 2010; Shih et al. 2012). These apparently contradictory results with regard to the function of Eα in mediating physical interactions within the 3′ end of the locus suggest that Eα-independent 3′ end contraction and Eα-dependent chromatin hub formation represent two distinct levels of organization. First, 3′ contraction would occur in an enhancer- and CTCF-independent fashion, and then this contraction would facilitate the contacts mediated by Eα-bound TFs and CTCF to form a chromatin hub involving the 3′ end of the locus.

It has been proposed that CTCF-associated Rad21 cohesin plays an essential role in the regulation of secondary rearrangements (Seitan et al. 2011). Because these experiments were performed in late DP thymocytes from Rad21 conditional knockout mice using a CD4-Cre transgene, the late Rad21 deletion did not allow for the evaluation of its effect on primary Vα-to-Jα rearrangements. Experiments performed with CTCF conditional knockout mice using Lck-Cre predict that the early deletion of Rad21 with Lck-Cre would also affect primary Vα-to-Jα rearrangements (Shih et al. 2012). Hence, CTCF/Rad21 are predicted to be relevant factors that facilitate general Vα-to-Jα rearrangements, with no specific role in primary versus secondary rearrangement. At present, no factor has been demonstrated to specifically regulate primary versus secondary Vα-to-Jα rearrangements.

In its role in the regulation of *Tcra*/*Tcrd* rearrangement, CTCF seems to specify the targets of Eα through its binding to Eα and specific promoters associated with distal Jα and proximal Vα gene segments, restricting Eα function within *Tcra* and inhibiting the possible Eα-dependent activation of *Tcrd* (Shih et al. 2012; Shih and Krangel 2013). This role of CTCF is different to that observed at the Ig loci, where it appears to function as an insulator of enhancer activity (Guo et al. 2011; Ribeiro de Almeida et al. 2011; Shih and Krangel 2013). What is common among the analyses of CTCF function in the *Tcra*/*Tcrd* and Ig loci is that CTCF is not responsible for the developmentally regulated AgR locus configuration (Guo et al. 2011; Ribeiro de Almeida et al. 2011; Lin et al. 2012; Shih et al. 2012; Shih and Krangel 2013). These results are consistent with the fact that CTCF is involved only in intra-domain but not in developmentally regulated inter-domain interactions (Lin et al. 2012). Currently, the molecular mechanisms for the contraction and decontraction of the AgR loci are not understood. Recent evidence suggests that p300 and developmentally regulated lineage-specific TFs, such as E2A and PU.1, are involved in regulated inter-domain interactions between enhancers during B lymphocyte development (Lin et al. 2012). These observations raise the possibility that lineage-specific TFs might be central players in establishing a developmentally regulated chromatin topology and locus configuration at AgR loci.

12.1.4.3 Epigenetic Regulation of Tcrb Recombination and Allelic Exclusion

Tcrb is fully rearranged and expressed in DN3a thymocytes (Fig. 12.2). *Tcrb* contains a single enhancer, Eβ, and a promoter that is associated with the Dβ gene segment present at each Dβ-Jβ cluster, Dβ1 and Dβ2 (Fig. 12.1). Eβ is T lineage specific and is activated very early during thymocyte development. Deletion of this enhancer inhibits V(D)J recombination at both Dβ-Jβ clusters, promoting the heterochromatinization of this region through H3/H4 hypoacetylation and CpG hypermethylation (Bories et al. 1996; Bouvier et al. 1996; Mathieu et al. 2000). The activity of Eβ seems to reside at binding sites for Ets-1, Runx1, and E2A/HEB (Sun et al. 1995; Tripathi et al. 2000). Deletion of the Dβ1 promoter inhibits the transcription and rearrangement of the Dβ1Jβ cluster but not the Dβ2Jβ cluster by regulating chromatin accessibility in a highly localized manner over a region of less than 450 bp surrounding the Dβ1 gene segment (Whitehurst et al. 2000; Oestreich et al. 2006). The loading of Sp1 and ATF/CREB to the Dβ1 promoter is mediated by Eβ (Spicuglia et al. 2002). The binding of these TFs to the Dβ1 promoter is important for its function. Hence, for the initial step of Dβ1-to-Jβ rearrangement, Eβ and the Dβ1 promoter are both required to remodel chromatin to permit RAG-1/2 to access the Dβ1 and Jβ gene segments, which likely occurs through the formation of a stable holocomplex between the promoter and enhancer (Oestreich et al. 2006).

Less is known about the control of the second step of *Tcrb* recombination, Vβ-to-DβJβ rearrangement. This step is controlled by allelic exclusion, which enforces the production of just a single functional protein from this locus by inhibiting recombination in one allele (Brady et al. 2010). Allelic exclusion also occurs during the rearrangement of the Ig loci but not the other TCR loci. Allelic exclusion at *Tcrb* is initiated by limiting Vβ-to-DβJβ rearrangements to one allele in DN thymocytes and is maintained by inhibiting further rearrangements in DP thymocytes. This process reduces the frequency of DP thymocytes that express two or more αβ TCRs and thereby facilitates central tolerance, limiting the generation of autoimmunity. It is now clear that the position of *Tcrb* in the nuclei of DN3a thymocytes has an essential role in directing its recombination in a single allele (Fig. 12.5) (Schlimgen et al. 2008; Chan et al. 2013). In experiments analyzing the position of the Ig loci during B lymphocyte development, directed monoallelic association of *Igh* and *Igk* with repressive nuclear compartments was determined to be responsible for the initiation and maintenance of allelic exclusion (Goldmit et al. 2005; Roldán et al. 2005). The mechanistic basis for this form of repression involves the sequestration of one RSS from RAG-1/2 to avoid the synapsis of two RSSs that are separated by large distances. In the case of *Tcrb*, the mechanism that establishes its allelic exclusion is somehow fundamentally different from that operating at the Ig loci because it is based on the high frequency of association of both alleles with repressive nuclear compartments in DN3a thymocytes (approximately 60 % association of two alleles versus 35 % association of a single allele) (Fig. 12.5) (Schlimgen et al. 2008). Interestingly, *Tcra/Tcrd* is rarely associated with repressive nuclear compartments in DN3a and DP thymocytes (Schlimgen et al. 2008), which is consistent with its active status in both developmental stages, allowing *Tcrd* rearrangements in DN3a cells and *Tcra* rearrangements in DP cells (Fig. 12.5). The mechanism by which these interactions inhibit Vβ-to-DβJβ recombination is not related to the inhibition of Vβ accessibility per se, as high levels of germline Vβ transcripts are detected in DN3a thymocytes from both *Tcrb* alleles (Jia et al. 2007). In fact, Eα insertion within the Vβ gene cluster, which leads to strong Vβ germline transcription in DP thymocytes, does not induce efficient Vβ rearrangement in DN3a cells, although a subtle reduction of the interactions of the locus with nuclear repressive compartments and a subtle reduction of allelic exclusion are observed (Jackson et al. 2005; Schlimgen et al. 2008).

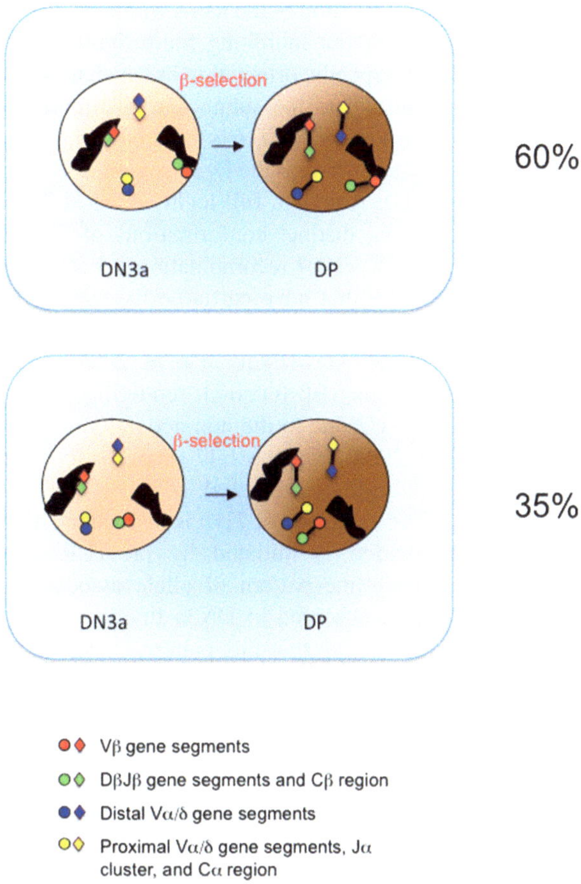

Fig. 12.5 Representation of the nuclear location and 3D architecture of *Tcra/Tcrd* and *Tcrb* during β-selection. *Red circles* and *diamonds* represent the Vβ gene segments of the two *Tcrb* alleles. *Green circles* and *diamonds* represent the DβJβ gene segments and the Cβ region of the two *Tcrb* alleles. *Blue circles* and *diamonds* represent the distal Vα/δ gene segments of the two *Tcra/Tcrd* alleles. *Yellow circles* and *diamonds* represent the proximal Vα/δ gene segments, the Jα gene segment cluster, and the Cα region of the two *Tcra* alleles. The *black areas* represent heterochromatin. In DN3a thymocytes, both of the *Tcra/Tcrd* alleles are fully contracted and located in the center of the nucleus. In addition, they are not associated with repressive nuclear compartments, allowing *Tcrd* rearrangements to occur. At the same time, both *Tcrb* alleles are associated with repressive nuclear compartments. They are tethered through both the Vβ and Cβ regions allowing sporadic Vβ-to-DβJβ rearrangements to occur (Chan et al. 2013). This is the basis for allelic exclusion at *Tcrb* in DN3a thymocytes. In DP thymocytes, both *Tcra* alleles are contracted at the 3′ end of the locus (including the proximal Vα/δ gene segments, the Jα cluster, and the Cα region), whereas the 5′ end of the locus is in an extended configuration to facilitate consecutive rounds of Vα-to-Jα recombination at *Tcra*. At this stage, both *Tcrb* alleles continue to be associated with repressive compartments, but their three-dimensional structure is changed to an extended configuration to enforce the allelic exclusion established in DN3a thymocytes. The frequencies of thymocytes containing both *Tcrb* alleles or one *Tcrb* allele in association with repressive nuclear compartments are indicated. The first diagram represents the percentage of the two *Tcrb* alleles being associated to repressive compartments, whereas the second diagram represents that of one *Tcrb* allele associated to repressive compartments and one *Tcrb* allele positioned away from these compartments. This figure is based on previously published data (Skok et al. 2007; Schlimgen et al. 2008; Chan et al. 2013)

The mechanism by which the association of the locus with repressive compartments inhibits V(D)J recombination without inhibiting transcription remains unknown, but it appears to be related to the specific orientation in which these loci are tethered to the repressive compartment and to the subnuclear distribution of RAG2 and RNAPII (Chan et al. 2013). It is known that *Igh* is tethered to the nuclear membrane through the distal V_H gene cluster, whereas the D_HJ_H region is located away (Kosak et al. 2002; Roldán et al. 2005). Although full leghts of most Tcrb alleles are in contact with the nuclear lamina, distinct conformations of Tcrb alleles can be distinguished (Chan et al. 2013). V(D)J recombination occurs preferently on the subset of peripheral Tcrb alleles that have partially dissociated from the lamina (Chan et al. 2013). The TFs Ets-1 and E47 have been involved in establishing allelic exclusion at *Tcrb* in DN3a thymocytes (Eyquem et al. 2004; Agata et al. 2007). Ets-1 appears to work by inhibiting Eβ function, rendering the rearranged DβJβ segments inaccessible to RAG-1/2 due to the non-activation of the Eβ-dependent Dβ promoters (Eyquem et al. 2004), whereas E47 supports *Tcrb* accessibility and recombination in DN3a lymphocytes (Agata et al. 2007).

In DP thymocytes, the allelic exclusion of *Tcrb* is maintained after the feedback inhibition of *Tcrb* recombination is initiated by pre-TCR signaling (Brady et al. 2010). Interestingly, the same pattern of allele association with nuclear repressive compartments that is observed in DN3a thymocytes is maintained in DP thymocytes (Fig. 12.5), indicating that this pattern is established early in the initiation of allelic exclusion (Schlimgen et al. 2008). These data suggest that the interactions of unrearranged alleles with nuclear repressive compartments are preserved in DN3a and DP thymocytes to ensure the maintenance of allelic exclusion. However, it seems that other mechanisms also contribute to the maintenance of the allelic exclusion of *Tcrb* in DP thymocytes because Vβ gene segment transcription and RAG-1/2 accessibility are reduced in the transition from DN3a to DP thymocytes, and both parameters correlate with allelic exclusion of Vβ-to-DβJβ recombination in DP thymocytes (Tripathi et al. 2002). In addition, *Tcrb* undergoes a process of gene contraction in DN3a thymocytes, which is reversed in DP thymocytes (Fig. 12.5) (Skok et al. 2007; Kondilis-Mangum et al. 2011). Hence, the suppression of further *Tcrb* recombination in DP thymocytes is associated with decontraction of the locus conformation (Fig. 12.5) as well as with epigenetic changes that reduce Vβ gene segment accessibility to RAG-1/2. These mechanisms of decontraction and reduced Vβ gene segment accessibility might operate together with the maintenance of the unrearranged *Tcrb* allele associated with nuclear repressive compartments to prevent further Vβ-to-DβJβ rearrangements in DP thymocytes and to assure monoallelic *Tcrb* expression.

The E2A factor E47 is the only TF that has been demonstrated to regulate *Tcrb* allelic exclusion in DP thymocytes (Agata et al. 2007). This TF is downregulated by pre-TCR signaling in the DN3a-to-DP transition, and its overexpression is able to override feedback inhibition and to promote Vβ-to-DβJβ recombination in DP thymocytes. Because the E47 dosage is rate limiting with respect to Vβ-to-DβJβ recombination and because forced E47 expression interferes with pre-TCR-mediated feedback inhibition, this TF is a good candidate to direct the interaction of *Tcrb* with nuclear repressive compartments. In addition, it is also possible that E47 mediates Vβ accessibility and recombination by inducing *Tcrb* contraction. This possibility is in agreement with recent data demonstrating that E2A factors are involved in the developmentally regulated inter-domain interactions between enhancers that occur during B lymphocyte development (Lin et al. 2012). Because the TFs HEB and E2A have specific and redundant roles during T-cell development (Zhuang et al. 1994; Bain et al. 1997; Barndt et al. 1999) and because HEB/E2A heterodimers are the major form of these factors in T cells (Sawada and Littman 1993), both TFs might have essential roles in controlling the different genomic conformations of the TCR loci during the differentiation of T lymphocytes.

12.1.4.4 Epigenetic Regulation of Tcrg Recombination

Most of our knowledge on the regulation of recombination at *Tcrg* is centered on the gene segments associated with Cγ1. One *Tcrg* enhancer that is positioned 3′ of Cγ1, Eγ, and another enhancer that is positioned between Vγ5 and Vγ2, HsA, do not affect *Tcrg* recombination, but they work together in the transcription of a rearranged *Tcrg* gene (Xiong et al. 2002). Similar to Eδ, c-Myb and Runx1 factors also appear to be involved in the regulation of Eγ (Redondo et al. 1992; Hsiang et al. 1993, 1995; Hernández-Munain and Krangel 1994; Hernández-Munain et al. 1996; Lauzurica et al. 1997; Tani-ichi et al. 2011). In addition, IL-7-induced TF STAT5 is recruited to Jγ promoters, Eγ, and HsA to induce histone acetylation and to activate germline transcription (Maki et al. 1996; Ye et al. 1999, 2001; Schlissel et al. 2000; Huang et al. 2001; Huang and Muegge 2001; Masui et al. 2008; Tani-ichi et al. 2009, 2010). Recruitment of STAT5 to these ACEs is necessary to establish chromatin accessibility to the Jγ and Vγ gene segments to RAG-1/2 (Schlissel et al. 2000; Huang et al. 2001). Similarly to *Tcrd* (Hernández-Munain et al. 1999), transcription of *Tcrg* is also repressed in DP thymocytes and αβ T lymphocytes, a phenomenon known as TCRγ silencing. The pre-TCR signaling induces transcriptional silencing of the *Tcrg* locus by reducing the recruitment of Runx1 and STAT5 to Eγ and HsA (Ferrero et al. 2006; Tani-ichi et al. 2011). We propose a similar mechanism for Eδ silencing in DP thymocytes because Runx1 functions as a structural factor in the formation of the Eδ enhanceosome (Hernández-Munain et al. 1999; Hernández-Munain and Krangel 2002).

As mentioned earlier, one interesting aspect of the regulation of Vγ-to-Jγ recombination is related to the rearrangements of specific Vγ gene segments during development that result in the generation of distinct waves of γδ T lymphocytes over the life of an organism with different Vγ/Vδ usage (Elliot et al. 1988; Korman et al. 1988). Rearrangements involving Vγ3 and Vγ4 predominate in fetal thymocytes, whereas those involving Vγ5 and Vγ2 are used in adult thymocytes (Garman et al. 1986; Havran and Allison 1988; Ito et al. 1989). These development-specific programmed rearrangements are controlled by promoters associated with the Vγ gene segments (Baker et al. 1998; Xiong et al. 2004). As described above for Vδ gene segment usage during development, E2A is involved in this regulation, as adult thymocytes from E2A$^{-/-}$ mice exhibit increased usage of fetal Vγ gene segments and suppressed rearrangements of the Vγ segments that are normally used in adults (Bain et al. 1999).

12.2 Consequences of Defects in V(D)J Recombination at the TCR Loci

Although beneficial, V(D)J recombination is a dangerous process. Defects in this process can cause immunodeficiencies and chromosomal translocations that lead to lethal leukemia (Aifantis et al. 2008; Van Vlierberghe and Ferrando 2012).

12.2.1 T-Cell Immunodeficiency Syndromes Derived from General Defects in V(D)J Recombination

12.2.1.1 Omenn Syndrome

V(D)J recombination defects are responsible for 20 % of severe combined immunodeficiency disease (SCID) in humans. The SCID phenotype that is derived from defects in V(D)J recombination results in T and B lymphocyte deficiencies but normal NK cells. Twenty to thirty percent of these patients exhibit null or hypomorphic mutations in *RAG1* or *RAG2*. Omenn syndrome (OS) is a SCID caused by hypomorphic mutations in *RAG1* and *RAG2* that produce mutant proteins with severely decreased recombination activity in vitro, resulting in the production of a restricted TCR repertoire (Villa et al. 1998). OS patients display elevated or normal numbers of T lymphocytes with a highly restricted oligoclonal repertoire and the absence of B lymphocytes. These patients normally suffer from recurrent and opportunistic infections by the age of 6 months. The reason why this constrained TCR repertoire may cause autoimmunity seems to be related to a relative deficiency in regulatory T lymphocytes, leading to impaired self-tolerance (Marrella et al. 2007). This disease is a rare autosomal recessive disease; since its description in 1965, less than 100 patients have been reported. In addition, OS is associated with erythrodermia, hepatosplenomegalia, lymphadenopathy, and alopecia and is accompanied by autoimmune and allergic symptoms with high levels of IgE. This disease is fatal without bone marrow transplantation.

12.2.1.2 Radiosensitive SCID

A subset of the remaining SCID patients exhibits mutations in genes that encode proteins involved in NHEJ, such as *Artemis*, *Cernunnos/XLF*, and *Lig4* (Moshous et al. 2001; Buck et al. 2006; van der Burg et al. 2006; Vera et al. 2013). These SCID patients additionally present increased cellular sensitivity to ionizing radiation due to a deficiency in DNA repair and predisposition to lymphomas. Null mutations in *Artemis* have high incidence among Athabascan-speaking Native Americans. These patients show complete absence of circulating mature T and B lymphocytes. Hypomorphic *Artemis* mutations cause a leaky SCID phenotype. Hypomorphic mutations in *Lig4* or *Cernunnos/LFX* result in leaky to severe SCID that is associated with microcephaly, developmental delay, the absence of B lymphocytes, and a residual number of T lymphocytes (Buck et al. 2006; van der Burg et al. 2006; Vera et al. 2013). This phenotype resembles that of Nijmegen breakage syndrome, which also includes microcephaly, developmental delay, and immunodeficiency but not SCID.

12.2.1.3 X-Linked SCID

Human X-linked SCID is a rare immunodeficiency disorder (1:100,000 live births) in which thymocyte development is arrested at the DN3a stage. In fact, the few T lymphocytes found in the peripheral blood of infants with this disease resemble DN3a thymocytes and lack mature T lymphocyte function. This disease results from a defect in the expression of the IL-2Rγ chain (Noguchi et al. 1993), which is a functional component of IL-7R. This defect causes an arrest of Vβ-to-DβJβ rearrangements, which leads to a block at the DN3a thymocyte stage due to their inability to express pre-TCRs (Sleasman et al. 1994). Although the defect that causes this SCID is not due to a primary defect in V(D)J recombination, its effect on the expression of IL-2Rγ causes a defect in *Tcrb* recombination that is itself responsible for the SCID phenotype.

12.2.1.4 Ataxia Telangiectasia

In addition to SCID, defects in V(D)J recombination can lead to leukemic processes that are associated with immunodeficiency. Ataxia telangiectasia (A-T) is a rare immunodeficiency disorder derived from chromosomal instability and defects in DNA repair (1:300,000 live births) (Taylor et al. 1996). These patients present primary immunodeficiency with reduced numbers of T lymphocytes, decreased levels of IgA, IgG2, and IgE, and thymic hypoplasia or an absent thymus. The major consequence of immunodeficiency in A-T patients is an increased likelihood of developing infections, and this defect is a significant cause of death. One of the most important features of A-T is the increased predisposition for T-cell acute lymphoblastic leukemia (T-ALL) and prolymphocytic leukemia. In fact, 10 % of all A-T patients develop a malignancy due to translocations and inversions involving chromosomes 7 and 14 at specific breakpoints at the *Tcrb* and *Tcra/Tcrd* loci, respectively. In addition to these problems, A-T patients also present neurologic defects resulting from cerebellar degeneration as well as abnormal eye movements, dysarthria, hypogonadism, and growth retardation.

This disease is caused by mutations in the ataxia telangiectasia mutated kinase (ATM), which, similar to the DNA-PKcs, functions during coding joint formation after RAG-1/2-induced cleavage. During NHEJ, these kinases phosphorylate proteins related to DNA repair, such as Ku70, Ku80, Xrcc4, Cernunnos/XLF, Artemis, DNA ligase IV, and the histone protein H2AX localized at double-strand breaks, as well as RAG-2 (Gapud et al. 2011). These two kinases have overlapping activities during chromosomal signal joint formation; they are important in preventing the aberrant resolution of RSSs that would produce potentially oncogenic chromosomal translocations (Gapud et al. 2011). Mice deficient in either DNA-PKcs or ATM are viable, whereas those that are deficient in both proteins exhibit early embryonic lethality (Gurley and Kemp 2001). Specifically, ATM deficiency or inactivation leads to increased genome instability and a highly elevated predisposition to lymphoid cancers due to aberrant translocations involving *Tcra/Tcrd* (Liyanage et al. 2000; Bredemeyer et al. 2006; Matei et al. 2007; Vacchio et al. 2007; Zha et al. 2010; Isoda et al. 2012). A lack in ATM also results in a severe defect in the DN3a-to-DN3b and DP-to-SP transitions. In fact, almost all $ATM^{-/-}$ mice die due to thymic lymphomas derived from aberrant *Tcrd* translocations and incorrect repair of RAG-1/2-induced double-strand breaks (Liyanage et al. 2000; Zha et al. 2010; Isoda et al. 2012). Experiments in ATM- and H2AX-deficient mice have demonstrated that these proteins cooperate to ensure normal V(D)J recombination, maintain cellular genomic stability, and suppress the generation of lymphomas derived from aberrant translocations of *Tcra/Tcrd* and *Tcrb* (Derheimer and Kastan 2010).

12.2.2 Generation of Leukemia Derived from Defects in the Process of V(D)J Recombination at the TCR Loci

V(D)J recombination is a dangerous process that can cause chromosomal translocations, leading to lethal leukemia. The most common T-cell leukemia, T-ALL, is derived from the transformation of precursor T cells. T-ALL is composed of a heterogeneous group of acute leukemias that are arrested at various stages of normal thymocyte development. Thirty-five percent of all human T-ALLs carry chromosomal translocations involving TCR loci in thymocytes, and most of them are found in children and adolescents (Aifantis et al. 2008; Van Vlierberghe and Ferrando 2012).

These aberrant translocations frequently involve the juxtaposition of a strong promoter and/or enhancer from a TCR locus with TF genes. The TF genes found to be translocated to TCR loci include those of the T-ALL factors *TAL1* (also known as *SCL*) and *TAL2*, the bHLH factor *BHLHB1*, the lymphoblastic leukemia-derived (LYL) factor *LYL1*, the factors encoded by the homeobox genes *HOX11* (*TLX1*) and

HOX11L2 (*TLX3*) and the *HOXA* cluster, the LIM-only factor domain genes *LMO1* and *LMO2*, c*MYC*, and c*MYB*, as well as other key genes involved in signaling for proliferation and differentiation, such as *LCK*, *NOTCH1*, *TAN1*, and *Cyclin-D2* (Aifantis et al. 2008; Van Vlierberghe and Ferrando 2012). These illegitimate TCR locus translocations lead to the aberrant expression of their corresponding proteins, resulting in abnormal proliferation and differentiation processes. For example, 5–10 % of pediatric and 30 % of adult T-ALLs show translocations of *TLX1* and *TLX3* into *Tcra/Tcrd*, 3 % of all childhood T-ALLs carry a translocation that places *TAL1* under the control of *Tcra/Tcrd*, approximately 3 % of T-ALL patients show translocations in the *HOXA* cluster into *Tcrb* and *Tcrg*, and in some cases, Notch1 is activated by the translocation of *NOTCH1* to *Tcrb*, driving the expression of a mutant form of Notch-1 that is truncated at its amino terminus.

The molecular mechanism for the development of T-ALLs as a consequence of the aberrant expression of these proteins during thymocyte development has been extensively investigated. The knowledge obtained from studies on V(D)J recombination and T lymphocyte development in normal cells has also been important for understanding the mechanisms that drive aberrant translocations that lead to the generation of leukemia. For example, the molecular mechanism for the T-cell maturation arrest at the DP stage in T-ALLs has been recently reported; this arrest was found to be derived from translocations of *TLX1* and *TLX3* into *Tcra/Tcrd* (Dadi et al. 2012). These translocations, which result in the overexpression of TLX1 and TLX3, represent the most frequent translocations observed in T-ALLs. The recruitment of TLX1/TLX3 to Eα was shown to result in reduced accessibility to the region influenced by the enhancer (increased levels of H3K27me3, a hallmark of silent chromatin), a drastic decrease in *Tcra* transcription, and strong inhibition of Vα-to-Jα recombination (Dadi et al. 2012). Arrest at the DP stage is a clear consequence of the inhibition of *Tcra* expression and Vα-to-Jα recombination, which is mediated via Eα repression because a functional TCRα is needed for TCRαβ expression and DP-to-SP thymocyte progression (Shinkai et al. 1993; Sleckman et al. 1997). Furthermore, Eα repression that is mediated by TLX1/3 factors was demonstrated to be controlled by their recruitment through enhancer-bound Ets-1 (Roberts et al. 1997; Hernández-Munain et al. 1998; del Blanco et al. 2009, 2012; Dadi et al. 2012).

12.3 Concluding Remarks

V(D)J recombination is subjected to a very tight regulation that involves nuclear dynamics and changes in higher-order chromatin architecture to provide RSS accessibility to RAG-1/2. RAG-1/2-mediated DNA cleavage at these loci is controlled by changes in the epigenome that are derived from chromatin modifications, transcriptional elongation, the location of the loci within the nucleus, and the three-dimensional architecture of the loci. This control is mediated by a functional interplay among ACEs, such as enhancers and promoters, and their bound TFs, which together orchestrate the accessibility of RAG-1/2 to chromatin through epigenetic mechanisms. Thus, the normal development of lymphocytes requires the stringent regulation of RAG-1/2 function to avoid illegitimate translocations that lead to the ectopic expression of TFs or signaling proteins involved in proliferation and differentiation. Future experiments are required to decipher the precise mechanisms by which the ACEs and their bound TFs that are present at a given TCR locus are activated and inactivated during T lymphocyte development and how they interact to form holocomplexes to regulate transcriptional elongation, recruit chromatin modifiers and remodeling complexes, and change the locus conformation and topology to ultimately activate V(D)J recombination. Studies performed in normal cells are fundamental to understand the deregulation of this process, which results in abnormal situations, leading to immunodeficiency, abnormal translocations, and the generation of leukemia.

Acknowledgments The authors are grateful for support from the Spanish goverment (BFU2009-08796), the Andalusian government (grants CTS-6587 and CVI-4526) and the European Regional Development Fund (ERDF/FEDER). We also thank Carlos Suñé for comments on the manuscript.

References

Abarrategui I, Krangel MS (2006) Regulation of T cell receptor-α gene recombination by transcription. Nat Immunol 7:1109–1115

Abarrategui I, Krangel MS (2007) Noncoding transcription controls downstream promoters to regulate T-cell receptor α recombination. EMBO J 26:4380–4390

Agata Y, Tamaki N, Sakamoto S, Ikawa T, Masuda K, Kawamoto H, Murre C (2007) Regulation of T cell receptor β gene rearrangement and allelic exclusion by the helix-loop-helix protein, E47. Immunity 27:871–884

Aifantis I, Raetz E, Buonamici S (2008) Molecular pathogenesis of T-cell leukemia and lymphoma. Nat Immunol 8:380–390

Bain G, Engel I, Robanus Maandag EC, te Riele HP, Voland JR, Sharp LL, Chun J, Huey B, Pinkel D, Murre C (1997) E2A deficiency leads to abnormalities in αβ T-cell development and to rapid development of T-cell lymphomas. Mol Cell Biol 17:4782–4791

Bain G, Romanow WJ, Albers K, Havran WL, Murre C (1999) Positive and negative regulation of V(D)J recombination by the E2A proteins. J Exp Med 189:289–300

Baker JE, Cado D, Raulet DH (1998) Developmentally programmed rearrangement of T cell receptor Vγ genes is controlled by sequences immediately upstream of the Vγ genes. Immunity 9:159–168

Balmelle N, Zamarreño N, Krangel MS, Hernández-Munain C (2004) Developmental activation of the TCR α enhancer requires functional collaboration among proteins bound inside and outside the core enhancer. J Immunol 173:5054–5063

Barndt R, Dai MF, Zhuang Y (1999) A novel role for HEB downstream or parallel to the pre-TCR signaling pathway during αβ thymopoiesis. J Immunol 163:3331–3343

Bassing CH, Alt FW, Hughes MM, D'Auteil M, Wehrly TD, Woodman BB, Gartner F, White JM, Davidson L, Sleckman BP (2000) Recombination signal sequences restrict chromosomal V(D) J recombination beyond the 12/23 rule. Nature 405:583–586

Bories JC, Demengeot J, Davidson L, Alt FW (1996) Gene-targeted deletion and replacement mutations of the T-cell receptor β-chain enhancer: the role of enhancer elements in controlling V(D)J recombination accesibility. Proc Natl Acad Sci U S A 93:7871–7876

Bosc N, Lefranc MP (2003) The mouse (Mus musculus) T cell receptor α (TCRA) and δ (TCRD) variable genes. Dev Comp Immunol 27:465–497

Bouvier G, Watrin F, Naspetti M, Verthuy C, Naquet P, Ferrier P (1996) Deletion of the mouse T-cell receptor β gene enhancer blocks αβ T-cell development. Proc Natl Acad Sci U S A 93: 7877–7881

Brady BL, Steinel NC, Bassing CH (2010) Antigen receptor allelic exclusion: an apdate and reappraisal. J Immunol 185:3801–3808

Bredemeyer AL, Sharma GG, Huang CY, Helmink BA, Walker LM, Khor KC, Nuskey B, Sullivan KE, Pandita TK, Bassing CH, Sleckman BP (2006) ATM stabilized DNA double-strand-break complexes during V(D)J recombination. Nature 442:466–470

Buck D, Moshous D, de Chasseval R, Ma Y, Le Deist F, Cavazzana-Calvo M, Fisher A, Casanova JL, Lieber MR, de Villartay JP (2006) Severe combined immunodeficiency and microcephaly in siblings with hypomorphic mutations in DNA ligase IV. Eur J Immunol 36:224–235

Callen E, Faryabi RB, Luckey M, Hao B, Daniel JA, Yang W, Sun H-W, Dressler G, Pen W, Chi H, Ge K, Krangel MS, Park J-H, Nussenzweig A (2012) The DNA damage- and transcription-associated protein Paxip1 controls thymocyte development and emigration. Immunity 37:1–15

Capone M, Watrin F, Fernex C, Hoevat H, Krippl B, Wu L, Scollay R, Ferrier P (1993) TCR β and TCR α gene enhancers confer tissue- and stage- specificity on V(D)J recombination events. EMBO J 12:4335–4346

Capone M, Hockett RD Jr, Zlotnik A (1998) Kinetics of T cell receptor β, γ, and δ rearrangements during adult thymic development: T cell receptor rearrangements are present in CD44+ CD25+ Pro-T thymocytes. Proc Natl Acad Sci U S A 95:12522–12527

Carabana J, Ortigoza E, Krangel MS (2005) Regulation of the murine Dδ2 promoter by upstream stimulatory factor 1, Runx1, and c-Myb. J Immunol 174:4144–4152

Chaumeil J, Micsinai M, Ntziachristos P, Deriano L, Wang JM-H, Ji Y, Nora EP, Rodesch MJ, Jeddeloh JA, Aifantis I, Kluger Y, Schatz DG, Skok JA (2013) Higher-order looping and nuclear organization of Tcra facilitate targeted RAG cleavage and regulated rearrangement in recombination centers. Cell Rep 3:359–370

Chan EA, Teng G, Corbett E, Choudhury KR, Bassing CH, Schatz DG, Krangel MS (2013) Peripheral subnuclear positioning suppresses Tcrb recombination and segregates Tcrb alleles from RAG2. Proc Natl Acad Sci U S A 110:E4628–E4637

Chen HT, Bahndoola A, Difilippantonio MJ, Zhu J, Brown MJ, Tai X, Rogakou EP, Brotz TM, Bonner WM, Ried T, Nussenzweig A (2000) Response to RAG-mediated V(D)J cleavage by NSB1 and γ-H2AX. Science 290:1962–1964

Ciofani M, Schmitt TM, Ciofani A, Michie AM, Cuburu N, Aublin A, Maryanski JL, Zúñiga-Pflucker JC (2004) Obligatory role for cooperative signaling by pre-TCR and Notch during thymocyte differentiation. J Immunol 172:5230–5239

Cobb RM, Oestreich KJ, Osipovich OA, Oltz EM (2006) Accessibility control of V(D)J recombination. Adv Immunol 91:45–109

Collins B, Clambey ET, Scott-Browne J, White J, Marrack P, Hagman J, Kappler JW (2013) Ikaros promotes rearrangement of TCR α genes in an ikaros null thymoma cell line. Eur J Immunol 43:1–12

Corthay A, Nandakumar KS, Holmdahl R (2001) Evaluation of the percentage of peripheral T cells with two differential T cell receptor a-chains and of their potential role in autoimmunity. J Autoimmun 16:423–429

Dadi S, Le Noir S, Payet-Bornet D, Lhermitte L, Zacarias-cabeza J, Bergeron J, Villarese P, Vachez E, Dik WA, Millien C, Radford I, Verhoeyen E, Cosset FL, Petit A, Ifrah N, Dombret H, Hermine O, Spicuglia S, Langerak AW, Macintyre EA, Nadel B, Ferrier P, Asnafi V (2012) TLX homoedomain oncogenes mediate T cell maturation arrest in T-ALL via interaction with ETS1 and suppresion of TCRα gene expression. Cancer Cell 21:563–576

del Blanco B, Roberts JL, Zamarreño N, Balmelle-Devaux N, Hernández-Munain C (2009) Flexible stereospecific interactions and composition within nucleoprotein complexes assembled on the TCRα gene enhancer. J Immunol 183:1871–1883
del Blanco B, García V, García-Mariscal A, Hernández-Munain C (2011) Control of V(D)J recombination through transcriptional elongation and changes in locus chromatin structure and nuclear organization. Genet Res Int 2011:970968. doi:10.4061/2011/970968
del Blanco B, García-Mariscal A, Wiest DL, Hernández-Munain C (2012) Tcra enhancer activation by inducible transcription factors downstream of pre-TCR signaling. J Immunol 188:3278–3293
Derheimer FA, Kastan MB (2010) Multiple roles of ATM in monitoring and maintaining DNA integrity. FEBS Lett 584:3675–3681
Diaz P, Cado D, Winoto A (1994) A locus control region in the T cell receptor α/δ locus. Immunity 1:207–217
Dik WA, Pike-Overzet K, Weerkamp F, de Ridder D, de Hass EFE, Baert MRM, van der Spek P, Koster EEL, Reinders MJT, van Dongen JJM, Langerak AW, Staal FJT (2005) New insights on human T cell development by quantitative T cell receptor gene rearrangement studing and gene expression profiling. J Exp Med 201:1715–1723
Egawa T, Tillman RE, Naoe Y, Taniuchi I, Littman DR (2007) The role of Runx transcription factors in thymocyte differentiation and in homeostasis of naïve T cells. J Exp Med 204: 1945–1957
Elliot JF, Rock EP, Patten PA, Davis MM, Chien Y-H (1988) The adult T-cell receptor δ-chain is deverse and distinct from that of fetal thymocytes. Nature 331:627–631
Eyquem S, Chemin K, Fasseu M, Bories J-C (2004) The Ets-1 transcription factor is required for complete pre-T cell receptor function and allelic exclusion at the T cell receptor β locus. Proc Natl Acad Sci U S A 101:15712–15717
Fehling HJ, Krotkova A, Saint-Ruf C, von Boehmer H (1995) Crucial role of the pre-T cell receptor α gene in development of αβ but not γδ T cells. Nature 375:795–798
Ferrero I, Mancini SJ, Grosjean F, Wilson A, Otten L, MacDonald HR (2006) TCRγ silencing during αβ T cell development depends upon pre-TCR-induced proliferation. J Immunol 177: 6038–6043
Ferrier P, Krippl B, Blackwell TK, Furley AJ, Suh H, Winoto A, Cook WD, Hood L, Constantini F, Alt FW (1990) Separate elements control DJ and VDJ rearrangement in a transgenic recombination substrate. EMBO J 9:117–125
Gapud EJ, Dorsett Y, Yin B, Bredemeyer AL, Mahowald GK, Omi KQ, Walker LM, Bednarski JJ, McKinnon PJ, Bassing CH, Nussenzweig A, Sleckman BP (2011) Ataxia telangiectasia mutated (Atm) and DNA-PKcs kinases have overlapping activities during chromosomal signal joint formation. Proc Natl Acad Sci U S A 108:2022–2027
Garman RD, Doherty PJ, Raulet DH (1986) Diversity, rearrangement and expression of murine T cell g genes. Cell 45:733–742
Genolet R, Stevenson BJ, Farinelli L, Osteras M, Luescher IF (2012) Highly diverse TCRa chain repertoire of pre-immune CD8+ cells reveals new insight in gene recombination. EMBO J 31:1666–1678
Ghosh S, Romanow WJ, Murre C (2001) Induction of a diverse T-cell receptor γ/δ repertoire by the helix-loop-helix proteins E2A and HEB in nonlymphoid cells. J Exp Med 193:769–775
Giese K, Kingsley C, Kirshner JR, Grosschedl R (1995) Assembly and function of a TCR α enhancer complex is dependent on LEF-1-induced DNA bending and multiple protein-protein interactions. Genes Dev 9:995–1008
Glusman G, Rowen L, Lee I, Boysen C, Roach JC, Smit AFA, Wang K, Koop BF, Hood L (2001) Comparative genomics of the human and mouse T cell receptor loci. Immunity 15:337–349
Godfrey DI, Kennedy J, Suda T, Zlotnik A (1993) A developmental pathway involving four phenotypically and functionally distinct subsets of $CD3^-CD4^-CD8^-$ triple-negative adult mouse thymocytes defined by CD44 and CD25 expression. J Immunol 150:4244–4252

Godfrey DI, Kennedy J, Mombaerts P, Tonegawa S, Zlotnik A (1994) Onset of TCRβ rearrangement and role of TCRβ expression during $CD3^{-}CD4^{-}CD8^{-}$ thymocyte differentiation. J Immunol 152:4783–4792

Golding A, Chandler S, Ballestar E, Wolffe AP, Schlissel MS (1999) Nucleosome structure completely inhibits in vitro cleavage by the V(D)J recombinase. EMBO J 18:3712–3723

Goldmit M, Ji Y, Skok J, Roldán E, Jung S, Cedar H, Bergman Y (2005) Epigenetic ontogeny of the Igκ locus during B cell development. Nat Immunol 6:198–203

Guo J, Hawwari A, Li H, Li H, Sun Z, Mahanta SK, Littman DR, Krangel MS, He Y-W (2002) Regulation of the TCRα repertoire by the survival window of $CD4^{+}CD8^{+}$ thymocytes. Nat Immunol 3:469–476

Guo C, Yoon HS, Franklin A, Jain S, Ebert A, Cheng H-L, Hansen E, Despo O, Bossen C, Vetterman C, Bates JG, Richard N, Myers D, Patel H, Gallagher M, Schlissel MS, Murre C, Busslinger M, Giallourakis C, Alt FW (2011) CTCF-binding elements mediate control of V(D) J recombination. Nature 477:424–431

Gurley KE, Kemp CJ (2001) Synthetic lethality between mutation in Atm and DNA-PKcs during murine embryogenesis. Curr Biol 11:191–194

Hao B, Krangel MS (2011) Long-distance regulation of fetal Vδ gene segment TRDV4 by the Tcrd enhancer. J Immunol 187:2484–2491

Havran WL, Allison JP (1988) Developmentally ordered appearance of thymocytes expression different T-cell antigen. Nature 335:443–445

Hawwari A, Krangel MS (2005) Regulation of TCR δ and α repertoires by local and long-distance control of variable gene segment chromatin structure. J Exp Med 202:467–472

Hawwari A, Krangel MS (2007) Role for rearranged variable gene segments in directing secondary T cell receptor a recombination. Proc Natl Acad Sci U S A 104:903–907

Hawwari A, Bock C, Krangel MS (2005) Regulation of T cell receptor α gene assembly by a complex hierarchy of germline Jα promoters. Nat Immunol 6:481–489

Hayday AC, Pennington DJ (2007) Key factors in the organized chaos of early T cell development. Nat Immunol 8:137–144

Helmink BA, Sleckman BP (2012) The response to and repair of RAG-mediated DNA double-strand breaks. Annu Rev Immunol 30:175–202

Hernández-Munain C, Krangel MS (1994) Regulation of the T-cell receptor δ enhancer by functional cooperation between c-Myb and core-binding factors. Mol Cell Biol 14:473–483

Hernández-Munain C, Krangel MS (2002) Distinct roles for c-Myb and core binding factor/ polyoma enhancer-binding protein 2 in the assembly and function of a multiprotein complex on the TCR δ enhancer in vivo. J Immunol 169:4362–4369

Hernández-Munain C, Lauzurica P, Krangel MS (1996) Regulation of T cell receptor δ gene rearrangement by c-Myb. J Exp Med 183:289–293

Hernández-Munain C, Roberts JL, Krangel MS (1998) Cooperation among multiple transcription factors is required for access to minimal T-cell receptor α-enhancer chromatin in vivo. Mol Cell Biol 18:3223–3233

Hernández-Munain C, Sleckman BP, Krangel MS (1999) A developmental switch from TCR δ enhancer to TCR α enhancer function during thymocyte maturation. Immunity 10:723–733

Hsiang YH, Spencer D, Wang S, Speck NA, Raulet DH (1993) The role of viral enhancer "core" motif-related sequences in regulating T cell receptor-γ and -δ gene expression. J Immunol 150:3905–3916

Hsiang YH, Goldman JP, Raulet DH (1995) The role of c-Myb or a related factor in regulating the T cell receptor γ gene enhancer. J Immunol 154:5195–5204

Huang J, Muegge K (2001) Control of chromatin accessibility for V(D)J recombination by interleukin-7. J Leukoc Biol 69:907–911

Huang CH, Sleckman BP (2007) Developmental stage-specific regulation of TCR-α-chain gene assembly by intrinsic features of the TEA promoter. J Immunol 179:449–454

Huang J, Durum SK, Muegge K (2001) Cutting edge: histone acetylation and recombination at the TCRγ locus follows IL-7 induction. J Immunol 167:6073–6077

Huang CY, Sleckman BP, Kanagawa O (2005) Revision of T cell receptor α chain genes is required for normal T lymphocyte development. Proc Natl Acad Sci U S A 102:14356–14361

Isoda T, Takagi M, Nakagama S, Sato M, Masuda K, Ikawa T, Azuma M, Morio T, Kawamoto H, Mizutani S (2012) Process for immune defect and chromosomal translocation during early thymocyte development lacking ATM. Blood 120:789–799

Ito K, Bonneville M, Takagani Y, Nakanishi N, Kanagawa O, Krecko EG, Tonegawa S (1989) Different gd-T cell receptors are expressed on thymocytes at different stages of development. Proc Natl Acad Sci U S A 86:631–635

Itohara S, Mombaerts P, Lafaille J, Iacomini J, Nelson A, Clarke AR, Hooper ML, Farr A, Tonegawa S (1993) T cell receptor de gene mutant mice: independent generation of αβ T cells and programmed rearrangements of γδ TCR genes. Cell 72:337–348

Jackson A, Kondilis HD, Khor B, Sleckman BP, Krangel MS (2005) Regulation of T cell receptor β allelic exclusion at a level beyond accessibility. Nat Immunol 6:189–197

Jackson AM, Krangel MS (2006) A role for MAPK in feedback of Tcrb recombination. J Immunol 176:6824–6830

Jaeger S, Fernández B, Ferrier P (2013) Epigenetic aspects of lymphocyte antigen receptor gene rearrangement or "when stochasticity completes randomness". Immunology 139:141–150

Jhunjhunwala S, van Zelm MC, Peak MM, Murre C (2009) Chromatin architecture and the generation of antigen receptor diversity. Cell 138:435–448

Ji Y, Little AJ, Banerjee JK, Hao B, Oltz EM, Krangel MS, Schatz DG (2010a) Promoters, enhancers, and transcription target RAG1 binding during V(D)J recombination. J Exp Med 207:2809–2816

Ji Y, Resch W, Corbett E, Yamane R, Casellas R, Schatz DG (2010b) The in vivo pattern of binding of RAG1 and RAG2 to antigen receptor loci. Cell 141:419–431

Jia J, Kondo M, Zhuang Y (2007) Germline transcription from T-cell receptor β gene is uncoupled from allelic exclusion. EMBO J 26:2387–2399

Jouvin-Marche E, Fuschiotti P, Marche PN (2009) Dynamic aspects of TCRα gene recombination: qualitative and quantitative assessment of the TCRα chain repertoire in man and mouse. Adv Exp Med Biol 650:82–92

Kohu K, Kubo M, Ichikawa H, Ohno S-I, Habu S, Sato T, Satake M (2008) Pleiotropic roles of Runx transcription factors in the differentiation and function of T lymphocytes. Curr Immunol Rev 4:101–115

Kondilis-Mangum HD, Cobb RM, Osipovich O, Srivatsan S, Oltz EM, Krangel MS (2010) Transcription-dependent mobilization of nucleosomes at accessible TCR gene segments in vivo. J Immunol 184:6970–6977

Kondilis-Mangum HD, Shih HY, Mahowald GK, Sleckman BP, Krangel MS (2011) Regulation of TCRβ allelic exclusion by gene segment proximity and accessibility. J Immunol 187: 6374–6381

Korman AJ, Galesic SM, Spencer D, Kruisbeek AM, Raulet D (1988) Predominant variable region usage by γ/δ T cell receptor-bearing cells in the adult thymus. J Exp Med 168:1021–1040

Kosak ST, Skok J, Medina KL, Riblet R, Le Beau MM, Fisher AG, Singh H (2002) Subnuclear comparmentalization of immunoglobulin loci during lymphocyte development. Science 296: 158–162

Krangel MS, Carabana J, Abbarategui I, Schlimgen R, Hawwari A (2004) Enforcing order within a complex locus: current perspectives on the control of V(D)J recombination at the murine T-cell receptor α/δ locus. Immunol Rev 200:224–232

Kwon J, Imbalzano AN, Matthews A, Oettinger MA (1998) Accessibility of nucleosomal DNA to V(D)J cleavage is modulated by RSS positioning and HMG1. Mol Cell 2:829–839

Kwon J, Morshead KB, Guyon JR, Kingston RE, Oettinger MA (2000) Histone acetylation and hSWI/SNF remodeling act in concert to stimulate V(D)J cleavage of nucleosomal DNA. Mol Cell 6:1037–1048

Lauzurica P, Krangel MS (1994a) Enhancer-dependent and -independent steps in the rearrangement of a human T cell receptor δ transgene. J Exp Med 179:43–55

Lauzurica P, Krangel MS (1994b) Temporal and lineage-specific control of T cell receptor α/δ gene rearrangement by T cell receptor α and δ enhancers. J Exp Med 179:1913–1921
Lauzurica P, Zhong XP, Krangel MS, Roberts JL (1997) Regulation of T cell receptor δ gene rearrangement by CBF/PEBP2. J Exp Med 185:1193–1201
Lieber MR (2010) The mechanism of double-strand DNA break repair by the nonhomologous DNA end-joining pathway. Annu Rev Biochem 79:181–211
Lin YC, Benner C, Mansson R, Heinz S, Miyazaki K, Miyazaki M, Chandra V, Bossen C, Glass CK, Murre C (2012) Global changes in the nuclear positioning of genes and intra- and interdomain genomic interactions that orchestrate B cell fate. Nat Immunol 13:1196–1204
Liu Y, Subrahmanyam R, Chakraborty T, Sen R, Desiderio S (2007) A plant homodomain in RAG-2 that binds hypermethylated lysine 4 of histone H3 is necessary for efficient antigen-receptor-gene rearrangement. Immunity 27:561–571
Livak F, Tourigny M, Schatz DG, Petrie HT (1999) Characterization of TCR gene rearrangements during adult murine T cell development. J Immunol 162:2575–2580
Liyanage M, Weaver Z, Barlow C, Coleman A, Pankratz DG, Anderson S, Wynshaw-Boris A, Ried T (2000) Abnormal rearrangement within the α/δ T-cell receptor locus in lymphomas from Atm-deficient mice. Blood 96:1940–1946
Maki K, Sunaga S, Ikuta K (1996) The V-J recombination of T cell receptor-γ genes is blocked in interleukin-7 receptor-deficient mice. J Exp Med 184:2423–2427
Malissen M, Trucy J, Jouvin-Marche E, Cazenave PA, Scollay R, Malissen B (1992) Regulation of TCR alpha and beta gene allelic exclusion during T-cell development. Immunol Today 13: 315–322
Marrella V, Poliani PL, Casati A, Rucci F, Frascoli L, Gougeon ML, Lemercier B, Bosticardo M, Ravatini M, Battaglia M, Roncarolo MG, Cavazzana-Calvo M, Fachetti F, Notarangelo MD, Vezzoni P, Grassi F, Villa A (2007) A hypomorphic R229Q Rag2 mouse mutant recapitulates Omenn syndrome. J Clin Invest 117:1260–1269
Masui N, Tani-ichi S, Maki K, Ikuta K (2008) Transcriptional activation of mouse TCR Jγ4 germline promoter by STAT5. Mol Immunol 45:849–855
Matei IR, Gladdy RA, Nutter LM, Canty A, Guidos CJ, Danska JS (2007) ATM deficiency disrupts Tcra locus integrity and the maturation of $CD4^{+}CD8^{+}$ thymocytes. Blood 109:1887–1896
Mathieu N, Hempel WM, Spicuglia S, Verthuy C, Ferrier P (2000) Chromatin remodeling by the T cell receptor (TCR)-β gene enhancer during early T cell development: implications for the control of TCR-β locus recombination. J Exp Med 192:625–636
Mauvieux L, Villey I, de Villartay JP (2001) T early α (TEA) regulates initial TCRVAJA rearrangements and leads to TCRJA coincidence. Eur J Immunol 31:2080–2086
McBlane F, Boyes J (2000) Stimulation of V(D)J recombination by histone acetylation. Curr Biol 10:483–486
McMillan RE, Sikes ML (2008) Differerential activation of dual promoters alters Dβ2 germline transcription during thymocyte development. J Immunol 180:3218–3228
McMillan RE, Sikes ML (2009) Promoter activity 5′ of Dβ2 is coordinated by E47, Runx1, and GATA-3. Mol Immunol 46:3009–3017
McMurry MT, Krangel MS (2000) A role for histone acetylation in the developmental regulation of VDJ recombination. Science 287:495–498
Monroe RJ, Sleckman BP, Monroe BC, Khor B, Claypool S, Ferrini R, Davidson L, Alt FW (1999) Developmental regulation of TCR δ locus accessibility and expression by the TCR δ enhancer. Immunity 10:503–513
Moshous D, Callebaut I, de Chasseval R, Corneo B, Cavazzana-Calvo M, Le Deist F, Tezcan I, Sanal O, Bertrand Y, Philippe N, Fisher A, de Villartay JP (2001) Artemis, a novel DNA double-strand break repair/V(D)J recombination protein, is mutated in human severe combined immune deficiency. Cell 105:177–186
Noguchi M, Huafang Y, Rosenblatt HM, Filipovich AH, Adelstein S, Modi WS, McBride OW, Leonard WJ (1993) Interleukin-2 receptor γ-chain mutation results in X-linked severe combined immunodeficiency in humans. Cell 73:147–157

Oestreich KJ, Cobb RM, Pierce S, Chen J, Ferrier P, Oltz EM (2006) Regulation of the TCRβ gene assembly by a promoter/enhancer holocomplex. Immunity 24:381–391
Ortiz BD, Cado D, Winoto A (1999) A new element within the T-cell receptor α locus required for tissue-specific locus control region activity. Mol Cell Biol 19:1901–1909
Ortiz BD, Harrow F, Cado D, Santoso B, Winoto A (2001) Function and factor interactions of a locus control region element in the mouse T cell receptor-α/Dad1 gene locus. J Immunol 167:3836–3845
Osipovich O, Oltz EM (2010) Regulation of antigen receptor gene assembly by genetic-epigenetic crosstalk. Semin Immunol 22:313–322
Osipovich O, Milley R, Meade A, Tachibana M, Shinkai Y, Krangel MS, Oltz EM (2004) Targeted inhibition of V(D)J recombination by a histone methyltransferase. Nat Immunol 5:309–315
Osipovich O, Cobb RM, Oestreich KJ, Pierce S, Ferrier P, Oltz EM (2007) Essential function for SWI-SNF chromatin-remodeling complexes in the promoter-directed assembly of Tcrb genes. Nat Immunol 8:809–816
Pasqual N, Gallagher M, Aude-Garcia C, Loiodice M, Thudersoz F, Demongeot J, Ceredig R, Marche PN, Jouvin-Marche E (2002) Quantitative and qualitative changes in V-J α rearrangement during mouse thymocytes differentiation: implication of a limited T cell receptor α chain repertoire. J Exp Med 196:1163–1173
Pearse M, Wu L, Egerton M, Wilson A, Shortman K, Scollay R (1989) A murine early thymocyte developmental sequence is marked by transient expression of the interleukin 2 receptor. Proc Natl Acad Sci U S A 86:1614–1618
Prinz I, Sansoni A, Kissenpfennig A, Ardouin L, Malissen M, Malissen B (2006) Visualization of the earliest steps of γδ T cell development in the adult thymus. Nat Immunol 7:995–1003
Ramón-Maiques S, Kuo AJ, Carney D, Matthews AG, Oettinger MA, Gozani O, Yang W (2007) The plant homodomain finger of RAG2 recognizes H3 methylated at both lysine-4 and arginine-2. Proc Natl Acad Sci U S A 104:18993–18998
Redondo JM, Pfohl JL, Hernández-Munain C, Wang S, Speck NA, Krangel MS (1992) Indistinguishable nuclear factor binding to functional core sites of the T-cell receptor δ and murine leukemia virus enhancers. Mol Cell Biol 12:4817–4823
Ribeiro de Almeida C, Stadhouders R, Bruijn MJW, Bergen IM, Thongjuea S, Lenhard B, van Ijcken W, Grosveld F, Galjart N, Soler E, Hendricks RW (2011) The DNA-binding protein CTCF limits proximal Vκ recombination and restricts κ enhancer interactions to the immunoglobulin κ light chain locus. Immunity 35:501–513
Roberts JL, Lauzurica P, Krangel MS (1997) Developmental regulation of VDJ recombination by the core fragment of the T cell receptor α enhancer. J Exp Med 185:131–140
Roldán E, Fuxa M, Chong W, Martinez D, Novatchkova M, Busslinger M, Skok JA (2005) Locus 'decontraction' and centromeric recruitment contribute to allelic exclusion of the immunoglobulin heavy-chain gene. Nat Immunol 6:31–41
Ryu CJ, Haines BB, Lee HR, Kang YH, Draganov DD, Lee M, Whitehurst CE, Hong HT, Chen J (2004) The T-cell receptor β variable gene promoter is required for efficient Vβ rearrangement but not allelic exclusion. Mol Cell Biol 24:7015–7023
Sato T, Ohno S-I, Hayashi T, Sato C, Kohu K, Satake M, Habu S (2005) Dual functions of Runx proteins for reactivating CD8 and silencing CD4 at the commitment process into CD8 thymocytes. Immunity 22:317–328
Sawada S, Littman DR (1993) A heterodimer of HEB and an E12-related protein interacts with the CD4 enhancer and regulate its activity in T-cell lines. Mol Cell Biol 13:5620–5628
Schlimgen RJ, Reddy KL, Singh H, Krangel MS (2008) Initiation of allelic exclusion by stochastic interaction of Tcrb alleles with repressive nuclear compartments. Nat Immunol 9:802–809
Schlissel MS, Durum SD, Muegge K (2000) The interleukin 7 receptor is required for T cell receptor γ locus accessibility to the V(D)J recombinase. J Exp Med 191:1045–1050
Seitan VC, Hao B, Tachibana-Konwalski K, Lavagnolli T, Mira-Bontenbal H, Brown KE, Teng G, Carroll T, Terry A, Horan K, Marks H, Adams DJ, Schatz DG, Aragón L, Fisher AG, Krangel MS,

Nasmyth K, Merkenschlager M (2011) A role for cohesin in T-cell-receptor rearrangement and thymocyte differentiation. Nature 476:467–471

Shih HY, Krangel MS (2010) Distinct contracted conformations of the Tcra/Tcrd locus during Tcra and Tcrd recombination. J Exp Med 207:1835–1841

Shih HY, Krangel MS (2013) Chromatin architecture, CCCTC-binding factor, and V(D)J recombination: managing long-distance relationships at antigen receptor loci. J Immunol 190: 4915–4921

Shih HY, Verma-Gaur J, Torkamani A, Feeney AJ, Galjart N, Krangel MS (2012) Tcra gene recombination is supported by a Tcra enhancer- and CTCF-dependent chromatin hub. Proc Natl Acad Sci U S A 109:3493–3502

Shinkai Y, Koyasu S, Nakayama K, Murphy KM, Loh DY, Reinherz EL, Alt FW (1993) Restoration of T cell development in RAG-2-deficient mice by functional TCR transgenes. Science 259:822–825

Sikes ML, Gomez RJ, Song J, Oltz EM (1998) A developmental stage-specific promoter directs germline transcription of Dβ Jβ gene segments in precursor T lymphocytes. J Immunol 161:1399–1405

Sikes ML, Meade A, Tripathi R, Krangel MS, Oltz EM (2002) Regulation of V(D)J recombination: a dominant role for promoter positioning in gene segment accessibilty. Proc Natl Acad Sci U S A 99:12309–12314

Skok JA, Gisler R, Novatchkova M, Farmer D, de Laat W, Busslinger M (2007) Reversible contraction by looping of the Tcra and Tcrb loci in rearranging thymocytes. Nat Immunol 8: 378–387

Sleasman JW, Harville TO, White GB, George JF, Barret DJ, Goodnow MM (1994) Arrested rearrangement of TCR Vβ genes in thymocytes from children with X-linked severe combined immunodeficiency disease. J Immunol 153:442–448

Sleckman BP, Bardon CG, Ferrini R, Davidson L, Alt FW (1997) Function of the TCR α enhancer in αβ and γδ T cells. Immunity 7:505–515

Spicuglia S, Payet D, Tripathi RK, Rameil P, Verthuy C, Imbert J, Ferrier P, Hempel WM (2000) TCRα enhancer activation occurs via a conformational change of a pre-assembled nucleo-protein complex. EMBO J 19:2034–2045

Spicuglia S, Kumar S, Yeh J-H, Vachez E, Chasson L, Gorbatch S, Cautres J, Ferrier P (2002) Promoter activation by enhancer-dependent and -independent loading of activator and coactivator complexes. Mol Cell 10:1479–1487

Stanhope-Baker P, Hudson KM, Shaffer AL, Constantinescu A, Schlissel MS (1996) Cell type-specific chromatin structure determines the targeting of V(D)J recombinase activity in vitro. Cell 85:887–897

Sun W, Graves BJ, Speck NA (1995) Transactivation of the Moloney murine leukemia virus and T-cell receptor β-chain enhancers by cbf and ets requires intact binding sites for both proteins. J Virol 69:4941–4949

Taghon T, Yui MA, Pant R, Diamond RA, Rothenberg EV (2006) Developmental and molecular characterization of emerging β- and γδ-selected pre-T cells in the adult mouse thymus. Immunity 24:53–64

Tani-ichi S, Satake M, Ikuta K (2009) Activation of the mouse TCRγ enhancers by STAT5. Int Immunol 21:1079–1088

Tani-ichi S, Lee HC, Ye SK, Ikuta K (2010) Accessibility control of TCR Vγ region by STAT5. Int Immunol 22:693–703

Tani-ichi S, Satake M, Ikuta K (2011) The pre-TCR signal induces transcriptioal silencing of the TCRγ locus by reducing the recruitment of STAT5 and Runx to transcriptional enhancers. Int Immunol 25:553–563

Taylor AMR, Metcalfe JA, Thick J, Mak Y-F (1996) Leukemia and lymphoma in a ataxia telangiectasia. Blood 87:423–438

Tourigny MR, Mazel S, Burtrum DB, Petrie HT (1997) The T cell receptor (TCR)-β gene recombination: dissociation from cell cycle regulation and developmental progression during T cell ontogeny. J Exp Med 185:1549–1556

Tripathi RK, Mathieu N, Spicuglia S, Payet D, Verthuy C, Bouvier G, Depetris D, Mattei MG, Hempel WM, Ferrier P (2000) Definition of a T-cell receptor β gene core enhancer of V(D)J recombination by transgenic mapping. Mol Cell Biol 20:42–53

Tripathi RK, Jackson A, Krangel MS (2002) A change in the structure of Vβ chromatin associated with TCR β allelic exclusion. J Immunol 168:2316–2324

Vacchio MS, Olaru A, Livak F, Hodes RJ (2007) ATM deficiency impairs thymocyte maturation because of defective resolution of T cell receptor α locus coding end breaks. Proc Natl Acad Sci U S A 104:6323–6328

van der Burg M, van Veelen RR, Verkaik NS, Wiegant WW, Hartwig NG, Barendregt BH, Brugmans L, Raams A, Jasper NG, Zdzienicka MZ, van Dongen JJ, van Gent DC (2006) A new type of radiosensitive $T^-B^-NK^+$ severe combined immunodeficiency caused by a LIG4 mutation. J Clin Invest 116:137–145

Van Vlierberghe P, Ferrando A (2012) The molecular basis of T cell acute lymphoblastic leukemia. J Clin Invest 122:3398–3406

Vera G, Rivera-Muñoz P, Abramowski V, Malivert L, Lim A, Bole-Feysot C, Martin C, Florkin B, Latour S, Revy P, de Villartay JP (2013) Cernunnos deficiency reduces thymocyte life span and alters the T cell repertoire in mice and humans. Mol Cell Biol 33:701–711

Villa A, Santagata S, Bozzi F, Giliani S, Frattini A, Imberti L, Benerini Gatta L, Ochs HD, Shwarz K, Notarangelo LD, Vezzoni P, Spanopoulou E (1998) Partial V(D)J recombination activity leads to Omenn syndrome. Cell 93:885–896

Villey I, Caillol D, Selz F, Ferrier P, de Villartay JP (1996) Defect in rearrangement of the most 5′ TCR-Jα following targeted deletion of T early α (TEA): implications for TCR α locus accessibility. Immunity 5:331–342

Wang Z, Zang C, Rosenfeld JA, Schones DE, Barski A, Cuddapah S, Cui K, Roh TY, Peng W, Zhang MQ, Zhao K (2008) Combinatorial patterns of histone acetylations and methylations in the human genome. Nat Genet 40:897–903

Whitehurst CE, Chattopadhyay S, Chen J (1999) Control of V(D)J recombinational accessibility of the Dβ1 gene segment at the TCRβ locus by a germline promoter. Immunity 10:313–322

Whitehurst CE, Schlissel MS, Chen J (2000) Deletion of germline promoter PDβ1 from the TCR β locus causes hypermethylation that impairs Dβ1 recombination by multiple mechanisms. Immunity 13:703–714

Wolfer A, Wilson A, Nemir M, MacDonald HR, Radtke F (2002) Inactivation of Notch1 impairs VDJβ rearrangement and allows pre-TCR-independent survival of early αβ lineage thymocytes. Immunity 16:869–879

Xiong N, Kang C, Raulet DH (2002) Redundant and unique roles of two enhancer elements in the TCRγ locus in gene regulation and γδ T cell development. Immunity 16:453–463

Xiong N, Baker JE, Kang C, Raulet DH (2004) The genomic arrangement of T cell receptor variable genes is determinant of the developmental rearrangement pattern. Proc Natl Acad Sci U S A 101:260–265

Yancopulos GD, Alt FW (1985) Developmentally controlled and tissue-specific expression of unrearranged V_H gene segments. Cell 40:271–281

Ye SK, Maki K, Kitamura T et al (1999) Induction of germline transcription in the TCRγ locus by Stat5: implications for accessibility control by the IL-7 receptor. Immunity 11:213–223

Ye SK, Agata Y, Lee HC, Kurooka H, Shimizu A, Honjo T, Ikuta K (2001) The IL-7 receptor controls the accessibility of the TCRγ locus by Stat5 and histone acetylation. Immunity 15: 813–823

Zha S, Bassing CH, Sanda T, Brush JW, Patel H, Goff PH, Murphy MM, Tepsuporn S, Gatti RA, Look AT, Alt FW (2010) ATM-deficient thymic lymphoma is associated with aberrant Tcrd rearrangement and gene amplification. J Exp Med 207:1369–1380

Zhou X-F, Yu J, Chang M, Zhang M, Zhou D, Cammas F, Sun SC (2012) TRIM28 mediates chromatin modifications at the TCRα enhancer and regulates the development of T and natural killer T cells. Proc Natl Acad Sci U S A 109:20083–20088

Zhuang Y, Soriano P, Weintraub H (1994) The helix-loop-helix gene E2A is required for B cell formation. Cell 79:875–884

Chapter 13
The Molecular Basis of B Cell Development and the Role of Deregulated Transcription and Epigenetics in Leukaemia and Lymphoma

Christopher M. Kirkham, James N. Scott, Joan Boyes, and Sarah Bevington

Abstract The development of B cells from their haematopoietic stem cell origins relies on a network of transcription factors that centre on PU.1, Ikaros and Pax5. These transcription factors cooperate to direct progenitor cells towards the early B cell lineage. Further maturation is then dependent on the process of V(D)J recombination, which creates a population of B cells expressing a hugely diverse repertoire of antigen receptors on their cell surface. When an antigen is bound by its cognate receptor, the antigen–antibody interaction is fine-tuned by somatic hypermutation (SHM) and the immune response is expanded by class switch recombination (CSR), which creates antibodies with different effector functions. The processes of V(D)J recombination, SHM and CSR all involve either the breaking or mutating of genomic DNA. Mistakes in any of these reactions can lead to chromosome translocations, which are thought to be the key event that triggers almost all lymphoid cancers. The first part of this review will discuss the transcriptional and epigenetic changes that lead to B cell lineage commitment, whilst the second part will cover the deregulation of these processes and their role in triggering B cell leukaemias and lymphomas. Lastly, we discuss recent advances in our understanding of the role of deregulated epigenetic and transcription factors in the development of B cell cancers.

Keywords V(D)J recombination • B cell development • Chromosome translocation • Somatic hypermutation • Class switch recombination

---

C.M. Kirkham • J.N. Scott • J. Boyes
School of Molecular and Cellular Biology, Faculty of Biological Sciences, University of Leeds, Leeds LS2 9JT, UK

S. Bevington (✉)
School of Immunity and Infection, Institute of Biomedical Research, University of Birmingham, Edgbaston, Birmingham, B15 2TT, UK
e-mail: s.l.bevington@bham.ac.uk

C. Bonifer and P.N. Cockerill (eds.), *Transcriptional and Epigenetic Mechanisms Regulating Normal and Aberrant Blood Cell Development*, Epigenetics and Human Health, DOI 10.1007/978-3-642-45198-0_13,

13.1 B Cell Commitment

B and T lymphocytes are central to the adaptive immune response, as they express highly specific and diverse receptors on the cell surface that are capable of recognising a vast range of antigens. Like all blood cells, B cells arise from haematopoietic stem cells (HSCs) in the bone marrow. Commitment of an HSC to the B cell lineage is regulated by a complex transcription factor network, at the centre of which are several key transcription factors, including PU.1, Ikaros, E2A and Pax5. The first step in HSC differentiation gives rise to multipotent progenitors (MPPs). Within this population, the more committed lymphoid-primed multipotent progenitors (LMPPs) have largely lost the potential to differentiate down the megakaryocyte or erythroid lineages and instead give rise to either myeloid or lymphoid cells (Laiosa et al. 2006). This lineage decision is governed by the levels of the transcription factor PU.1: high levels promote myeloid differentiation, whereas lower levels steer the cell towards the lymphoid lineage and differentiation to the common lymphoid progenitor (CLP) (DeKoter and Singh 2000). The importance of PU.1 in early haemopoietic development was identified in the 1990s when PU.1 knockout mice were shown to be embryonic lethal due to a block in differentiation at the MPP stage (McKercher et al. 1996; Scott et al. 1994).

The level of PU.1 expression is regulated by the transcription factors Ikaros and Gfi1. Ikaros positively regulates Gfi1, which in turn antagonises the expression of PU.1, thus promoting commitment to the lymphoid lineage (Fig. 13.1) (Spooner et al. 2009; Wang et al. 1996). Ikaros is also important in the transition between CLPs and committed B cells. At the CLP stage, a lack of Ikaros leads to development of natural killer cells at the expense of B and T cells (Wang et al. 1996). Consistent with this, in a global study of Ikaros target genes, more than half of the genes up-regulated during commitment from the CLP stage to the B lineage were regulated by Ikaros (Ferreiros Vidal et al. 2013).

Other transcription factors that play an instrumental role in B cell commitment are E2A, Ebf-1 and Foxo1. The helix–loop–helix factor E2A is required to induce the expression of Ebf-1 and Foxo1. These three proteins then cooperate to regulate key genes involved in differentiation (Lin et al. 2010; Seet et al. 2004; Welinder et al. 2011). Included in these is *Pax5*, which plays a crucial role in B cell development by activating B cell-specific genes and repressing genes associated with other lineages, and as such it is considered to be the guardian of B cell identity (Delogu et al. 2006; Schebesta et al. 2007). The importance of Pax5 in B cell commitment was demonstrated in a striking study in which the *Pax5* gene was inactivated in mature B cells, thus causing them to dedifferentiate back to early uncommitted progenitors which then had the potential to generate functional T lymphocytes (Cobaleda et al. 2007). The differentiation of an HSC into a B cell is also guided by the c-kit, Flt3 and IL-7 receptor signalling pathways, as discussed elsewhere (Laiosa et al. 2006).

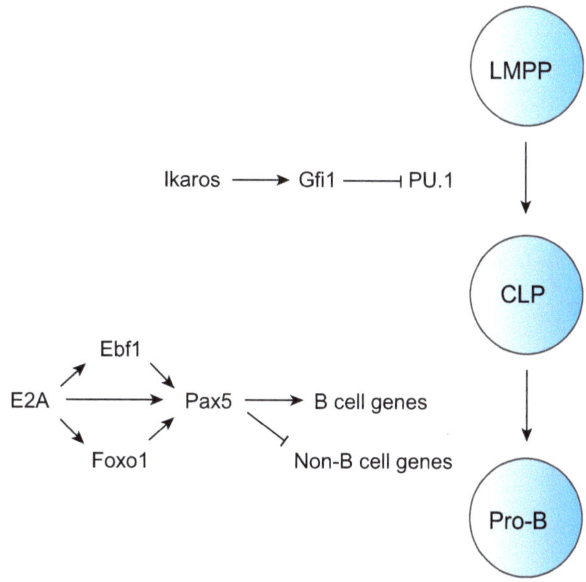

Fig. 13.1 Transcriptional regulation of early B cell development. *Lines with an arrowhead* indicate positive regulation, *lines with a flat head* indicate repression

13.2 B Cell Development

Commitment to the B cell lineage gives rise to the progenitor B (pro-B) cell population. Further development at this stage is dependent on the process of V(D)J recombination, which generates the antigen receptors of the adaptive immune system by excising large segments of the immunoglobulin loci. The heavy chain locus (IgH) is rearranged in pro-B cells, firstly by recombining D and J gene segments, followed by the stochastic joining of the recombined DJ to one of the ~150 V gene segments (Fig. 13.2b).

Once a productive VDJ_H rearrangement has occurred, the resulting μ chain is expressed on the cell surface, in a complex with the surrogate light chains, λ5 and VpreB (Bassing et al. 2002). Expression of this complex, termed the pre-B cell receptor (pre-BCR), causes a cascade of signalling events that triggers proliferation of the large pre-B cell population, followed by development to the small pre-B cell stage (Geier and Schlissel 2006). In these cells, light chain recombination is initiated at one of the two light chain isotypes, kappa (Igκ) or lambda (Igλ) (Gorman and Alt 1998), where recombination occurs between V and J gene segments only. The rearranged light chain product is then expressed with the rearranged heavy chain to form the B cell receptor molecule, IgM.

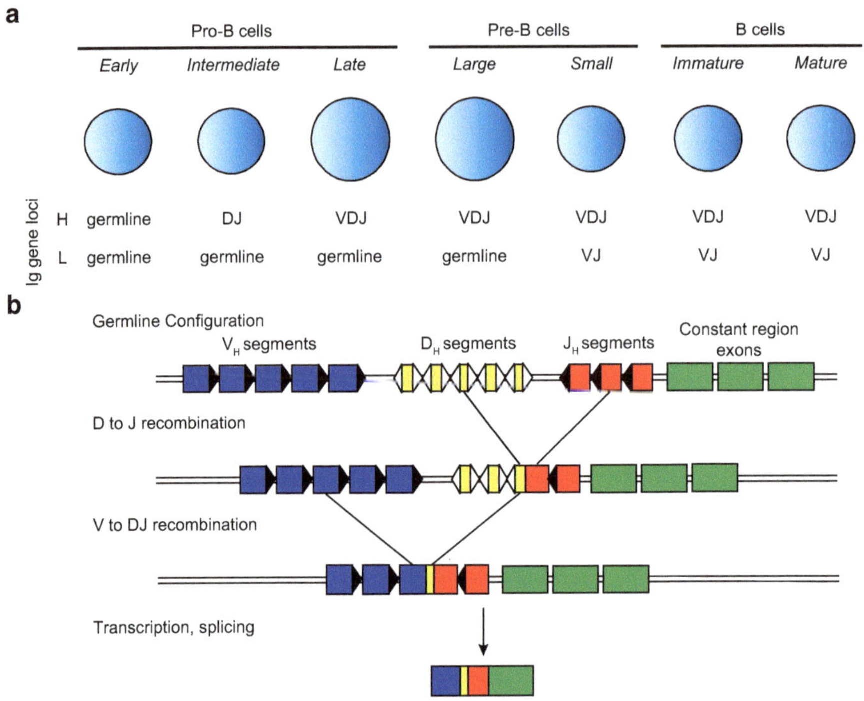

Fig. 13.2 (**a**) Stages of B cell development. The extent of recombination at the heavy (H) and light (L) chain loci is indicated underneath each developmental stage. (**b**) Schematic representation of a simplified IgH locus and V(D)J recombination. Note that a real locus has many more copies of each gene segment than shown here. *Open triangles* are 12-RSSs; *black triangles* are 23-RSSs

13.3 V(D)J Recombination

In B cells, recombination only occurs at the IgH, Igκ and Igλ immunoglobulin loci. Each locus contains multiple discontinuous variable (V), joining (J) and in some cases diversity (D) gene segments, separated by non-coding intergenic DNA (Gellert 2002). During V(D)J recombination, one of each of the V, D and J gene segments is joined to create an exon that encodes the antigen-binding portion of the receptor; the large number of different possible combinations of V, D and J gene segments is responsible for generating about half of the observed antigen receptor diversity (Fig. 13.2b).

Each gene segment is flanked by a recombination signal sequence (RSS), consisting of a conserved heptamer and an A–T rich nonamer, separated by a non-conserved spacer of either 12 ± 1 nucleotides (12-RSS) or 23 ± 1 nucleotides (23-RSS). Only the gene segments that are flanked by RSSs with dissimilar spacer lengths undergo efficient recombination, a limitation referred to as the '12/23 rule' (Tonegawa 1983). Notably, 12 and 23 RSSs are arranged in a way that prevents

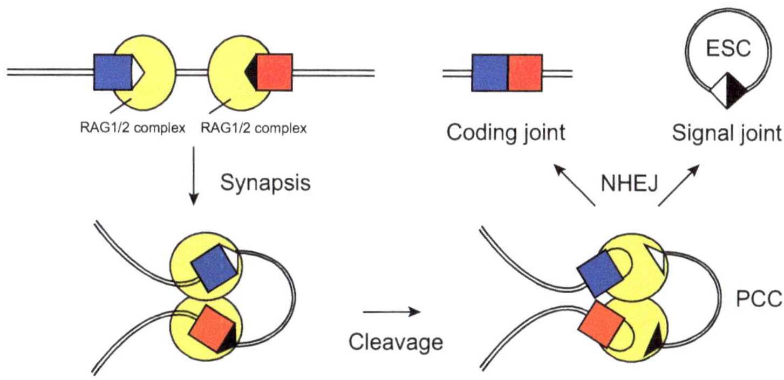

Fig. 13.3 Schematic of V(D)J recombination. RAG proteins bind to a pair of RSSs and bring them together in a synaptic complex. Double-strand breaks are made at the RSS/coding segment boundary, which are then resolved via the NHEJ pathway to form a coding joint and signal joint. *Blue and red squares* are coding segments, *triangles* are RSSs, and *circles* are RAGs. PCC is the Post Cleavage Complex; ESC is the Excised Signal Circle

non-productive rearrangements, such as the joining of two V_H segments. During V(D)J recombination, the protein products of the recombination-activating genes, *RAG1* and *RAG2*, bind to the RSSs to form a synaptic complex between a 12- and a 23-RSS. The RAG proteins create a single-strand nick precisely at the heptamer/RSS boundary, leaving a free 3′ hydroxyl group, which then attacks the opposite DNA strand in a direct trans-esterification reaction. This creates two coding ends sealed in a hairpin structure and two blunt 5′ phosphorylated signal ends. The DNA-dependent protein kinase (DNA-PK) and the non-homologous end joining (NHEJ) machinery then process and join the DNA to produce a coding joint to form the exon of the antigen receptor gene and a signal joint in the DNA that is excised from the genome. This latter piece of DNA is the recombination by-product, the excised signal circle (ESC; Fig. 13.3 and see re-integration below).

13.4 Regulation of V(D)J Recombination

13.4.1 Rearrangement is Ordered

Despite the huge payoff of having a diverse immune system, V(D)J recombination is an inherently dangerous process that creates double-stranded breaks (DSBs) in the genome of developing B cells and as such it is a tightly controlled process with strict regulations imposed at a number of different levels. RAG1 and RAG2 are the only lymphoid-specific proteins required for V(D)J recombination; these proteins bind to the RSSs found within *both* the immunoglobulin and the T cell receptor loci. Nevertheless, lineage-specific regulation ensures that the immunoglobulin genes are only fully recombined in B cells, whereas T lymphocytes exclusively rearrange the T cell receptor loci. Furthermore, different Ig loci are targeted for recombination at defined cell stages within each lineage. This strict regulation can be explained by the 'accessibility hypothesis', first proposed by Yancopoulos and Alt in the 1980s. This suggests that specific changes in chromatin structure increase antigen receptor locus accessibility to allow the RAGs to initiate rearrangement only in the correct cell type and at the correct stage of development (Alt et al. 1984; Yancopoulos and Alt 1985). Consistent with this, a number of studies demonstrated that the packaging of RSSs into nucleosomes renders them inaccessible to RAG binding and consequently inhibits V(D)J recombination (Golding et al. 1999; Kwon et al. 1998; McBlane and Boyes 2000). Furthermore, nucleosomes are preferentially positioned over RSSs (Baumann et al. 2003), implying that they must be remodelled before V(D)J recombination can be initiated. However, exactly which chromatin changes are required to increase accessibility and how they are regulated remain open questions.

13.4.2 Histone Acetylation and V(D)J Recombination

Many studies have highlighted correlations between the developmental regulation of V(D)J recombination and the acetylation of histones H3 and H4 (Espinoza and Feeney 2005; Huang and Muegge 2001; Maes et al. 2001; Roth and Roth 2000; Ye et al. 2001). Indeed, McMurry and Krangel provided a striking association between histone acetylation, accessibility and V(D)J recombination at the TCRαγ locus (McMurry and Krangel 2000). Further analyses spanning the entire IgH locus demonstrated that histone acetylation was regulated both globally and locally. Large active domains were acetylated but within these regions peaks of acetylation were localised to the RSSs (Johnson et al. 2003; Morshead et al. 2003). Morshead et al. also showed that BRG1, the catalytic subunit of hSWI/SNF, is bound to the same regions that are enriched for histone acetylation at both the IgH and TCRβ loci (Morshead et al. 2003). Consistent with this, elevated levels of histone acetylation were shown to increase the accessibility of chromatin to nucleosome remodelling complexes, which subsequently allowed the RAG proteins to bind at RSSs (Nightingale et al. 2007). Despite the strong correlation between histone acetylation and the initiation of V(D)J recombination, this mark alone is insufficient for full locus activation (Hesslein et al. 2003; Sikes et al. 2002; Tripathi et al. 2002).

13.4.3 Histone Methylation and V(D)J Recombination

Trimethylation of histone H3 lysine 4 (H3K4me3) is also critical for V(D)J recombination. Increased H3K4me3 was first observed at the IgH and TCRβ loci in peaks at the ends of regions undergoing recombination (Morshead et al. 2003). More comprehensive studies at the Igκ locus, performed in inducible cell lines, primary B cells and in cell lines representing distinct developmental stages, uncovered a correlation between increased levels of this modification and the initiation of V(D)J recombination (Fitzsimmons et al. 2007; Goldmit et al. 2005; Perkins et al. 2004; Xu and Feeney 2009). Subsequently, the essential role for H3K4me3 was demonstrated by the discovery that the PHD domain of RAG2 interacts with this modification (Liu et al. 2007b; Matthews et al. 2007; Ramon-Maiques et al. 2007). Consistent with this, knockdown of WDR5, a co-factor of the H3K4 methyltransferase MLL, led to a reduction in the level of V(D)J rearrangements (Matthews et al. 2007).

The interaction between RAG2 and H3K4me3 was further verified in an elegant genome-wide study by the Schatz laboratory which showed high levels of RAG2 binding at the J gene segment RSSs, consistent with the previous findings that H3K4me3 is predominantly targeted to these regions (Ji et al. 2010). However, this study also detected RAG2 binding at other sites that are enriched for H3K4me3 throughout the genome. In contrast, the binding of RAG1, which is not dependent on H3K4me3, was more specifically targeted to the antigen receptor loci (Ji et al. 2010).

13.4.4 Sterile Transcription

Studies in the mid-1980s identified sterile, or non-coding, transcripts of the unrearranged antigen receptor loci (Blackwell et al. 1986; Lennon and Perry 1985; Schlissel and Baltimore 1989; Yancopoulos and Alt 1985). Subsequently, these transcripts were found to be up-regulated at the same stage at which V(D)J recombination was initiated (Duber et al. 2003; Engel et al. 1999) suggesting that transcription could play a role in regulating V(D)J recombination. However, other studies at the time cast doubt on the requirement of sterile transcription for rearrangement but instead proposed that increased transcription was merely a consequence of a more open chromatin structure. Indeed, transcripts were detected without the activation of recombination (Angelin-Duclos and Calame 1998; Fernex et al. 1995; Tripathi et al. 2000) and, conversely, recombination was shown to occur in the absence of transcription (Sikes et al. 2002). More recently, the Krangel laboratory demonstrated that transcription plays a critical role in regulating V(D)J recombination of TCR genes. Transcription through specific RSSs at the TCRα locus was blocked by the insertion of a transcriptional terminator downstream of the T early α (TEA) promoter (Abarrategui and Krangel 2006, 2007); this suppressed Vα to Jα recombination and greatly diminished levels of H3K4me2, H3K4me3, H3K36me3 and H3 acetylation at the downstream Jα gene segments. These studies suggest that sterile transcription controls chromatin structure by activating immediately downstream RSSs and regulating the addition of histone modifications. Consistent with this, transcription is known to regulate chromatin changes associated with gene activation, such as histone acetylation, H2B ubiquitination (Workman 2006) and notably H3K4me3. The latter occurs via association of the histone methyltransferase, Set1, with RNA polymerase II; this results in deposition of H3K4me3 during transcription initiation (Bernstein et al. 2005; Ng et al. 2003; Pokholok et al. 2005), thereby helping to recruit RAG2.

Other studies have shown that during the transcription of chromatin templates, H2A/H2B dimers are transiently evicted from nucleosomes to enable the passage of the RNA polymerase (Belotserkovskaya et al. 2003; Orphanides et al. 1999). Notably, a recent study found that even when all histone modifications previously associated with recombination are present, recombination is not fully activated. Instead, a strong correlation was found between the level of recombination and RSS accessibility; this accessibility was found to rely on the transcription-mediated eviction of H2A–H2B dimers at RSSs (Bevington and Boyes 2013). The authors argue that by making the RSSs available only transiently, this may play an important role in reducing the number of RAG-mediated double-stranded DNA breaks, thereby helping to maintain genomic stability.

13.5 Enhancers and Transcription Factors at the Immunoglobulin Loci

V(D)J recombination is controlled by *cis*-acting regulatory elements. Each antigen receptor locus contains at least one recombination enhancer and in some cases these elements share redundant functions, maintaining efficient V(D)J recombination when another element is removed or compromised (Chowdhury and Sen 2004; Gorman et al. 1996; Inlay et al. 2002; Schlissel 2004; Takeda et al. 1993; Xu et al. 1996). The function of the enhancers is to regulate the key changes required for V(D)J recombination, including stimulating sterile transcription, increasing the level of activating histone modifications and opening up the chromatin to increase accessibility. Moreover, the enhancers of the antigen receptor loci are able to stimulate these changes over large distances consistent with more recent findings in embryonic stem cells in which functionally interacting domains can extend up to a megabase (Dixon et al. 2012). Indeed, some of the immunoglobulin enhancers are located hundreds of kilobases from the gene segments that are targeted for rearrangement. Analysis of these enhancers identified several transcription factors that appear to play a key role in the regulation of antigen receptor rearrangement, including E2A, IRF4 and PU.1.

13.5.1 E2A

E2A was the first E-protein identified in mammals and was shown to bind to the immunoglobulin heavy and light chain enhancers at conserved sequences defined as E-boxes (Atchison 1988; Jones and Zhuang 2009). More recently, additional E-boxes have been located in close proximity to the Vκ gene segments (Sakamoto et al. 2012). Consistent with the idea that these factors activate recombination via the enhancers, mutation of the E-boxes at the heavy chain intronic enhancer (Fernex et al. 1995) and the kappa intronic enhancer (Inlay et al. 2004) impairs V(D)J rearrangement of the targeted allele.

One function of the E2A proteins is to recruit histone-modifying activities. E2A can interact with the histone acetyltransferases, p300/CBP and SAGA (Eckner et al. 1996; Ogryzko et al. 1996) and E2A-deficient pre-B cells have reduced levels of histone acetylation at both of the kappa recombination enhancers (Lazorchak et al. 2006). Moreover, when expressed in a non-B cell line, E2A was shown to bind at the Vκ gene segments, recruit p300 and induce sterile transcription and V(D)J recombination (Sakamoto et al. 2012). Since genomic interaction studies found that the binding of E2A, PU.1 and p300 overlap, it was suggested that these factors could regulate V(D)J recombination by anchoring enhancers and/or promoters into the same complex (Lin et al. 2012).

13.5.2 PU.1

PU.1 plays an important role in the regulation of immunoglobulin gene activation with binding sites located at the heavy chain intronic enhancer (Nelsen et al. 1993; Rivera et al. 1993), the kappa and lambda light chain enhancers (Eisenbeis et al. 1993; Pongubala et al. 1992) and at specific promoter regions (Schwarzenbach et al. 1995; Shin and Koshland 1993). Notably, in cooperation with other transcription factors, PU.1 synergistically activates IgH transcription and increases locus accessibility (Nelsen et al. 1993; Nikolajczyk et al. 1999; Rivera et al. 1993).

13.5.3 IRF4

IRF4 and IRF8 function redundantly to control pre-B cell development (Lu et al. 2003) and, consistent with a role for IRF4 in light chain activation, binding sites are present at the kappa 3′ enhancer (κE3′) and the lambda enhancers ($E\lambda_{3-1}/E\lambda_{2-4}$) (Eisenbeis et al. 1993; Rudin and Storb 1992; Schlissel 2004). Importantly, IRF4 binding to these sites increases in pre-B cells, concomitant with an increase in IgL recombination (Bevington and Boyes 2013; Johnson et al. 2008; Muljo and Schlissel 2003).

The binding of IRF4 to κE3′ and $E\lambda_{3-1}/E\lambda_{2-4}$ is dependent on DNA-bound PU.1 (Eisenbeis et al. 1995; Pongubala et al. 1992). Moreover, IRF4 interacts with E2A to stimulate significant transcriptional synergy at κE3′ (Nagulapalli and Atchison 1998; Nagulapalli et al. 2002) and knockdown of IRF4 impairs the recruitment of E2A to the enhancer (Lazorchak et al. 2006). Therefore it seems likely that interactions between IRF4 and other proteins are important for stimulation of the light chain loci.

Binding sites for E2A and IRF4 have also been identified at the IgH intronic enhancer, where binding of the complex induces heavy chain sterile transcription (Nagulapalli and Atchison 1998). However, further studies suggest that this binding plays a more important role later in B cell development during class switch recombination (Klein et al. 2006).

13.5.4 STAT5

STAT5 plays a critical role in regulating stage-specific activation of the immunoglobulin loci. In pro-B cells, IL-7 signalling activates the Jak-STAT pathway, stimulating phosphorylation of STAT5 which promotes dimer formation and transcriptional activation. STAT5 is then recruited to the distal V_H gene segments where it is thought to activate recombination via increased histone acetylation and sterile transcription of the large V_HJ558 gene family (Bertolino et al. 2005). Conversely, also in pro-B cells, STAT5 acts as a repressor, inhibiting kappa light chain activation by binding to both the kappa intronic enhancer (κEi) and the kappa 3′ enhancer (κE3′). When STAT5 is bound, the activators E2A and IRF4/PU.1 are unable to access their binding sites at κEi and κE3′, respectively (Hodawadekar et al. 2012; Malin et al. 2010; Mandal et al. 2009). Consistent with this, sterile transcription and V(D)J rearrangements at the Igκ locus in pro-B cells are inhibited (Malin et al. 2010; Mandal et al. 2009). Successful rearrangement of the heavy chain locus and pre-BCR signalling reduces the cells' responsiveness to IL-7, resulting in a loss of STAT5 activation. Consequently, STAT5 binding is diminished at the kappa regulatory regions, enabling the initiation of light chain recombination.

13.5.5 PAX5

More recently, Pax5-activated intergenic repeat (PAIR) elements have been identified at the 5′ region of the V_H gene segment array (Ebert et al. 2011). The elements are marked by an active chromatin signature and antisense transcription. In Pax5-deficient pro-B cells, antisense transcription is reduced suggesting a regulatory role for Pax5 in V_H to DJ_H recombination. In addition to Pax5-binding sites, these regions contain sites for the transcription factors CTCF and E2A. Both CTCF and Pax5 have been shown to play key roles in locus contraction and during formation of the long-range interactions required for V(D)J recombination (see below).

13.6 Long-Range Interactions Between Regulatory Elements

Rearrangements occur over huge distances, up to 3 Mb, and for this to occur, the antigen receptor loci undergo substantial large-scale genomic re-organisation during lymphocyte development. Fluorescence in situ hybridisation (FISH) studies demonstrated that the IgH and Igκ loci are in an extended configuration in T cells but they become contracted in pro-B cells and pre-B cells, respectively (Kosak et al. 2002; Roldan et al. 2005; Sayegh et al. 2005; Skok et al. 2007). IgH locus contraction and distal V_H to DJ_H rearrangements are facilitated by the transcription factors, Pax5 and YY1 (Fuxa et al. 2004; Hesslein et al. 2003; Liu et al. 2007a). FISH analyses demonstrated that the distance between the V_H and C_H regions was increased in Pax5-deficient pro-B cells, affecting distal V_H gene usage (Fuxa et al. 2004). In YY1-deficient mice, development is blocked prior to the onset of IgH V(D)J recombination (Gordon et al. 2003; Liu et al. 2007a). Similar to the absence of Pax5, the defect mainly affects recombination of the distal V_H gene segments and, furthermore, the distance separating the V_H gene segments from the C_H region is also increased.

More recently, a key role for CTCF in facilitating long-range interactions and chromosomal loops at the immunoglobulin loci has been uncovered. Developmentally regulated binding of CTCF and the cofactor cohesin was shown at sites within the IgH and Igκ loci (Degner et al. 2011) and these sites provide an anchor for long-range interactions and the formation of chromosomal loops. Notably, YY1-binding sites were also found at the base of these looped structures (Guo et al. 2011a). At the heavy chain locus two forms of loops have been identified: Eμ-dependent loops mediate interactions between the D_H and J_H gene segments, whereas E-μ-independent loops are created between specific V_H gene families. Subsequently, the looped domains are dynamically brought together to compact the locus and facilitate V(D)J recombination (Guo et al. 2011a). The importance of the CTCF sites in regulating stage-specific recombination was demonstrated when a region adjacent to the IgH DFL16.1 gene segment containing two conserved CTCF-binding sites was deleted. The mutated alleles were able to recombine V_H to D_H gene segments before D_H to J_H rearrangement. Lineage specification was also affected as V_H to DJ_H recombination was detected in developing thymocytes (Guo et al. 2011b). This and other studies suggested that these sites were acting as a physical barrier, preventing the activation of V_H genes before D_H to J_H recombination had initiated (Featherstone et al. 2010; Guo et al. 2011b).

13.7 Mature B Cell Development and Plasma Cell Differentiation

Upon encountering an antigen, naïve B cells that express a cognate IgM receptor go on to proliferate and differentiate into either memory B cells or antibody-secreting plasma cells in the germinal centres (GCs) of secondary lymphoid organs such as the lymph nodes (Kelsoe 1996; MacLennan 2005). Here, B cells undergo clonal expansion, producing a large number of cells expressing the same immunoglobulin. The affinity of the B cell receptor–antigen interaction is fine-tuned by the process of somatic hypermutation (SHM), where an extremely high rate of localised point mutation occurs at IgH and IgL variable exons (Peled et al. 2008). These mutations are clustered around specific nucleotides corresponding to the complementarity-determining regions of the receptor, thus creating a huge number of B cells that express subtle variations of the same immunoglobulin. Some of these will bind to the antigen with greater affinity and will be subjected to positive selective pressure and proliferate further, whereas those cells containing mutations that weaken the interaction will be lost (Casali et al. 2006; Di Noia and Neuberger 2007; Li et al. 2004; Teng and Papavasiliou 2007).

13.8 AID and Somatic Hypermutation

Activation of B cells by an antigen triggers Pax5 and E2A to upregulate expression of activation induced cytidine deaminase (AID); this initiates SHM by deaminating deoxycytidine (dC) residues in single-stranded DNA to deoxyuridine (dU), thus creating a uracil:guanine mismatch (Fig. 13.4). Because AID can only act on single-stranded (ss)DNA, transcription through the immunoglobulin locus is essential to create single-stranded transcription bubbles before SHM can occur (Peled et al. 2008).

Since deoxyuridine is not normally found in DNA, these residues are repaired in one of three ways: DNA replication, base-excision repair (BER), or mismatch repair (MMR). Repair by DNA replication simply involves interpretation of the uracil residue as deoxythymidine, resulting in a C-to-T mutation or a G-to-A mutation in the opposite strand. In base excision repair, uracil DNA glycosylase (UNG) removes the uracil residue to create an abasic site. The abasic site is converted to a single-strand DNA break by apurinic/apyrimidinic endonuclease 1 (APE1), which can then be filled in with any of the four bases by a translesional error-prone DNA polymerases such as REV1 or Pol μ (Peled et al. 2008). In MMR, the G:U mismatch is recognised by a different set of proteins, but also results in the mutation of residues surrounding the U:G mismatch. Either the Msh2–Msh3 or Msh2–Msh6 heterodimer recognises the U:G mismatch, which then recruit (amongst others) MLH1 and PMS2 leading to a single-strand nick near the mismatch. Exonuclease 1 excises a stretch of the surrounding DNA strand, and

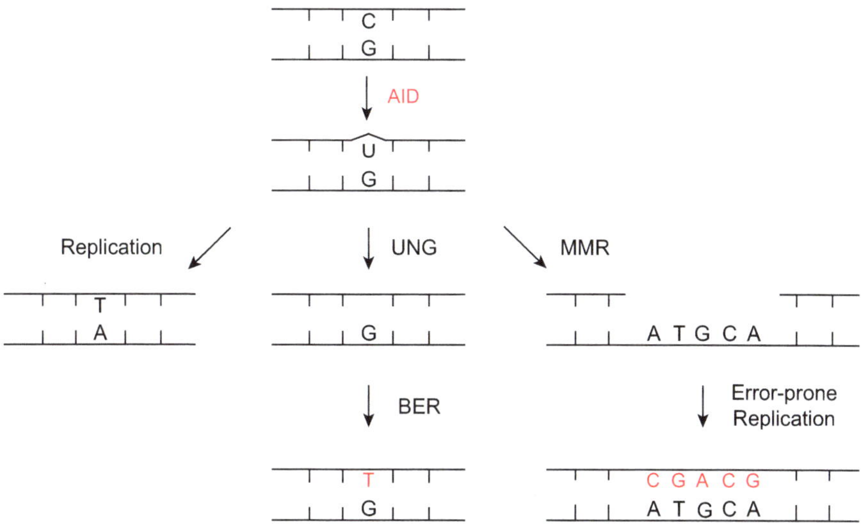

Fig. 13.4 Somatic hypermutation. AID deaminates cytidine to uridine, which is then processed in one of three ways: replication, where the U is recognised as a T, base excision repair (BER) or mismatch repair (MMR). In BER, only one residue is mutated, whereas in MMR, bases surrounding the lesion are also mutated

PCNA recruits error-prone polymerases to repair the gap, resulting in several new mutations surrounding the site of the original G:U mismatch (Peled et al. 2008). Intriguingly, MMR and base excision repair are normally beneficial repair pathways, but in SHM they are hijacked and made to introduce mutations in the V region. In both cases it is thought that mono-ubiquitinylated PCNA is responsible for recruiting the error-prone polymerases (Poltoratsky et al. 2000).

13.9 Class Switch Recombination

Before stimulation by a complementary antigen, naïve B cells express either the IgM or IgD isotypes by alternative splicing of the D or M constant exons onto the variable gene exon. Changing to any of the other isotypes (IgG, IgA or IgE) is mediated by class switch recombination (CSR, also known as isotype switching). CSR is an intrachromosomal deletion recombination reaction that occurs between G-rich tandem-repeated DNA sequences called switch (S) regions that are located upstream of all heavy chain constant C_H regions except IgD (Fig. 13.5). Each naïve B cell has the potential to switch to any isotype, although the switch is directed by various cytokines that activate sterile transcription through the specific switch regions.

The initial steps of CSR are similar to SHM. However, instead of generating a mutation, a double-stranded DNA break is made, which is subsequently repaired by

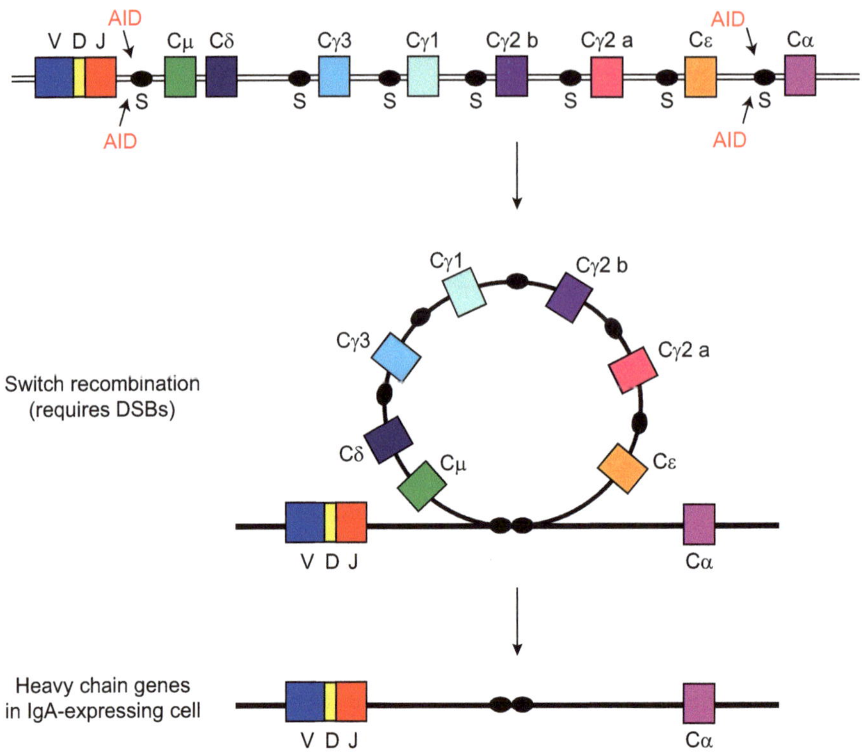

Fig. 13.5 Mechanism of class switch recombination. AID creates double-strand breaks at switch regions (*black ovals* labelled "s"). Recombination occurs between the two double strand breaks, bringing the alternative constant region adjacent to the rearranged variable region. A complete heavy chain transcript is made by mRNA splicing

the NHEJ pathway. First, sterile transcription through the C_H locus creates localised regions of ssDNA at S regions, and AID introduces several dU residues in the S regions on both strands of DNA. UNG recognises the U:G mismatch and removes the deoxyuracil to create an apyrimidinic ribose residue, which is excised by APE1 or APE2, thus forming a single-stranded nick. These single-stranded breaks can spontaneously form a DSB if two are created close to each other on opposite strands. If not, it is possible that proteins from the MMR pathway can convert single-strand breaks that are not near each other into DSBs. The NHEJ pathway then joins DSBs at donor and acceptor switch regions via the same mechanism as in V(D)J recombination, and the intervening DNA is excised (Stavnezer et al. 2008).

13.10 Terminal Differentiation to Plasma Cells

Terminal differentiation of B cells into antibody-secreting plasma cells is regulated by BLIMP1, a zinc-finger transcription factor, considered to be the primary trigger for plasma cell differentiation. BLIMP1 orchestrates this transition by repression of *Pax5*, which has two major effects. The first is the repression of the germinal centre programme by removing the positive regulation of *Aicda* and *Bcl6* by Pax5 (Sciammas and Davis 2004; Shaffer et al. 2002), and the second is the derepression of *Xbp1*. Xbp1 expression results in a dramatic expansion of the endoplasmic reticulum and activation of genes involved in the protein secretion pathway, thus allowing the production and secretion of large amounts of antibody (Shaffer et al. 2004). In addition, BLIMP1 regulates the differential 3′ end processing of the immunoglobulin heavy chain pre-mRNA, leading to increased levels of antibody secretion (Sciammas and Davis 2004; Shapiro-Shelef et al. 2003). Together, these processes produce highly efficient, immunoglobulin-secreting machines.

In addition to its role in regulating recombination, IRF4 plays a central role in coordinating exit from the GC programme. Whilst downregulated throughout the GC reaction, IRF4 is upregulated during exit from the germinal centres and coordinates plasma cell development by terminating the GC programme (De Silva et al. 2012). This is achieved by derepression of BLIMP1 expression and downregulation of *Bcl6* expression, which is the master regulator of the GC programme (De Silva et al. 2012). The central role of IRF4 in regulating plasma cell development is exemplified by the fact that deregulation of IRF4 alone can trigger multiple myeloma in plasma cells (see "Sect. 13.16").

13.11 B Cell Leukaemia and Lymphoma

Normal B cell development involves the programmed formation of multiple genomic DSBs and mutations, through the mechanisms of V(D)J recombination, SHM and class switch recombination. Whilst these processes are fundamental to the generation of a diverse immune receptor repertoire, the formation of double-strand breaks presents a major threat to the genomic integrity of developing B cells. Indeed, chromosome translocations arising from aberrant recombination are a hallmark of lymphoid malignancies (Kuppers 2005). These translocations alone are generally not enough to fully trigger carcinogenesis. For instance the *BCL2/IgH* translocation can be found in circulating pre-leukemic blood cells of healthy individuals (Limpens et al. 1991; Janz et al. 2003). However, they are a defining feature of many B cell malignancies and are thought to be the critical initiating event (Kuppers 2005). Full progression of these malignancies requires additional genetic mutations in order to further deregulate the transcriptional and epigenetic programmes that direct B cell development. The following section will discuss the sources of aberrant B cell development and how they can trigger malignancy.

13.12 Chromosome Translocations

There are three general mechanisms by which the recurrent translocations found in B cell tumours are thought to promote malignancy (Nussenzweig and Nussenzweig 2010). The first involves upregulation of a proto-oncogene or down-regulation of a tumour suppressor gene by the juxtaposition of one of these genes with a strong regulator element from one of the immunoglobulin loci (Adams et al. 1985; Vanasse et al. 1999). For example, the t(14;18) translocation, found in nearly all follicular lymphomas (Raghavan et al. 2001), brings a strong enhancer from the J_H locus on chromosome 14 within range of the *BCL2* promoter on chromosome 18. In a normal physiological setting, the role of BCL2 is to prevent caspase 9 and 3-mediated apoptosis by inhibiting the release of mitochondrial cytochrome c (Kridel et al. 2012). Therefore, upregulation of BCL2 expression by the J_H enhancer promotes survival of these cells by reducing their capacity to undergo apoptosis. Another classic example is the *c-myc/IgH* translocation, which is found nearly all cases of Burkitt's lymphoma (see "Sect. 13.15") (Robbiani and Nussenzweig 2013).

The second mechanism by which translocations can promote B cell leukaemogenesis is via formation of a fusion protein. The classic example of this is the t(9;22)(q34;q11) Philadelphia chromosome, which encodes the BCR-ABL1 kinase, and whilst being a hallmark of chronic myeloid leukaemia, it is also found in 20–40 % of adults with B cell acute lymphoblastic leukaemia (B-ALL). The BCR-ABL kinase is a constitutively active kinase that promotes transformation by aberrantly enhancing the RAS-MAPK, JAK-STAT and PI3K-AKT pathways (Teitell and Pandolfi 2009).

The final mechanism involves the deregulation of microRNA expression. This is a relatively recent discovery that is less well characterised than the first two mechanisms, but includes the microRNA families miR-15 and miR-16 which act as tumour suppressors by down-regulating *BCL2* (Calin and Croce 2007; Calin et al. 2004).

The first step in any chromosomal translocation is the formation of two DSBs. Nearly all B cell translocations that deregulate a proto-oncogene involve a break at an immunoglobulin locus, caused by V(D)J recombinase or AID activity. There are several possible sources of the second break, including the mistargeting of RAG activity through the recognition of cryptic RSSs and non-standard DNA structures, mistargeted AID activity, or another process such as collapse of a replication fork or exposure to ionising radiation (Tsai and Lieber 2010).

It should be noted that translocations that create a fusion protein (e.g., BCR-ABL, TEL-AML1) do not involve antigen receptor loci. The sources of these breaks are less well-defined and potentially more diverse, with similar mechanisms operating widely in non-lymphoid cancers. There is some evidence that they could include the presence of Alu repeats, for example at the *Bcr* and *Abl1* breakpoint cluster regions (Elliott et al. 2005), palindromic-mediated genomic instability (Kurahashi et al. 2006), or an imbalance of metabolic pathways, based on epidemiological evidence in infant leukaemias (Rabkin and Janz 2008). Nevertheless, the TEL-AML1 fusion protein, also known as ETV6-RUNX1, is the most common genetic lesion in childhood ALL (Shurtleff et al. 1995). Expression of this fusion protein is thought to promote transformation by overexpressing the erythropoietin receptor and downstream activation of the JAK/STAT signalling pathway (Torrano et al. 2011). In some cases the fusion partner simply directs mis-expression of the oncogene (e.g. BCR-ABL).

13.12.1 Mistargeted RAG Activity

The first indication that V(D)J recombination might be involved in chromosome translocations arose through analysis of translocation breakpoints in B and T cell malignancies, which revealed that many of them involve cryptic RSSs, namely sequences that bear homology to true RSSs (Dalla-Favera et al. 1982; Kirsch et al. 1982). This led to the popular idea that all translocations occur because the RAG complex mistakenly identifies these cryptic RSSs as suitable recombination partners, but otherwise carries out normal V(D)J recombination (Fig. 13.6a).

There are over ten million cryptic RSSs scattered throughout the entire genome, thus creating huge scope for translocations by this process (Lewis et al. 1997). However, in vivo recombination experiments using extra-chromosomal plasmid substrates indicated that whilst some cryptic RSSs are used as recombination substrates very efficiently, such as the *LMO2*, *TAL1* and *TAL2* cryptic RSSs, others are used very inefficiently, if at all, including those found at common translocation breakpoints such as the *BCL1* and *BCL2* loci (Marculescu et al. 2002; Raghavan et al. 2001). Furthermore, many of the sequences found at breakpoints such as the *BCL2* major breakpoint region (Mbr) do not contain any of the heptamer/nonamer sequences that are required for RAG binding, indicating that many of the translocations found in B cell malignancies are not caused by recombination with a cryptic RSS (Raghavan et al. 2004b). If the *BCL2* Mbr does not contain an efficient recombination signal, then why is the t(14;18) translocation so prevalent? One possible explanation lies in the ability of the RAG proteins to bind and cleave non-B form DNA. Analysis of the Mbr revealed that it reacts with bisulphite, a chemical probe for ssDNA, suggesting that the Mbr forms a stable region of single-stranded, non-B form DNA, and this is efficiently recognised and cleaved by the RAG complex in transfected cell lines (Raghavan et al. 2004a, b, 2005). However, it remains to be seen whether or not RAGs can cleave the *BCL2* Mbr in developing B cells.

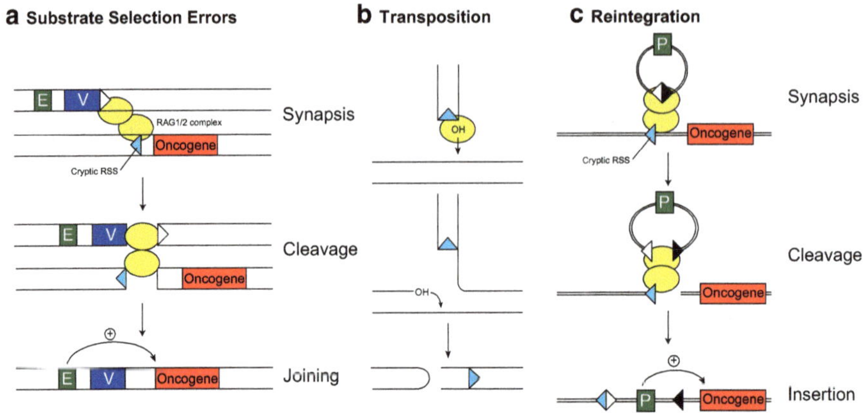

Fig. 13.6 Mechanisms of aberrant RAG activity (**a**) Substrate selection error. RAGs carry out normal recombination but mistakenly recognise a cryptic RSS or non-standard DNA structure. End donation occurs in the same way, except that RAGs do not make one of the breaks. (**b**) In transposition, the 3′-OH group of a broken RSS attacks the phosphodiester backbone at a random genomic locus. (**c**) In reintegration, RAGs catalyse normal recombination, albeit between an excised signal circle and a genomic RSS. *Blue triangles* cryptic RSS, *E* immunoglobulin locus enhancer, *P* promoter, *V* variable gene segment

13.12.2 End Donation

Some translocations, where the non-immunoglobulin partner does not bear the hallmarks of either RAG or AID activity, arise via another unrelated process, such as the collapse of a replication fork or a random DNA break caused by ionising radiation or oxidative free radicals (Fig. 13.6a) (Kuppers and Dalla-Favera 2001; Lewis 1994; Tycko and Sklar 1990). Analysis of translocation junctions in follicular lymphomas showed duplications of sequences at the breakpoint, indicative of a staggered DNA break, which is a feature that is not consistent with RAG cutting (Bakhshi et al. 1987). These observations led to the proposal of the end donation model, which are estimated to be responsible for 30–40 % of the translocations found in follicular and mantle cell lymphomas (Jager et al. 2000; Welzel et al. 2001). However, the DSBs created by RAG cutting should be held in place in a post-cleavage complex until they are resolved by the NHEJ pathway, so it is unclear how they would become free to join with other DSBs (Roth 2003). Notably, these translocations are increased in the absence of Ataxia Telangiectasia Mutated (ATM) kinase or the NHEJ machinery, highlighting the critical role of these complexes in preventing translocations (Roth 2003).

13.13 Reinsertion of V(D)J Recombination By-Products

In addition to gross chromosome translocations, RAG proteins can mediate two other types of DNA rearrangements, often involving the recombination by-products, which can lead to oncogene activation. In vitro, the RAG complex is able to catalyse transposition reactions, where the reactive 3′ hydroxyl group of a signal end attacks a phosphodiester bond at any location, causing random insertion of the signal ends in a joining reaction that is independent of NHEJ factors (Agrawal et al. 1998; Hiom et al. 1998; Roth 2003). The source of these signal ends can be either the cleavage of genomic RSSs adjacent to a coding segments or the signal joint of an ESC (the by-product of V(D)J recombination) which can reopened by the RAG complex to give two signal ends (Fig. 13.6b) (Neiditch et al. 2002).

Many studies have highlighted the similarity between V(D)J recombination and viral transposition, leading to the hypothesis that the RAG proteins evolved from a viral transposase which integrated into the genome of jawed vertebrates (Hansen and McBlane 2000). Like RAGs, transposases specifically recognise short sequences flanking the mobile DNA segment and introduce double-strand breaks between the recognition sequences and the DNA segment by a direct, one-step trans-esterification mechanism (van Gent et al. 1996). Indeed, the RAG complex catalyses transposition in vitro with a high efficiency (Agrawal et al. 1998; Hiom et al. 1998). However, there have been surprisingly few cases reported in vivo. One study in a human T cell line found an excised signal-end fragment from the T cell receptor locus which had integrated into an intron of the *HPRT* gene (Messier et al. 2003), but importantly there has been no description of a case of leukaemia or lymphoma caused by transposition. The initial in vitro studies were carried out using core RAG proteins, where the regions dispensable for the basic recombination reaction are removed for ease of purification. However, the non-core C-terminus of RAG2 is capable of almost entirely inhibiting the transposition reaction in vivo (Elkin et al. 2003), thus making transposition an unlikely source of translocations in B cells.

13.14 Reintegration

In contrast to transposition, the reintegration of ESCs has been shown to be a potential major contributor to genomic instability (Curry et al. 2007; Vanura et al. 2007). Reintegration is distinct from transposition in that a transposition reaction shows no target specificity, whereas reintegration joins the ESCs with RSSs located on chromosomes. These could be an authentic RSS or one of the ten million cryptic RSSs littered throughout the human genome (Fig. 13.6c).

A genome-wide assessment of signal end integrations in developing thymocytes (T cell precursors) found that half of the insertions out of a total of 43 observations were caused by reintegration of an ESC (Curry et al. 2007). Furthermore, they could not detect any obvious translocations or transpositions. Therefore, current evidence indicates that the main pathway for insertion of an ESC back into the genome is actually reintegration and not the extensively studied transposition reaction (Ramsden et al. 2010).

Vanura et al. (2007) cloned the cryptic RSSs adjacent to the *TAL2* and *LMO2* oncogenes into plasmid substrates and found that an excised signal circle could reintegrate into the cryptic RSSs with an efficiency equal to trans-recombination between authentic RSSs. To test the physiological relevance of this finding, they examined the level of pseudo-hybrid joint formation in mouse thymocytes and found that they occurred at the same frequency as chromosomal translocations. Recent gene profiling studies have shown that a substantial number of lymphoma and leukaemia cases display oncogene activation in the absence of any detectable cytogenetic abnormalities (Vanura et al. 2007). These studies suggest that reintegration of ESCs into cryptic RSSs adjacent to oncogenes is a possible potent cause of T cell leukaemias and follicular lymphomas. Moreover, the relatively small size of excised signal circles (<1 Mb) means they would be undetectable by traditional cytogenetic methods. This would go some way to explain why translocation is well documented, but reintegration has only been recently described. The existence of ESC reintegration in vivo is reinforced by a report in which an ESC was found inserted in the cryptic RSS in *HPRT* exon 1, a region that is commonly involved in V(D)J recombination errors (Messier et al. 2006). It has been estimated that reintegration of episomal signal circles occurs following 1–10 % of V(D)J-mediated recombination reactions and that in humans there are around 5,000 reintegrations per human per day (Vanura et al. 2007), making this reaction a very significant threat to genomic stability.

13.15 Mistargeting of AID Activity

Many human lymphomas involve mature B cells, which have already undergone V(D)J recombination and therefore no longer express RAG proteins. Nevertheless, these cancers are usually associated with translocations involving immunoglobulin genes, such as the *Bcl2/IgH* translocation in follicular lymphoma and the *c-myc/IgH* translocation in sporadic Burkitt's lymphoma (Kuppers 2005). In contrast, mature T cell lymphomas are very rare, probably because these cells do not undergo further genetic alteration or double-strand break generation. Analysis of the translocations in mature B cells indicates that most of the immunoglobulin locus breakpoints are in the V (indicating SHM error) or switch (indicating CSR error) regions. Thus, it appears that AID activity plays a central role in many of these translocations (Robbiani and Nussenzweig 2013).

Two mouse plasmacytoma models have been used to test the possibility that translocation involving switch regions are the result of mistargeted AID activity. Plasmacytomas with *c-myc-IgH* translocations can be induced by injection of mineral oil or by overexpression of interleukin-6 (Kovalchuk et al. 2002; Potter and Wiener 1992), and most of the translocations are between the 5′ UTR of the *c-myc* gene and switch regions at the *IgH* locus. AID-deficient mice do not acquire *c-myc-IgH* translocations, suggesting that AID is essential to initiate the process (Ramiro et al. 2006; Unniraman et al. 2004). UNG is also necessary to create the DSB for these translocations (Ramiro et al. 2006), and since UNG creates DSBs in the initial steps of CSR, it appears that the source of the *c-myc/IgH* translocation is aberrant CSR and not SHM.

The cause of the *c-myc* locus break remained elusive for some time with the favoured hypothesis being that AID activity was responsible for this break as well, based on the observations that cleavage at both loci occurs at the same stage of B cell development and that transcription through the *c-myc* locus generates G-loops that can be bound by AID (Duquette et al. 2005). This prediction was borne out by experiments by Robbiani et al. (2008), who showed that when an artificial DSB is made at the *IgH* locus, the *c-myc/IgH* translocation is detected only in AID-proficient cells, thus demonstrating the necessity of AID in creating the break at *c-myc* (Robbiani et al. 2008).

In addition to initiating translocations, AID can introduce mutations at a number of non-immunoglobulin loci in a hypermutation-like manner, including at the proto-oncogenes *PIM1*, *MYC*, *RhoH/TTF (ARHH)* and *PAX5* (Pasqualucci et al. 1998, 2001). Mutations in these four oncogenes can be found in 50 % of diffuse large B cell lymphomas (DLBCLs), but not in normal germinal centre B cells, indicating that mistargeted AID activity can promote malignancy development not only by translocation-mediated mechanisms but also by a hypermutation-mediated mechanism (Pasqualucci et al. 2001).

13.16 Deregulation of Transcription and Epigenetics in B Cell Leukaemia/Lymphoma

Whilst chromosomal translocations are thought to be the key event that triggers most cancers of the B cell lineage, progression of these malignancies requires the deregulation of the transcriptional and epigenetic mechanisms that govern the normal development of B cells.

Indeed, 39 % of diffuse large B-cell lymphomas (DLBCLs) and 41 % of follicular lymphomas carry inactivating mutations in the histone acetyltransferase genes *CREBBP* (CBP) and/or *EP300* (p300) (Pasqualucci et al. 2011a). CBP and p300 are transcription coactivators that enhance gene transcription through multiple mechanisms, including the acetylation of histones (Bannister and Kouzarides 1996; Ogryzko et al. 1996), acetylation of transcriptional activators (for example p53 and GATA-1) (Blobel et al. 1998; Gu and Roeder 1997) and through the acetylation of transcriptional repressors (Bereshchenko et al. 2002). They can also act as scaffolding proteins, building bridges between other proteins and the transcriptional apparatus to form enhancersomes (Chan and La Thangue 2001).

MLL2, a histone methyltransferase gene, is also frequently mutated in lymphoid cancers, as it has been found in over one-third of DLBCL cases, and notably, these mutations occur in conserved regions of *MLL2* that are critical for methyltransferase activity. Moreover, in most cases these mutations only affect one allele, suggesting that a reduction in the activity of these chromatin-modifying enzymes is important for the progression of these lymphomas and that these genes act as haploinsufficient tumour supressors (Pasqualucci et al. 2011b).

Translocations involving a related histone methyltransferase, mixed lineage leukaemia (*MLL*), are implicated in a number of acute leukaemias. *MLL* is critical for normal haematopoiesis, and more than 50 translocation partners have been found to date. The most common *MLL* translocations in ALL, t(4;11) and t(11;19), lead to the expression of the fusion proteins MLL-AF4 and MLL-ENL, respectively (Liedtke and Cleary 2009). AF4 and ENL are normally involved in the maintenance of transcription elongation regulation by forming higher order complexes with the H3K79 methylase hDOT1-L (human DOT1-like), and in the case of AF4, with P-TEFb (RNA polymerase II transcription elongation factor b) (Liedtke and Cleary 2009). Combining this activity with the DNA-binding domain of MLL ultimately leads to the aberrant expression of MLL target genes, which include the *HOX* genes, thus contributing to the development of ALL.

DNA-modifying enzymes have also been implicated in lymphoma development, with inactivating mutations of *TET2* being found in 2 % of B cell lymphomas. The Ten Eleven Translocation (TET) family of proteins were initially identified from chromosome translocations involved in myeloid leukaemias (Lorsbach et al. 2003; Ono et al. 2002), but subsequent biochemical analyses showed that they catalyse the conversion of 5-methyl-cytosine (5mC), to 5-hydroxymethyl-cytosine (5hmC) (Tahiliani et al. 2009). *TET2* mutations have now been found to occur early in the development of a large number of myeloid cancers, and more recently, in a number of human lymphoid cancers. In a mouse model where one or both copies of *TET2* were inactivated, HSC homeostasis and differentiation were affected in both the early and late stages of haematopoiesis, including lymphoid differentiation. This gave these cells a competitive advantage that led to the development of malignancies (Quivoron et al. 2011). Since epigenetic modifiers such as *TET2* and *MLL* have global effects on the genome, perturbation of their normal function can alter the expression profiles of huge numbers of genes, ultimately leading to malignancy.

In addition to epigenetic factors, many of the transcription factors that modulate B cell development can become activated by translocations as a result of aberrant recombination during B cell development. For example, Pax5, the guardian of B cell identity, is itself a target of translocation where the *PAX5* gene and a strong enhancer from the *IgH* locus are brought together. This translocation is associated with aggressive B-cell non-Hodgkin lymphomas, and the major effect is likely to be the prevention of *PAX5* repression at the start of plasma cell differentiation (Thomas-Tikhonenko and Cozma 2008). Likewise, E2A is involved in a translocation that produces a chimeric transcription factor, E2A-HLF, which is found in

B-precursor ALL. In addition to translocations, aberrantly spliced variants of Ikaros, which are deficient in DNA-binding activity, lead to childhood ALL (Sun et al. 1999).

A particularly notable example of a B-cell regulator that is activated by translocations is IRF4. This factor becomes over-expressed in following its translocation to the IgH locus (Iida et al. 1997), leading to multiple myeloma, a cancer of plasma cells. Subsequent innovative studies showed that this single factor coordinates an aberrant transcription network that sustains the malignant growth of the multiple myeloma cells (Shaffer et al. 2008). Specifically, IRF4 directly activates expression of *MYC*, which is central to an aberrant transcriptional network of over 100 genes, and notably, MYC directly activates *IRF4*, forming a positive feedback loop in myeloma cells that reinforces an aberrant expression profile that promotes malignancy (Shaffer et al. 2008). Importantly, knockdown of *IRF4* alone causes death of the multiple myeloma cells, providing strong evidence for the growing paradigm of oncogene addiction in certain cancers (Shaffer et al. 2008; Weinstein and Joe 2008).

13.17 Conclusion

B cells undergo multiple breakage and rejoining reactions as an essential step in their development, but a mistake in the regulation of any of these reactions can lead to malignancy-promoting translocations. Ironically, quite a number of the genes that are activated via translocations are the very transcription factors that specify B cell development. Any imbalance in these factors can perturb B cell development and potentially contribute to leukaemia. Thus, greater understanding of translocations mechanisms and their effect on B cell development transcription networks seems essential to guide the development of more specific and powerful leukaemia therapies.

Acknowledgments This work was funded by an EPSRC studentship (to CMK) a Lady Tata Memorial Trust studentship (to JNS), a MRC grant G0801101 (to JB) and the BBSRC (SB).

References

Abarrategui I, Krangel MS (2006) Regulation of T cell receptor-alpha gene recombination by transcription. Nat Immunol 7:1109–1115. doi:10.1038/ni1379, ni1379 [pii]

Abarrategui I, Krangel MS (2007) Noncoding transcription controls downstream promoters to regulate T-cell receptor alpha recombination. EMBO J 26:4380–4390

Adams JM, Harris AW, Pinkert CA, Corcoran LM, Alexander WS, Cory S, Palmiter RD, Brinster RL (1985) The c-myc oncogene driven by immunoglobulin enhancers induces lymphoid malignancy in transgenic mice. Nature 318:533–538

Agrawal A, Eastman QM, Schatz DG (1998) Transposition mediated by RAG1 and RAG2 and its implications for the evolution of the immune system. Nature 394:744–751. doi:10.1038/29457

Alt FW, Yancopoulos GD, Blackwell TK, Wood C, Thomas E, Boss M, Coffman R, Rosenberg N, Tonegawa S, Baltimore D (1984) Ordered rearrangement of immunoglobulin heavy chain variable region segments. EMBO J 3:1209–1219

Angelin-Duclos C, Calame K (1998) Evidence that immunoglobulin VH-DJ recombination does not require germ line transcription of the recombining variable gene segment. Mol Cell Biol 18:6253–6264

Atchison ML (1988) Enhancers: mechanisms of action and cell specificity. Annu Rev Cell Biol 4:127–153. doi:10.1146/annurev.cb.04.110188.001015

Bakhshi A, Wright JJ, Graninger W, Seto M, Owens J, Cossman J, Jensen JP, Goldman P, Korsmeyer SJ (1987) Mechanism of the t(14;18) chromosomal translocation: structural analysis of both derivative 14 and 18 reciprocal partners. Proc Natl Acad Sci USA 84:2396–2400

Bannister AJ, Kouzarides T (1996) The CBP co-activator is a histone acetyltransferase. Nature 384:641–643. doi:10.1038/384641a0

Bassing CH, Swat W, Alt FW (2002) The mechanism and regulation of chromosomal V(D)J recombination. Cell 109(Suppl):S45–S55, S009286740200675X [pii]

Baumann M, Mamais A, McBlane F, Xiao H, Boyes J (2003) Regulation of V(D)J recombination by nucleosome positioning at recombination signal sequences. EMBO J 22:5197–5207. doi:10.1093/emboj/cdg487

Belotserkovskaya R, Oh S, Bondarenko VA, Orphanides G, Studitsky VM, Reinberg D (2003) FACT facilitates transcription-dependent nucleosome alteration. Science 301:1090–1093. doi:10.1126/science, 1085703 301/5636/1090 [pii]

Bereshchenko OR, Gu W, Dalla-Favera R (2002) Acetylation inactivates the transcriptional repressor BCL6. Nat Genet 32:606–613. doi:10.1038/ng1018, ng1018 [pii]

Bernstein BE, Kamal M, Lindblad-Toh K, Bekiranov S, Bailey DK, Huebert DJ, McMahon S, Karlsson EK, Kulbokas EJ 3rd, Gingeras TR, Schreiber SL, Lander ES (2005) Genomic maps and comparative analysis of histone modifications in human and mouse. Cell 120:169–181

Bertolino E, Reddy K, Medina KL, Parganas E, Ihle J, Singh H (2005) Regulation of interleukin 7-dependent immunoglobulin heavy-chain variable gene rearrangements by transcription factor STAT5. Nat Immunol 6:836–843. doi:10.1038/ni1226, ni1226 [pii]

Bevington S, Boyes J (2013) Transcription-coupled eviction of histones H2A/H2B governs V(D)J recombination. EMBO J. doi:10.1038/emboj.2013.42, emboj201342 [pii]

Blackwell TK, Moore MW, Yancopoulos GD, Suh H, Lutzker S, Selsing E, Alt FW (1986) Recombination between immunoglobulin variable region gene segments is enhanced by transcription. Nature 324:585–589. doi:10.1038/324585a0

Blobel GA, Nakajima T, Eckner R, Montminy M, Orkin SH (1998) CREB-binding protein cooperates with transcription factor GATA-1 and is required for erythroid differentiation. Proc Natl Acad Sci USA 95:2061–2066

Calin GA, Croce CM (2007) Chromosomal rearrangements and microRNAs: a new cancer link with clinical implications. J Clin Invest 117:2059–2066. doi:10.1172/JCI32577

Calin GA, Sevignani C, Dumitru CD, Hyslop T, Noch E, Yendamuri S, Shimizu M, Rattan S, Bullrich F, Negrini M, Croce CM (2004) Human microRNA genes are frequently located at fragile sites and genomic regions involved in cancers. Proc Natl Acad Sci USA 101:2999–3004. doi:10.1073/pnas.0307323101, 0307323101 [pii]

Casali P, Pal Z, Xu Z, Zan H (2006) DNA repair in antibody somatic hypermutation. Trends Immunol 27:313–321. doi:10.1016/j.it.2006.05.001, S1471-4906(06)00143-8 [pii]

Chan HM, La Thangue NB (2001) p300/CBP proteins: HATs for transcriptional bridges and scaffolds. J Cell Sci 114:2363–2373

Chowdhury D, Sen R (2004) Regulation of immunoglobulin heavy-chain gene rearrangements. Immunol Rev 200:182–196. doi:10.1111/j.0105-2896.2004.00177.x, IMR177 [pii]

Cobaleda C, Schebesta A, Delogu A, Busslinger M (2007) Pax5: the guardian of B cell identity and function. Nat Immunol 8:463–470. doi:10.1038/ni1454, ni1454 [pii]

Curry JD, Schulz D, Guidos CJ, Danska JS, Nutter L, Nussenzweig A, Schlissel MS (2007) Chromosomal reinsertion of broken RSS ends during T cell development. J Exp Med 204:2293–2303. doi:10.1084/jem.20070583

Dalla-Favera R, Bregni M, Erikson J, Patterson D, Gallo RC, Croce CM (1982) Human c-myc onc gene is located on the region of chromosome 8 that is translocated in Burkitt lymphoma cells. Proc Natl Acad Sci USA 79:7824–7827

De Silva NS, Simonetti G, Heise N, Klein U (2012) The diverse roles of IRF4 in late germinal center B-cell differentiation. Immunol Rev 247:73–92. doi:10.1111/j.1600-065X.2012.01113.x

Degner SC, Verma-Gaur J, Wong TP, Bossen C, Iverson GM, Torkamani A, Vettermann C, Lin YC, Ju Z, Schulz D, Murre CS, Birshtein BK, Schork NJ, Schlissel MS, Riblet R, Murre C, Feeney AJ (2011) CCCTC-binding factor (CTCF) and cohesin influence the genomic architecture of the Igh locus and antisense transcription in pro-B cells. Proc Natl Acad Sci USA 108:9566–9571. doi:10.1073/pnas.1019391108, 1019391108 [pii]

DeKoter RP, Singh H (2000) Regulation of B lymphocyte and macrophage development by graded expression of PU.1. Science 288:1439–1441, 8531 [pii]

Delogu A, Schebesta A, Sun Q, Aschenbrenner K, Perlot T, Busslinger M (2006) Gene repression by Pax5 in B cells is essential for blood cell homeostasis and is reversed in plasma cells. Immunity 24:269–281. doi:10.1016/j.immuni.2006.01.012, S1074-7613(06)00111-7 [pii]

Di Noia JM, Neuberger MS (2007) Molecular mechanisms of antibody somatic hypermutation. Annu Rev Biochem 76:1–22. doi:10.1146/annurev.biochem.76.061705.090740

Dixon JR, Selvaraj S, Yue F, Kim A, Li Y, Shen Y, Hu M, Liu JS, Ren B (2012) Topological domains in mammalian genomes identified by analysis of chromatin interactions. Nature 485:376–380. doi:10.1038/nature11082

Duber S, Engel H, Rolink A, Kretschmer K, Weiss S (2003) Germline transcripts of immunoglobulin light chain variable regions are structurally diverse and differentially expressed. Mol Immunol 40:509–516

Duquette ML, Pham P, Goodman MF, Maizels N (2005) AID binds to transcription-induced structures in c-MYC that map to regions associated with translocation and hypermutation. Oncogene 24:5791–5798. doi:10.1038/sj.onc.1208746, 1208746 [pii]

Ebert A, McManus S, Tagoh H, Medvedovic J, Salvagiotto G, Novatchkova M, Tamir I, Sommer A, Jaritz M, Busslinger M (2011) The distal V(H) gene cluster of the Igh locus contains distinct regulatory elements with Pax5 transcription factor-dependent activity in pro-B cells. Immunity 34:175–187. doi:10.1016/j.immuni.2011.02.005, S1074-7613(11)00043-4 [pii]

Eckner R, Ludlow JW, Lill NL, Oldread E, Arany Z, Modjtahedi N, DeCaprio JA, Livingston DM, Morgan JA (1996) Association of p300 and CBP with simian virus 40 large T antigen. Mol Cell Biol 16:3454–3464

Eisenbeis CF, Singh H, Storb U (1993) PU.1 is a component of a multiprotein complex which binds an essential site in the murine immunoglobulin lambda 2-4 enhancer. Mol Cell Biol 13:6452–6461

Eisenbeis CF, Singh H, Storb U (1995) Pip, a novel IRF family member, is a lymphoid-specific, PU.1-dependent transcriptional activator. Genes Dev 9:1377–1387

Elkin SK, Matthews AG, Oettinger MA (2003) The C-terminal portion of RAG2 protects against transposition in vitro. EMBO J 22:1931–1938. doi:10.1093/emboj/cdg184

Elliott B, Richardson C, Jasin M (2005) Chromosomal translocation mechanisms at intronic alu elements in mammalian cells. Mol Cell 17:885–894. doi:10.1016/j.molcel.2005.02.028, S1097-2765(05)01144-5 [pii]

Engel H, Rolink A, Weiss S (1999) B cells are programmed to activate kappa and lambda for rearrangement at consecutive developmental stages. Eur J Immunol 29:2167–2176

Espinoza CR, Feeney AJ (2005) The extent of histone acetylation correlates with the differential rearrangement frequency of individual VH genes in pro-B cells. J Immunol 175:6668–6675

Featherstone K, Wood AL, Bowen AJ, Corcoran AE (2010) The mouse immunoglobulin heavy chain V-D intergenic sequence contains insulators that may regulate ordered V(D)J recombination. J Biol Chem 285:9327–9338. doi:10.1074/jbc.M109.098251, M109.098251 [pii]
Fernex C, Capone M, Ferrier P (1995) The V(D)J recombinational and transcriptional activities of the immunoglobulin heavy-chain intronic enhancer can be mediated through distinct protein-binding sites in a transgenic substrate. Mol Cell Biol 15:3217–3226
Ferreiros Vidal I, Carroll T, Taylor B, Terry A, Liang Z, Bruno L, Dharmalingam G, Khadayate S, Cobb BS, Smale ST, Spivakov M, Srivastava P, Petretto E, Fisher AG, Merkenschlager M (2013) Genome-wide identification of Ikaros targets elucidates its contribution to mouse B cell lineage specification and pre-B cell differentiation. Blood. doi:10.1182/blood-2012-08-450114, blood-2012-08-450114 [pii]
Fitzsimmons SP, Bernstein RM, Max EE, Skok JA, Shapiro MA (2007) Dynamic changes in accessibility, nuclear positioning, recombination, and transcription at the Ig kappa locus. J Immunol 179:5264–5273, 79/8/5264 [pii]
Fuxa M, Skok J, Souabni A, Salvagiotto G, Roldan E, Busslinger M (2004) Pax5 induces V-to-DJ rearrangements and locus contraction of the immunoglobulin heavy-chain gene. Genes Dev 18:411–422
Geier JK, Schlissel MS (2006) Pre-BCR signals and the control of Ig gene rearrangements. Semin Immunol 18:31–39. doi:10.1016/j.smim.2005.11.001, S1044-5323(05)00086-2 [pii]
Gellert M (2002) V(D)J recombination: RAG proteins, repair factors, and regulation. Annu Rev Biochem 71:101–132. doi:10.1146/annurev.biochem.71.090501.150203, 090501.150203 [pii]
Golding A, Chandler S, Ballestar E, Wolffe AP, Schlissel MS (1999) Nucleosome structure completely inhibits in vitro cleavage by the V(D)J recombinase. EMBO J 18:3712–3723
Goldmit M, Ji Y, Skok J, Roldan E, Jung S, Cedar H, Bergman Y (2005) Epigenetic ontogeny of the Igk locus during B cell development. Nat Immunol 6:198–203
Gordon SJ, Saleque S, Birshtein BK (2003) Yin Yang 1 is a lipopolysaccharide-inducible activator of the murine 3′ Igh enhancer, hs3. J Immunol 170:5549–5557
Gorman JR, Alt FW (1998) Regulation of immunoglobulin light chain isotype expression. Adv Immunol 69:113–181
Gorman JR, van der Stoep N, Monroe R, Cogne M, Davidson L, Alt FW (1996) The Ig(kappa) enhancer influences the ratio of Ig(kappa) versus Ig(lambda) B lymphocytes. Immunity 5:241–252
Gu W, Roeder RG (1997) Activation of p53 sequence-specific DNA binding by acetylation of the p53 C-terminal domain. Cell 90:595–606, S0092-8674(00)80521-8 [pii]
Guo C, Gerasimova T, Hao H, Ivanova I, Chakraborty T, Selimyan R, Oltz EM, Sen R (2011a) Two forms of loops generate the chromatin conformation of the immunoglobulin heavy-chain gene locus. Cell 147:332–343. doi:10.1016/j.cell.2011.08.049, S0092-8674(11)01082-8 [pii]
Guo C, Yoon HS, Franklin A, Jain S, Ebert A, Cheng HL, Hansen E, Despo O, Bossen C, Vettermann C, Bates JG, Richards N, Myers D, Patel H, Gallagher M, Schlissel MS, Murre C, Busslinger M, Giallourakis CC, Alt FW (2011b) CTCF-binding elements mediate control of V(D)J recombination. Nature 477:424–430. doi:10.1038/nature10495, nature10495 [pii]
Hansen JD, McBlane JF (2000) Recombination-activating genes, transposition, and the lymphoid-specific combinatorial immune system: a common evolutionary connection. Curr Top Microbiol Immunol 248:111–135
Hesslein DG, Pflugh DL, Chowdhury D, Bothwell AL, Sen R, Schatz DG (2003) Pax5 is required for recombination of transcribed, acetylated, 5′ IgH V gene segments. Genes Dev 17:37–42
Hiom K, Melek M, Gellert M (1998) DNA transposition by the RAG1 and RAG2 proteins: a possible source of oncogenic translocations. Cell 94:463–470, S0092-8674(00)81587-1 [pii]
Hodawadekar S, Park K, Farrar MA, Atchison ML (2012) A developmentally controlled competitive STAT5-PU.1 DNA binding mechanism regulates activity of the Ig kappa E3′ enhancer. J Immunol 188:2276–2284. doi:10.4049/jimmunol.1102239, jimmunol.1102239 [pii]

Huang J, Muegge K (2001) Control of chromatin accessibility for V(D)J recombination by interleukin-7. J Leukoc Biol 69:907–911

Iida S, Rao PH, Butler M, Corradini P, Boccadoro M, Klein B, Chaganti RS, Dalla-Favera R (1997) Deregulation of MUM1/IRF4 by chromosomal translocation in multiple myeloma. Nat Genet 17:226–230. doi:10.1038/ng1097-226

Inlay M, Alt FW, Baltimore D, Xu Y (2002) Essential roles of the kappa light chain intronic enhancer and 3′ enhancer in kappa rearrangement and demethylation. Nat Immunol 3:463–468

Inlay MA, Tian H, Lin T, Xu Y (2004) Important roles for E protein binding sites within the immunoglobulin kappa chain intronic enhancer in activating Vkappa Jkappa rearrangement. J Exp Med 200:1205–1211

Jager U, Bocskor S, Le T, Mitterbauer G, Bolz I, Chott A, Kneba M, Mannhalter C, Nadel B (2000) Follicular lymphomas' BCL-2/IgH junctions contain templated nucleotide insertions: novel insights into the mechanism of t(14;18) translocation. Blood 95:3520–3529

Ji Y, Resch W, Corbett E, Yamane A, Casellas R, Schatz DG (2010) The in vivo pattern of binding of RAG1 and RAG2 to antigen receptor loci. Cell 141:419–431. doi:10.1016/j.cell.2010.03.010, S0092-8674(10)00246-1 [pii]

Janz S, Potter M, Rabkin CS (2003) Lymphoma- and leukemia-associated chromosomal translocations in healthy individuals. Genes Chromosomes Cancer 36:211–223

Johnson K, Angelin-Duclos C, Park S, Calame KL (2003) Changes in histone acetylation are associated with differences in accessibility of V(H) gene segments to V-DJ recombination during B-cell ontogeny and development. Mol Cell Biol 23:2438–2450

Johnson K, Hashimshony T, Sawai CM, Pongubala JM, Skok JA, Aifantis I, Singh H (2008) Regulation of immunoglobulin light-chain recombination by the transcription factor IRF-4 and the attenuation of interleukin-7 signaling. Immunity 28:335–345. doi:10.1016/j.immuni.2007.12.019, S1074-7613(08)00044-7 [pii]

Jones ME, Zhuang Y (2009) Regulation of V(D)J recombination by E-protein transcription factors. Adv Exp Med Biol 650:148–156

Kelsoe G (1996) The germinal center: a crucible for lymphocyte selection. Semin Immunol 8:179–184. doi:10.1006/smim.1996.0022, S1044-5323(96)90022-6 [pii]

Kirsch IR, Morton CC, Nakahara K, Leder P (1982) Human immunoglobulin heavy chain genes map to a region of translocations in malignant B lymphocytes. Science 216:301–303

Klein U, Casola S, Cattoretti G, Shen Q, Lia M, Mo T, Ludwig T, Rajewsky K, Dalla-Favera R (2006) Transcription factor IRF4 controls plasma cell differentiation and class-switch recombination. Nat Immunol 7:773–782. doi:10.1038/ni1357, ni1357 [pii]

Kosak ST, Skok JA, Medina KL, Riblet R, Le Beau MM, Fisher AG, Singh H (2002) Subnuclear compartmentalization of immunoglobulin loci during lymphocyte development. Science 296:158–162

Kovalchuk AL, Kim JS, Park SS, Coleman AE, Ward JM, Morse HC 3rd, Kishimoto T, Potter M, Janz S (2002) IL-6 transgenic mouse model for extraosseous plasmacytoma. Proc Natl Acad Sci USA 99:1509–1514. doi:10.1073/pnas.022643999, 022643999 [pii]

Kridel R, Sehn LH, Gascoyne RD (2012) Pathogenesis of follicular lymphoma. J Clin Invest 122:3424–3431. doi:10.1172/JCI63186, 63186 [pii]

Kuppers R (2005) Mechanisms of B-cell lymphoma pathogenesis. Nat Rev Cancer 5:251–262. doi:10.1038/nrc1589, nrc1589 [pii]

Kuppers R, Dalla-Favera R (2001) Mechanisms of chromosomal translocations in B cell lymphomas. Oncogene 20:5580–5594. doi:10.1038/sj.onc.1204640

Kurahashi H, Inagaki H, Ohye T, Kogo H, Kato T, Emanuel BS (2006) Palindrome-mediated chromosomal translocations in humans. DNA Repair (Amst) 5:1136–1145. doi:10.1016/j.dnarep.2006.05.035, S1568-7864(06)00175-3 [pii]

Kwon J, Imbalzano AN, Matthews A, Oettinger MA (1998) Accessibility of nucleosomal DNA to V(D)J cleavage is modulated by RSS positioning and HMG1. Mol Cell 2:829–839

Laiosa CV, Stadtfeld M, Graf T (2006) Determinants of lymphoid-myeloid lineage diversification. Annu Rev Immunol 24:705–738. doi:10.1146/annurev.immunol.24.021605.090742

Lazorchak AS, Schlissel MS, Zhuang Y (2006) E2A and IRF-4/Pip promote chromatin modification and transcription of the immunoglobulin kappa locus in pre-B cells. Mol Cell Biol 26:810–821

Lennon GG, Perry RP (1985) C mu-containing transcripts initiate heterogeneously within the IgH enhancer region and contain a novel 5′-nontranslatable exon. Nature 318:475–478

Lewis SM (1994) The mechanism of V(D)J joining: lessons from molecular, immunological, and comparative analyses. Adv Immunol 56:27–150

Lewis SM, Agard E, Suh S, Czyzyk L (1997) Cryptic signals and the fidelity of V(D)J joining. Mol Cell Biol 17:3125–3136

Li Z, Woo CJ, Iglesias-Ussel MD, Ronai D, Scharff MD (2004) The generation of antibody diversity through somatic hypermutation and class switch recombination. Genes Dev 18:1–11. doi:10.1101/gad.1161904, 18/1/1 [pii]

Liedtke M, Cleary ML (2009) Therapeutic targeting of MLL. Blood 113:6061–6068. doi:10.1182/blood-2008-12-197061

Limpens J, de Jong D, van Krieken JH, Price CG, Young BD, van Ommen GJ, Kluin PM (1991) Bcl-2/JH rearrangements in benign lymphoid tissues with follicular hyperplasia. Oncogene 6:2271–2276

Lin YC, Jhunjhunwala S, Benner C, Heinz S, Welinder E, Mansson R, Sigvardsson M, Hagman J, Espinoza CA, Dutkowski J, Ideker T, Glass CK, Murre C (2010) A global network of transcription factors, involving E2A, EBF1 and Foxo1, that orchestrates B cell fate. Nat Immunol 11:635–643. doi:10.1038/ni.1891, ni.1891 [pii]

Lin YC, Benner C, Mansson R, Heinz S, Miyazaki K, Miyazaki M, Chandra V, Bossen C, Glass CK, Murre C (2012) Global changes in the nuclear positioning of genes and intra- and interdomain genomic interactions that orchestrate B cell fate. Nat Immunol 13:1196–1204. doi:10.1038/ni.2432, ni.2432 [pii]

Liu H, Schmidt-Supprian M, Shi Y, Hobeika E, Barteneva N, Jumaa H, Pelanda R, Reth M, Skok J, Rajewsky K, Shi Y (2007a) Yin Yang 1 is a critical regulator of B-cell development. Genes Dev 21:1179–1189

Liu Y, Subrahmanyam R, Chakraborty T, Sen R, Desiderio S (2007b) A plant homeodomain in Rag-2 that binds hypermethylated lysine 4 of histone H3 is necessary for efficient antigen-receptor-gene rearrangement. Immunity 27(4):561–571

Lorsbach RB, Moore J, Mathew S, Raimondi SC, Mukatira ST, Downing JR (2003) TET1, a member of a novel protein family, is fused to MLL in acute myeloid leukemia containing the t(10;11)(q22;q23). Leukemia 17:637–641. doi:10.1038/sj.leu.2402834, 2402834 [pii]

Lu R, Medina KL, Lancki DW, Singh H (2003) IRF-4,8 orchestrate the pre-B-to-B transition in lymphocyte development. Genes Dev 17:1703–1708

MacLennan IC (2005) Germinal centers still hold secrets. Immunity 22:656–657. doi:10.1016/j.immuni.2005.06.002, S1074-7613(05)00167-6 [pii]

Maes J, O'Neill LP, Cavelier P, Turner BM, Rougeon F, Goodhardt M (2001) Chromatin remodeling at the Ig loci prior to V(D)J recombination. J Immunol 167:866–874

Malin S, McManus S, Cobaleda C, Novatchkova M, Delogu A, Bouillet P, Strasser A, Busslinger M (2010) Role of STAT5 in controlling cell survival and immunoglobulin gene recombination during pro-B cell development. Nat Immunol 11:171–179. doi:10.1038/ni.1827, ni.1827 [pii]

Mandal M, Powers SE, Ochiai K, Georgopoulos K, Kee BL, Singh H, Clark MR (2009) Ras orchestrates exit from the cell cycle and light-chain recombination during early B cell development. Nat Immunol 10:1110–1117. doi:10.1038/ni.1785, ni.1785 [pii]

Marculescu R, Le T, Simon P, Jaeger U, Nadel B (2002) V(D)J-mediated translocations in lymphoid neoplasms: a functional assessment of genomic instability by cryptic sites. J Exp Med 195:85–98

Matthews AG, Kuo AJ, Ramon-Maiques S, Han S, Champagne KS, Ivanov D, Gallardo M, Carney D, Cheung P, Ciccone DN, Walter KL, Utz PJ, Shi Y, Kutateladze TG, Yang W, Gozani O, Oettinger MA (2007) RAG2 PHD finger couples histone H3 lysine 4 trimethylation

with V(D)J recombination. Nature 450:1106–1110. doi:10.1038/nature06431, nature06431 [pii]
McBlane F, Boyes J (2000) Stimulation of V(D)J recombination by histone acetylation. Curr Biol 10:483–486
McKercher SR, Torbett BE, Anderson KL, Henkel GW, Vestal DJ, Baribault H, Klemsz M, Feeney AJ, Wu GE, Paige CJ, Maki RA (1996) Targeted disruption of the PU.1 gene results in multiple hematopoietic abnormalities. EMBO J 15:5647–5658
McMurry MT, Krangel MS (2000) A role for histone acetylation in the developmental regulation of VDJ recombination [see comments]. Science 287:495–498
Messier TL, O'Neill JP, Hou SM, Nicklas JA, Finette BA (2003) In vivo transposition mediated by V(D)J recombinase in human T lymphocytes. EMBO J 22:1381–1388. doi:10.1093/emboj/cdg137
Messier TL, O'Neill JP, Finette BA (2006) V(D)J recombinase mediated inter-chromosomal HPRT alterations at cryptic recombination signal sequences in peripheral human T cells. Hum Mutat 27:829. doi:10.1002/humu.9440
Morshead KB, Ciccone DN, Taverna SD, Allis CD, Oettinger MA (2003) Antigen receptor loci poised for V(D)J rearrangement are broadly associated with BRG1 and flanked by peaks of histone H3 dimethylated at lysine 4. Proc Natl Acad Sci USA 100:11577–11582. doi:10.1073/pnas.1932643100, 1932643100 [pii]
Muljo SA, Schlissel MS (2003) A small molecule Abl kinase inhibitor induces differentiation of Abelson virus-transformed pre-B cell lines. Nat Immunol 4:31–37
Nagulapalli S, Atchison ML (1998) Transcription factor Pip can enhance DNA binding by E47, leading to transcriptional synergy involving multiple protein domains. Mol Cell Biol 18:4639–4650
Nagulapalli S, Goheer A, Pitt L, McIntosh LP, Atchison ML (2002) Mechanism of e47-Pip interaction on DNA resulting in transcriptional synergy and activation of immunoglobulin germ line sterile transcripts. Mol Cell Biol 22:7337–7350
Neiditch MB, Lee GS, Huye LE, Brandt VL, Roth DB (2002) The V(D)J recombinase efficiently cleaves and transposes signal joints. Mol Cell 9:871–878, S109727650200494X [pii]
Nelsen B, Tian G, Erman B, Gregoire J, Maki R, Graves B, Sen R (1993) Regulation of lymphoid-specific immunoglobulin mu heavy chain gene enhancer by ETS-domain proteins. Science 261:82–86
Ng HH, Robert F, Young RA, Struhl K (2003) Targeted recruitment of Set1 histone methylase by elongating Pol II provides a localized mark and memory of recent transcriptional activity. Mol Cell 11:709–719, S1097276503000923 [pii]
Nightingale KP, Gendreizig S, White DA, Bradbury C, Hollfelder F, Turner BM (2007) Cross-talk between histone modifications in response to histone deacetylase inhibitors: MLL4 links histone H3 acetylation and histone H3K4 methylation. J Biol Chem 282:4408–4416
Nikolajczyk BS, Sanchez JA, Sen R (1999) ETS protein-dependent accessibility changes at the immunoglobulin mu heavy chain enhancer. Immunity 11:11–20
Nussenzweig A, Nussenzweig MC (2010) Origin of chromosomal translocations in lymphoid cancer. Cell 141:27–38. doi:10.1016/j.cell.2010.03.016, S0092-8674(10)00289-8 [pii]
Ogryzko VV, Schiltz RL, Russanova V, Howard BH, Nakatani Y (1996) The transcriptional coactivators p300 and CBP are histone acetyltransferases. Cell 87:953–959, S0092-8674(00)82001-2 [pii]
Ono R, Taki T, Taketani T, Taniwaki M, Kobayashi H, Hayashi Y (2002) LCX, leukemia-associated protein with a CXXC domain, is fused to MLL in acute myeloid leukemia with trilineage dysplasia having t(10;11)(q22;q23). Cancer Res 62:4075–4080
Orphanides G, Wu WH, Lane WS, Hampsey M, Reinberg D (1999) The chromatin-specific transcription elongation factor FACT comprises human SPT16 and SSRP1 proteins. Nature 400:284–288
Pasqualucci L, Migliazza A, Fracchiolla N, William C, Neri A, Baldini L, Chaganti RS, Klein U, Kuppers R, Rajewsky K, Dalla-Favera R (1998) BCL-6 mutations in normal germinal center B

cells: evidence of somatic hypermutation acting outside Ig loci. Proc Natl Acad Sci USA 95:11816–11821
Pasqualucci L, Neumeister P, Goossens T, Nanjangud G, Chaganti RS, Kuppers R, Dalla-Favera R (2001) Hypermutation of multiple proto-oncogenes in B-cell diffuse large-cell lymphomas. Nature 412:341–346. doi:10.1038/35085588, 35085588 [pii]
Pasqualucci L, Dominguez-Sola D, Chiarenza A, Fabbri G, Grunn A, Trifonov V, Kasper LH, Lerach S, Tang H, Ma J, Rossi D, Chadburn A, Murty VV, Mullighan CG, Gaidano G, Rabadan R, Brindle PK, Dalla-Favera R (2011a) Inactivating mutations of acetyltransferase genes in B-cell lymphoma. Nature 471:189–195. doi:10.1038/nature09730, nature09730 [pii]
Pasqualucci L, Trifonov V, Fabbri G, Ma J, Rossi D, Chiarenza A, Wells VA, Grunn A, Messina M, Elliot O, Chan J, Bhagat G, Chadburn A, Gaidano G, Mullighan CG, Rabadan R, Dalla-Favera R (2011b) Analysis of the coding genome of diffuse large B-cell lymphoma. Nat Genet 43:830–837. doi:10.1038/ng.892, ng.892 [pii]
Peled JU, Kuang FL, Iglesias-Ussel MD, Roa S, Kalis SL, Goodman MF, Scharff MD (2008) The biochemistry of somatic hypermutation. Annu Rev Immunol 26:481–511. doi:10.1146/annurev.immunol.26.021607.090236
Perkins EJ, Kee BL, Ramsden DA (2004) Histone 3 lysine 4 methylation during the pre-B to immature B-cell transition. Nucleic Acids Res 32:1942–1947
Pokholok DK, Harbison CT, Levine S, Cole M, Hannett NM, Lee TI, Bell GW, Walker K, Rolfe PA, Herbolsheimer E, Zeitlinger J, Lewitter F, Gifford DK, Young RA (2005) Genome-wide map of nucleosome acetylation and methylation in yeast. Cell 122:517–527
Poltoratsky V, Goodman MF, Scharff MD (2000) Error-prone candidates vie for somatic mutation. J Exp Med 192:F27–F30
Pongubala JM, Nagulapalli S, Klemsz MJ, McKercher SR, Maki RA, Atchison ML (1992) PU.1 recruits a second nuclear factor to a site important for immunoglobulin kappa 3′ enhancer activity. Mol Cell Biol 12:368–378
Potter M, Wiener F (1992) Plasmacytomagenesis in mice: model of neoplastic development dependent upon chromosomal translocations. Carcinogenesis 13:1681–1697
Quivoron C, Couronne L, Della Valle V, Lopez CK, Plo I, Wagner-Ballon O, Do Cruzeiro M, Delhommeau F, Arnulf B, Stern MH, Godley L, Opolon P, Tilly H, Solary E, Duffourd Y, Dessen P, Merle-Beral H, Nguyen-Khac F, Fontenay M, Vainchenker W, Bastard C, Mercher T, Bernard OA (2011) TET2 inactivation results in pleiotropic hematopoietic abnormalities in mouse and is a recurrent event during human lymphomagenesis. Cancer Cell 20:25–38. doi:10.1016/j.ccr.2011.06.003, S1535-6108(11)00225-X [pii]
Rabkin CS, Janz S (2008) Mechanisms and consequences of chromosomal translocation. Cancer Epidemiol Biomarkers Prev 17:1849–1851. doi:10.1158/1055-9965.EPI-07-2902, 17/8/1849 [pii]
Raghavan SC, Kirsch IR, Lieber MR (2001) Analysis of the V(D)J recombination efficiency at lymphoid chromosomal translocation breakpoints. J Biol Chem 276:29126–29133. doi:10.1074/jbc.M103797200, M103797200 [pii]
Raghavan SC, Houston S, Hegde BG, Langen R, Haworth IS, Lieber MR (2004a) Stability and strand asymmetry in the non-B DNA structure at the bcl-2 major breakpoint region. J Biol Chem 279:46213–46225. doi:10.1074/jbc.M406280200, M406280200 [pii]
Raghavan SC, Swanson PC, Wu X, Hsieh CL, Lieber MR (2004b) A non-B-DNA structure at the Bcl-2 major breakpoint region is cleaved by the RAG complex. Nature 428:88–93. doi:10.1038/nature02355, nature02355 [pii]
Raghavan SC, Swanson PC, Ma Y, Lieber MR (2005) Double-strand break formation by the RAG complex at the bcl-2 major breakpoint region and at other non-B DNA structures in vitro. Mol Cell Biol 25:5904–5919. doi:10.1128/MCB.25.14.5904-5919.2005, 25/14/5904 [pii]
Ramiro AR, Jankovic M, Callen E, Difilippantonio S, Chen HT, McBride KM, Eisenreich TR, Chen J, Dickins RA, Lowe SW, Nussenzweig A, Nussenzweig MC (2006) Role of genomic instability and p53 in AID-induced c-myc-Igh translocations. Nature 440:105–109. doi:10.1038/nature04495, nature04495 [pii]

Ramon-Maiques S, Kuo AJ, Carney D, Matthews AG, Oettinger MA, Gozani O, Yang W (2007) The plant homeodomain finger of RAG2 recognizes histone H3 methylated at both lysine-4 and arginine-2. Proc Natl Acad Sci USA 104:18993–18998. doi:10.1073/pnas.0709170104, 0709170104 [pii]
Ramsden DA, Weed BD, Reddy YV (2010) V(D)J recombination: born to be wild. Semin Cancer Biol. doi:10.1016/j.semcancer.2010.06.002, S1044-579X(10)00044-1 [pii]
Rivera RR, Stuiver MH, Steenbergen R, Murre C (1993) Ets proteins: new factors that regulate immunoglobulin heavy-chain gene expression. Mol Cell Biol 13:7163–7169
Robbiani DF, Nussenzweig MC (2013) Chromosome translocation, B cell lymphoma, and activation-induced cytidine deaminase. Annu Rev Pathol 8:79–103. doi:10.1146/annurev-pathol-020712-164004
Robbiani DF, Bothmer A, Callen E, Reina-San-Martin B, Dorsett Y, Difilippantonio S, Bolland DJ, Chen HT, Corcoran AE, Nussenzweig A, Nussenzweig MC (2008) AID is required for the chromosomal breaks in c-myc that lead to c-myc/IgH translocations. Cell 135:1028–1038. doi:10.1016/j.cell.2008.09.062, S0092-8674(08)01310-X [pii]
Roldan E, Fuxa M, Chong W, Martinez D, Novatchkova M, Busslinger M, Skok JA (2005) Locus 'decontraction' and centromeric recruitment contribute to allelic exclusion of the immunoglobulin heavy-chain gene. Nat Immunol 6:31–41
Roth DB (2003) Restraining the V(D)J recombinase. Nat Rev Immunol 3:656–666
Roth DB, Roth SY (2000) Unequal access: regulating V(D)J recombination through chromatin remodeling. Cell 103:699–702
Rudin CM, Storb U (1992) Two conserved essential motifs of the murine immunoglobulin lambda enhancers bind B-cell-specific factors. Mol Cell Biol 12:309–320
Sakamoto S, Wakae K, Anzai Y, Murai K, Tamaki N, Miyazaki M, Miyazaki K, Romanow WJ, Ikawa T, Kitamura D, Yanagihara I, Minato N, Murre C, Agata Y (2012) E2A and CBP/p300 act in synergy to promote chromatin accessibility of the immunoglobulin kappa locus. J Immunol 188:5547–5560. doi:10.4049/jimmunol.1002346, jimmunol.1002346 [pii]
Sayegh CE, Jhunjhunwala S, Riblet R, Murre C (2005) Visualization of looping involving the immunoglobulin heavy-chain locus in developing B cells. Genes Dev 19:322–327. doi:10.1101/gad.1254305, 19/3/322 [pii]
Schebesta A, McManus S, Salvagiotto G, Delogu A, Busslinger GA, Busslinger M (2007) Transcription factor Pax5 activates the chromatin of key genes involved in B cell signaling, adhesion, migration, and immune function. Immunity 27:49–63. doi:10.1016/j.immuni.2007.05.019, S1074-7613(07)00331-7 [pii]
Schlissel MS (2004) Regulation of activation and recombination of the murine Igkappa locus. Immunol Rev 200:215–223
Schlissel MS, Baltimore D (1989) Activation of immunoglobulin kappa gene rearrangement correlates with induction of germline kappa gene transcription. Cell 58:1001–1007
Schwarzenbach H, Newell JW, Matthias P (1995) Involvement of the Ets family factor PU.1 in the activation of immunoglobulin promoters. J Biol Chem 270:898–907
Sciammas R, Davis MM (2004) Modular nature of Blimp-1 in the regulation of gene expression during B cell maturation. J Immunol 172:5427–5440
Scott EW, Simon MC, Anastasi J, Singh H (1994) Requirement of transcription factor PU.1 in the development of multiple hematopoietic lineages. Science 265:1573–1577
Seet CS, Brumbaugh RL, Kee BL (2004) Early B cell factor promotes B lymphopoiesis with reduced interleukin 7 responsiveness in the absence of E2A. J Exp Med 199:1689–1700. doi:10.1084/jem.20032202, jem.20032202 [pii]
Shaffer AL, Lin KI, Kuo TC, Yu X, Hurt EM, Rosenwald A, Giltnane JM, Yang L, Zhao H, Calame K, Staudt LM (2002) Blimp-1 orchestrates plasma cell differentiation by extinguishing the mature B cell gene expression program. Immunity 17:51–62, S1074761302003357 [pii]
Shaffer AL, Shapiro-Shelef M, Iwakoshi NN, Lee AH, Qian SB, Zhao H, Yu X, Yang L, Tan BK, Rosenwald A, Hurt EM, Petroulakis E, Sonenberg N, Yewdell JW, Calame K, Glimcher LH, Staudt LM (2004) XBP1, downstream of Blimp-1, expands the secretory apparatus and other

organelles, and increases protein synthesis in plasma cell differentiation. Immunity 21:81–93. doi:10.1016/j.immuni.2004.06.010, S1074761304001670 [pii]
Shaffer AL, Emre NC, Lamy L, Ngo VN, Wright G, Xiao W, Powell J, Dave S, Yu X, Zhao H, Zeng Y, Chen B, Epstein J, Staudt LM (2008) IRF4 addiction in multiple myeloma. Nature 454:226–231. doi:10.1038/nature07064, nature07064 [pii]
Shapiro-Shelef M, Lin KI, McHeyzer-Williams LJ, Liao J, McHeyzer-Williams MG, Calame K (2003) Blimp-1 is required for the formation of immunoglobulin secreting plasma cells and pre-plasma memory B cells. Immunity 19:607–620, S107476130300267X [pii]
Shin MK, Koshland ME (1993) Ets-related protein PU.1 regulates expression of the immunoglobulin J-chain gene through a novel Ets-binding element. Genes Dev 7:2006–2015
Shurtleff SA, Buijs A, Behm FG, Rubnitz JE, Raimondi SC, Hancock ML, Chan GC, Pui CH, Grosveld G, Downing JR (1995) TEL/AML1 fusion resulting from a cryptic t(12;21) is the most common genetic lesion in pediatric ALL and defines a subgroup of patients with an excellent prognosis. Leukemia 9:1985–1989
Sikes ML, Meade A, Tripathi R, Krangel MS, Oltz EM (2002) Regulation of V(D)J recombination: a dominant role for promoter positioning in gene segment accessibility. Proc Natl Acad Sci USA 99:12309–12314
Skok JA, Gisler R, Novatchkova M, Farmer D, de Laat W, Busslinger M (2007) Reversible contraction by looping of the Tcra and Tcrb loci in rearranging thymocytes. Nat Immunol 8:378–387. doi:10.1038/ni1448, ni1448 [pii]
Spooner CJ, Cheng JX, Pujadas E, Laslo P, Singh H (2009) A recurrent network involving the transcription factors PU.1 and Gfi1 orchestrates innate and adaptive immune cell fates. Immunity 31:576–586. doi:10.1016/j.immuni.2009.07.011, S1074-7613(09)00415-4 [pii]
Stavnezer J, Guikema JE, Schrader CE (2008) Mechanism and regulation of class switch recombination. Annu Rev Immunol 26:261–292. doi:10.1146/annurev.immunol.26.021607.090248
Sun L, Goodman PA, Wood CM, Crotty ML, Sensel M, Sather H, Navara C, Nachman J, Steinherz PG, Gaynon PS, Seibel N, Vassilev A, Juran BD, Reaman GH, Uckun FM (1999) Expression of aberrantly spliced oncogenic ikaros isoforms in childhood acute lymphoblastic leukemia. J Clin Oncol 17:3753–3766
Tahiliani M, Koh KP, Shen Y, Pastor WA, Bandukwala H, Brudno Y, Agarwal S, Iyer LM, Liu DR, Aravind L, Rao A (2009) Conversion of 5-methylcytosine to 5-hydroxymethylcytosine in mammalian DNA by MLL partner TET1. Science 324:930–935. doi:10.1126/science.1170116, 1170116 [pii]
Takeda S, Zou YR, Bluethmann H, Kitamura D, Muller U, Rajewsky K (1993) Deletion of the immunoglobulin kappa chain intron enhancer abolishes kappa chain gene rearrangement in cis but not lambda chain gene rearrangement in trans. EMBO J 12:2329–2336
Teitell MA, Pandolfi PP (2009) Molecular genetics of acute lymphoblastic leukemia. Annu Rev Pathol 4:175–198. doi:10.1146/annurev.pathol.4.110807.092227
Teng G, Papavasiliou FN (2007) Immunoglobulin somatic hypermutation. Annu Rev Genet 41:107–120. doi:10.1146/annurev.genet.41.110306.130340
Thomas-Tikhonenko A, Cozma D (2008) PAX5 and B-cell neoplasms: transformation through presentation. Future Oncol 4:5–9. doi:10.2217/14796694.4.1.5
Tonegawa S (1983) Somatic generation of antibody diversity. Nature 302:575–581
Torrano V, Procter J, Cardus P, Greaves M, Ford AM (2011) ETV6-RUNX1 promotes survival of early B lineage progenitor cells via a dysregulated erythropoietin receptor. Blood 118:4910–4918. doi:10.1182/blood-2011-05-354266, blood-2011-05-354266 [pii]
Tripathi RK, Mathieu N, Spicuglia S, Payet D, Verthuy C, Bouvier G, Depetris D, Mattei MG, Hempe LW, Ferrier P (2000) Definition of a T-cell receptor beta gene core enhancer of V(D)J recombination by transgenic mapping. Mol Cell Biol 20:42–53
Tripathi R, Jackson A, Krangel MS (2002) A change in the structure of Vbeta chromatin associated with TCR beta allelic exclusion. J Immunol 168:2316–2324
Tsai AG, Lieber MR (2010) Mechanisms of chromosomal rearrangement in the human genome. BMC Genomics 11 Suppl 1: S1. doi:10.1186/1471-2164-11-S1-S1, 1471-2164-11-S1-S1 [pii]

Tycko B, Sklar J (1990) Chromosomal translocations in lymphoid neoplasia: a reappraisal of the recombinase model. Cancer Cells 2:1–8

Unniraman S, Zhou S, Schatz DG (2004) Identification of an AID-independent pathway for chromosomal translocations between the Igh switch region and Myc. Nat Immunol 5:1117–1123. doi:10.1038/ni1127, ni1127 [pii]

van Gent DC, Mizuuchi K, Gellert M (1996) Similarities between initiation of V(D)J recombination and retroviral integration. Science 271:1592–1594

Vanasse GJ, Concannon P, Willerford DM (1999) Regulated genomic instability and neoplasia in the lymphoid lineage. Blood 94:3997–4010

Vanura K, Montpellier B, Le T, Spicuglia S, Navarro JM, Cabaud O, Roulland S, Vachez E, Prinz I, Ferrier P, Marculescu R, Jager U, Nadel B (2007) In vivo reinsertion of excised episomes by the V(D)J recombinase: a potential threat to genomic stability. PLoS Biol 5:e43. doi:10.1371/journal.pbio.0050043, 1544-9173-5-3-e43 [pii]

Wang JH, Nichogiannopoulou A, Wu L, Sun L, Sharpe AH, Bigby M, Georgopoulos K (1996) Selective defects in the development of the fetal and adult lymphoid system in mice with an Ikaros null mutation. Immunity 5:537–549, S1074-7613(00)80269-1 [pii]

Weinstein IB, Joe A (2008) Oncogene addiction. Cancer Res 68:3077–3080. doi:10.1158/0008-5472.CAN-07-3293, discussion 3080, 68/9/3077 [pii]

Welinder E, Mansson R, Mercer EM, Bryder D, Sigvardsson M, Murre C (2011) The transcription factors E2A and HEB act in concert to induce the expression of FOXO1 in the common lymphoid progenitor. Proc Natl Acad Sci USA 108:17402–17407. doi:10.1073/pnas.1111766108, 1111766108 [pii]

Welzel N, Le T, Marculescu R, Mitterbauer G, Chott A, Pott C, Kneba M, Du MQ, Kusec R, Drach J, Raderer M, Mannhalter C, Lechner K, Nadel B, Jaeger U (2001) Templated nucleotide addition and immunoglobulin JH-gene utilization in t(11;14) junctions: implications for the mechanism of translocation and the origin of mantle cell lymphoma. Cancer Res 61:1629–1636

Workman JL (2006) Nucleosome displacement in transcription. Genes Dev 20:2009–2017

Xu CR, Feeney AJ (2009) The epigenetic profile of Ig genes is dynamically regulated during B cell differentiation and is modulated by pre-B cell receptor signaling. J Immunol 182:1362–1369, 182/3/1362 [pii]

Xu Y, Davidson L, Alt FW, Baltimore D (1996) Deletion of the Ig kappa light chain intronic enhancer/matrix attachment region impairs but does not abolish V kappa J kappa rearrangement. Immunity 4:377–385

Yancopoulos GD, Alt FW (1985) Developmentally controlled and tissue-specific expression of unrearranged VH gene segments. Cell 40:271–281

Ye SK, Agata Y, Lee HC, Kurooka H, Kitamura T, Shimizu A, Honjo T, Ikuta K (2001) The IL-7 receptor controls the accessibility of the TCRgamma locus by Stat5 and histone acetylation. Immunity 15:813–823

Part IV
Epigenetic Control of Immune Cell Function

Chapter 14
Epigenetic Control of Immune T Cell Memory

Atsushi Onodera, Damon J. Tumes, and Toshinori Nakayama

Abstract The main role of the immune system is to protect against infections caused by invading pathogens. The adaptive immune system is particularly important for protection against repeated exposure to pathogens, due to its ability to memorize antigens during an initial infection and then respond rapidly and strongly to subsequent antigen challenges from the same or a related pathogen. This system forms immunological memory. Most epigenetic studies in immunology have focused on analysis of differentiation of CD4 T subsets, key players in the adaptive immune system. However, the relationship between immunological memory and epigenetics has not been as well studied. In recent years, with the advancement of technology such as ChIP-seq or RNA-seq methods, the importance of epigenetic mechanisms in immunological memory is becoming apparent. This review outlines our understanding of how CD4 T cells acquire and maintain function during or after differentiation, using Th2 cells as a model. In addition, we summarize the general characteristics of memory T cells from the perspective of epigenetics and discuss the possibility of clinical application of epigenetic studies in immunology.

Keywords Immunological memory • CD4 T cells • Th2 cells • GATA3 • Polycomb • Trithorax

14.1 Introduction

The immune system is an extremely important system for maintaining in vivo homeostasis, and small changes in the balance of the immune system can cause the onset of several diseases. Decreased immune function increases the risk of infectious disease and the development of malignant tumors, whereas excess

---

A. Onodera • D.J. Tumes • T. Nakayama (✉)
Department of Immunology, Graduate School of Medicine, Chiba University, 1-8-1 Inohana, Chuo-ku, Chiba 260-8670, Japan
e-mail: tnakayama@faculty.chiba-u.jp

C. Bonifer and P.N. Cockerill (eds.), *Transcriptional and Epigenetic Mechanisms Regulating Normal and Aberrant Blood Cell Development*, Epigenetics and Human Health, DOI 10.1007/978-3-642-45198-0_14,

immune responses cause allergy and autoimmune diseases. Immunological memory has been at the core of our understanding of disease protection for more than 200 years, since Edward Jenner first proved in 1796 that vaccination against smallpox could provide protection from disease. It is now becoming clear that immunological memory is regulated not only by immune processes but also by general cellular memory mechanisms (Zediak et al. 2011b). The ability to maintain function after differentiation into the different types of cells needed to make up our body can be referred to as cellular memory. "Epigenetics" is an important keyword in understanding the concept of the cellular memory (Goldberg et al. 2007). Histone modifications and DNA methylation form the molecular basis of the epigenetic regulation of cellular memory.

Epigenetic studies have made a substantial contribution to the elucidation of mechanisms controlling immunological memory, particularly in the field of T cells. This review aims to summarize the role that epigenetics plays in the establishment of immunological memory in T cells. In addition, we introduce the latest progress in the field of immunological memory, mainly based on our data in which Th2 cells, one of the CD4 T cell subsets, are used as a model.

14.2 How Do Epigenetic Approaches Contribute to Studies in the Immunological Memory Field?

The immune system has the capacity to remember previous exposure to antigens that occurred during a primary infection. This memory enables the immune system to launch a more rapid and stronger response to the second antigenic challenge than to the first exposure. We will not describe the process of generation of immunological memory here due to length constraints. For a good review of this topic, see (Seder and Ahmed 2003). However, in very broad terms, immunological memory can be considered as being formed by a combination of both internal and external factors (Fig. 14.1). Internal factors include acquisition of high-affinity antigen receptors by gene rearrangement and high-level expression of certain transcription factors, cytokines, and cell surface molecules (Cuddapah et al. 2010; Zediak et al. 2011b). In addition, the external environment, such as the concentration of physiologically active substances including cytokines, intercellular interactions, and the extra-cellular matrix, can also influence memory cell function and formation (Shinoda et al. 2012; Tokoyoda et al. 2010). Epigenetic studies have focused on internal factors that regulate memory cells and have begun to elucidate the mechanisms controlling cellular memory of adaptive immune cells (Nakayama and Yamashita 2009). The concept of epigenetics is considered important because epigenetic information can dictate the transcriptional state of a gene (Goldberg et al. 2007). Typical examples of epigenetic changes are chemical modifications of histone proteins around which the genomic DNA is wound (histone modification) and DNA methylation involving the addition of a methyl group to the number

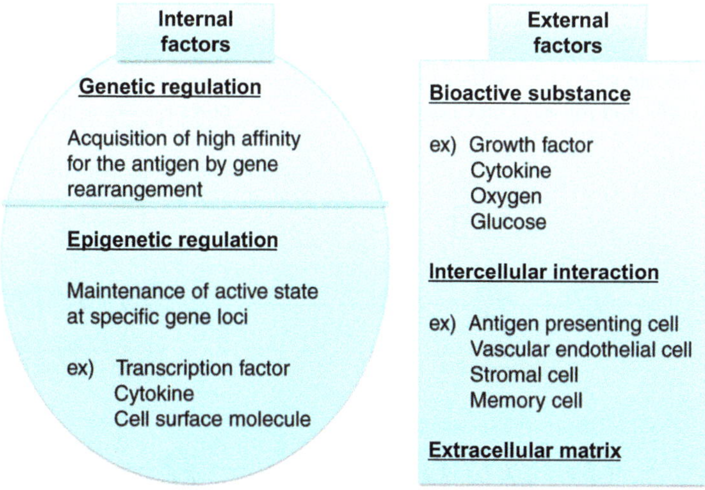

Fig. 14.1 Internal and external factors in immunological memory. Mechanisms of immunological memory are classified into internal factors and external factors. The internal factors are subdivided into genetically controlled mechanisms and epigenetically controlled mechanisms and shown with some examples (*left*). The external factors are subdivided into three factors: physiologically active substances, cell–cell interactions, and extracellular matrix (examples are shown in *right panel*)

5 position of the cytosine pyrimidine ring (DNA methylation) (Turner 2002; Wu and Zhang 2010). As shown in Fig. 14.2a, methylation of histone H3K27, methylation of H3K9, and DNA methylation are indicators of transcriptional repression. On the other hand, methylation of H3K4 and acetylation H3K9 and H3K27 are indicators of transcriptional activation. Here, symbols such as K27 indicate the number of a particular amino acid in the N-terminus of histone H3. The transcriptional repression and transcriptional activation states are commonly referred to as closed chromatin structure and open chromatin structures, respectively. The term chromatin refers to a complex of DNA and protein existing in a eukaryotic cell nucleus.

The epigenetic concept can thus explain the question "how can the form and function of specific cells or tissues be different while they each have the same underlying DNA sequence?" Figure 14.2b shows two types of cells, cell type A and cell type B. The two cell types have the same genes (i.e., same genetic information), as the two cell types are of the same origin. However, gene A is expressed only in type A cells, and gene B is expressed only in type B cells. This mechanism can be explained as follows: in type A cells, the region of DNA encompassing the A gene has an open chromatin structure, in which active histone modifications such as H3K9ac and H3K4me3 are abundant. However, in type B cells, the chromatin structure at the A gene region contains repressive epigenetic marks such as H3K27me3 and DNA methylation and is closed. The open chromatin structure increases accessibility for molecules such as RNA polymerase and transcription factors that can control transcription. Thus, these molecules can easily accumulate

a

	Repressive mark	Active mark
Histone modification	Histone H3K27 methylation Histone H3K9 methylation	Histone H3K4 methylation Histone H3K9 acetylation
DNA methylation	DNA methylation	DNA 5-hydroxy methylation

b

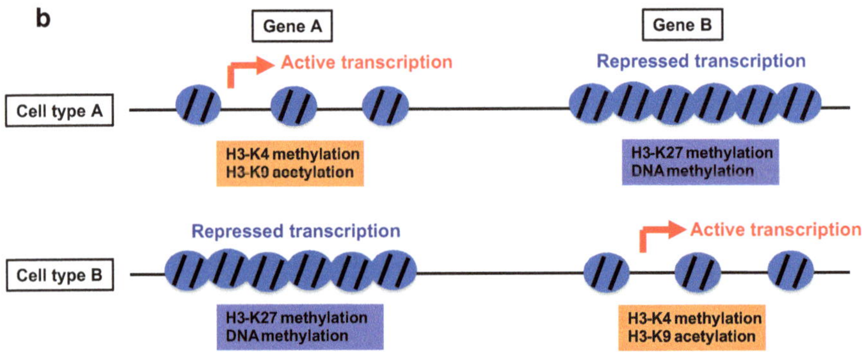

Fig. 14.2 Examples of epigenetic regulation. (**a**), Some typical examples of epigenetic information are shown. Histone modifications and DNA methylation are divided into two classes: active marks (*left*) and repressive marks (*right*). (**b**), Difference in epigenetic states at differentially expressed genes between two distinct cell types. In type A cells, the A gene has open chromatin structure and is highly expressed. On the other hand, in type B cells, the B gene has open chromatin structure and is highly expressed

at genes located in areas of open chromatin and activate transcription of these genes. Conversely, the closed chromatin structure decreases accessibility to transcriptional regulatory molecules and represses gene transcription. This also applies to the B gene. Chromatin structure around the B gene is closed in type A cells, whereas chromatin around the B gene contains open structure in type B cells. A complex combination of these epigenetic mechanisms allows cells with the same genetic information to acquire specific functions, according to the tissue each cell belongs to.

These internal factors regulating immunological memory are an example of how we can use our understanding of the concepts of genetics and epigenetics to understand how the immune system functions. In contrast to the epigenetic changes described above, the antigen receptor genes of lymphocytes such as B cells and the T cells are known to acquire high affinity to antigen by gene rearrangement (Hozumi and Tonegawa 1976), i.e., by direct changes to the genetic code within the cell that are largely irreversible. This process is one of the very few examples of normal physiological genetic rearrangements. Therefore, to avoid the need to directly rearrange the genetic code, epigenetic mechanisms are utilized for the acquisition of memory cell-specific gene expression profiles: high-level expression of certain transcription factors, cytokines, and cell surface molecules.

14.3 The Factors That Control Epigenetic States

The epigenetic modifications such as those described above are introduced and interpreted by epigenetic regulators which include Polycomb (PcG) and Trithorax (TrxG) proteins (Margueron and Reinberg 2011; Nakayama and Yamashita 2009; Schuettengruber et al. 2007, 2011). These complexes will only be described briefly here because they are discussed at length in other chapters in this volume. PcG and TrxG proteins were originally identified in *Drosophila*, but they also play major roles in defining mammalian gene expression programs during cell differentiation and in cancer cells. The Mammalian counterparts of TrxG proteins are the SET/MLL family of proteins. Figure 14.3 shows a schematic view of the principle PcG and TrxG complexes (Mohan et al. 2012). Previous studies report that the Polycomb-repressive complex 2 (PRC2) maintains a transcriptionally repressive state, primarily via EZH1/2 which has methyl transferase activity specific for H3K27. This state is then reinforced by the Polycomb-repressive complex 1 (PRC1) containing RING1 which has a repressive histone H2AK119 ubiquitin ligase activity. Conversely, the TrxG complex participates in the maintenance of a transcriptionally active state. The SET domain of MLL1/2 methylates H3K4 and induces open chromatin. In addition, some reports show that the TrxG complex interacts with histone acetyl transferase (HAT) and directly induces transcriptional activation (Schuettengruber et al. 2011). In these ways, it is thought that PcG and TrxG complexes antagonize each other and regulate the ON/OFF state of their target genes.

14.4 Experimental Methods Used to Determine Epigenetic States

The chromatin immunoprecipitation (ChIP) assay is an experimental technique used to investigate the chromatin state (i.e., epigenetic state) within cells (Solomon et al. 1988). Firstly, target cells are collected and DNA–protein is cross-linked inside the cell using formaldehyde. Next, the DNA–protein complex is fragmented by sonication or a restriction enzyme. After performing immunoprecipitation by specific antibody to a target protein, DNA fragments are purified and collected. The collected DNA is analyzed by quantitative PCR using specific primers to detect the genomic regions of interest (Fig. 14.4a). Various kinds of histone modifications and transcription factor-binding patterns are thus detected by selecting protein-specific antibodies.

More recently, chromatin immunoprecipitation with high-throughput sequencing methods (ChIP-Seq) has been enabled due to the development of advanced next-generation sequencers (Fig. 14.4b) (Barski et al. 2007). ChIP-Seq is a method that can rapidly read the nucleotide sequences of the DNA fragments recovered from ChIP experiments, followed by mapping where these DNA fragments are

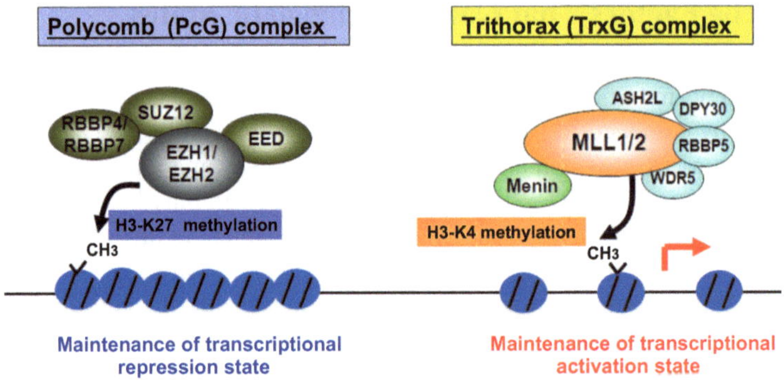

Fig. 14.3 Polycomb (PcG) and Trithorax (TrxG) complex. A schematic representation of the Polycomb PRC2 complex (*left*) and the Trithorax MLL complex (*right*) is shown (Mohan et al. 2012). Menin protein is encoded by the *MEN1* (multiple endocrine neoplasia 1) gene, whose mutation is involved in multiple endocrine neoplasia type 1 in human

derived from in the genome. Thus, this technique enables us to analyze histone modification states and binding sites of particular transcription factors on a genome-wide scale. Figure 14.4c shows an example of conventional ChIP analysis of Bmi1 (PcG)- and Menin (TrxG)-binding patterns in Splenic B, Naïve CD4 T, and fully developed Th2 cells, demonstrating that the activation of the *Gata3* gene during Th2 cell differentiation is accompanied by decreased association of the Bmi1 PcG protein and increased association of the Menin TrxG protein.

14.5 Differentiation of CD4 T Cell Subsets

Naive CD4 T cells that receive antigen stimulation are known to differentiate into various subsets, depending on their surrounding environment and the type of cells that provide the antigenic signals for activation. The signals from cytokines, transmitted via their receptors expressed on the cell surface, are particularly important factors in deciding the type of differentiation that a cell will undergo (Fig. 14.5) (Kanno et al. 2012; Mosmann et al. 1986; Wilson et al. 2009; Zhou et al. 2009; Zhu et al. 2010). Th1 cells produce IFNγ and direct cell-mediated immunity against intracellular pathogens. The interleukin (IL)-12-STAT4 (signal transducer and activator of transcription 4) signaling pathway induces upregulation of the transcription factor T-bet and is required for Th1 cell differentiation. Th2 cells produce IL-4, IL-5, and IL-13 (Th2 cytokines) and are involved in humoral immunity and allergic reactions, which can include allergies and asthma. Upregulation of the IL-4-STAT6-dependent transcription factor GATA3 is crucial for Th2 cell differentiation. Th17 cells produce IL-17A, IL-17 F, and IL-22 and play an important role in immunity against bacteria and fungus. In mice, experimental autoimmune encephalomyelitis (EAE), a model of multiple sclerosis, is

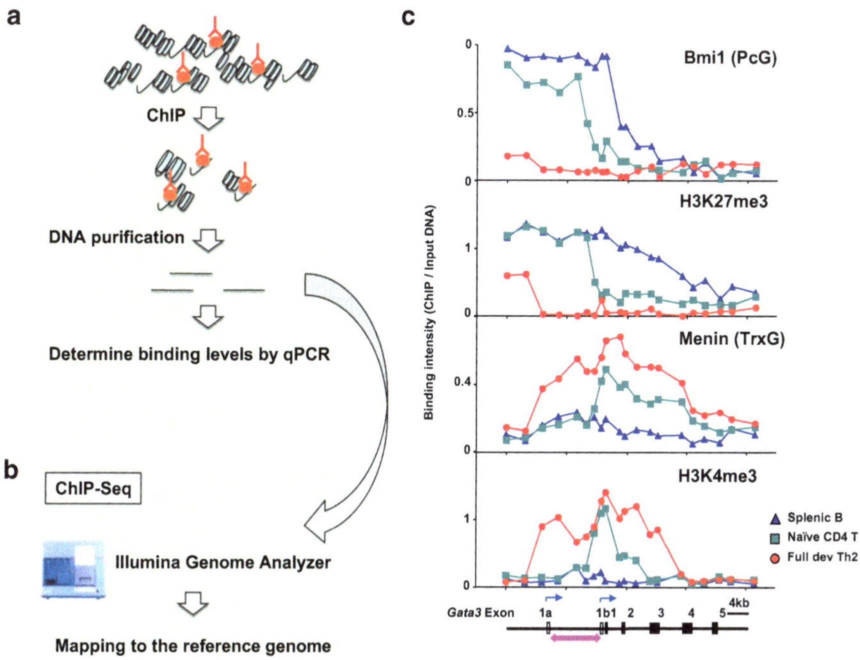

Fig. 14.4 ChIP and ChIP-Seq method. (**a–b**), A summary of ChIP and ChIP-Seq method is shown. Performing ChIP by using a specific antibody against the protein a researcher wants to examine binding to DNA is common in both conventional ChIP and ChIP-Seq analysis. In conventional ChIP assay, purified DNA is quantified by quantitative PCR (**a**). In ChIP-Seq analysis, DNA sequences of purified and pretreated samples are read by next-generation sequencer and mapped on the reference genome (**b**). (**c**), An example of conventional ChIP analysis. Bmi1 (PcG)- and Menin (TrxG)-binding patterns and histone modifications were analyzed by ChIP with semiquantitative PCR. © 2010 Onodera et al. Journal of Experimental Medicine. 207:2493-2506. doi:10.1084/jem.20100760 Results in splenic B cells (*navy*), CD4 T cells (*green*) and fully developed Th2 cells (*red*) are shown

known as a Th17 cell-mediated disorder. The master transcription factor for Th17 cell differentiation is RORγt, whose expression is upregulated by cytokines including TGF-β, IL-6, IL-21, and IL-23. STAT3 is a signal transmission molecule located downstream of the receptors of IL-6, IL-21, and IL-23. Regulatory T cells (Treg), a subset with immune suppressive function, are characterized as cells highly expressing Foxp3 protein. Tregs are classified into two types: naturally occurring Treg (nTreg) that differentiate in the thymus, and inducible Treg (iTreg) whose differentiation is induced by TGF-β in peripheral tissues. In addition, other subsets have also been reported: Tfh cells assist in the maturation of B cells within the follicles of lymphoid organs, and Th9 cells have the ability to produce IL-9.

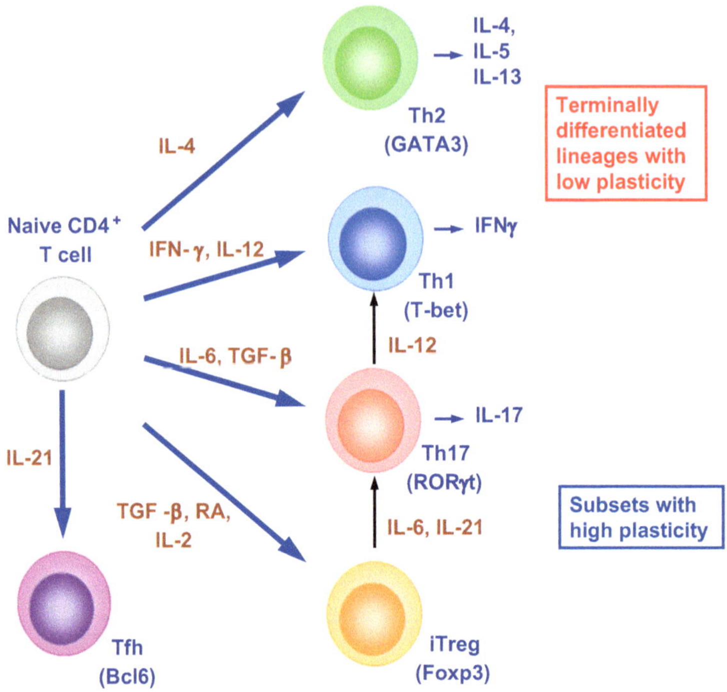

Fig. 14.5 CD4 T cell subset differentiation and features. Naïve CD4 T cells are subjected to antigen presentation from antigen presenting cells (APC) and differentiate into Th1, Th2, Th17, iTreg, and Tfh subsets. Transcription factors and cytokines necessary for the differentiation of each subset are shown (Zhou et al. 2009)

14.6 Acquisition and Maintenance of Th2 Cell Identity

Here we briefly describe mechanisms of induction of Th2 cell differentiation and then outline data generated in mouse models showing how Th2 cell identity is maintained during the memory phase. Figure 14.6 (left) shows how the differentiation of Th2 cells is induced. Efficient Th2 cell differentiation requires two signaling pathways: the TCR (T cell receptor)-signaling pathway and the IL-4 and IL-4 receptor-signaling pathway (Nakayama and Yamashita 2010). GATA3, whose expression is upregulated by combined signaling through these two signaling pathways, is regarded as the master transcription factor responsible for forming active (accessible) chromatin at the Th2 cytokine gene loci (Amsen et al. 2009; Ho et al. 2009). STAT6-deficient CD4 T cells display impaired GATA3 upregulation, resulting in impaired Th2 cell differentiation and IL-4 production. This shows that the identity of Th2 cells is gained in an IL-4-STAT6-dependent manner along with the existence of signals from the TCR. However, until recently, the detailed mechanisms by which STAT6 induces the upregulation of *Gata3* had been unknown. Recent research revealed that STAT6 directly bound to the *Gata3* gene

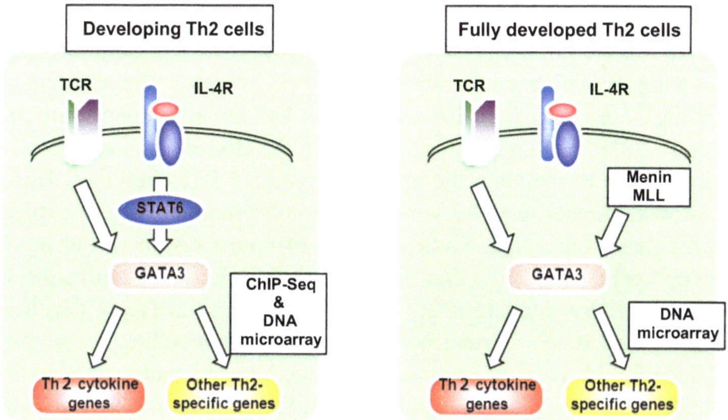

Fig. 14.6 Acquisition and maintenance of Th2 cell identity. Developing Th2 cells, represented in the *left panel*, refer to CD4 T cells cultured under the Th2 condition for <1 week. Developed Th2 cells, shown in the *right panel*, refer to Th2 cells cultured for 2–4 weeks in vitro or Th2 cells transferred into congenic mice and maintained for more than 4 weeks in vivo (memory Th2 cells). ChIP-Seq and DNA microarray datasets are deposited in the Gene Expression Omnibus (GEO) database (http://www.ncbi.nlm.nih.gov/geo) under accession number GSE28292 and GSE46185

locus and induced epigenetic changes mediated by displacement of PcG proteins by TrxG proteins (Onodera et al. 2010). While there have been many reports demonstrating that high-level expression of GATA3 is indispensable for induction of Th2 cytokine genes, it was also unknown if GATA3 participated in the regulation of the expression of other Th2-specific genes. Recently, ChIP-Seq analysis was successfully utilized to identify genome-wide targets of GATA3 in Th2 cells (Horiuchi et al. 2011). Furthermore, by combining a *Gata3* knockdown system with DNA microarray, the functional target genes of GATA3 were comprehensively identified (Sasaki et al. 2013). In addition, GATA3 function was shown to be controlled via protein–protein interactions. The Sox4 protein was also shown to antagonize GATA3 function (Kuwahara et al. 2012), and Chd4 was shown to coordinately regulate the formation of active and suppressive GATA3 complexes in T helper cells (Hosokawa et al. 2013).

Another important issue to address is how the identity of developed Th2 cells is maintained after differentiation (Fig. 14.6, right). Interestingly, IL-4 is not necessary for the maintenance of developed Th2 cell function. This observation is true for in vitro-developed Th2 cells and in vivo-generated memory Th2 cells (Yamashita et al. 2004a). Th2 function is maintained in an IL-4-independent manner, and this maintenance mechanism requires GATA3 in vitro (Yamashita et al. 2004b). Additionally, Pai et al. have reported that GATA3 is required for the functional maintenance of Th2 cells in vitro (Pai et al. 2004). However, the IL-4-STAT6-independent mechanism for maintaining *Gata3* expression in developed Th2 cells remains unclear. Two reports clearly proved that MLL and Menin, both members of the TrxG complex, were important for the functional maintenance of developed

Th2 cells (Onodera et al. 2010; Yamashita et al. 2006). The same mechanism was also recently reported in human Th2 cells (Nakata et al. 2010). On the other hand, Bmi1 and Ring1B, which are members of the PcG complex (the counterpart of the TrxG complex), are important for survival of Th2 cells by controlling apoptosis (Suzuki et al. 2010; Yamashita et al. 2008). In developed Th2 cells, GATA3 was also indispensable to maintain the ability to produce Th2 cytokines. However, it was not known whether GATA3 is required for the maintenance of expression of Th2-specific genes other than Th2 cytokines. From the results of our in vitro and in vivo experiments, we found that maintenance of high-level expression GATA3 was indispensable for keeping many Th2-specific genes active in developed Th2 cells (Sasaki et al. 2013). Furthermore, we reported that pathogenic memory Th2 cells produced IL-5 in vivo and were involved in the pathogenesis of airway inflammation (Endo et al. 2011) and that NKT cells controlled the pool size of memory Th2 cells (Iwamura et al. 2012). In these and other similar ways, researchers are currently expending considerable effort trying to unravel the mechanisms of immune memory from the viewpoint of maintenance of Th2 cell identity.

14.7 General Transcriptional and Epigenetic Features of Memory T Cells

14.7.1 Memory T Cell Classification and Their Transcriptional Features

We have discussed immunological memory in terms of maintenance of Th2 cell phenotype until this section. An increasing number of groups are beginning to investigate epigenetic regulation of immune cell memory, most of which are using the approach of separating and classifying cells by surface molecules and then analyzing the epigenetic states of these populations. In this section, we describe the currently known epigenetic characteristics of immune memory cells defined by cell surface molecules, focusing on T cells.

In murine T cells, high expression of the cell surface molecule CD44 is used as a marker of memory cells (Dutton et al. 1998). Memory T cells can be further subdivided into central memory T cells (TCM; CD62L high) and effector memory T cells (TEM; CD62L low) according to the expression levels of CD62L (Sallusto et al. 2004). CD45RA and CD45RO are often used as markers of naïve and memory human T cells, respectively (Dutton et al. 1998), and like murine memory cells, human memory cells can also be subdivided into TCM and TEM based on expression of CD62L. T cells are classified into helper T cells expressing CD4 and cytotoxic T cells expressing CD8; however, memory cells generated from these two populations display a high degree of transcriptional similarity (Seder and Ahmed 2003), as do TCM and TEM cells (Weng et al. 2012). Therefore, we will discuss here general gene expression profiles and epigenetic signatures that are shared by all memory T cells. Naïve and memory T cells show less than a 5 % difference in their overall gene expression profiles when assessing the number of genes that are differentially expressed (Araki et al. 2009; Kaech et al. 2002). Weng et al. reported a list of highly expressed genes in human memory T cells (Weng et al. 2012). Kinetic analysis before and after T cell activation has revealed that there are activation-induced genes that are upregulated more rapidly in activated memory T cells than in activated naïve T cells, in addition to genes that are expressed at higher levels by activated memory T cells than by activated naïve T cells (Araki et al. 2009). When comparing memory T cells to naïve T cells, genes that are highly expressed in either resting or activated memory T cells are likely to be important for memory T cell function. Genes that are highly expressed by memory cells include those with immune functions, such as cytokines, chemokines, and receptors (Weng et al. 2012). Memory T cells also express genes that promote T cell survival and homeostasis and other genes with multiple or undefined functions (Weng et al. 2012).

14.7.2 Epigenetic Features of Memory T Cells

Although it is easy to imagine that epigenetic mechanisms are important for the control of the expression of these genes, there are technical limitations in our ability to efficiently analyze memory T cell populations. First, the number of memory T cells is so small in vivo that it is difficult to collect sufficient cell number for analysis. Several detailed studies have overcome this problem and found direct evidence that epigenetic processes are involved. Analyses of histone modifications in memory T cells have revealed that the more rapid response of memory T cells is associated with histone H3 acetylation and H3K4 methylation (Araki et al. 2008; Fann et al. 2006; Zediak et al. 2011b). For example, histone acetylation at the *Ifng* gene locus provides a molecular basis for the enhanced responsiveness of memory CD8 T cells (Northrop et al. 2006). Functions of memory T cells are also associated with reduced histone H3K27 methylation at numerous genes (Araki et al. 2009; Zediak et al. 2011a). In addition, Bernstein et al. have reported that in embryonic stem (ES) cells, many key developmental genes display both histone H3K4me3 and H3K27me3 marks, displaying a so-called bivalent phenotype (Bernstein et al. 2006). Resolution of these bivalent genes was also observed in differentiated cells (Mikkelsen et al. 2007). This is also the case in memory T cells, indicating that bivalent domains mark and regulate a select group of important genes in T cells (Araki et al. 2009). The second difficulty encountered when analyzing epigenetic modifications in memory T cells is that this group of cells exist as a heterogeneous population, so population analysis is not sufficient to fully understand the mechanisms regulating the phenotype of these cells. Dispirito and Shen (2010) have shown a global increase in histone H3 acetylation levels in single cells using a flow cytometric assay. The development of techniques that enable single-cell genome-wide analysis should shed light on this complex issue (Kalisky et al. 2011).

Finally, the timing of when the features of memory T cells are completely established is unclear. Are memory T cells programmed during the effector phase, in which T cells encounter a pathogen for the first time, or during the contraction phase, while the levels of pathogen is rapidly declining and the infection is being eliminated? Additionally, it is possible that epigenetic states in T cells change during the memory phase. Many questions about the process of memory T cell formation remain unsolved. Some studies provide evidence for stable epigenetic marks that are established during the effector phase and persist in the memory phase (Fann et al. 2006; Mirabella et al. 2010; Yamashita et al. 2004a). The persistence of an active chromatin signature at relevant gene loci is observed in resting memory cells, even when those genes are transcriptionally inactive (Zediak et al. 2011b). The epigenetic study of memory T cells is at a nascent stage, and further research development along with progress in technology to analyze rare cell populations will continue to advance this field.

14.8 Perspective

We are conducting epigenetic research of immune cell memory using Th2 cells as a model. Th2 cells are considered to be responsible for causing allergic diseases, so elucidating the mechanisms controlling Th2 cell differentiation and maintenance of their function is relevant to the development of potential therapies. In particular, it is thought that the maintenance of functions after development of Th2 function is closely related to the pathogenesis of chronic disease and repetitive allergic reactions. For example, several studies have found that in pollinosis patients, antigen-specific immune cells remain a long time and repeatedly respond to cedar pollen scattered in early spring each year (Horiguchi et al. 2008). That is, as a treatment for patients who suffer from allergic diseases, suppressing the initial differentiation of Th2 cells is not sufficient. Instead, approaches to suppress the function of developed Th2 cells or to change them into other subsets are required. Of course, most memory T cells are not pathogenic and are instead important for defense against infection. Th2 cells assist in antibody production by B cells and are crucial for the elimination of pathogens by humoral immunity. Th1 cells are indispensable to the cell-mediated immunity mediated mainly by CD8 T cells. The specific defense mechanisms against infection that are regulated by memory T cells remain unclear and is one of the major unresolved issues related to immunological memory. For many immunology researchers, a major goal is to unravel the mystery of the immune memory mechanism and to establish new and efficient methods for vaccine development (Pulendran and Ahmed 2011). In today's modern society, preventive medicine is highly desired. However, there are many kinds of pathogens against which we cannot successfully generate preventive immune responses using current vaccine technology. In addition, the establishment of a rapid vaccine development system for emerging infectious diseases that pose a threat to humans such as avian influenza (Gao et al. 2013) and MERS (Middle East respiratory syndrome)

(Memish et al. 2013) is an urgent unmet need. Suppressing pathogenic memory T cells and increasing beneficial memory T cells are two essential aspects of immunological memory study when considering the application to clinical medicine. Moreover, it is hoped that we can apply the knowledge acquired using Th2 cells as a model to further studies of immunological memory in other CD4 T cell subsets, CD8 T cells, and B cells.

As discussed in this review, immunological memory is an important concept that forms the basis of long-lasting immunity. However, the mechanisms controlling memory cell development and functional maintenance have not yet been fully elucidated. Certainly, the nature of immunological memory is controlled by epigenetic mechanisms. However, there are few immunology researchers who have studied it from such a viewpoint. Elucidation of the molecular mechanism controlling immunological memory in terms of epigenetic regulation will lead to further development of immunology and new applications to clinical medicine. We consider there to be at least two important points with respect to investigations of the factors regulating immunological memory. The first is delineation of the mechanisms regulating memory cell survival, and the second is to determine the regulatory pathways by which memory cells maintain their phenotype. Survival of nonfunctional memory cells is likely avoided to conserve essential energy and nutrients. In contrast, survival of functional memory cells that retain the ability to react quickly upon secondary antigen encounter is central to immunological memory. With respect to the survival of memory cells, is likely that only the cells able to provide the most efficient responses to secondary antigen encounter are allowed to survive and become memory cells. However, how this process is regulated is almost completely unknown (Mueller et al. 2013). If the general concept of immunological memory was investigated in the same manner as the memory in a brain, research of functional maintenance would be similar to investigating the mechanism regulating "how can we remember." In contrast, research of memory cell survival would be more similar to investigating "why only a part of memory remains" or "why do we have selective memory."

Ultimately, through our epigenetic research on Th2 cells, we would like to find the fundamental principles that can be applied generally to cell differentiation and maintenance of differentiated cell identity. Recent reports about the generation of induced pluripotent stem (iPS) cells reversed the old thinking that once cells underwent functional specialization that they could not return to an undifferentiated state (Takahashi et al. 2007; Takahashi and Yamanaka 2006). Research on the differentiation of specific tissues from iPS cells for the purpose of regenerative medicine is a good example of the study of cell differentiation that relies on our understanding of epigenetic regulation. The epigenetic mechanisms defined using Th2 cell differentiation systems combined with analytic methods of ChIP-Seq and bioinformatics technology that we have developed and utilized may therefore also benefit the development of new regenerative medicine studies. In the near future, technical progress will enable us to perform single-cell ChIP-Seq analysis or high-throughput ChIP-Seq analysis with clinical samples (Northrup and Zhao 2011). The bioinformatic techniques for ChIP-Seq analysis which we and others have developed will also be useful when analyzing these samples.

Finally, the effects of radiation on epigenetic states should be described. Since the accident at the Fukushima nuclear power plant in 2011, people have made arguments about how radiation influences the human body (Christodouleas et al. 2011). It is likely that not only genetic changes but also epigenetic changes are induced by radiation. Thus, epigenetic research will come to take on more and more importance in the future.

References

Amsen D, Spilianakis CG, Flavell RA (2009) How are T(H)1 and T(H)2 effector cells made? Curr Opin Immunol 21:153–160

Araki Y, Fann M, Wersto R et al (2008) Histone acetylation facilitates rapid and robust memory CD8 T cell response through differential expression of effector molecules (eomesodermin and its targets: perforin and granzyme B). J Immunol 180:8102–8108

Araki Y, Wang Z, Zang C et al (2009) Genome-wide analysis of histone methylation reveals chromatin state-based regulation of gene transcription and function of memory CD8+ T cells. Immunity 30:912–925

Barski A, Cuddapah S, Cui K et al (2007) High-resolution profiling of histone methylations in the human genome. Cell 129:823–837

Bernstein BE, Mikkelsen TS, Xie X et al (2006) A bivalent chromatin structure marks key developmental genes in embryonic stem cells. Cell 125:315–326

Christodouleas JP, Forrest RD, Ainsley CG et al (2011) Short-term and long-term health risks of nuclear-power-plant accidents. N Engl J Med 364:2334–2341

Cuddapah S, Barski A, Zhao K (2010) Epigenomics of T cell activation, differentiation, and memory. Curr Opin Immunol 22:341–347

Dispirito JR, Shen H (2010) Histone acetylation at the single-cell level: a marker of memory CD8+ T cell differentiation and functionality. J Immunol 184:4631–4636

Dutton RW, Bradley LM, Swain SL (1998) T cell memory. Annu Rev Immunol 16:201–223

Endo Y, Iwamura C, Kuwahara M et al (2011) Eomesodermin controls interleukin-5 production in memory T helper 2 cells through inhibition of activity of the transcription factor GATA3. Immunity 35:733–745

Fann M, Godlove JM, Catalfamo M et al (2006) Histone acetylation is associated with differential gene expression in the rapid and robust memory CD8(+) T-cell response. Blood 108:3363–3370

Gao R, Cao B, Hu Y et al (2013) Human infection with a novel avian-origin influenza A (H7N9) virus. N Engl J Med 368:1888–1897

Goldberg AD, Allis CD, Bernstein E (2007) Epigenetics: a landscape takes shape. Cell 128: 635–638

Ho IC, Tai TS, Pai SY (2009) GATA3 and the T-cell lineage: essential functions before and after T-helper-2-cell differentiation. Nat Rev Immunol 9:125–135

Horiguchi S, Tanaka Y, Uchida T et al (2008) Seasonal changes in antigen-specific T-helper clone sizes in patients with Japanese cedar pollinosis: a 2-year study. Clin Exp Allergy 38:405–412

Horiuchi S, Onodera A, Hosokawa H et al (2011) Genome-wide analysis reveals unique regulation of transcription of Th2-specific genes by GATA3. J Immunol 186:6378–6389

Hosokawa H, Tanaka T, Suzuki Y et al (2013) Functionally distinct Gata3/Chd4 complexes coordinately establish T helper 2 (Th2) cell identity. Proc Natl Acad Sci USA 110:4691–4696

Hozumi N, Tonegawa S (1976) Evidence for somatic rearrangement of immunoglobulin genes coding for variable and constant regions. Proc Natl Acad Sci USA 73:3628–3632

Iwamura C, Shinoda K, Endo Y et al (2012) Regulation of memory CD4 T-cell pool size and function by natural killer T cells in vivo. Proc Natl Acad Sci USA 109:16992–16997

Kaech SM, Hemby S, Kersh E et al (2002) Molecular and functional profiling of memory CD8 T cell differentiation. Cell 111:837–851

Kalisky T, Blainey P, Quake SR (2011) Genomic analysis at the single-cell level. Annu Rev Genet 45:431–445

Kanno Y, Vahedi G, Hirahara K et al (2012) Transcriptional and epigenetic control of T helper cell specification: molecular mechanisms underlying commitment and plasticity. Annu Rev Immunol 30:707–731

Kuwahara M, Yamashita M, Shinoda K et al (2012) The transcription factor Sox4 is a downstream target of signaling by the cytokine TGF-beta and suppresses T(H)2 differentiation. Nat Immunol 13:778–786

Margueron R, Reinberg D (2011) The Polycomb complex PRC2 and its mark in life. Nature 469:343–349

Memish ZA, Zumla AI, Al-Hakeem RF et al (2013) Family cluster of middle east respiratory syndrome coronavirus infections. N Engl J Med 368(26):2487–2494

Mikkelsen TS, Ku M, Jaffe DB et al (2007) Genome-wide maps of chromatin state in pluripotent and lineage-committed cells. Nature 448:553–560

Mirabella F, Baxter EW, Boissinot M et al (2010) The human IL-3/granulocyte-macrophage colony-stimulating factor locus is epigenetically silent in immature thymocytes and is progressively activated during T cell development. J Immunol 184:3043–3054

Mohan M, Herz HM, Shilatifard A (2012) SnapShot: Histone lysine methylase complexes. Cell 149(498–498):e1

Mosmann TR, Cherwinski H, Bond MW et al (1986) Two types of murine helper T cell clone I. Definition according to profiles of lymphokine activities and secreted proteins. J Immunol 136:2348–2357

Mueller SN, Gebhardt T, Carbone FR et al (2013) Memory T cell subsets, migration patterns, and tissue residence. Annu Rev Immunol 31:137–161

Nakata Y, Brignier AC, Jin S et al (2010) c-Myb, Menin, GATA-3, and MLL form a dynamic transcription complex that plays a pivotal role in human T helper type 2 cell development. Blood 116:1280–1290

Nakayama T, Yamashita M (2009) Critical role of the Polycomb and Trithorax complexes in the maintenance of CD4 T cell memory. Semin Immunol 21:78–83

Nakayama T, Yamashita M (2010) The TCR-mediated signaling pathways that control the direction of helper T cell differentiation. Semin Immunol 22:303–309

Northrop JK, Thomas RM, Wells AD et al (2006) Epigenetic remodeling of the IL-2 and IFN-gamma loci in memory CD8 T cells is influenced by CD4 T cells. J Immunol 177: 1062–1069

Northrup DL, Zhao K (2011) Application of ChIP-Seq and related techniques to the study of immune function. Immunity 34:830–842

Onodera A, Yamashita M, Endo Y et al (2010) STAT6-mediated displacement of polycomb by trithorax complex establishes long-term maintenance of GATA3 expression in T helper type 2 cells. J Exp Med 207:2493–2506

Pai SY, Truitt ML, Ho IC (2004) GATA-$_3$ deficiency abrogates the development and maintenance of T helper type 2 cells. Proc Natl Acad Sci USA 101:1993–1998

Pulendran B, Ahmed R (2011) Immunological mechanisms of vaccination. Nat Immunol 12:509–517

Sallusto F, Geginat J, Lanzavecchia A (2004) Central memory and effector memory T cell subsets: function, generation, and maintenance. Annu Rev Immunol 22:745–763

Sasaki T, Onodera A, Hosokawa H et al (2013) Genome-wide gene expression profiling revealed a critical role for GATA3 in the maintenance of the Th2 cell identity. PLoS One 8:e66468

Schuettengruber B, Chourrout D, Vervoort M et al (2007) Genome regulation by polycomb and trithorax proteins. Cell 128:735–745
Schuettengruber B, Martinez AM, Iovino N et al (2011) Trithorax group proteins: switching genes on and keeping them active. Nat Rev Mol Cell Biol 12:799–814
Seder RA, Ahmed R (2003) Similarities and differences in CD4+ and CD8+ effector and memory T cell generation. Nat Immunol 4:835–842
Shinoda K, Tokoyoda K, Hanazawa A et al (2012) Type II membrane protein CD69 regulates the formation of resting T-helper memory. Proc Natl Acad Sci USA 109:7409–7414
Solomon MJ, Larsen PL, Varshavsky A (1988) Mapping protein-DNA interactions in vivo with formaldehyde: evidence that histone H4 is retained on a highly transcribed gene. Cell 53: 937–947
Suzuki A, Iwamura C, Shinoda K et al (2010) Polycomb group gene product Ring1B regulates Th2-driven airway inflammation through the inhibition of Bim-mediated apoptosis of effector Th2 cells in the lung. J Immunol 184:4510–4520
Takahashi K, Yamanaka S (2006) Induction of pluripotent stem cells from mouse embryonic and adult fibroblast cultures by defined factors. Cell 126:663–676
Takahashi K, Tanabe K, Ohnuki M et al (2007) Induction of pluripotent stem cells from adult human fibroblasts by defined factors. Cell 131:861–872
Tokoyoda K, Hauser AE, Nakayama T et al (2010) Organization of immunological memory by bone marrow stroma. Nat Rev Immunol 10:193–200
Turner BM (2002) Cellular memory and the histone code. Cell 111:285–291
Weng NP, Araki Y, Subedi K (2012) The molecular basis of the memory T cell response: differential gene expression and its epigenetic regulation. Nat Rev Immunol 12:306–315
Wilson CB, Rowell E, Sekimata M (2009) Epigenetic control of T-helper-cell differentiation. Nat Rev Immunol 9:91–105
Wu SC, Zhang Y (2010) Active DNA demethylation: many roads lead to Rome. Nat Rev Mol Cell Biol 11:607–620
Yamashita M, Shinnakasu R, Nigo Y et al (2004a) Interleukin (IL)-4-independent maintenance of histone modification of the IL-4 gene loci in memory Th2 cells. J Biol Chem 279:39454–39464
Yamashita M, Ukai-Tadenuma M, Miyamoto T et al (2004b) Essential role of GATA3 for the maintenance of type 2 helper T (Th2) cytokine production and chromatin remodeling at the Th2 cytokine gene loci. J Biol Chem 279:26983–26990
Yamashita M, Hirahara K, Shinnakasu R et al (2006) Crucial role of MLL for the maintenance of memory T helper type 2 cell responses. Immunity 24:611–622
Yamashita M, Kuwahara M, Suzuki A et al (2008) Bmi1 regulates memory CD4 T cell survival via repression of the *Noxa* gene. J Exp Med 205:1109–1120
Zediak VP, Johnnidis JB, Wherry EJ et al (2011a) Cutting edge: persistently open chromatin at effector gene loci in resting memory CD8+ T cells independent of transcriptional status. J Immunol 186:2705–2709
Zediak VP, Wherry EJ, Berger SL (2011b) The contribution of epigenetic memory to immunologic memory. Curr Opin Genet Dev 21:154–159
Zhou L, Chong MM, Littman DR (2009) Plasticity of CD4+ T cell lineage differentiation. Immunity 30:646–655
Zhu J, Yamane H, Paul WE (2010) Differentiation of effector CD4 T cell populations. Annu Rev Immunol 28:445–489

Chapter 15
The Macrophage Epigenome and the Control of Inflammatory Gene Expression

Sara Polletti, Alessia Curina, Gioacchino Natoli, and Serena Ghisletti

Abstract The combination of large-scale genomic studies and computational tools has revealed the complexity and dynamics of the mechanisms that regulate the inflammatory response. The specificity of the inflammatory gene expression program in innate immune cells, such as macrophages, reflects a simple underlying mechanism: transcription factors controlling macrophage differentiation generate a unique, cell type-specific repertoire of accessible genomic regions that enable the recruitment of stimulus-induced transcription factors. This mechanism constrains the activity of non-cell type-specific inflammatory transcription factors at macrophage-specific regulatory elements. In this chapter, we provide an overview of transcriptional regulation in macrophages and we discuss the recent progresses on how the interplay between genomic and epigenomic information results in a fine-tuned inflammatory response.

Keywords Chromatin • Epigenome • Inflammation • Enhancers • Macrophages • PU.1

15.1 Introduction

Inflammation is a complex physiological response to infection, tissue stress, and injury that involves both immune system cells, notably macrophages and neutrophils, and parenchymal cells (Medzhitov 2008).

Macrophages are resident phagocytic cells that play central roles in the innate and adaptive immune defense as well as in tissue homeostasis (Geissmann et al. 2010). They originate from blood monocytes and they are widely distributed in tissues, where they are active both as immune effector cells with a broad

S. Polletti • A. Curina • G. Natoli • S. Ghisletti (✉)
Department of Experimental Oncology, European Institute of Oncology (IEO), Via Adamello 16, 20139 Milan, Italy
e-mail: serena.ghisletti@ieo.eu

C. Bonifer and P.N. Cockerill (eds.), *Transcriptional and Epigenetic Mechanisms Regulating Normal and Aberrant Blood Cell Development*, Epigenetics and Human Health, DOI 10.1007/978-3-642-45198-0_15,

microbial recognition capacity and as housekeeping phagocytes responsible for maintenance of tissue and organism integrity. Macrophages are highly heterogeneous cells in terms of phenotype and gene expression and can rapidly change their function in response to microenvironmental inputs (Gordon and Taylor 2005; Lawrence and Natoli 2011; Mosser and Edwards 2008). This striking heterogeneity reflects the specialization of tissue macrophages in the microenvironments of different tissues such as liver (Kupffer cells), brain (microglial cells), lung (alveolar macrophages), and skin (Langerhans cells). Moreover, marked dynamic changes in macrophage physiology occur after they come in contact with microbial stimuli or endogenous stress signals (such as cell debris). Therefore, the generic definition of macrophages comprises a variety of cells with many different functions and functional states that are specified by the complex interplay between microenvironmental and tissue-specific signals.

The complexity of the inflammatory response reflects the coordinated action of proinflammatory mediators (such as bioactive lipids, cytokines, and chemokines) whose controlled production requires the deployment of a tightly enforced gene expression program regulated in a stimulus- and cell type-specific manner during the different phases of the inflammatory process. Genes encoding proteins with antimicrobial and proinflammatory activities must be rapidly and robustly induced in the presence of microbial stimuli or danger signals. At the same time, inflammatory genes must be maintained in a transcriptionally repressed state under normal homeostatic conditions. In addition, some inducible mediators of inflammation, such as antimicrobial peptides and complement factors, directly target infectious microorganisms, while others, including cytokines and chemokines, activate endothelial cells and other cells of both the innate and adaptive immune systems to the site of infection.

Responses to microbial pathogens are initiated when intracellular pattern-recognition receptors (PRRs) encounter pathogen-associated molecular patterns (PAMPs) (Takeuchi and Akira 2010). The Toll-like receptors (TLRs) are the first-described and best-characterized class of PRRs mainly involved in the induction of acute inflammation. The PRR–PAMP interaction promotes the activation of several signal transduction pathways, which in turn regulate the expression and/or the activity of a limited set of transcription factors (TFs) belonging to a few major families: NF-κB (Nuclear Factor κB), IRFs (Interferon-Regulatory Factors), STATs (Signal Transducers and Activators of Transcription), and AP-1 (Activator Protein-1) (Kawai and Akira 2010). These TFs bind to specific DNA recognition elements and directly or indirectly control the expression of hundreds of genes involved in different phases of inflammation (Bhatt et al. 2012; Medzhitov and Horng 2009; Smale 2010). As a first step, the inflammatory stimulus directly activates signaling pathways and TFs downstream of the corresponding PRR and induces the expression of primary response genes. Some primary response genes encode cytokines and chemokines, which immediately contribute to microbial protection, but many others encode TFs and signaling molecules, which in turn activate secondary response genes in a transcriptional cascade.

15.2 The Transcriptional Regulatory Repertoire of Macrophages

Several experimental data, mainly referring to macrophages activated with lipopolysaccharide (LPS), the prototypical Toll-like Receptor 4 (TLR4) agonist, have suggested that inducible recruitment of NF-kB and IRFs to target gene promoters is influenced by their preexisting nucleosomal organization (Ramirez-Carrozzi et al. 2009).

It is useful to distinguish in this context primary response genes (PRGs) that are usually rapidly activated after stimulus and secondary response genes (SRGs) that display delayed activation kinetics. PRGs are formally defined as genes that can be induced without de novo protein synthesis, while SRGs require new protein synthesis for inducible expression (Herschman 1991). The promoters of most PRGs—such those encoding tumor necrosis factor alpha (TNFα), superoxide dismutase 2 (SOD2), and prostaglandin G/H synthase 2 (PTGS2)—contain a CpG island (Deaton and Bird 2009; Hargreaves et al. 2009; Ramirez-Carrozzi et al. 2009). The very high CG content of CpG islands tends to directly interfere with the assembly of stable nucleosomes (Fenouil et al. 2012; Ramirez-Carrozzi et al. 2009); moreover, protein complexes associated with CpG islands further contribute to nucleosomal depletion and provoke a comparatively higher nucleosome loss in vivo than in vitro (Valouev et al. 2011). Therefore, a distinct property of CpG islands is their relative depletion of nucleosomes, which allows an unrestricted access to these regions of both TFs expressed in a constitutive fashion and TFs activated by stimulation. In addition, multimolecular complexes containing proteins with CXXC domains, such as SET1 and MLL, are able to recognize the unmethylated CpG dinucleotides present at high density at CpG islands and to catalyze the deposition of H3K4me3, a histone mark associated with active promoters. Although the precise functional role of H3K4me3 is still a matter of debate, this is probably one of the events that contribute to the organization of an active promoter (Ayton et al. 2004; Deaton and Bird 2009; Lee and Skalnik 2005, 2008). While the impact, if any, of H3K4me3 at inflammatory genes is still unknown, a recent study indicated a specific role for the histone H3K4 methyltransferase MLL4 (Wbp7) in maintaining the expression of the enzyme that catalyzes the first step of glycosylphosphatidylinositol (GPI) anchor synthesis. Loss of GPI anchor-dependent loading of proteins on the macrophage membrane also affected CD14, the coreceptor for lipopolysaccharide (LPS) and other bacterial molecules, eventually causing an attenuation of LPS-triggered intracellular signals (Austenaa et al. 2012).

More importantly, CpG islands are constitutively associated with RNA polymerase II (Pol II) as well as carry acetylated histones and are both transcribed at a low basal level, poising them for further induction as soon as cells are stimulated (Ramirez-Carrozzi et al. 2009). The recruitment of inducible TFs to the promoter is followed by phosphorylation of serine 2 at the Pol II C-terminal domain repeats, which allows productive transcription and its coordination with the splicing machinery (Hargreaves et al. 2009; Ramirez-Carrozzi et al. 2009). In summary, CpG island-containing promoters are in a poised chromatin conformation, which allows fast binding of activated TFs (such as NF-κB, AP1, and IRFs) and rapid transcription in response to stimulation.

Secondary response genes, such as those encoding interleukin 6 (*Il6*), nitric oxide synthase (*Nos2*), and interleukin-12p40 (*Il12p40*), as well as some primary response genes with delayed activation kinetics such as *Ccl5*, are subjected to regulatory mechanisms involving the nucleosome remodeling activities of the SWI/SNF complex (Ramirez-Carrozzi et al. 2009). Promoters of SRGs are characterized by well-positioned nucleosomes and repressive chromatin features and become more accessible due to the recruitment of chromatin-remodeling complexes and chromatin modifiers (such as histone methyltransferases and acetyltransferases), which allow binding of inflammatory TFs and Pol II (Hargreaves et al. 2009; Ramirez-Carrozzi et al. 2009). These features were initially described based on the results collected on a limited number of representative genes. A recent systematic analysis of LPS-induced gene expression carried out by RNA-seq on different subcellular RNA fractions has provided a comprehensive, genome-scale, picture of these events (Bhatt et al. 2012). By this approach, nascent (chromatin-associated), nucleoplasmic and cytoplasmic transcripts expressed in macrophages following the stimulation with an inflammatory stimulus were identified. The analysis of nascent transcripts allowed understanding the transcriptional response at high resolution since their level exclusively reflects the transcriptional activity of the gene. This analysis has revealed that CpG islands are in fact present also at some SRG promoters. SRGs with a CpG island lack constitutive transcriptional activity probably because some TFs necessary for their activation are not expressed in the basal state. Therefore, the presence or absence of a CpG island does not distinguish between PRGs and SRGs. However, taken collectively, CpG island genes differ from those without a CpG island in that they have on average a comparatively lower fold induction in response to stimulation. The high dynamic range of non-CpG island genes can probably be explained by the tight regulation conferred by the presence of stable and strategically positioned nucleosomes that prevent any basal transcriptional activity.

15.3 The Genomic Organization of Macrophages

During the past few years, technologies enabling genome-wide analyses of regulatory regions have revealed organizational features characteristic of the macrophage genome and have provided an increasingly clear picture of how the inflammatory response is regulated in a specific cellular context. Studies of regulatory elements in macrophages have indicated that competence for responses to an inflammatory stimulus is programmed at an early stage of differentiation by factors involved in lineage commitment and macrophage identity, which are responsible for the organization of the macrophage-specific *cis*-regulatory repertoire (Ghisletti and Natoli

2013; Lichtinger et al. 2012; Natoli 2010; Natoli et al. 2011). In other words, the convergence at *cis*-regulatory regions of lineage-determining TFs and the classic inflammatory TFs (like NF-kB and IRFs) explain the role of the cellular context in modulating the response to inflammatory stimuli. In macrophages, the same genomic location includes binding sites for macrophage-specific lineage-determining TFs and for ubiquitously expressed TFs, which are recruited on these regions upon stimulation.

15.3.1 Chromatin Features of Cis*-regulatory Regions*

Generally, the regulatory information contained in mammalian genomes is located right upstream of the transcription start site (TSS) both at gene promoters and at distal regulatory elements, specifically enhancers (Bulger and Groudine 2011). These *cis*-regulatory elements are relatively nucleosome depleted, as demonstrated by DNAse I-hypersensitivity-based approaches, or by FAIRE (Formaldehyde-Assisted Isolation of Regulatory Elements) assay, which allows the recovery of the soluble (i.e., nucleosome-free) fraction of the chromatin (Cockerill 2011; Ernst et al. 2011; Giresi et al. 2007; Neph et al. 2012; Sabo et al. 2006; Schones et al. 2008; Song et al. 2011; Thurman et al. 2012). Importantly, nucleosome positioning at *cis*-regulatory regions influences TFs occupancy because most TFs are unable to bind their cognate sites when embedded in a nucleosomal context. Because of the presence of clustered recognition sites for multiple TFs, enhancers can be functionally considered as platforms that recruit cooperating TFs (Spitz and Furlong 2012). Analysis of the genomic distribution of histone modifications revealed a specific chromatin signature of enhancers, characterized by high levels of monomethylation of histone H3 Lysine 4 (H3K4me1) in the absence of significant levels of trimethylation of the same residue (H3K4me3), which is instead highly enriched at promoters (Barski et al. 2007; Heintzman et al. 2007, 2009; Zhou et al. 2011). Additional marks associated with active enhancers include binding of histone acetyltransferases such as p300 and CBP and histone acetylation (particularly but not exclusively H3K27Ac) (Ghisletti et al. 2010; Heintzman et al. 2009; Visel et al. 2009). Interestingly, enhancers are also characterized by a distinctive nucleosomal composition, they being enriched in non-canonical histone variants, mainly the H2A variant H2A.Z and H3.3 (Barski et al. 2007; Calo and Wysocka 2013; Zlatanova and Thakar 2008). H2A.Z deposition may create domains of nucleosomal instability that probably facilitate initial TF binding events (Hu et al. 2012; Li et al. 2012). Notably, the annotation of *cis*-regulatory regions in many different cell types enabled by recent epigenomic profiling technologies clearly indicates that enhancers are the most dynamically utilized part of the genome, playing a crucial role in driving cell type-specific gene expression and being capable of activating transcription from great distances (Blow et al. 2010; Ernst et al. 2011; Heintzman et al. 2009; Pennacchio et al. 2007; Shen et al. 2012; Xi et al. 2007). Moreover, a recent large-scale genomic mapping across multiple cell lines revealed an enormous number of putative enhancer elements, indicating a great complexity in the way genomic regulatory information is used in different cells (Dunham et al. 2012). Almost the entire repertoire of enhancers in macrophages is constitutively bound by PU.1, a TF constantly expressed at high levels in macrophages and required to induce and to maintain their differentiation (Olson et al. 1995).

15.3.2 PU.1 Is the Myeloid Master Regulator

PU.1 *(Purine-rich box 1)* is exclusively expressed in cells of the hematopoietic lineage and belongs to the ETS family of TFs, one of the largest families of winged helix-loop-helix DNA-binding proteins. DNA- and protein-binding assays showed that ETS-binding profiles cluster into four distinct classes (I–IV) (Wei et al. 2010). PU.1 (SPI1) and its paralogs SPIB and SPIC recognize both in vitro and in vivo highly specific sequences that differ at a few critical positions (mainly at the 5' of the binding site) from the binding sites of all other ETS proteins (Wei et al. 2010). Several studies with PU.1 gene-disrupted mice indicate that PU.1 is a critical regulator of differentiation within the hematopoietic system and is particularly important for myeloid and B lymphocyte lineage development (Nerlov and Graf 1998). $Pu.1^{-/-}$ mice, which are born alive but die of severe septicemia within 48 h, are characterized by a normal amount of erythrocytes and megakaryocytes, but they lack mature myeloid and B cells (McKercher et al. 1996; Scott et al. 1994). More recently, conditional knockout mouse models indicated that PU.1 is not essential for myeloid and lymphoid lineage commitment, but it is absolutely required for the normal differentiation of most myeloid lineages and B cells (Carotta et al. 2010; Iwasaki et al. 2005). Importantly, development and function of B cells and myeloid cells are dependent on the precise regulation of PU.1 protein concentration (Rosenbauer and Tenen 2007). The expression of PU.1 is low at the beginning of differentiation but rises when progenitors become more lineage restricted, such as in common myeloid progenitors (CMPs) and common lymphoid progenitors (CLPs) (Dakic et al. 2005; Nutt et al. 2005). Upon further lineage differentiation, PU.1 is expressed at different levels in mature blood cells. Precisely, high levels favor macrophage differentiation, whereas about tenfold lower levels of PU.1 are associated with B-cell development (Bakri et al. 2005; Dahl et al. 2003; DeKoter and Singh 2000). Conversely, Pu.1 expression is downregulated during early erythroid and T-cell differentiation (Rosenbauer et al. 2006; Kueh et al. 2013).

15.3.3 PU.1 Controls the Genomic Regulatory Landscape in Macrophages

Genome-wide mapping of PU.1 binding revealed that its distribution is widespread in the macrophage genome (Ghisletti et al. 2010; Heinz et al. 2010). PU.1 binds dozens of thousands of genomic sites and it is constitutively associated with nearly all enhancers marked by H3K4me1 (Ghisletti et al. 2010; Heinz et al. 2010). Interestingly, PU.1 binding is able to promote the deposition of H3K4me1 and to create small open regions of accessible DNA that can be bound by other TFs, such as those activated by inflammatory stimuli. In this context, several experimental evidences have suggested that PU.1 might act as a pioneer factor during macrophage differentiation. Pioneer factors are functionally defined as sequence-specific DNA-binding proteins able to bind to their target sites when embedded in a nucleosomal context that is not permissive for binding of other TFs (Zaret and Carroll 2011). PU.1 expression in non-myeloid cells or in PU.1-negative myeloid progenitors is sufficient to induce nucleosome-free DNA sequences at the same genomic regions identified as enhancers in macrophages (Ghisletti et al. 2010; Heinz et al. 2010). In this context, PU.1 must be able to bind condensed chromatin and to attract chromatin-remodeling factors. However, biochemical evidence that PU.1 has the ability to invade inaccessible nucleosomal chromatin is still unavailable. Therefore, PU.1 is actively involved in determining the baseline accessible chromatin landscape of macrophages, thus enabling the recruitment of other TFs unable to invade nucleosomes, such as many of those responsive to environmental stimuli. In this regard, ChIP-seq technology was used to identify, throughout the genome, enhancers involved in the inflammatory response (Ghisletti et al. 2010). In this study, LPS-induced genomic occupancy of the transcription co-activator/protein acetyltransferase p300 was used to identify inducible enhancers in mouse macrophages, thereby revealing common features of enhancers that contribute to the inflammatory response. These genomic regions occupied by PU.1 in resting macrophages are bound after stimulation by inducible TFs (such as NF-kB and IRF family members) and recruit p300. These enhancer-bound factors are then able to interact with proteins recruited on the inflammatory gene promoters via DNA looping to induce the transcription of inflammatory response genes.

In this context, from a mechanistic point of view, PU.1 may have a role not only in controlling the establishment of the enhancer repertoire but also in the formation of intricate patterns of specific three-dimensional connections among promoters and sets of regulatory elements located up to hundreds of kilobases from each other in the linear genome (Fig. 15.1). Chromosome conformation capture (3C) has been used in HSC to understand to what extent PU.1 is involved in physical interactions between promoters and enhancers. PU.1 autoregulation through a distal regulatory element—the 14 kB URE (Upstream Regulatory Element)—has been demonstrated to be due to the formation of a chromosome loop that allows promoter–enhancer interactions and consequently gene activation (Staber et al. 2013) (Leddin et al. 2011). Also in dendritic cells, PU.1 was recently reported to control long-

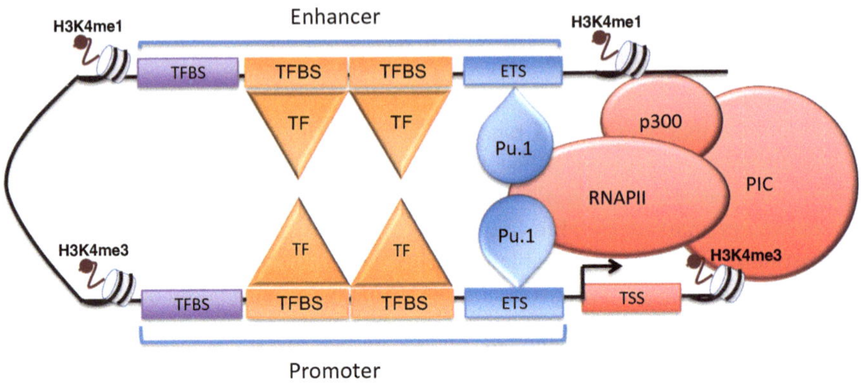

Fig. 15.1 Macrophage specific *cis*-regulatory repertoire. Enhancers and promoters are nucleosome-depleted regions characterized by multiple transcription factor binding sites (TFBSs). In macrophages, PU.1 marks cell type-specific *cis*-regulatory regions (characterized by chromatin marks permissive for transcription—e.g., H3K4me3 and H3K4me1) and recruits to the chromatin other TFs. Cooperative interactions with partner TFs might play an important role in specifying sites of PU.1 binding in macrophages. PU.1 may have a role not only in controlling the establishment of the enhancer repertoire but also in the formation of three-dimensional interactions between enhancer- and promoter-bound PU.1, by regulating DNA looping to promote the recruitment of the transcriptional machinery (RNAPII and PIC) to TSS

distant contacts between regulatory elements and the irf8 gene (Schonheit et al. 2013).

However, to what extent PU.1 is involved in globally controlling 3D chromatin contacts and to what extent these interactions are dynamically modified in response to stimulation remains to be defined.

15.3.4 Role of PU.1 as Pioneer Factor

The activation and the maintenance of a set of regulatory regions depend on the chromatin context, which represents a critical barrier affecting access of most TFs. A restricted class of TFs, pioneer factors, allows the regulation of gene expression in a specific nucleosomal context (Magnani et al. 2011; Zaret and Carroll 2011). Pioneer factors are functionally defined as sequence-specific DNA-binding proteins able to bind to nucleosomal target sites that are not permissive for binding of other TFs (Zaret and Carroll 2011). The role of pioneer factors is not restricted to the control of inducible genes in mature cells, but they are also essential for developmental and differentiation programs, acting as placeholders that will be replaced by other TFs at later stages of development (Cirillo and Zaret 1999; McPherson et al. 1993; Zaret and Carroll 2011). Recent experimental evidence suggests that PU.1 might act as a pioneer factor during macrophage differentiation. First of all, PU.1 expression in non-myeloid cells or in PU.1-negative myeloid progenitors is sufficient to induce nucleosome-free DNA sequences at the same genomic regions identified as enhancers in macrophages (Ghisletti et al. 2010; Heinz et al. 2010). In the majority of cases, binding of PU.1 alone is indeed sufficient to promote chromatin changes, whereas in some cases the generation of cell type-specific open regions requires the collaborative interaction between PU.1 and other TFs, such as C/EBPβ (Heinz et al. 2010). Therefore, PU.1 binding is able to promote the deposition of H3K4me1 and to create small open regions of accessible DNA that can be bound by other TFs, such as those activated by inflammatory stimuli. In this context, PU.1 must be able to bind condensed chromatin and to attract chromatin-remodeling factors. However, biochemical evidence that PU.1 has the ability to invade inaccessible nucleosomal chromatin is still lacking. In conclusion, PU.1 is actively involved in determining the baseline accessible chromatin landscape of macrophages, thus enabling the recruitment of other TFs unable to invade nucleosomes, such as many of those responsive to environmental stimuli.

15.3.5 The Enhancer Repertoire Dynamically Changes Upon Stimulation

Altogether, the experimental data discussed above suggest that macrophages have a specific enhancer repertoire that is predetermined and established prior to receiving a stimulus, since *cis*-regulatory regions controlling inducible genes become associated with TFs involved in lineage commitment and maintenance of cell identity at an early stage of differentiation (Lichtinger et al. 2012; Natoli 2010). However, several studies have reported dynamic changes in chromatin marks in response to external inputs that usually correlate with transcriptional activity. In this scenario, H3K4me1 represents a general mark of distal regulatory regions and additional modifications (mainly acetylation) can distinguish between those enhancers that are active and those that are poised and can subsequently be activated during developmental transitions or in response to stimulation. Therefore, enhancers have been classified as active, poised, and intermediate depending on their chromatin signature (Creyghton et al. 2010; Rada-Iglesias et al. 2011). Enhancers associated with active genes are characterized by high level of H3K4me1 and H3K27ac, as opposed to inactive and poised enhancers containing H3K4me1 only. Poised enhancers display high levels of H3K4me1, but they are negative for H3K27Ac (and in some cases may also be positive for the repressive mark H3K27me3) (Zentner et al. 2011). Therefore, enhancers can be considered dynamic regulatory regions that gradually assemble through developmental stages and whose function is modulated by environmental stimuli. In the inflammatory response, external inputs have a broad impact on chromatin organization and marking at *cis*-regulatory elements. In macrophages, the p300 histone acetyltransferase is recruited to a specific subset of enhancers that are activated by LPS (Ghisletti et al. 2010). Moreover, a recent genome-wide analysis confirmed that thousands of *cis*-

regulatory regions gain H4 acetylation upon the proinflammatory stimulus LPS (Chen et al. 2012). These studies also suggested a specific and non-redundant role of the histone deacetylase HDAC3 in controlling acetylation levels at a subset of genomic regions (Chen et al. 2012).

In spite of such dynamic changes in acetylation, the repertoire of enhancers was thought to be fixed in a specific cell type because of the activity of TFs involved in cell specification. It was recently reported that in macrophages proinflammatory stimuli cause more than simple changes in acetylation and in fact result in the reorganization of a subset of enhancers (Ostuni et al. 2013). In this study, genomic regions that gained enhancers marks in response to stimulation were identified and termed *latent* enhancers (Ostuni et al. 2013). These results indicated that macrophages can acquire new genomic regulatory properties in response to stimulation. These latent enhancers are unmarked in the unperturbed state, but they recruit the master regulator PU.1 and stimulus-specific TFs only after stimulation. Therefore, PU.1 can bind latent enhancers only by cooperating with other TFs activated by stimulation (e.g., STAT1 and STAT6), likely because these regulatory regions have a low-affinity binding site for PU.1(Ostuni et al. 2013). Once the stimulus has ceased, many of these latent enhancers do not return to the original state, but retain some enhancer marks and remain poised for subsequent stimulation (Ostuni et al. 2013). Therefore, environmental stimuli can reorganize in a selective manner the enhancer repertoire.

15.3.6 Combinatorial Control at Cis*-regulatory Regions*

As discussed above, in macrophages PU.1 controls the establishment and the maintenance of the enhancer repertoire. However, how it cooperates at a genomic level with other TFs to define or activate specific subsets of enhancers is still poorly understood. From a mechanistic point of view, TFs acting as lineage-restricted organizers of the genomic regulatory information need additional restricted or non-restricted TFs to bind and activate specific subsets of enhancers. To what extent PU.1 distribution in macrophages requires cooperative binding is still unknown and specific rules controlling functional cooperation between TFs remain to be defined. In this context, relevant insights have been obtained in a model of hematopoietic differentiation, by measuring genome-wide dynamics of TF assembly on their target genes at a genome-wide scale (Lichtinger et al. 2012). In this model, the expression of the TF RUNX1 in hemogenic endothelium (HE) has been demonstrated to regulate the assembly of a hematopoiesis-specific global pattern of transcription factor binding. Similarly, PU.1 recruitment should regulate the assembly of specific TFs to regulatory regions in different cell types. In B cells, where the concentration of PU.1 is several fold lower than in macrophages as previously discussed, PU.1 binding has been shown to depend on cooperative interactions with partner TFs (like E2A and EBF) that are B cell specific (Heinz et al. 2010). Thus, PU.1 distribution in B cells displays little overlap with macrophages, where it might cooperate with different TFs to activate specific subsets of enhancers. Therefore, cooperative interactions with partner TFs likely play an important role also in specifying sites of PU.1 binding in macrophages, but its higher levels of expression may both reduce dependence on other TFs and determine the selection of a different genomic repertoire of sites.

According to one recent study performed in dendritic cells, three conceptually distinguishable broad classes of TFs cooperate to activate *cis*-active regulatory elements. A high-throughput chromatin immunoprecipitation method (HT-ChIP) was applied to build genome-wide dynamic maps of 25 TFs at different time points following LPS stimulation (Garber et al. 2012). The first class included TFs with a very pervasive association with almost all regulatory elements in the genome, such as PU.1 and C/EBPβ. The broad distribution of these TFs is compatible with their role as chromatin openers that facilitate access of a second group of TFs that were termed *primers* and included JUNB, IRF4, and ATF3. This name is justified by the notion that they are able to prime for activation regions that are associated with stimulus-dependent gene induction (Garber et al. 2012). A third set of TFs (that includes NF-kB and STATs family members) bind dynamically to specific set of genes in a stimulus-dependent manner and control the induction of gene expression (Garber et al. 2012).

15.4 Conclusions

Inflammation is a physiological and fundamental response of the organism to counteract infections and invasive events, to activate lymphocytes, to promote wound healing, and to repair damaged tissues (Murray and Smale 2012). On the other hand, the inflammatory response must be tightly regulated to prevent molecular, cellular, and organ damage (Okin and Medzhitov 2012). For example, reactive oxygen species and nitrogen intermediates, amongst others, produced by activated macrophages are highly toxic for the invading agents, but they are also potentially harmful for tissues and organs where the infection takes place (Serbina et al. 2008). As a consequence, when the response becomes excessive in magnitude or duration, such as in sepsis or chronic inflammation, the normally protective role of inflammation becomes detrimental. In addition to chronic inflammatory and autoimmune disease, such as Crohn's disease, multiple sclerosis, and rheumatoid arthritis, chronic inflammation has been associated with diseases such as obesity, cardiovascular and neurodegeneration, as well as cancer (Kawane et al. 2006; Murphy et al. 2003; Smith et al. 2009). The pathogenesis of chronic autoimmune diseases and cancer for example is often associated with sustained production of inflammatory cytokines that are normally transiently expressed. For these reasons, the inflammatory response must be finely regulated in order to suppress genes responsible for inflammation-associated pathologies while maintaining a robust response to microbial infection. Therefore, a full understanding of the mechanisms by which macrophages achieve the transcriptional selectivity of the inflammatory response

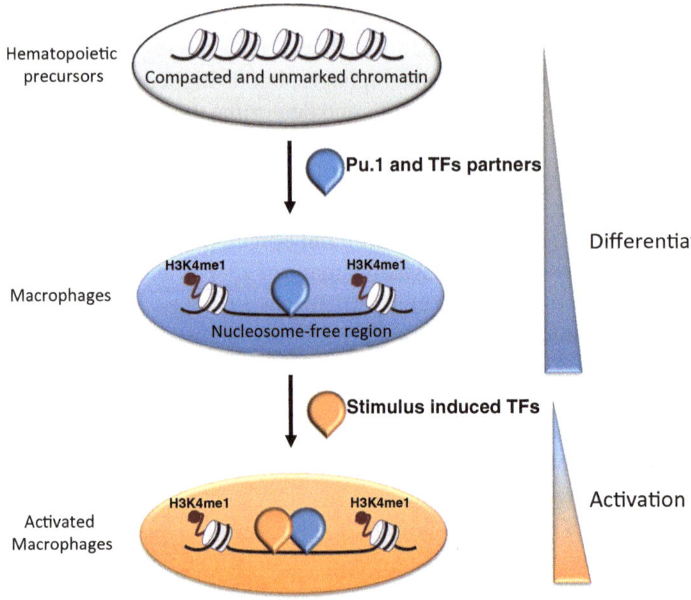

Fig. 15.2 Enhancer landscape in macrophages is achieved during terminal differentiation and activation. The enhancer repertoire is established during differentiation by the combined activity of transcription factors involved in lineage determination, such as PU.1 and its partners. PU.1 binding is sufficient to promote the deposition of H3K4me1 in a nucleosome-free region. Environmental stimuli (e.g., response to LPS) trigger the recruitment of stimulus-induced TFs, such as NF-kB, AP-1, IRFs, and STATs family members, to a specific set of proinflammatory genes

becomes crucial. Recent advances in genome-scale methods have allowed to uncover that different chromatin organization can contribute to the selectivity of the inflammatory response. As we have summarized here, the more complete understanding of the genomic regulatory elements used for transcriptional regulation in macrophages and the detailed description of the genomic distribution of TFs involved in macrophage specification and function have provided clues on how transcriptional selectivity in the inflammatory response is achieved.

The emerging picture is that competence for the response to inflammatory stimuli is preprogrammed by TFs involved in macrophage commitment and maintenance of the differentiated state; the molecular counterpart of this preprogramming is represented by the generation of regions of stably open chromatin that enable the recruitment of TFs activated by stimulation (a simplified model is represented in Fig. 15.2). Nonetheless, insights into the complex interplay between TFs, epigenomic modifications, and genome organization are still lacking, and new technologies and approaches will be required to obtain them. Specifically, it will be critical to delineate in the future the complete panel of TFs that contribute to the organization and maintenance of macrophage-specific regulatory networks; it will also be crucial to understand if and how they control the three-dimensional network of interactions between distal regulatory elements and target genes.

Noncoding RNAs generated at genomic regulatory regions as well as long noncoding RNAs may provide additional regulatory layers impacting macrophage specification and function (De Santa et al. 2010; Kim et al. 2010; Natoli and Andrau 2011).

References

Austenaa L et al (2012) The histone methyltransferase Wbp7 controls macrophage function through GPI glycolipid anchor synthesis. Immunity 36:572–585

Ayton PM, Chen EH, Cleary ML (2004) Binding to nonmethylated CpG DNA is essential for target recognition, transactivation, and myeloid transformation by an MLL oncoprotein. Mol Cell Biol 24:10470–10478

Bakri Y et al (2005) Balance of MafB and PU.1 specifies alternative macrophage or dendritic cell fate. Blood 105:2707–2716

Barski A et al (2007) High-resolution profiling of histone methylations in the human genome. Cell 129:823–837

Bhatt DM et al (2012) Transcript dynamics of proinflammatory genes revealed by sequence analysis of subcellular RNA fractions. Cell 150:279–290

Blow MJ et al (2010) ChIP-Seq identification of weakly conserved heart enhancers. Nat Genet 42:806–810

Bulger M, Groudine M (2011) Functional and mechanistic diversity of distal transcription enhancers. Cell 144:327–339

Calo E, Wysocka J (2013) Modification of enhancer chromatin: what, how, and why? Mol Cell 49:825–837

Carotta S et al (2010) The transcription factor PU.1 controls dendritic cell development and Flt3 cytokine receptor expression in a dose-dependent manner. Immunity 32:628–641

Chen X et al (2012) Requirement for the histone deacetylase Hdac3 for the inflammatory gene expression program in macrophages. Proc Natl Acad Sci U S A 109:E2865–E2874

Cirillo LA, Zaret KS (1999) An early developmental transcription factor complex that is more stable on nucleosome core particles than on free DNA. Mol Cell 4:961–969

Cockerill PN (2011) Structure and function of active chromatin and DNase I hypersensitive sites. FEBS J 278:2182–2210

Creyghton MP et al (2010) Histone H3K27ac separates active from poised enhancers and predicts developmental state. Proc Natl Acad Sci U S A 107:21931–21936

Dahl R et al (2003) Regulation of macrophage and neutrophil cell fates by the PU.1:C/EBPalpha ratio and granulocyte colony-stimulating factor. Nat Immunol 4:1029–1036

Dakic A et al (2005) PU.1 regulates the commitment of adult hematopoietic progenitors and restricts granulopoiesis. J Exp Med 201:1487–1502

De Santa F et al (2010) A large fraction of extragenic RNA pol II transcription sites overlap enhancers. PLoS Biol 8:e1000384

Deaton AM, Bird A (2009) CpG islands and the regulation of transcription. Genes Dev 25:1010–1022

DeKoter RP, Singh H (2000) Regulation of B lymphocyte and macrophage development by graded expression of PU.1. Science 288:1439–1441

Dunham I et al (2012) An integrated encyclopedia of DNA elements in the human genome. Nature 489:57–74

Ernst J et al (2011) Mapping and analysis of chromatin state dynamics in nine human cell types. Nature 473:43–49

Fenouil R et al (2012) CpG islands and GC content dictate nucleosome depletion in a transcription-independent manner at mammalian promoters. Genome Res 22:2399–2408

Garber M et al (2012) A high-throughput chromatin immunoprecipitation approach reveals principles of dynamic gene regulation in mammals. Mol Cell 47:810–822
Geissmann F et al (2010) Development of monocytes, macrophages, and dendritic cells. Science 327:656–661
Ghisletti S, Natoli G (2013) Deciphering cis-regulatory control in inflammatory cells. Philos Trans R Soc Lond B Biol Sci 368:20120370
Ghisletti S et al (2010) Identification and characterization of enhancers controlling the inflammatory gene expression program in macrophages. Immunity 32:317–328
Giresi PG et al (2007) FAIRE (Formaldehyde-assisted isolation of regulatory elements) isolates active regulatory elements from human chromatin. Genome Res 17:877–885
Gordon S, Taylor PR (2005) Monocyte and macrophage heterogeneity. Nat Rev Immunol 5:953–964
Hargreaves DC, Horng T, Medzhitov R (2009) Control of inducible gene expression by signal-dependent transcriptional elongation. Cell 138:129–145
Heintzman ND et al (2007) Distinct and predictive chromatin signatures of transcriptional promoters and enhancers in the human genome. Nat Genet 39:311–318
Heintzman ND et al (2009) Histone modifications at human enhancers reflect global cell-type-specific gene expression. Nature 459:108–112
Heinz S et al (2010) Simple combinations of lineage-determining transcription factors prime cis-regulatory elements required for macrophage and B cell identities. Mol Cell 38:576–589
Herschman HR (1991) Primary response genes induced by growth factors and tumor promoters. Annu Rev Biochem 60:281–319
Hu G et al (2012) H2A.Z facilitates access of active and repressive complexes to chromatin in embryonic stem cell self-renewal and differentiation. Cell Stem Cell 12:180–192
Iwasaki H et al (2005) Distinctive and indispensable roles of PU.1 in maintenance of hematopoietic stem cells and their differentiation. Blood 106:1590–1600
Kawai T, Akira S (2010) The role of pattern-recognition receptors in innate immunity: update on Toll-like receptors. Nat Immunol 11:373–384
Kawane K et al (2006) Chronic polyarthritis caused by mammalian DNA that escapes from degradation in macrophages. Nature 443:998–1002
Kim TK et al (2010) Widespread transcription at neuronal activity-regulated enhancers. Nature 465:182–187
Kueh HY et al (2013) Positive feedback between PU.1 and the cell cycle controls myeloid differentiation. Science 341:670–673
Lawrence T, Natoli G (2011) Transcriptional regulation of macrophage polarization: enabling diversity with identity. Nat Rev Immunol 11:750–761
Leddin M et al (2011) Two distinct auto-regulatory loops operate at the PU.1 locus in B cells and myeloid cells. Blood 117:2827–2838
Lee JH, Skalnik DG (2005) CpG-binding protein (CXXC finger protein 1) is a component of the mammalian Set1 histone H3-Lys4 methyltransferase complex, the analogue of the yeast Set1/COMPASS complex. J Biol Chem 280:41725–41731
Lee JH, Skalnik DG (2008) Wdr82 is a C-terminal domain-binding protein that recruits the Setd1A Histone H3-Lys4 methyltransferase complex to transcription start sites of transcribed human genes. Mol Cell Biol 28:609–618
Li Z et al (2012) Foxa2 and H2A.Z mediate nucleosome depletion during embryonic stem cell differentiation. Cell 151:1608–1616
Lichtinger M et al (2012) RUNX1 reshapes the epigenetic landscape at the onset of haematopoiesis. EMBO J 31:4318–4333
Magnani L, Eeckhoute J, Lupien M (2011) Pioneer factors: directing transcriptional regulators within the chromatin environment. Trends Genet 27:465–474
McKercher SR et al (1996) Targeted disruption of the PU.1 gene results in multiple hematopoietic abnormalities. EMBO J 15:5647–5658

McPherson CE et al (1993) An active tissue-specific enhancer and bound transcription factors existing in a precisely positioned nucleosomal array. Cell 75:387–398
Medzhitov R (2008) Origin and physiological roles of inflammation. Nature 454:428–435
Medzhitov R, Horng T (2009) Transcriptional control of the inflammatory response. Nat Rev Immunol 9:692–703
Mosser DM, Edwards JP (2008) Exploring the full spectrum of macrophage activation. Nat Rev Immunol 8:958–969
Murphy CA et al (2003) Divergent pro- and antiinflammatory roles for IL-23 and IL-12 in joint autoimmune inflammation. J Exp Med 198:1951–1957
Murray PJ, Smale ST (2012) Restraint of inflammatory signaling by interdependent strata of negative regulatory pathways. Nat Immunol 13:916–924
Natoli G (2010) Maintaining cell identity through global control of genomic organization. Immunity 33:12–24
Natoli G, Andrau JC (2011) Noncoding transcription at enhancers: general principles and functional models. Annu Rev Genet 46:1–19
Natoli G, Ghisletti S, Barozzi I (2011) The genomic landscapes of inflammation. Genes Dev 25:101–106
Neph S et al (2012) An expansive human regulatory lexicon encoded in transcription factor footprints. Nature 489:83–90
Nerlov C, Graf T (1998) PU.1 induces myeloid lineage commitment in multipotent hematopoietic progenitors. Genes Dev 12:2403–2412
Nutt SL et al (2005) Dynamic regulation of PU.1 expression in multipotent hematopoietic progenitors. J Exp Med 201:221–231
Okin D, Medzhitov R (2012) Evolution of inflammatory diseases. Curr Biol 22:R733–R740
Olson MC et al (1995) PU. 1 is not essential for early myeloid gene expression but is required for terminal myeloid differentiation. Immunity 3:703–714
Ostuni R et al (2013) Latent enhancers activated by stimulation in differentiated cells. Cell 152:157–171
Pennacchio LA et al (2007) Predicting tissue-specific enhancers in the human genome. Genome Res 17:201–211
Rada-Iglesias A et al (2011) A unique chromatin signature uncovers early developmental enhancers in humans. Nature 470:279–283
Ramirez-Carrozzi VR et al (2009) A unifying model for the selective regulation of inducible transcription by CpG islands and nucleosome remodeling. Cell 138:114–128
Rosenbauer F, Tenen DG (2007) Transcription factors in myeloid development: balancing differentiation with transformation. Nat Rev Immunol 7:105–117
Rosenbauer F et al (2006) Lymphoid cell growth and transformation are suppressed by a key regulatory element of the gene encoding PU.1. Nat Genet 38:27–37
Sabo PJ et al (2006) Genome-scale mapping of DNase I sensitivity in vivo using tiling DNA microarrays. Nat Methods 3:511–518
Schones DE et al (2008) Dynamic regulation of nucleosome positioning in the human genome. Cell 132:887–898
Schonheit J et al (2013) PU.1 level-directed chromatin structure remodeling at the Irf8 gene drives dendritic cell commitment. Cell Rep 3:1617–1628
Scott EW et al (1994) Requirement of transcription factor PU.1 in the development of multiple hematopoietic lineages. Science 265:1573–1577
Serbina NV et al (2008) Monocyte-mediated defense against microbial pathogens. Annu Rev Immunol 26:421–452
Shen Y et al (2012) A map of the cis-regulatory sequences in the mouse genome. Nature 488:116–120
Smale ST (2010) Selective transcription in response to an inflammatory stimulus. Cell 140:833–844

Smith AM et al (2009) Disordered macrophage cytokine secretion underlies impaired acute inflammation and bacterial clearance in Crohn's disease. J Exp Med 206:1883–1897
Song L et al (2011) Open chromatin defined by DNaseI and FAIRE identifies regulatory elements that shape cell-type identity. Genome Res 21:1757–1767
Spitz F, Furlong EE (2012) Transcription factors: from enhancer binding to developmental control. Nat Rev Genet 13:613–626
Staber PB et al (2013) Sustained PU.1 levels balance cell-cycle regulators to prevent exhaustion of adult hematopoietic stem cells. Mol Cell 49:934–946
Takeuchi O, Akira S (2010) Pattern recognition receptors and inflammation. Cell 140:805–820
Thurman RE et al (2012) The accessible chromatin landscape of the human genome. Nature 489:75–82
Valouev A et al (2011) Determinants of nucleosome organization in primary human cells. Nature 474:516–520
Visel A et al (2009) ChIP-seq accurately predicts tissue-specific activity of enhancers. Nature 457:854–858
Wei GH et al (2010) Genome-wide analysis of ETS-family DNA-binding in vitro and in vivo. EMBO J 29:2147–2160
Xi H et al (2007) Identification and characterization of cell type-specific and ubiquitous chromatin regulatory structures in the human genome. PLoS Genet 3:e136
Zaret KS, Carroll JS (2011) Pioneer transcription factors: establishing competence for gene expression. Genes Dev 25:2227–2241
Zentner GE, Tesar PJ, Scacheri PC (2011) Epigenetic signatures distinguish multiple classes of enhancers with distinct cellular functions. Genome Res 21:1273–1283
Zhou VW, Goren A, Bernstein BE (2011) Charting histone modifications and the functional organization of mammalian genomes. Nat Rev Genet 12:7–18
Zlatanova J, Thakar A (2008) H2A.Z: view from the top. Structure 16:166–179

Glossary

5-Aza-2-deoxy-cytidine (Decitabine) A cytosine in which the 5 carbon of the cytosine ring has been replaced with nitrogen. Decitabine is exclusively incorporated into DNA, inhibiting mammalian DNA methyltransferases.

5-Azacytidine (AZA) A cytidine RNA analog in which the 5 carbon of the cytosine ring has been replaced with nitrogen. 5-Azacytidine can be incorporated into RNA, and after metabolic activation also into DNA, where it functions as an inhibitor of mammalian DNA methyltransferases.

Acetylation (ac) The enzymatic introduction of an acetyl group to an organic compound, for instance, histones.

Acquired immune deficiency syndrome (AIDS) A disease of the human immune system caused by the human immunodeficiency virus (HIV). Presently, there is no cure or vaccine for AIDS; however, antiretroviral treatment can slow the course of the disease and can lead to a near-normal life expectancy.

Acute lymphocytic leukemia (ALL) A cancer of lymphocyte lineage whose rapid growth interferes with the production of normal blood cells in the bone marrow. ALL is a common acute leukemia affecting both children and adults.

Acute myeloid leukemia (AML) A cancer of the myeloid line of white blood cells whose rapid growth interferes with the production of normal blood cells in the bone marrow. AML is the most common acute leukemia affecting adults, and its incidence increases with age.

Acute promyelocytic leukemia (APL) A subtype of AML, a cancer of the blood and bone marrow. Since there is an abnormal accumulation of immature granulocytes called promyelocytes in APL, it is also known as acute progranulocytic leukemia. APL is responsive to all-trans retinoic acid therapy.

Alleles Different variants or copies of a gene. For most genes on the chromosomes, there are two copies: one copy inherited from the mother and the other from the father. The DNA sequence of each of these copies may be different because of genetic polymorphisms.

Alpha-thalassemia/mental retardation syndrome X-linked (ATRX) A protein that belongs to the switch/sucrose nonfermentable (SWI/SNF) family of chromatin-remodeling molecules that facilitates gene expression by allowing

C. Bonifer and P.N. Cockerill (eds.), *Transcriptional and Epigenetic Mechanisms Regulating Normal and Aberrant Blood Cell Development*, Epigenetics and Human Health, DOI 10.1007/978-3-642-45198-0, © Springer-Verlag Berlin Heidelberg 2014

transcription factors to gain access to their targets in chromatin. Mutations in *ATRX* alter DNA methylation and are associated with an X-linked mental retardation syndrome that is often accompanied by ATRX syndrome.

Antigen (Ag) A region of a molecule that is recognized by a specific Ig or TCR. Ags allow the immune system to recognize specific pathogens, and different individual B cells or T cells have the ability to acquire different Ag specificities.

Ataxia telangiectasia mutated (ATM) A serine/threonine protein kinase that is recruited and activated by DNA double-strand breaks. It phosphorylates several key proteins, thereby activating DNA damage checkpoint delay. This results in cell cycle arrest and subsequent DNA repair or apoptosis.

Basic helix-loop-helix (bHLH) The basic helix-loop-helix motif is characterized by two α-helices connected by a loop. bHLH proteins normally bind to a consensus sequence called an E-box. The canonical E-box, CACGTG, is palindromic; however, some bHLH transcription factors bind to related non-palindromic sequences that are similar to the E-box.

Bisulfite sequencing (BS) A procedure in which sodium bisulfite is used to deaminate cytosine to uracil in genomic DNA. Conditions are chosen so that 5-methylcytosine is not changed. PCR amplification and subsequent DNA sequencing then reveal the exact position of cytosines that are methylated in genomic DNA.

Bivalent chromatin A chromatin region that is modified by a combination of histone modifications such that it represses gene transcription, but at the same time retains the potential of acquiring gene expression.

Bromo domain Protein motif found in a variety of nuclear proteins, including transcription factors and HATs involved in transcriptional activation. Bromodomains bind to histone tails carrying acetylated lysine residues.

cAMP response element binding protein (CREB) A transcriptional activator for many immediate early genes.

Cell fate The programmed path of cell differentiation. Although all cells have the same DNA, their cell fate can be different. Some cells develop into the brain, whereas others are the precursors of blood. Cell fate is determined in part by the organization of chromatin—DNA and the histone proteins—in the nucleus.

Cellular memory (epigenetic) Specific active and repressive organizations of chromatin can be maintained from one cell to its daughter cells; this is called epigenetic inheritance. It ensures that specific states of gene expression are inherited over many cell generations.

ChIP-chip After chromatin immunoprecipitation, DNA is purified from the immunoprecipitated chromatin fraction and hybridized on arrays of short DNA fragments representing specific regions of the genome.

ChIP-seq Sequencing of the totality of DNA fragments obtained by ChIP using next-generation sequencing to quantify patterns of enrichment across the genome.

Chromatid In each somatic cell generation, the genomic DNA is replicated in order to make two copies of each individual chromosome. During the M phase of the cell cycle, these copies—called chromatids—are microscopically visible and next to each other before they get distributed to the daughter cells.

Chromatin The nucleo-protein-complex constituting the chromosomes in eukaryotic cells. Structural organization of chromatin is complex and involves different levels of compaction. The lowest level of compaction is represented by an extended array of nucleosomes.

Chromatin conformation capture assay (3C and Hi-C) 3C is a cross-linking technique that is useful for better understanding gene regulation which identifies interactions between regulatory elements such as enhancers and promoters. Hi-C is high-throughput next-generation sequencing version of this technique used to analyze the long-range organization of chromosomes and their interactions.

Chromatin immunoprecipitation (ChIP) This is a method for examining protein–DNA interactions occurring in the cell. DNA-binding proteins are cross-linked to the DNA and enriched using antibodies with specific affinity to particular proteins (e.g., histones) or covalent modifications on proteins. After ChIP, the genomic DNA is purified from the chromatin fragments brought down by the antiserum and analyzed by qPCR, microarray (ChIP-chip), or next-generation sequencing (ChIP-seq).

Chromatin remodeling Locally, the organization and compaction of chromatin can be altered by different enzymatic machineries. This is called chromatin remodeling. Several chromatin-remodeling proteins move nucleosomes along the DNA and require ATP for their action.

Chromodomain (chromatin organization modifier domain) A protein–protein interaction motif first identified in *Drosophila melanogaster* HP1 and polycomb group proteins. It is also found in other nuclear proteins involved in transcriptional silencing and heterochromatin formation. Chromodomains consist of approximately 50 amino acids that bind to histone tails methylated at certain lysine residues.

Chromosomal domain It is often observed in higher eukaryotes that chromatin is organized (e.g., by histone methylation) the same way across hundreds to thousands of kilobases of DNA. These "chromosomal domains" can comprise multiple genes that are similarly expressed. Some chromosomal domains are controlled by genomic imprinting.

Copy number variation (CNV) Alterations in the DNA of a genome that results in a cell having an increased or decreased number of copies of one or more sections of the DNA. These variations range from kilobases to megabases in size.

CpG dinucleotide A cytosine followed by a guanine in the sequence of bases of the DNA. Cytosine methylation in mammals occurs primarily at CpG dinucleotide positions.

CpG island (CGI) A small stretch of DNA, several hundred bases up to several kilobases in size, that is particularly rich in CpG dinucleotides, and is also relatively enriched in cytosines and guanines. Most CpG islands comprise promoter sequences that drive the expression of genes.

CpG island methylator phenotype (CIMP) Cancers can be classified according to the degree of methylation in their genome. Those with high degrees of methylation are referred to as having a CpG island methylator phenotype and are characterized by epigenetic instability.

CREB-binding protein (CBP) A protein involved in transcriptional regulation that is often associated with histone acetyltransferases such as p300.

Cytosine methylation DNA methylation in mammals occurs at cytosines that are part of CpG dinucleotides. As a consequence of the palindromic nature of the CpG sequence, methylation is symmetrical and affects both strands of DNA at a methylated target site. When present at promoters, it is usually associated with transcriptional repression.

Deacetylation The removal of acetyl groups from proteins. Deacetylation of histones is often associated with gene repression and is mediated by histone deacetylases (HDACs).

de novo DNA methylation The addition of methyl groups to a stretch of DNA that is not yet methylated.

Deoxyribonucleic acid (DNA) A molecule encoding the genetic instructions used in the development and function of all known living organisms and many viruses.

Differentially methylated region (DMR) A segment of DNA generally rich in cytosine and guanine nucleotides, with the cytosine nucleotides methylated on only one parental allele. DNA methylation of these regulatory elements is parent-of-origin dependent when they regulate the mono-allelic expression of imprinted genes.

Disomy The occurrence in the cell of two copies of a chromosome, or part of a chromosome, that are identical and of the same parental origin (i.e., uniparental disomy).

DNA demethylation Removal of methyl groups from the DNA. This can occur actively by an enzymatically mediated process, or passively when methylation is not maintained after DNA replication.

DNA methylation A biochemical modification of DNA resulting from the addition of a methyl group to either adenine or cytosine bases. Methylation in mammals is essentially confined to cytosines that are in CpG dinucleotides. Methyl groups can be removed from DNA by DNA demethylation.

DNA methyltransferase The enzyme that adds new (de novo) methylation to the DNA or maintains existing patterns of DNA methylation.

Double-strand break (DSB) A break in double-stranded DNA in which both strands are cleaved can result in mutagenic events or cell death if left unrepaired or repaired inappropriately.

Double-stranded RNA (dsRNA) RNA with two complementary strands; it is similar to the DNA found in all cells. dsRNA forms the genetic material of double-stranded RNA viruses.

Down syndrome A chromosomal condition caused by the presence of all or part of a third copy of chromosome 21. This syndrome is named after John Langdon

Down, the British physician who described it in 1866. Down syndrome is the most common chromosome abnormality in humans. It is typically associated with a delay in cognitive ability and physical growth and a particular set of facial characteristics.

Eight-twenty-one (ETO) This gene derives its name from its association with many cases of acute myelogenous leukemia (AML) in which a reciprocal translocation, t(8;21), brings together a large portion of the *ETO* gene from chromosome eight and part of the *AML1* gene from chromosome 21.

Embryo (EMB) A multicellular diploid eukaryote in its earliest stage of development. In humans, it is called an embryo until about 8 weeks after fertilization, and then it is called a fetus.

Embryonic stem (ES) cells Cultured cells obtained from the inner cell mass of the blastocyst. These cells are totipotent and can be differentiated into all of the different somatic cell lineages.

Enhancer A small, specialized sequence of DNA which, when recognized by specific regulatory proteins, can enhance the activity of the promoter of a gene (s) located in close proximity.

Enhancer RNA (eRNA) Enhancer regions can produce their own RNA or eRNA that can intensify the ability of cells to produce specific protein coding transcripts.

Epigenetic code Patterns of DNA methylation and histone modifications can modify the way genes on the chromosomes are expressed. This led to the idea that combinations of epigenetic modifications constitute a code on top of the genetic code that modulates gene expression and can be recognized by specific non-histone proteins.

Epigenetic inheritance The somatic inheritance, or inheritance through the germ line, of epigenetic information (i.e., changes that affect gene function without the occurrence of an alteration in the DNA sequence).

Epigenetic marks Regional modifications of DNA and chromatin proteins. This includes DNA methylation and histone methylation that can be maintained from one cell generation to the next and may affect the way genes are expressed.

Epigenetic reprogramming The resetting of epigenetic marks on the genome so that they become like those of another cell type or of another developmental stage. Epigenetic reprogramming occurs in primordial germ cells brought back to a "ground state." Epigenetic reprogramming and dedifferentiation also occur after somatic cell nuclear transfer.

Epigenetics The study of heritable changes in gene function that arise without an apparent change in the genomic DNA sequence. Epigenetic mechanisms are involved in the formation and maintenance of cell lineages during development and in X-inactivation and genomic imprinting; they are frequently perturbed in diseases.

Epigenome The epigenome is the overall epigenetic state of a particular cell. In the developing embryo, each cell type has a different epigenome. Epigenome maps represent the presence of DNA methylation, histone modification, and other chromatin modifications along the chromosomes.

Epigenome-wide association studies (EWAS) The principle of epigenome-wide association studies involves scanning cases and controls to identify epigenetic variations associated with a specific trait or disease.

Epigenotype The totality of epigenetic marks that are found along the DNA sequence of the genome in a particular cell lineage or at a particular developmental stage.

Epimutation A change in the normal epigenetic marking of a gene or regulatory DNA sequence (e.g., DNA methylation) that affects gene expression.

Escape of X-inactivation Regions and genes on the X-chromosomes that are not affected by the dosage compensation/X-inactivation mechanism and remain active on both X-chromosomes in females.

Euchromatin A type of chromatin that appears lightly stained when observed through the microscope at interphase. Euchromatic chromosomal domains are loosely compacted and relatively rich in genes. The opposite type of chromatin organization is heterochromatin.

Fluorescent in situ hybridization (FISH) A cytogenetic technique that uses fluorescent probes to detect and localize the presence or absence of specific DNA sequences on chromosomes.

Genome The entirety of an organism's hereditary information that is encoded either in DNA or in RNA for many types of viruses. The genome includes both the genes and the noncoding sequences of the DNA.

Genome-wide association study (GWAS) An examination of all or most of the genes in groups of individuals different for a specific trait or disease in order to identify DNA sequence-based factors that contribute to the origin of such phenotypes.

Genomic imprinting An epigenetic phenomenon that affects a small subset of genes in the genome of Therian mammals and results in mono-allelic gene expression in a parent-of-origin-dependent manner.

Glucocorticoid receptor (GR) A receptor encoded by *NR3C1* that glucocorticoids (e.g., cortisol) bind to it. The GR regulates genes that modulate development, metabolism, immune functions, and stress response.

Glucocorticoids Steroid hormones that bind to the glucocorticoid receptor (GR) and affect development, immunological functions, metabolic processes, and stress response.

Green fluorescent protein (GFP) A protein composed of 238 amino acids, and first isolated from the jellyfish, *Aequorea victoria*. It exhibits bright green fluorescence when exposed to light in the blue to ultraviolet range. GFP is frequently used as a reporter of gene expression.

Heterochromatin A type of chromatin that appears dark when observed through the microscope at interphase. Heterochromatic chromosomal domains, found in all cell types, are highly compacted, are rich in repeat sequences, and show little or no gene expression. Extended regions of heterochromatin are found close to centromeres and at telomeres.

Histone acetylation Posttranslational modification of the ε-amino group of lysine residues in histones that is catalyzed by a family of enzymes called histone acetyltransferases (HATs). Acetylation contributes to the formation of decondensed, transcriptionally permissive chromatin structures and facilitates interaction with proteins containing bromo domains.

Histone acetyltransferase (HAT) An enzyme that acetylates specific lysine amino acids on histone proteins.

Histone code A theory that specific (combinations of) histone modifications are recognized in by non-histone proteins (through specific protein domains, such as bromo- and chromodomains) and thereby bring about a specific chromatin configuration and expression state (see epigenetic code).

Histone deacetylase (HDAC) An enzyme that removes acetyl groups from histone proteins (and for some, from specific non-histone proteins). This increases the positive charge of histones and enhances their attraction to the negatively charged phosphate groups in DNA, resulting in chromatin condensation.

Histone deacetylase inhibitor (HDACi) A class of compounds that interferes with the function of histone deacetylases. These compounds are used in psychiatry and neurology as mood stabilizers and anti-epileptics. They are also being investigated as possible treatments for cancer and inflammatory disease.

Histone demethylase (HDM) Proteins catalyzing the active enzymatic removal of methyl groups from either lysine or arginine residues of histones. Prominent examples are LSD1 and Jumonji proteins.

Histone methylation Posttranslational methylation of amino acid residues in histones catalyzed by histone methyltransferases (HMTs). Histone methylation is found at arginine as mono- or dimethylation and lysine as mono-, di-, or trimethylation. Different types of methylation can be found in either open transcriptionally active or closed transcriptionally silent chromatin. Methylated lysine residues are recognized by proteins containing chromodomains.

Histone methyltransferase (HMT) Enzymes catalyzing the transfer of methyl groups from S-adenosyl-methionine (SAM) to lysine or arginine residues in histones.

Histone variants Canonical histones with distinct amino acid changes accumulating at specific chromatin regions associated with the activating or silencing of transcription.

Immunoglobulins (Ig) Proteins that represent a central component of the adaptive immune system. Igs are generated via a multitude of different combinations Ig gene rearrangement and SHM in different individual B cells to develop different Ag recognition specificities.

Imprinted genes Genes that show a parent-of-origin-specific gene expression pattern controlled by epigenetic marks that originate from the germ line.

Imprinting See genomic imprinting

Imprinting control region (ICR) Region of the DNA that shows germ-line-derived, parent-of-origin-dependent epigenetic marks that control the parental-specific allelic expression of neighboring imprinted genes.

Induced pluripotent stem cells (iPS) Cells derived from differentiated somatic cells by in vitro reprogramming. Reprogramming is triggered by the activation of pluripotency factor genes and cultivation in ES cell medium. iPS cells are capable of generating all cell types of an embryo.

Inner cell mass (ICM) In early embryogenesis, the inner cell mass of cells will eventually give rise to the fetus. This structure forms before implantation into the endometrium of the uterus. The ICM lies within the blastocyst cavity and is entirely surrounded by a single layer of cells called the trophoblast.

Intracisternal A particle (IAP) A family of retrovirus-like elements that encode for virus-like particles found regularly in early rodent embryos. They are also transcribed in a wide variety of neoplasms because of DNA hypomethylation.

Large intervening noncoding RNA (lincRNA) A molecule of RNA 200 to many thousands of nucleotides in length that is transcribed by nonprotein coding areas of DNA. These ribonucleotides may play a role in a variety of biological processes, such as cancer formation.

Long interspersed elements (LINE) Highly repeated sequences, 6,000–8,000 base pairs in length, that contain RNA polymerase II promoters. They also have an open reading frame that is related to the reverse transcriptase of retroviruses, but they do not contain LTRs (long terminal repeats). Copies of the LINE1 family form about 15 % of the human genome. LINE elements are usually transcriptionally silent and marked by DNA methylation.

Long noncoding RNA (lncRNA) Nonprotein coding transcripts longer than 200 nucleotides. This limit distinguishes long ncRNAs from microRNAs (miRNAs), short interfering RNAs (siRNAs), Piwi-interacting RNAs (piRNAs), and small nucleolar RNAs (snoRNAs).

Long terminal repeat (LTR) Sequences of DNA that repeat hundreds or thousands of times. They are found in retroviral DNA and in retrotransposons that flank functional genes. They are used by viruses to insert their genetic sequences into the host genome.

Lymphoid primed multipotent progenitor cell (LMPP) A partially committed hematopoietic progenitor cell that retains the ability to give rise to both T and B lymphoid lineage cells and to granulocyte–macrophage lineage cells. These cells are not thought to give rise to erythroid/megakaryocytic lineage cells.

Messenger RNA (mRNA) A large family of RNA molecules that convey genetic information from DNA to the ribosome, where they specify the amino acid sequence of the protein products of gene expression.

Methyl-CpG-binding protein 2 (MeCP2) A protein that is essential for the normal function of nerve cells; mutations in this gene cause Rett syndrome.

Methylated DNA immunoprecipitation-microarray (MeDIP-chip) A genome-wide, high-resolution approach to detect DNA methylation in the whole genome or CpG islands. The method utilizes anti-methylcytosine antibody to immunoprecipitate DNA that contains highly methylated CpG sites. The enriched methylated DNA is then interrogated using DNA microarrays.

Methylated DNA immunoprecipitation-sequencing (MeDIP-seq) A genome-wide, high-resolution approach to detect DNA methylation in the whole genome or CpG islands. The method utilizes anti-methylcytosine antibody to immunoprecipitate DNA that contains highly methylated CpG sites. The enriched methylated DNA is then interrogated using massive parallel sequencing techniques.

Methyl-CpG binding domain (MBD) Protein domain in methyl-CpG-binding proteins (MBPs) responsible for recognizing and binding to methylated cytosine residues in DNA. Proteins containing MBDs form a specific family of proteins with various molecular functions.

Methyl-CpG-binding proteins (MBPs) Proteins containing domains (such as MBD) that bind to 5-methyl-cytosine in the context of CpG dinucleotides. MBPs mostly act as mediators for molecular functions such as transcriptional control or DNA repair.

Methyl-DNA-binding domain capture-sequencing (MethylCap-seq) A recently developed technique for the genome-wide profiling of DNA methylation. This technique consists of capturing the methylated DNA fragments by their methyl-CpG-binding domains (MBDs) and the subsequent deep sequencing of eluted DNA.

Methyl tetrahydrofolate reductase (MTHFR) A key enzyme in the folate S-adenosylmethionine (SAM) pathway.

microRNA (miRNA) A small noncoding RNA molecule about 22 nucleotides in length found in plants and animals. It functions in transcriptional and posttranscriptional regulation of gene expression.

Mitogen-activated protein kinase (MAPK) A protein in a cellular signaling pathway that transduces signals from the cell surface to the nucleus and modifies gene expression by affecting the activities of transcription factors.

Mixed-lineage leukemia (MLL) A type of childhood leukemia in which a piece of chromosome 11 is translocated to another chromosome. Children with this type of leukemia have a particularly poor prognosis. The name comes from the gene expression profiles in this disease being different than those seen in ALL and AML.

Mixed-lineage leukemia gene (MLL) A gene identified in mixed-lineage leukemia which is a member of the TRX group of SET domain histone modifying proteins. The MLL gene is a common site of chromosomal translocations in Mixed-Lineage Leukemia.

Myelodysplastic syndrome (MDS) A hematopoietic disorder characterized by an imbalance in the production of myeloid lineage cells in the bone marrow.

Myelo-proliferative disorder (MDS) A hematopoietic disorder similar to MDS which is characterized by excessive production of myeloid lineage cells in the bone marrow.

Next-generation sequencing (NGS) A technology similar to capillary electrophoresis-based Sanger sequencing where the bases of a small fragment of DNA are sequentially identified from signals emitted as each fragment is resynthesized from a DNA template strand. NGS extends this process across millions of reactions in a massively parallel fashion, rather than being limited to a single or a few DNA fragments.

Noncoding RNA (ncRNA) RNA transcripts that do not encode for a protein. ncRNA generation frequently involves RNA processing.

Nucleosome Fundamental organizational unit of chromatin consisting of 147 base pairs of DNA wound around a histone octamer.

Nucleosome Free Region (NFR) Regions in the DNA with an increased accessibility to micrococcal nuclease digestion. Thus, NFR refers to a deficiency in experimentally determined nucleosomes, but it does not imply a complete lack of histones. NFRs at the 5′ and 3′ ends of genes are sites of transcription initiation for mRNA and noncoding RNA.

Open reading frame (ORF) An open reading frame is a portion of a DNA molecule that, when translated into amino acids, contains no stop codons.

Plant homeodomain (PHD) The PHD finger is a Cys_4-His-Cys_3 zinc-finger-like motif found in nuclear proteins thought to be involved in epigenetics and chromatin-mediated transcriptional regulation.

Polycomb group proteins (PCG) A family of proteins initially discovered in fruit flies that can remodel chromatin such that epigenetic silencing of genes takes place. Polycomb group proteins are well known for silencing *Hox* genes through modulation of chromatin structure during embryonic development.

Polycomb response elements (PREs) *cis*-regulatory DNA elements that recruit both the Polycomb group (PcG) and Trithorax group (TrxG) proteins that are required for gene silencing and activation, respectively.

Position effect variegation (PEV) Cell/tissue-specific variability of gene expression controlled by the temporal inheritance of certain epigenetic states. PEV is a consequence of variable formation of heterochromatin across the respective gene. A classical example of PEV is found in certain mutations leading to variegated eye pigmentation in fruit flies.

Posttranslational modification (PTM) Proteins are created by ribosomes translating mRNA into polypeptide chains that then undergo posttranslational modifications such as folding and cutting before becoming mature proteins.

Promyelocytic leukemia (PML) A subtype of acute myelogenous leukemia (AML). It is a cancer of the blood and bone marrow with an abnormal accumulation of immature granulocytes called promyelocytes. The disease is characterized by a chromosomal translocation involving the *retinoic acid receptor alpha (RARA)* gene and is unique from other forms of AML in its responsiveness to all-trans retinoic acid therapy.

Quantitative real-time polymerase chain reaction (qPCR) A laboratory technique based on PCR that is used to amplify and simultaneously quantify a targeted DNA molecule.

Reduced Representation Bisulfite Sequencing (RRBS) A technique that couples bisulfite conversion and next-generation sequencing. It is an innovative method that enriches genomic regions with a high density of potential methylation sites and allows for the determination of DNA methylation at a single-nucleotide resolution.

Regions of altered methylation (RAMs) Persistent RAMs seen in precancerous tissues are thought to play a critical role in the genesis of cancer.

Reverse transcriptase (RT) An enzyme used to generate complementary DNA (cDNA) from an RNA template, a process termed reverse transcription. RT is needed for the replication of retroviruses, and RT inhibitors are widely used as antiretroviral drugs. Reverse transcriptase was discovered independently by Howard Temin at the University of Wisconsin–Madison and David Baltimore at MIT; a discovery for which they shared the 1975 Nobel Prize in Physiology or Medicine.

Ribonucleic acid (RNA) A ubiquitous family of large biological molecules that perform multiple vital roles in the coding, decoding, regulation, and expression of genes. RNA is assembled as a chain of nucleotides, but it is usually single stranded.

RNA-directed DNA methylation (RdDM) An epigenetic process first elucidated in plants whereby small double-stranded RNA (dsRNA) is processed to guide methylation to complementary DNA loci.

RNA-induced silencing complex (RISC) A multiprotein complex that incorporates one strand of a small interfering RNA (siRNA) or microRNA (miRNA). RISC uses the siRNA or miRNA as a template for recognizing complementary mRNA, which is then cleaved by activating RNase. This process is important in both gene regulation and the defense against viral infections.

RNA interference (RNAi) Posttranscriptional regulatory effects on mRNAs (i.e., control of translation or stability) triggered by processed dsRNA and ssRNA. Effects are propagated by enzymatic complexes such as RISC containing the small RNAs bound by Argonaute proteins.

Recombination signal sequence (RSS) DNA sequences recognized by the recombination apparatus that fuse together different gene segments of Ig and TCR genes to generate different Ag specificities in lymphocytes.

S-Adenosyl methionine (SAM) A cofactor for all DNA methyltransferases (DNMTs) and histone methyltransferases (HMTs), providing the methyl group added to either cytosines (DNA) or histones (arginine or lysine).

S-Adenosylhomocysteine (SAH) Hydrolyzed product formed after the methylation reaction catalyzed by DNA and histone methyltransferases using SAM as a methyl group donor. SAH is a competitive inhibitor of SAM for most methyltransferases.

SET domain A domain found in virtually all lysine-specific histone methyltransferases (HMTs). A protein–protein interaction domain required for HMT activity and modulation of chromatin structure that is frequently associated with cysteine-rich Pre-SET and Post-SET domains.

Short interspersed nuclear element (SINE) Non-long terminal repeat retrotransposons are highly abundant and heterogeneous; their length is about 300 base pairs. The most abundant SINEs in humans are in the Alu family.

Single nucleotide polymorphism (SNP) A DNA sequence variation occurring when a single nucleotide in the genome differs between members of a biological species or paired chromosomes.

Small interfering RNAs (siRNAs) RNAs that range in the size between 21 and 24 nucleotides and are derived from double-stranded long RNAs cleaved by Dicer. siRNAs are incorporated into the RISC complex to be targeted to complementary RNAs to promote cleavage of these mRNAs.

Somatic Hyper-mutation (SMH) Enzyme-mediated process by which the DNA sequence of TCR and Ig genes is altered to generate different Ag specificities.

snoRNAs Small nucleolar RNAs involved in processing of small RNAs such as ribosomal RNAs.

Stem cell Noncommitted cell that has the capacity to self-renew. Stem cells also have the capacity to differentiate into specialized cells.

Sumoylation Addition of a Small Ubiquitin-like Modifier or SUMO group to histone residues that is associated with transcriptional modification.

T-cell antigen receptor (TCR) Receptors that allow T cells to recognize specific antigens. Individual T cells acquire different Ag specificities by a process of gene rearrangement that varies greatly among individual T cells.

Tetrahydrofolate (THF) A coenzyme in many reactions, especially in the metabolism of amino acids and nucleic acids. It is produced from dihydrofolic acid by dihydrofolate reductase. It acts as the donor of a group with one carbon atom. A shortage of THF can cause megaloblastic anemia.

Totipotency Capacity of stem cells to produce all cell types required to form a mammalian embryo, i.e., embryonic and extraembryonic cells. Totipotent cells are formed during the first cleavages of the embryo.

Transcriptional gene silencing (TGS) The stable repression of transcription that mainly affects transposons, chromosomal repeats, and transgenic inserts; however, it can also involve protein encoding genes. It results from epigenetic modifications of DNA and histones that create an environment of heterochromatin around a gene, making it inaccessible to transcriptional machinery.

Transcriptome The set of all RNA molecules, including mRNA, rRNA, tRNA, and other noncoding RNA produced in a cell.

Trichostatin A (TSA) An inhibitor of certain types of histone deacetylases.

Trithorax group proteins (TRX) Proteins containing a trithorax-like bromodomain: They are usually involved in recognizing histone modifications marking transcriptionally active regions and contributing to the maintenance of activity.

Trithorax response elements (TRE) Chromosomal regions, a few hundred base pairs long, that maintain the active or silent transcriptional state of their associated genes after the initial determining activators and repressors have disappeared.

Trophoblasts (TB) Cells forming the outer layer of a blastocyst that provide nutrients to the embryo; they develop into the placenta.

Ultrabithorax (Ubx) A member of the homeobox gene family. In fruit flies, it is expressed in the third thoracic and first abdominal segments where it represses wing formation.

Untranslated region (UTR) The sections on each side of a coding sequence on a strand of mRNA. It is called the 5′ UTR if it is the leader sequence and the 3′ UTR if it is trailer sequence.

X-chromosome inactivation Epigenetically controlled form of dosage compensation in female mammals resulting in transcriptional silencing of genes on the surplus X-chromosome. X-chromosome inactivation is triggered by the noncoding RNA *Xist*, and it is manifested by various epigenetic modifications, including histone methylation, histone deacetylation, and DNA methylation.

X-inactivation center (XIC) Region at which the XIST-mediated inactivation starts. Allelic differences in the XIC may lead to skewed X-chromosome inactivation.

X-inactive specific transcript (XIST) The mammalian XIST gene encodes for a nonprotein encoding RNA that coats the inactive X-chromosome.

X trisomy A form of chromosomal variation characterized by the presence of an extra X chromosome in each cell of a female. There is usually no distinguishable difference between women with triple X and the rest of the female population.

Yolk sac (YS) A membranous sac attached to the embryo, providing early nourishment in the form of yolk in bony fishes, sharks, reptiles, birds, and primitive mammals. It functions as the developmental circulatory system of the human embryo before internal circulation begins.

Zinc finger (ZNF) A small protein structural motif that is formed by the coordination of one or more zinc ions in order to stabilize the fold. The vast majority of zinc finger proteins function as interaction modules that bind DNA, RNA, proteins, or other small molecules.

Index

C. Bonifer and P.N. Cockerill (eds.), *Transcriptional and Epigenetic Mechanisms Regulating Normal and Aberrant Blood Cell Development*, Epigenetics and Human Health, DOI 10.1007/978-3-642-45198-0, © Springer-Verlag Berlin Heidelberg 2014

MIX
Papier aus verantwortungsvollen Quellen
Paper from responsible sources
FSC® C105338

FSC
www.fsc.org

Printed by Libri Plureos GmbH
in Hamburg, Germany